Lecture Notes in Artificial Intelligence   11055

Subseries of Lecture Notes in Computer Science

LNAI Series Editors

Randy Goebel
University of Alberta, Edmonton, Canada
Yuzuru Tanaka
Hokkaido University, Sapporo, Japan
Wolfgang Wahlster
DFKI and Saarland University, Saarbrücken, Germany

LNAI Founding Series Editor

Joerg Siekmann
DFKI and Saarland University, Saarbrücken, Germany

More information about this series at http://www.springer.com/series/1244

Ngoc Thanh Nguyen · Elias Pimenidis
Zaheer Khan · Bogdan Trawiński (Eds.)

Computational Collective Intelligence

10th International Conference, ICCCI 2018
Bristol, UK, September 5–7, 2018
Proceedings, Part I

 Springer

Editors
Ngoc Thanh Nguyen
Faculty of Information Technology
Ton Duc Thang University
Ho Chi Minh City
Vietnam

and

Faculty of Computer Science
  and Management
Wrocław University of Science
  and Technology
Wrocław
Poland

Elias Pimenidis
Department of Computer Science
  and Creative Technologies
University of the West of England
Bristol
UK

Zaheer Khan
Department of Computer Science
  and Creative Technologies
University of the West of England
Bristol
UK

Bogdan Trawiński
Faculty of Computer Science
  and Management
Wrocław University of Science
  and Technology
Wrocław
Poland

ISSN 0302-9743   ISSN 1611-3349 (electronic)
Lecture Notes in Artificial Intelligence
ISBN 978-3-319-98442-1   ISBN 978-3-319-98443-8 (eBook)
https://doi.org/10.1007/978-3-319-98443-8

Library of Congress Control Number: 2018950468

LNCS Sublibrary: SL7 – Artificial Intelligence

This Springer imprint is published by the registered company Springer Nature Switzerland AG
The registered company address is: Gewerbestrasse 11, 6330 Cham, Switzerland

Preface

This volume contains the proceedings of the 10th International Conference on Computational Collective Intelligence (ICCCI 2018), held in Bristol, UK, September 5–7, 2018. The conference was co-organized by the University of the West of England, Bristol, UK, and the Wrocław University of Science and Technology, Poland. The conference was run under the patronage of the IEEE SMC Technical Committee on Computational Collective Intelligence.

Following the successes of the First ICCCI (2009) held in Wrocław, Poland, the Second ICCCI (2010) in Kaohsiung, Taiwan, the Third ICCCI (2011) in Gdynia, Poland, the 4th ICCCI (2012) in Ho Chi Minh City, Vietnam, the 5th ICCCI (2013) in Craiova, Romania, the 6th ICCCI (2014) in Seoul, South Korea, the 7th ICCCI (2015) in Madrid, Spain, the 8th ICCCI (2016) in Halkidiki, Greece, and the 9th ICCCI (2017) in Nicosia, Cyprus, this conference continued to provide an internationally respected forum for scientific research in the computer-based methods of collective intelligence and their applications.

Computational collective intelligence (CCI) is most often understood as a sub-field of artificial intelligence (AI) dealing with soft computing methods that facilitate group decisions or processing knowledge among autonomous units acting in distributed environments. Methodological, theoretical, and practical aspects of CCI are considered as the form of intelligence that emerges from the collaboration and competition of many individuals (artificial and/or natural). The application of multiple computational intelligence technologies such as fuzzy systems, evolutionary computation, neural systems, consensus theory, etc., can support human and other collective intelligence, and create new forms of CCI in natural and/or artificial systems. Three subfields of the application of computational intelligence technologies to support various forms of collective intelligence are of special interest but are not exclusive: the Semantic Web (as an advanced tool for increasing collective intelligence), social network analysis (as the field targeted at the emergence of new forms of CCI), and multi-agent systems (as a computational and modeling paradigm especially tailored to capture the nature of CCI emergence in populations of autonomous individuals).

The ICCCI 2018 conference featured a number of keynote talks and oral presentations, closely aligned to the theme of the conference. The conference attracted a substantial number of researchers and practitioners from all over the world, who submitted their papers for the main track and four special sessions.

The main track, covering the methodology and applications of CCI, included: knowledge engineering and Semantic Web, social network analysis, recommendation methods and recommender systems, agents and multi-agent systems, text processing and information retrieval, data mining methods and applications, decision support and control systems, sensor networks and Internet of Things, as well as computer vision techniques. The special sessions, covering some specific topics of particular interest, included: cooperative strategies for decision-making and optimization, complex decision systems, machine learning in real-world data, as well as intelligent sustainable smart cities.

We received over 240 submissions from 39 countries all over the world. Each paper was reviewed by two to four members of the international Program Committee (PC) of either the main track or one of the special sessions. Finally, we selected 98 best papers for oral presentation and publication in two volumes of the *Lecture Notes in Artificial Intelligence* series.

We would like to express our thanks to the keynote speakers: Andrew Adamatzky from the University of the West of England, UK, Anthony Pipe from the Bristol Robotics Laboratory, UK, Tadeusz Szuba from the AGH University of Science and Technology, Poland, and Jan Treur from the Vrije Universiteit Amsterdam, The Netherlands, for their world-class plenary speeches.

Many people contributed toward the success of the conference. First, we would like to recognize the work of the PC co-chairs and special sessions organizers for taking good care of the organization of the reviewing process, an essential stage in ensuring the high quality of the accepted papers. The workshop and special session chairs deserve a special mention for the evaluation of the proposals and the organization and coordination of the work of seven special sessions. In addition, we would like to thank the PC members, of the main track and of the special sessions, for performing their reviewing work with diligence. We thank the local Organizing Committee chairs, publicity chair, Web chair, and technical support chair for their fantastic work before and during the conference. Finally, we cordially thank all the authors, presenters, and delegates for their valuable contribution to this successful event. The conference would not have been possible without their support.

Our special thanks are also due to Springer for publishing the proceedings and sponsoring awards, and to all the other sponsors for their kind support.

It is our pleasure to announce that the ICCCI conference series continues to have a close cooperation with the Springer journal *Transactions on Computational Collective Intelligence*, and the IEEE SMC Technical Committee on Transactions on Computational Collective Intelligence.

Finally, we hope that ICCCI 2018 contributed significantly to the academic excellence of the field and will lead to the even greater success of ICCCI events in the future.

September 2018

Ngoc Thanh Nguyen
Elias Pimenidis
Zaheer Khan
Bogdan Trawiński

Organization

Organizing Committee

Honorary Chairs

Pierre Lévy	University of Ottawa, Canada
Cezary Madryas	Wroclaw University of Science and Technology, Poland
Paul Olomolaiye	University of the West of England, UK

General Chairs

Ngoc Thanh Nguyen	Wroclaw University of Science and Technology, Poland
Larry Bull	University of the West of England, UK

Program Chairs

Zaheer Khan	University of the West of England, UK
Costin Badica	University of Craiova, Romania
Edward Szczerbicki	University of Newcastle, Australia
Gottfried Vossen	University of Münster, Germany

Special Session Chairs

Bogdan Trawinski	Wroclaw University of Science and Technology, Poland
Mehmet Aydin	University of the West of England, UK

Doctoral Chair

Emmanuel Ogunshile	University of the West of England, UK

Organizing Chair

Elias Pimenidis	University of the West of England, UK

Publicity Chair

Nikolaos Polatidis	University of Brighton, UK

Local Organizing Committee

Stewart Green	University of the West of England, UK
Kamran Soomro	University of the West of England, UK
Marcin Jodlowiec	Wroclaw University of Science and Technology, Poland
Marek Krotkiewicz	Wroclaw University of Science and Technology, Poland
Marcin Maleszka	Wroclaw University of Science and Technology, Poland
Krystian Wojtkiewicz	Wroclaw University of Science and Technology, Poland

Web Chair

Jake Hallam   University of the West of England, UK

Keynote Speakers

Andrew Adamatzky   University of the West of England, UK
Anthony Pipe   Bristol Robotics Laboratory, UK
Tadeusz Szuba   AGH University of Science and Technology, Poland
Jan Treur   Vrije Universiteit Amsterdam, The Netherlands

Special Session Organizers

CSDMO 2018: Special Session Cooperative Strategies for Decision-Making and Optimization

Piotr Jedrzejowicz   Gdynia Maritime University, Poland
Dariusz Barbucha   Gdynia Maritime University, Poland

CDS 2018: Special Session on Complex Decision Systems

Alicja Wakulicz–Deja   University of Silesia, Poland
Agnieszka Nowak–   University of Silesia, Poland
   Brzezinska
Malgorzata   University of Silesia, Poland
   Przybyła-Kasperek

MLRWD 2018: Special Session on Machine Learning in Real-World Data

Krzysztof Kania   University of Economics in Katowice, Poland
Przemyslaw Juszczuk   University of Economics in Katowice, Poland
Jan Kozak   University of Economics in Katowice, Poland
Bogna Zacny   University of Economics in Katowice, Poland

ISSC 2018: Special Session on Intelligent Sustainable Smart Cities

Libuse Svobodova   University of Hradec Kralove, Czech Republic
Ali Selamat   Universiti Teknologi Malaysia, Malaysia
Petra Maresova   University of Hradec Kralove, Czech Republic
Arkadiusz Kawa   Poznan University of Economics and Business, Poland
Bartlomiej Pieranski   Poznan University of Economics and Business, Poland
Peter Brida   University of Zilina, Slovakia

Program Committee

Muhammad Abulaish   South Asian University, India
Sharat Akhoury   University of Cape Town, South Africa
Bashar Al-Shboul   University of Jordan, Jordan
Ana Almeida   GECAD-ISEP-IPP, Portugal

Tzung-Pei Hong	National University of Kaohsiung, Taiwan
Mong-Fong Horng	National Kaohsiung University of Applied Sciences, Taiwan
Frédéric Hubert	Laval University, Canada
Maciej Huk	Wroclaw University of Science and Technology, Poland
Zbigniew Huzar	Wroclaw University of Science and Technology, Poland
Dosam Hwang	Yeungnam University, South Korea
Lazaros Iliadis	Democritus University of Thrace, Greece
Agnieszka Indyka-Piasecka	Wroclaw University of Science and Technology, Poland
Dan Istrate	Universite de Technologie de Compiegne, France
Mirjana Ivanovic	University of Novi Sad, Serbia
Jaroslaw Jankowski	West Pomeranian University of Technology, Szczecin, Poland
Joanna Jedrzejowicz	University of Gdansk, Poland
Piotr Jedrzejowicz	Gdynia Maritime University, Poland
Gordan Jezic	University of Zagreb, Croatia
Geun Sik Jo	Inha University, South Korea
Kang-Hyun Jo	University of Ulsan, South Korea
Jason Jung	Chung-Ang University, South Korea
Tomasz Kajdanowicz	Wroclaw University of Science and Technology, Poland
Petros Kefalas	University of Sheffield, Greece
Rafal Kern	Wroclaw University of Science and Technology, Poland
Zaheer Khan	University of the West of England, UK
Marek Kisiel-Dorohinicki	AGH University of Science and Technology, Poland
Attila Kiss	Eötvös Loránd University, Hungary
Marek Kopel	Wroclaw University of Science and Technology, Poland
Leszek Koszalka	Wroclaw University of Science and Technology, Poland
Leszek Kotulski	AGH University of Science and Technology, Poland
Ivan Koychev	University of Sofia St. Kliment Ohridski, Bulgaria
Jan Kozak	University of Economics in Katowice, Poland
Adrianna Kozierkiewicz	Wroclaw University of Science and Technology, Poland
Bartosz Krawczyk	Virginia Commonwealth University, USA
Ondrej Krejcar	University of Hradec Kralove, Czech Republic
Dalia Kriksciuniene	Vilnius University, Lithuania
Dariusz Krol	Wroclaw University of Science and Technology, Poland
Marek Krotkiewicz	Wroclaw University of Science and Technology, Poland
Elzbieta Kukla	Wroclaw University of Science and Technology, Poland
Julita Kulbacka	Wroclaw Medical University, Poland
Marek Kulbacki	Polish-Japanese Academy of Information Technology, Poland
Piotr Kulczycki	Polish Academy of Science, Systems Research Institute, Poland
Kazuhiro Kuwabara	Ritsumeikan University, Japan
Halina Kwasnicka	Wroclaw University of Science and Technology, Poland

Mark Last	Ben-Gurion University of the Negev, Israel
Hoai An Le Thi	Universite de Lorraine, France
Florin Leon	Gheorghe Asachi Technical University of Iasi, Romania
Edwin Lughofer	Johannes Kepler University Linz, Austria
Juraj Machaj	University of Zilina, Slovakia
Bernadetta Maleszka	Wroclaw University of Science and Technology, Poland
Marcin Maleszka	Wroclaw University of Science and Technology, Poland
Yannis Manolopoulos	Aristotle University of Thessaloniki, Greece
Urszula Markowska-Kaczmar	Wroclaw University of Science and Technology, Poland
Adam Meissner	Poznan University of Technology, Poland
Héctor Menéndez	University College London, UK
Mercedes Merayo	Universidad Complutense de Madrid, Spain
Jacek Mercik	WSB University in Wroclaw, Poland
Radoslaw Michalski	Wroclaw University of Science and Technology, Poland
Peter Mikulecky	University of Hradec Kralove, Czech Republic
Javier Montero	Universidad Complutense de Madrid, Spain
Ahmed Moussa	Universite Abdelmalek Essaadi, Morocco
Dariusz Mrozek	Silesian University of Technology, Poland
Kazumi Nakamatsu	University of Hyogo, Japan
Grzegorz J. Nalepa	AGH University of Science and Technology, Poland
Fulufhelo Nelwamondo	Council for Scientific and Industrial Research, South Africa
Filippo Neri	University of Naples Federico II, Italy
Linh Anh Nguyen	University of Warsaw, Poland
Loan T. T. Nguyen	Nguyen Tat Thanh University, Vietnam
Adam Niewiadomski	Lodz University of Technology, Poland
Agnieszka Nowak-Brzezinska	University of Silesia, Poland
Alberto Núñez	Universidad Complutense de Madrid, Spain
Manuel Núñez	Universidad Complutense de Madrid, Spain
Tarkko Oksala	Aalto University, Finland
Mieczyslaw Owoc	Wroclaw University of Economics, Poland
Marcin Paprzycki	Systems Research Institute, Polish Academy of Sciences, Poland
Marek Penhaker	VSB -Technical University of Ostrava, Czech Republic
Isidoros Perikos	University of Patras, Greece
Marcin Pietranik	Wroclaw University of Science and Technology, Poland
Elias Pimenidis	University of the West of England, UK
Nikolaos Polatidis	University of Brighton, UK
Piotr Porwik	University of Silesia, Poland
Radu-Emil Precup	Politehnica University of Timisoara, Romania
Ales Prochazka	University of Chemistry and Technology, Czech Republic
Paulo Quaresma	Universidade de Evora, Portugal
Mohammad Rashedur Rahman	North South University, Bangladesh

Ewa Ratajczak-Ropel	Gdynia Maritime University, Poland
Tomasz M. Rutkowski	University of Tokyo, Japan
Virgilijus Sakalauskas	Vilnius University, Lithuania
Jose L. Salmeron	University Pablo de Olavide, Spain
Ali Selamat	Universiti Teknologi Malaysia, Malaysia
Andrzej Sieminski	Wroclaw University of Science and Technology, Poland
Dragan Simic	University of Novi Sad, Serbia
Vladimir Sobeslav	University of Hradec Kralove, Czech Republic
Stanimir Stoyanov	University of Plovdiv Paisii Hilendarski, Bulgaria
Yasufumi Takama	Tokyo Metropolitan University, Japan
Zbigniew Telec	Wroclaw University of Science and Technology, Poland
Diana Trandabat	University Alexandru Ioan Cuza of Iasi, Romania
Bogdan Trawinski	Wroclaw University of Science and Technology, Poland
Jan Treur	Vrije Universiteit Amsterdam, The Netherlands
Maria Trocan	Institut Superieur d'Electronique de Paris, France
Krzysztof Trojanowski	Cardinal Stefan Wyszynski University in Warsaw, Poland
Ualsher Tukeyev	Al-Farabi Kazakh National University, Kazakhstan
Olgierd Unold	Wroclaw University of Science and Technology, Poland
Bay Vo	Ho Chi Minh City University of Technology, Vietnam
Lipo Wang	Nanyang Technological University, Singapore
Izabela Wierzbowska	Gdynia Maritime University, Poland
Krystian Wojtkiewicz	Wroclaw University of Science and Technology, Poland
Slawomir Zadrozny	Systems Research Institute, Polish Academy of Sciences, Poland
Danuta Zakrzewska	Lodz University of Technology, Poland
Constantin-Bala Zamfirescu	Lucian Blaga University of Sibiu, Romania
Katerina Zdravkova	St. Cyril and Methodius University, Macedonia
Aleksander Zgrzywa	Wroclaw University of Science and Technology, Poland
Adam Ziebinski	Silesian University of Technology, Poland
Drago Zagar	University of Osijek, Croatia

Special Session Program Committees

CSDMO 2018: Special Session Cooperative Strategies for Decision - Making and Optimization

Dariusz Barbucha	Gdynia Maritime University, Poland
Vincenzo Cutello	University of Catania, Italy
Ireneusz Czarnowski	Gdynia Maritime University, Poland
Joanna Jedrzejowicz	Gdansk University, Poland
Piotr Jedrzejowicz	Gdynia Maritime University, Poland
Edyta Kucharska	AGH University of Science and Technology, Poland
Antonio D. Masegosa	University of Deusto, Spain
Javier Montero	Complutense University of Madrid, Spain
Ewa Ratajczak-Ropel	Gdynia Maritime University, Poland

Iza Wierzbowska	Gdynia Maritime University, Poland
Mahdi Zargayouna	IFSTTAR, France

CDS 2018: Special Session on Complex Decision Systems

Alicja Wakulicz-Deja	University of Silesia, Katowice, Poland
Grzegorz Baron	Silesian University of Technology, Gliwice, Poland
Rafal Deja	Academy of Business in Dabrowa Gornicza, Poland
Michal Draminski	Institute of Computer Science, Polish Academy of Sciences, Warsaw, Poland
Agnieszka Duraj	Lodz University of Technology, Poland
Katarzyna Harezlak	Silesian University of Technology, Gliwice, Poland
Michal Kozielski	Silesian University of Technology, Gliwice, Poland
Dariusz Mrozek	Silesian University of Technology, Gliwice, Poland
Bozena Malysiak-Mrozek	Silesian University of Technology, Gliwice, Poland
Agnieszka Nowak-Brzezinska	University of Silesia, Katowice, Poland
Malgorzata Przybyla-Kasperek	University of Silesia, Katowice, Poland
Roman Siminski	University of Silesia, Katowice, Poland
Urszula Stanczyk	Silesian University of Technology, Gliwice, Poland
Beata Zielosko	University of Silesia, Katowice, Poland
Tomasz Xieski	University of Silesia, Katowice, Poland

MLRWD 2018: Special Session on Machine Learning in Real-World Data

Franciszek Bialas	University of Economics in Katowice, Poland
Grzegorz Dziczkowski	University of Economics in Katowice, Poland
Marcin Grzegorzek	University of Siegen, Germany
Ignacy Kaliszewski	Systems Research Institute, Polish Academy of Sciences, Poland
Krzysztof Kania	University of Economics in Katowice, Poland
Jan Kozak	University of Economics in Katowice, Poland
Przemyslaw Juszczuk	University of Economics in Katowice, Poland
Janusz Miroforidis	Systems Research Institute, Polish Academy of Sciences, Poland
Agnieszka Nowak-Brzezinska	University of Silesia, Poland
Dmitry Podkopaev	Systems Research Institute, Polish Academy of Sciences, Poland
Malgorzata Przybyla-Kasperek	University of Silesia, Poland
Tomasz Stas	University of Economics in Katowice, Poland
Magdalena Tkacz	University of Silesia, Poland
Bogna Zacny	University of Economics in Katowice, Poland

ISSC 2018: Special Session on Intelligent Sustainable Smart Cities

Costin Badica	University of Craiova, Romania
Peter Bracinik	University of Zilina, Slovakia
Peter Brida	University of Zilina, Slovakia
Davor Dujak	University of Osijek, Croatia
Martina Hedvicakova	University of Hradec Kralove, Czech Republic
Marek Hoger	University of Zilina, Slovakia
Petra Maresova	University of Hradec Kralove, Czech Republic
Hana Mohelska	University of Hradec Kralove, Czech Republic
Arkadiusz Kawa	Poznan University of Economics and Business, Poland
Waldemar Koczkodaj	Laurentian University, Canada
Ondrej Krejcar	University of Hradec Kralove, Czech Republic
Martina Latkova	University of Zilina, Slovakia
Juraj Machaj	University of Zilina, Slovakia
Miroslava Mikusova	University of Zilina, Slovakia
Jaroslaw Olejniczak	Wroclaw University of Economics, Poland
Pawel Piatkowski	Poznan University of Technology, Poland
Bartlomiej Pieranski	Poznan University of Economics and Business, Poland
Petra Poulova	University of Hradec Kralove, Czech Republic
Michal Regula	University of Zilina, Slovakia
Marek Roch	University of Zilina, Slovakia
Carlos Andres Romano	Polytechnic University of Valencia, Spain
Ali Selamat	Universiti Teknologi Malaysia, Malaysia
Marcela Sokolova	University of Hradec Kralove, Czech Republic
Libuse Svobodova	University of Hradec Kralove, Czech Republic
Emese Tokarcikova	University of Zilina, Slovakia
Hana Tomaskova	University of Hradec Kralove, Czech Republic
Marek Vokoun	Institute of Technology and Business in Ceske Budejovice, Czech Republic

Contents – Part I

Recommendation Methods and Recommender Systems

Agents and Multi-Agent Systems

Text Processing and Information Retrieval

Sensor Networks and Internet of Things

Data Mining Methods and Applications

Contents – Part II

Cooperative Strategies for Decision Making and Optimization

Complex Decision Systems

Machine Learning in Real-World Data

Intelligent Sustainable Smart Cities

Computer Vision Techniques

Knowledge Engineering and Semantic Web

ViewpointS: Towards a Collective Brain

Philippe Lemoisson[1,2] and Stefano A. Cerri[3(✉)]

[1] CIRAD, UMR TETIS, 34398 Montpellier, France
philippe.lemoisson@cirad.fr
[2] TETIS, Univ Montpellier, AgroParisTech, CIRAD, CNRS,
IRSTEA, Montpellier, France
[3] LIRMM, Univ Montpellier, CNRS, 161 Rue Ada, 34095 Montpellier, France
cerri@lirmm.fr

Abstract. Tracing knowledge acquisition and linking learning events to interaction between peers is a major challenge of our times. We have conceived, designed and evaluated a new paradigm for constructing and using collective knowledge by Web interactions that we called ViewpointS. By exploiting the similarity with Edelman's Theory of Neuronal Group Selection (TNGS), we conjecture that it may be metaphorically considered a Collective Brain, especially effective in the case of trans-disciplinary representations. Far from being without doubts, in the paper we present the reasons (and the limits) of our proposal that aims to become a useful integrating tool for future quantitative explorations of individual as well as collective learning at different degrees of granularity. We are therefore challenging each of the current approaches: the logical one in the semantic Web, the statistical one in mining and deep learning, the social one in recommender systems based on authority and trust; not in each of their own preferred field of operation, rather in their integration weaknesses far from the holistic and dynamic behavior of the human brain.

Keywords: Collective brain · Collective intelligence · Knowledge graph
Human learning · Knowledge acquisition · Semantic web · Social web

1 Introduction

On one side, today's research on the human brain allows us to visualize and trace the activity along the beams connecting the neural maps. When publishing the Theory of Neuronal Group Selection (TNGS) more than thirty years ago, G.M. Edelman emphasized the observation/action loop and the social interactions loop. Both loops continuously evolve the beams connecting the neural maps under the supervision of our homeostatic internal systems also called system of values, and generate learning. On the other side, we live a digital revolution where the Web plays an increasing role in the collective construction of knowledge; this happens through the semantic Web and its ontologies, via the indexing and mining techniques of the search engines and via the social Web and its recommender systems based on authority and trust.

The goal of our approach is twofold: (i) to exploit the metaphor of the brain for improving the collective construction of knowledge and (ii) to better exploit our digital traces in order to refine the understanding of our learning processes. We have designed

© Springer Nature Switzerland AG 2018
N. T. Nguyen et al. (Eds.): ICCCI 2018, LNAI 11055, pp. 3–12, 2018.
https://doi.org/10.1007/978-3-319-98443-8_1

and prototyped a Knowledge Graph where resources (human or artificial agents, documents and descriptors) are dynamically interlinked by beams of digital connections called viewpoints (human viewpoints or artificial viewpoints issued from algorithms). We-as-agents endlessly exploit and update this graph, so that by similarity with the TNGS, we conjecture that it may be metaphorically considered a Collective Brain evolving under the supervision of all our individual systems of values.

In Sect. 2, we present a schematic view of the biological bases of cognition, starting by the "three worlds" of Popper (1978) where the interaction between minds can be studied. We then re-visit the TNGS and the role played by our system of values (internal drives, instinct, intentionality…). We finally illustrate "learning through interaction" as exposed by D. Laurillard and J. Piaget.

In Sect. 3, we explore the collective construction of knowledge in the Web paradigm, assuming that a large proportion of the traces we produce and consume today are digital ones. We distinguish three paradigms, respectively governed by logics, by statistics and by authority and trust. Thus our challenge is to integrate these paradigms and describe how individual systems of values participate to learning events.

Section 4 is dedicated to the ViewpointS approach, as a candidate for answering the challenge. The metaphor of "neural maps interconnected by beams of neurons" led to the design of a graph of "knowledge resources interconnected by beams of viewpoints", where each agent can exploit the traces of others and react to them by adding new traces. As a result, the combination of all individual "system of values" regulates the evolution of knowledge. We conjecture that it may be metaphorically considered a Collective Brain.

We conclude by recapitulating our proposal which has the limits inherent to any integrator: we are not yet sure if the collective knowledge emerging from our proposed Collective Brain will perform competitively with the existing separate paradigms respectively governed by logics, by statistics and by authority and trust. Nevertheless, if our proposal does not ensure scientific discovery about learning, we hope it represents a progress toward its comprehension.

2   A Schematic View of the Biological Bases of Cognition

In this section, we start by adopting a well-known philosophical position where the questions of cognition and interaction can be addressed. Then we draft a schematic view of the lessons learned from Edelman about the biological mechanisms supporting cognition, and finally we use this representation within D. Laurillard's conversational learning scenario in order to test it against the question of knowledge acquisition through interaction.

2.1   The Three Worlds

To start with our analysis about minds in interaction, we need some philosophical default position; "the three worlds" of Popper [1] provides a relevant framework. Such a framework had already found an expression in the semantic triangle of Odgen and Richards [2]. The strong interconnection of the three worlds is developed in [3] where

J. Searle explains how the interpretation of repeated collective experiences by individuals bears the emergence of an institutional reality founded on the use of language. In the following, we shall refer to the three worlds as W_1, W_2 and W_3, with the following definitions:

W_1 is the bio-physical world where objects and events exist independently from us, from our perceptions, our thoughts and our languages. Causal relations, insofar we are not directly implied by some event, are also considered independent from us.

W_2 is the internal world of subjectivity, where the perception of objects and events of W_1 leave traces in memory that are combined in order to participate to the construction of our own knowledge, our consciousness about the world, where intentions appear and the emotions that will be the trigger for our actions.

W_3 is the world of the cultures and languages, made of interpretable traces: signs, symbols, rules of behavior and rules for representing objects and events of W_1. W_3 is the support of communication among individuals. Within W_3, we find all specialized languages of the scientific disciplines, as well as the language of emotions and feelings, for instance represented by smileys. Digital images such as satellite images or scanned documents are also part of W_3.

W_1 is where it happens, W_3 is where we can communicate about what happens, and W_2 is where the links and the learning events are. For this reason, we are going to pay special attention to the internal world W_2.

2.2   The Internal World of the Mind

This section pays a heavy tribute to the work of Edelman [4, 5], founder of the Theory of Neuronal Group Selection (TNGS), and one the firsts to emphasize that the brain is not a computer, but a highly dynamic, distributed and complex system, maybe the most complex "object" of the known universe. There is neither correlation between our personality and the shape of our skull (despite the teachings of phrenology), nor localized coding of information; no autopsy will ever reveal any single chunk of knowledge available in the brain.

According to the TNGS, every brain is twice unique: first because its cellular organization results from the laws of morphogenesis. Most important, however, is Edelman's second reason for the brain uniqueness: the brain is a set of "neural maps" continuously selected according to the individual's experiences. These cards, or adaptive functional units, are bi-directionally linked one-another by a fundamental integrating mechanism: the "re-entry". This crucial hypothesis allows a functional integration requiring neither any "super-card" nor any "supervising program": the neural maps are like "musicians of an orchestra linked one-another by wires in the absence of a unique conductor". The bi-directional re-entry links are the result of a selective synaptic reinforcement among groups of neurons; similarly: the cards result from a synaptic reinforcement internal to each group of neurons composing them. These reinforcements are triggered and managed by the homeostatic internal systems, also called "system of values" of each individual.

Figure 1 (/left part of the figure) shows an observation-action loop that highlights several brain cards re-entering when grasping an apple. This type of loops originates the perceptual categorization event, common to all highly evolved organisms, decisive for adapting the behavior to the likelihood of benefits or dangers.

In humans as well as in some higher mammals, there is a second level of categorization, supported by cards situated in the temporal, frontal and parietal areas. Beyond the immediate cartography of the world, humans may shape some durable concepts (conceptual categorization) that consider the past and/or the future.

Finally, the human brain parts specialized in language (the Wernicke and Broca areas) play a major role in the emergence of a consciousness of a higher level, enabling the human subject to "map" his-her own experience and study him-herself.

The basic principles of the TNGS (selective reinforcement and re-entry according to the advantages offered to the subject) potentially explain any learning process, from simple memorization to skill acquisition and knowledge acquisition. All these processes are regulated by our system of values.

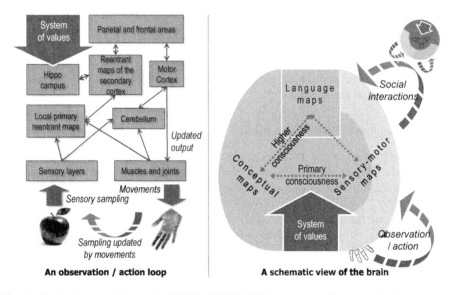

Fig. 1. The brain according to the TNGS of G.M. Edelman: a complex network of re-entrant maps in interaction loops with the world

A kernel element of the TNGS is quite relevant for us: knowledge is supported by a "physiological complex and adaptive network of neural maps"; the metaphor of "knowledge graph" is therefore justified. It induces to search for a topology allowing to define distances and proximity, like it was conjectured by the "zone of proximal development" of Vygotsky [6, 7]; such a topology will be presented in chapter 4.

2.3 Minds in Social Interaction

According to the two loops at the right of Fig. 1, we always learn through interaction: observation/action versus social interaction. These two loops clearly appear in D. Laurillard's work [8] when analyzing the acquisition of knowledge in higher education.

In her scenario, a student and his teacher simultaneously experiment and discuss. In [9] we extended this scenario to interactions within a group of peers co-constructing a representation of a shared territory.

In this multi-peers scenario, interactions occur at two levels: (i) peers act in the shared territory (in W_1/objects and events) and (ii) peers exchange personal views of the shared territory (in W_3/language). Doing so, the assimilation/adaptation processes described by Piaget in [10] are activated (in W_2/mind), which can be interpreted in terms of series of re-entry loops according to the TNGS. In [11] we proposed a roadmap for understanding and forecasting cognitive and emotional events linked to serendipitous learning.

As a consequence of all above processes, inner views tend to synchronize and yield a shared representation. We propose that what happens on the Web is a generalization of this prototypical scenario, and can be called collective knowledge acquisition. Our approach aims at tracing it; this will be exemplified in chapter 4.

3   Humans in Web Interaction

The change in our lives that we have been experiencing since when Internet has gained a significant place, often called the digital revolution, has been theoretically addressed by several authors, among which Vial [12] and Cardon [13] who respectively provide a philosophical and a sociological approach. This revolution has suggested a significant hope: Internet as a space of shared knowledge able to bring in new levels of under-standing in the sense given by Gruber in [14]. Internet is a support for a huge set of digital traces interpretable by humans but also by machines; if we refer to the con-ceptual framework above, it is part of W_3/language.

This space is far from being homogeneous however, and the approaches to co-build shared knowledge are multiples; hereafter, we consider three paradigms.

The first paradigm is governed by the *logical* evidence: we usually call semantic Web this logically structured part of Internet where humans interact with databases encoding the knowledge of experts according to consensual conceptual schemes such as ontologies. This allows logical responses (provided by reliable algorithms) to cor-rectly formulated questions (and only to questions with such a property). But there are problems and limits. Firstly, ontologies only represent a fragment of the reality, and the consensus they reflect is necessarily local and temporary. Secondly, formal query languages assume a closed world – what is rarely the case. Thirdly, formal query languages require a learning effort in order to be used properly. Finally, interconnecting ontologies and supporting their evolution with time in a rapidly changing world are very heavy and costly processes. Various approaches based on automatic alignment [15] or machine learning [16] exist, but the task is endless since each ontology's evolution is domain-dependent.

The second paradigm is governed by the *statistical* evidence. The issue is to exploit techniques of data mining, i.e.: scan without too many assumptions a corpus, also called data set, of tweets, sequences, clicks, documents, … and detect regularities, frequencies, co-occurrences of items or terms. In other words: to feed suitable algo-rithms with the big data in order to reveal regularities. By reducing the complexity of the digital world W_3, the mining algorithms make it readable for humans. However, the simplicity of these descriptions must pay a price to the expressiveness or even to the effectiveness: we are just shown the surface and not the depth, the "meaning". Today, a simple question with three independent keywords on Google may give very disappointing results. What is even worse, inferential statistics – the only one allowing us to take significant decisions - require selecting the data according to a predefined goal, and this remains hidden from the user. Interpreting the results in order to build chunks of science is therefore heavily biased.

The third paradigm is based on authority and trust, which rely on *emotions* and *feelings*. The algorithms of the social Web provide information search and recommendations by graph analysis of the various personal, subjective and spontaneous light traces such as 'likes', 'bookmarks' and 'tweets'. They clearly operate in an open world; the limits are firstly the impossibility to logically assess the quality of responses and secondly the absence of guarantee concerning their stability along time.

Coming back to the dream of Gruber and many others to fuse the three paradigms, a first attempt is the semantic Web project [17] which somehow aims at subsuming them within the *logical* one; after a first wave of enthusiasm it seems that the limits listed above resist, even if they are daily pushed forward. The ViewpointS approach discussed in chapter 4 aims to offer a potential step forward in the direction of subsuming the three paradigms within the third one i.e., building up upon *trust* towards 'peers', would they be humans, databases or mining algorithms.

4   The ViewpointS Approach Discussed and Exemplified

This section first briefly recalls the ViewpointS framework and formalism for building collective knowledge in the metaphor of the brain - a detailed description can be found in [18, 19] - and then illustrates them through an imaginary case.

In the ViewpointS approach, the "neural maps interconnected by beams of neurons" are transposed into a graph of "knowledge resources (agents, documents, topics) interconnected by beams of viewpoints". The "systems of values" of the agents influence not only the viewpoints they emit, but also the way they interpret the graph.

We call *knowledge resources* all the resources contributing to knowledge: agents, documents and topics. We call *viewpoints* the links between *knowledge resources*. Each *viewpoint* is a subjective connection established by an agent (Human or Artificial) between two *knowledge resources*; the *viewpoint* $(a_1, \{r_2, r_3\}, \theta, \tau)$ stands for: the agent a_1 believes at time τ that r_2 and r_3 are related according to the emotion carried by θ. We call Knowledge Graph the bipartite graph consisting of *knowledge resources* and *viewpoints*. Given two *knowledge resources*, the aggregation of the beam of all connections (*viewpoints*) linking them can be quantified and interpreted as a proximity. We call *perspective* the set of rules implementing this quantification by evaluating each viewpoint and then aggregating all these evaluations into a single value. The *perspective* is tuned by the "consumer" of the information, not by third a part "producer" such as Google or Amazon algorithms; each time an agent wishes to exploit the knowledge of the community, he does so through his own subjective *perspective* which acts as an interpreter.

Tuning a perspective may for instance consist in giving priority to trustworthy *agents*, or to the most recent *viewpoints*, or to the *viewpoints* issued from the logical paradigm. This clear separation between the storing of the traces (the *viewpoints*) and their subjective interpretation (through a *perspective*) protects the human agents involved in sharing knowledge against the intrusion of third-part algorithms reifying external system of values, such as those aiming at invading our psyche, influencing our actions [20], or even computing bankable profiles exploitable by brands or opinion-makers [21]. Adopting a *perspective* yields a tailored *knowledge map* where distances can be computed between knowledge resources, i.e. where the semantics emerge both from the topology of the *knowledge graph* and from our own system of values expressed by the tuned perspective.

The shared semantics emerge from the dynamics of the observation/action loops. Agents browse the shared knowledge through the *perspectives* they adopt (observation), and reversely update the graph by adding new *viewpoints* expressing their feedback (action). Along these exploitation/feedback cycles, shared knowledge is continuously elicited against the systems of values of the agents in a selection process.

To illustrate this, we develop below (Fig. 2) an imaginary case where learners have to select resources inside an Intelligent Tutoring System (ITS) to which a Knowledge Graph is associated. They wish to learn about the topic 'apple' and from step1 to step4 the learners adopt a 'neutral' perspective which puts in balance all types of viewpoints (issued from the logical or mining paradigms, or from the emotions and feelings of the learners). However at step5 B chooses a perspective discarding his own viewpoints in order to discover new sources of knowledge.

Step 1 illustrates the initial state of the knowledge. A, B and C are co-learners in the ITS (linked as such within the logical paradigm); the big arrows within the icons represent their respective systems of values, which play a key role both in the choice of *perspectives* and in the emission of *viewpoints*. D_1, D_2 and D_3 are documents that a mining algorithm has indexed by the topic/tag 'apple'.

Step 2: A is a calm person who has time; she browses through D_1, D_2 and D_3 and has a positive feeling about D_1 and D_2 (she likes both and finds them relevant with respect to 'apple'); the capture of her feedbacks results in linking D_1 and D_2 to her and reinforcing the links between the documents and the topic 'apple'. B is always in a hurry; he asks the Knowledge Graph the question "which is the shortest path between me and the topic 'apple'?" According to the paths in the diagram, he gets a double answer: B-A-D_1-'apple'and B-A-D_2-'apple'.

Step 3: B's feedback to D_1 is positive; this results in reinforcing the path B-A-D_1-'apple'. If he would re-ask his question, he would now get only D_1.

Step 4: C likes to explore; rather than taking a short path she browses through D_1, D_2 and D_3 and has a positive emotion about D_3 (she likes it and finds it relevant with respect to 'apple'); this results in linking D_3 to her and reinforcing the linking between D_3 and the topic 'apple'. At this stage, if A, B and C would ask for the shortest path to 'apple', they would respectively get D_1, D_1 and D_3.

Step 5: B is not fully satisfied by D_1. He asks for a novel short path between him and the topic 'apple' using a new perspective: he discards the viewpoints expressing his own feelings in order to discover new sources of knowledge. According to this new perspective, B-A-D_1-'apple', B-A-D_2-'apple'and B-C-D_3-'apple' have the same length i.e., D_1, D_2 and D_3 are equidistant from him. B may now discard D_1 (already visited) and D_2 (already rejected) and read D_3.

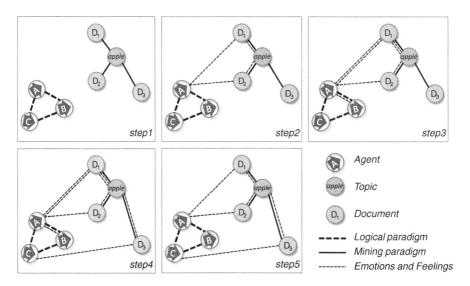

Fig. 2. The network of interlinked resources evolves along the attempts of the learners A, B and C to "catch" the topic 'apple' through performing the modules D1, D2 or D. What is figured in the schemas is not the Knowledge Graph itself, but the views (also called Knowledge Maps) resulting from the perspectives; in these maps, the more links between two resources, the closer they are.

Along the five steps of this imaginary case, the evolution of "knowledge paths" follows the metaphor of the selective reinforcement of neural beams, except that this reinforcement is not regulated by a single system of values, rather by a collaboration/ competition between the three systems of values of A, B and C. The three co-learners learn as a whole, in a trans-disciplinary way: the dynamics are governed by a topology mixing information and emotions, not by pure logics.

5   Conclusion

Starting from the three worlds proposed by K. Popper (the external world of objects and events, the internal world of mind and the world of language), we have browsed through the TNGS of G.M. Edelman and learnt how the learning events are supported by an adaptive neural network, are regulated by our systems of values and occur mainly through social interactions. We have then focused on Web interactions and reformulated the question of the emergence of collective knowledge partially supported by algorithms.

The ViewpointS approach and formalism offer to integrate most if not all these elements in the metaphor of a collective brain; we illustrate through an imaginary case how to trace and enhance collective knowledge acquisition. Within ViewpointS, the three paradigms for knowledge acquisition (logical inferences of the semantic Web, statistical recommendations of the mining community, authority and trust of the social Web) are merged into a knowledge graph of digital traces interpretable by human and artificial agents. Within this graph, the beams of connections are regulated by the individual systems of values that support affect i.e., culture, personality traits, as defined in [22].

What we gain in the integration may be lost with respect to the advantages of each of the three knowledge acquisition paradigms taken individually. For this reason, we are not yet sure if the collective knowledge emerging from ViewpointS graphs and maps (our proposed Collective Brain) will perform competitively with a similar wisdom emerging from each of the three crowds.

Nevertheless: as it has been always the case in the synergies between technological developments and scientific progress, the developments do not ensure scientific discovery, rather may facilitate the process. For instance: Galileo's telescopes did not directly produce the results of modern astronomy, but enabled a significant progress. We hope and believe that our proposed Collective Brain will have a positive impact in understanding and enhancing some aspects of human cognition.

References

1. Popper, K.: The Tanner Lectures on Human Values, pp. 143–167. University of Michigan, Michigan (1978)
2. Hampton, J.A.: Concepts in the Semantic Triangle. In: The Conceptual Mind : New Directions in the Study of Concepts (2016)
3. Searle, J.: Speech acts: an essay in the philosophy of language. Cambridge University Press, Cambridge (1969)
4. Edelman, G.M.: Neural Darwinism: The theory of neuronal group selection. Basic Books, New York (1989)
5. Edelman, G.M., Tononi, G.: Comment la matière devient conscience | Éditions Odile Jacob. Sciences, Paris (2000)
6. Vygotski, L.: Apprentissage et développement: tensions dans la zone proximale. Paris La Disput. 2ème éd. Augment. 233 (1933)
7. Vygotsky, L.S.: Mind in Society: The Development of Higher Psychological Processes, vol. 1978 (1978)
8. Laurillard, D.: A conversational framework for individual learning applied to the 'learning organisation' and the 'learning society'. Syst. Res. Behav. Sci. 16(2), 113–122 (1999)
9. Lemoisson, P., Passouant, M.: Un cadre pour la construction collaborative de connaissances lors de la conception d'un observatoire des pratiques territoriales. Cah. Agric. 21(1), 11–17 (2012)
10. Piaget, J.: La construction du réel chez l'enfant. In: La construction du réel chez l'enfant, Fondation Jean Piaget, Ed. Lonay, Suisse: Delachaux et Niestlé, pp. 307–339 (1937)
11. Cerri, S.A., Lemoisson, P.: Tracing and enhancing serendipitous learning with ViewpointS. Brain Function Assessment in Learning. LNCS (LNAI), vol. 10512, pp. 36–47. Springer, Cham (2017). https://doi.org/10.1007/978-3-319-67615-9_3
12. Vial, S.: L'être et l'écran - Comment le numérique change la perception (2013)

13. Cardon, D., Smyrnelis, M.-C.: La démocratie Internet. Transversalités, 123(3), 65–73, 13-Sep-2012. Institut Catholique de Paris
14. Gruber, T.: Collective knowledge systems: where the social web meets the semantic web. Web Semant. Sci. Serv. Agents World Wide Web **6**(1), 4–13 (2008)
15. Jain, P., et al.: Contextual ontology alignment of LOD with an upper ontology: a case study with proton, pp. 80–92. Springer, Heidelberg (2011)
16. Zhao, L., Ichise, R.: Ontology integration for linked data. J. Data Semant. **3**(4), 237–254 (2014)
17. Berners-Lee, T., Hendler, J., Lassila, O.: The Semantic Web. Sci. Am. **284**(5), 34–43 (2001)
18. Lemoisson, P., Surroca, G., Jonquet, C., Cerri, S.A.: ViewpointS: when social ranking meets the semantic web. In: AAAI Publications, The Thirtieth International Flairs Conference, Special Track on Artificial Intelligence for Big Social Data Analysis, Florida Artificial Intelligence Research Society Conference, North America (2017)
19. Lemoisson, P., Surroca, G., Jonquet, C., Cerri, S.A.: ViewPointS: capturing formal data and informal contributions into an evolutionary knowledge graph. Int. J. Knowl. Learn. **12**(2), 119–145 (2018)
20. Duchatelle, V.: Hacker son auto-prophétie, Paris (2017)
21. Kosinski, M., Bachrach, Y., Kohli, P., Stillwell, D., Graepel, T.: Manifestations of user personality in website choice and behaviour on online social networks. Mach. Learn. **95**(3), 357–380 (2014)
22. Bamidis, P.D.: Affective learning: principles, technologies, practice. Brain Function Assessment in Learning. LNCS (LNAI), vol. 10512, pp. 1–13. Springer, Cham (2017). https://doi.org/10.1007/978-3-319-67615-9_1

Intelligent Collectives: Impact of Diversity on Susceptibility to Consensus and Collective Performance

Van Du Nguyen[1,3(✉)], Hai Bang Truong[2], Mercedes G. Merayo[3],
and Ngoc Thanh Nguyen[1]

[1] Department of Information Systems,
Faculty of Computer Science and Management,
Wroclaw University of Science and Technology, Wrocław, Poland
{van.du.nguyen, Ngoc-Thanh.Nguyen}@pwr.edu.pl
[2] Faculty of Computer Science, University of Information Technology,
Vietnam National University Ho Chi Minh City (VNU-HCM),
Ho Chi Minh City, Vietnam
bangth@uit.edu.vn
[3] Dept. Sistemas Informáticos y Computación,
Universidad Complutense de Madrid, Madrid, Spain
mgmerayo@ucm.es

Abstract. In this paper, we present an approach to analyzing the impact of diversity, one of the most crucial determinants of intelligent collectives, on susceptibility to consensus and collective performance. In the common understanding, susceptibility to consensus refers to the situation in which the obtained collective prediction determined on the basis of individual predictions can be accepted as the representative for the collective as a whole. Computational experiments have indicated that when collectives are small, it is difficult to obtain a high probability of susceptibility to consensus. For large collectives, however, diversity seems not to matter susceptibility to consensus. Furthermore, the findings have also shown that higher collective performances can be the consequence of more diverse collectives. In other words, diversity is positively correlated with collective performance.

Keywords: Susceptibility to consensus · Wisdom of crowds
Collective intelligence

1 Introduction

Currently, along with the rapid development of information technologies, very often solving a common problem is entrusted to a large number of autonomous units (i.e. humans, agent systems). The effectiveness of such an approach has been widely applied to improve predictions on the outcomes of future events or estimating unknown quantities [2, 3, 9, 19, 21]. Moreover, the authors have found that the predictions based on collectives of uninformed individuals are equal to or even better than those

© Springer Nature Switzerland AG 2018
N. T. Nguyen et al. (Eds.): ICCCI 2018, LNAI 11055, pp. 13–22, 2018.
https://doi.org/10.1007/978-3-319-98443-8_2

generated by traditional forecasting approaches [2, 9]. Nofer et al. [18] have found that, on average, the collective performance is 0.59% points higher than that of experts.

By using a large number of autonomous units, one can tap into the so-called the wisdom of crowds (WoC) that is often understood as a situation in which collective predictions obtained by aggregating individual predictions are often superior to most (sometimes all) of individual predictions [21]. In the effectiveness of the WoC, diversity has often considered as the most crucial determinant [1, 3, 7, 19–21]. Nguyen et al. [14] have classified diversity into two kinds: diversity in the composition of members of a collective and diversity of their predictions on a given problem. However, most of research results so far have mainly based on the former kind of diversity. Only a few of them are based on the diversity of individual predictions. For instance, some results [19, 21] have suggested that a collective as a whole can give more accurate collective prediction if individual predictions are more diverse. It is due to the phenomenon of correlated error will be reduced in the process of aggregation [6, 10, 11].

In the common understanding, susceptibility to consensus refers to the situation in which the obtained collective prediction determined on the basis of individual predictions can be accepted as the representative for the collective as a whole. Consider the following example in which groups of people are asked for giving their responses to the question "*What is the minimum temperature for tomorrow in Wroclaw, Poland?*" Let us suppose that each collective provides three estimates (in degree) as follows: $X = \{-3, 1, 2\}$, $Y = \{-1, 0, 1\}$. Considering arithmetic mean, these collectives have the same collective estimate that is $0°$ In contrast to the collective X, the estimation of collective Y would be more reliable because the individual estimates of the collective Y are more consistent than those of the first one. Moreover, it can be seen that collective Y is more likely to be susceptible to consensus than collective X. This phenomenon leads to a natural question: "*How diversity affects susceptibility to consensus?*"

Apart from the impact of diversity on susceptibility to consensus, this paper also aims at analyzing the impact of diversity on collective performance. Suppose that a group of people is asked for providing predictions on an unknown quantity. This kind of problems has the following characteristics:

- The proper value is not known in advance;
- The proper value is independent of individual predictions;
- The collective prediction reflects the proper value to some degree.

After individual predictions are given, they are classified into two groups. The diversity value of the first one is maximal. For simplicity, we named this case as *MaxDiv* to differentiate it from the second case in which the diversity value of the collective is minimal, *MinDiv*. Intuitively, from the theory of consensus, we will be happy with the second group because its individual predictions are more consistent with each other. However, in this paper, we will show a paradox in which more diverse collectives will produce better collective performances. The measures of collective performance are based on the accuracy of collective prediction and the ratio of collective error to individual errors.

The rest of the paper is organized as follows. Section 2 presents some basic notions used in the current study. In Sect. 3, some related works are briefly discussed. The research model, computational experiments and their evaluation are presented in Sect. 4. The paper is ended with some conclusions and future research.

2 Preliminaries

This section briefly presents some basic notions from collective prediction, susceptibility to consensus and diversity measures, followed by the measures of collective performance.

2.1 Collective Prediction

Collective prediction, which is determined by means of the aggregation of individual predictions, is often considered as a representative of the collective as a whole [13]. In the current study, the individual predictions are given as follows:

$$X = \{x_1, x_2, \ldots, x_n\}$$

where n presents the cardinality of the collective and $X \in \prod(U)$. $\prod(U)$ is the set of all non-empty finite multisets of U, the set of values representing the potential predictions on a given problem.

In the present study we assume that the representation of each prediction is a single value and a criterion is used for collective prediction determination. Collective prediction is defined as follows (using arithmetic mean):

$$x^* = \frac{1}{n}\sum_{i=1}^{n} x_i$$

2.2 Diversity Measure

In this work, we define diversity as the variety of individual predictions on a specific problem. In order to measure diversity, we apply the functions defined in [13]. The first function is based on the pairwise distances between individual predictions.

$$c_3(X) = \begin{cases} \frac{1}{n(n-1)}\sum_{x_i,x_j \in X} d(x_i, x_j), & \text{for } n > 1 \\ 0, & \text{otherwise} \end{cases}$$

where n represents the cardinality of the collective X and $d(x_i, x_j)$ represents the distance between x_i and x_j.

According to $c_3(X)$, the more different individual predictions of a collective the higher level of diversity.

The second function is based on the distances from collective prediction to individual predictions.

$$c_5(X) = \frac{1}{n}\sum_{x_i \in X} d(x^*, x_i)$$

2.3   Susceptibility to Consensus

As stated earlier, we understand susceptibility to consensus as the collective prediction, which is obtained from the aggregation of the underlying predictions, can be accepted as the representative for the collective as a whole. In other words, if the individual predictions of a collective are dense enough, then the collective prediction will be more reliable. The criterion of susceptibility to consensus has been workout out by Nguyen in [13].

Given a collective X, by susceptibility to consensus, collective X should satisfy the following condition:

$$d_{t\_mean}(X) \geq c_5(X)$$

where $d_{t\_mean}(X) = \frac{1}{n(n+1)} \sum_{x_i, x_j \in X} d(x_i, x_j)$ represents the average distance of all distances between predictions of collective X.

2.4   Collective Performance

Given a collective X, let x^* be its collective prediction and r be the proper value. We use two different functions to define the measure of collective performance.

- $Diff(X) = 1 - d(r, x^*)$
- $QIC(X) = 1 - \frac{d^2(r, x^*)}{(d^2(r, X)/n)}$

The first function aims at reflecting the accuracy of collective prediction. The higher collective performance, the closer collective prediction to the proper value. Meanwhile, the second one represents the capability of a collective as a whole in comparison with its individuals in solving a given problem.

3   Related Work

This section briefly introduces recent research on susceptibility to consensus as well as the role of diversity in improving collective performance.

The problem of susceptibility to consensus was introduced by Nguyen [13]. As its name implies, a collective whose individual predictions are dense enough will be called susceptible to consensus. Initially, it has been mainly applied in the context of the problem of knowledge integration for evaluating if a representative of a collective, which is determined on the basis of knowledge states of a collective, is reliable or not. In [8], the author has investigated susceptibility to consensus with some representations of knowledge including real numbers, ordered partitions, ordered coverings, and binary vectors [8]. Recently, susceptibility criterion has been taken into account in determining collective predictions of large collectives via two-stage consensus choice [17]. In this work, it is analyzed the relationship between diversity and susceptibility to consensus with different scenarios of collective cardinality.

The research results on the important role of diversity are mainly based on the diversity of collective members [1, 3, 7, 19–21]. With this kind of diversity, the more diverse are the members of the collective, the better is the collective prediction determined from individual predictions. It has been shown [19, 21] that this kind of diversity is essential for obtaining more accurate collective prediction because it effectively reduces correlated errors among individual predictions of a collective [6, 10, 11].

To date, the diversity of a collective can be enhanced by expanding its cardinality [4]. Indeed, it can be seen that maintaining diversity in a large collective is easier than such a task in a small collective. Regarding cardinality, the experiments conducted by Wagner et al. [22] have revealed that the higher collective performance is a consequence of higher collective cardinality. In [4] the authors have also found similar results.

By means of computational experiments [15, 16], we have found that the larger the collective we use, the higher the quality of collective prediction we get.

4 Computational Experiments and Their Evaluation

As stated earlier, this paper presents an analysis of the impact of diversity on susceptibility to consensus as well as on collective performance with different scenarios of cardinality. In particular, we intend to analyze, how diversity affects susceptibility to consensus and collective performance when collectives have the same cardinality. For this aim, first, a general research model is presented. Then, we introduce the simulation design, followed by experimental results and their evaluation.

4.1 Research Model and Settings

The general research model of the current study is depicted in Fig. 1. Notice that we only take into account diversity of individual predictions.

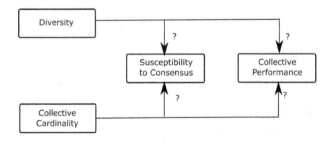

Fig. 1. General research model

In this study, U represents a set of predictions such that the difference between each of them and the proper value does not exceed a predefined threshold ∂ (that is, $(\forall x_i \in U) : d(r,x) \leq \partial)$.

The conducted computational experiment is similar to the task of estimating an unknown quantity. For this assumption, we use $\partial = 1000$, and $r = 1000$. Accordingly, the individual predictions of each collective are randomly generated from the interval [0, 2000] (uniform distribution). Additionally, we use a large collective that is divided into two smaller ones of the same cardinality:

- maximal diverse collective (*MaxDiv*);
- minimal diverse collective (*MinDiv*).

Considering these settings, the computational results have indicated that the cardinality needed to obtain a reliable prediction (plus or minus 100) with 95% confidence interval must be around 129. Therefore, in the following experiments, we use three scenarios in which the collectives have different cardinalities:

- Small collectives: cardinality of 9;
- Medium collectives: cardinality of 129;
- Large collectives: cardinality of 229.

4.2   Impact of Diversity on Susceptibility to Consensus

Figure 2 depicts the results obtained from the experiments. It shows that when collectives are small (M09), in the case of c_3-based diversity, only 72% *MaxDiv* collectives are susceptible to consensus, while in the case of c_5-based diversity they reach 76%. Conversely, all *MinDiv* collectives satisfy the condition of susceptibility to consensus. In the case of large collectives (M129 and M229), however, *MaxDiv and MinDiv* collectives are susceptible to consensus.

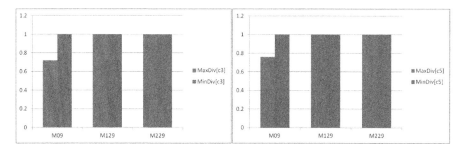

Fig. 2. Impact of diversity on susceptibility to consensus

From these results, we can conclude that diversity is negatively associated with susceptibility to consensus in small collectives (M09) but not so in large collectives (M129 and M229). An important question that needs to be addressed at this point is which collective will provide higher collective performance? We address this research problem in the following section.

4.3   The Relationship Between Diversity and Collective Performance

According to Fig. 3, in the case of *Diff* measure, collective performances of *MaxDiv* collectives are higher than those of *MinDiv* collectives. In particular, in the case of c_3-based diversity, *Diff* values range from 0.8893 (M09) to 0.9848 (M229) for *MaxDiv*

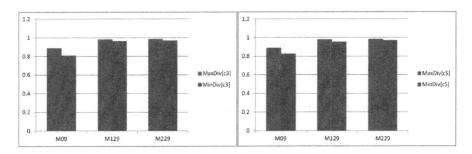

Fig. 3. Simulation results using *Diff* measure

collectives. Meanwhile, for *MinDiv* collectives, *Diff* values range from 0.8100 (M09) to 0.9730 (M229).

Similarly, in the case of c_5-based diversity, *Diff* values range from 0.8914 (M09) to 0.9837 (M229) for *MaxDiv* collectives. Meanwhile, for *MinDiv* collectives, *Diff* values range from 0.8280 (M09) to 0.9720 (M229).

According to Fig. 4, in the case of c_3-based diversity, *QIC* values range from 0.9591 (M09) to 0.9993 (M229) for *MaxDiv* collectives. Meanwhile, for *MinDiv* collectives, *QIC* values range from 0.7320 (M09) to 0.9820 (M229). Similarly, in the case of c_5-based diversity, *QIC* values range from 0.9572 (M09) to 0.9994 (M229) for *MaxDiv* collectives. Meanwhile, for *MinDiv* collectives, *QIC* values range from 0.7392 (M09) to 0.9864 (M229).

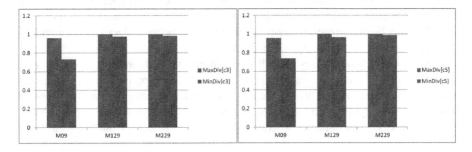

Fig. 4. Simulation results using *QIC* measure

As it is shown in above figures, collectives with higher diversity levels give higher collective performances. However, the difference between collective performances with previous settings is statistically significant. We will perform statistical tests to determine whether such differences are statistically significant by using the nonparametric Mann-Whitney U Test. For all statistical tests, the significance level 0.05 is used. The results of statistical tests are reported in Tables 1 and 2.

According to above tables, the differences between *Diff*-based (as well as *QIC*-based) collective performances are statistically significant. Therefore, it can be reasonable to conclude that diversity of individual predictions is positively correlated with

Table 1. Results of statistical analysis [c_3-based diversity]

	M09	M129	M229
Diff	5.46E − 05	3.95E − 08	2.04E − 06
QIC	0.0	0.0	0.0

Table 2. Results of statistical analysis [c_5-based diversity]

	M09	M129	M229
Diff	5.35E − 07	1.59E − 07	4.63E − 05
QIC	0.0	0.0	0.0

collective performance. From these results, it can be seen that maintaining the diversity of individual predictions is an important task in the use of WoC. Recently, an approach has proposed [12] to fostering diversity by introducing an incentive scheme for rewarding accurate minority predictions. Moreover, this kind of diversity can be the consequence of the existence of the former kind of diversity. These issues will be the subject of our future research.

5   Conclusions and Future Works

Diversity has been proven effective in leading to more accurate collective predictions based on the use of the WoC [1, 3, 5, 7, 20]. The current study has presented an approach to analyzing the relationships between diversity and collective performance as well as between diversity and susceptibility to consensus. The results have indicated that collectives having higher diversity levels will give better collective performances in comparison with those having lower diversity levels.

Regarding its impact on susceptibility to consensus, we found that when the cardinality is small, diversity is negatively associated with the probability of being susceptible to consensus. However, when collectives are large enough, diversity seems do not matter the susceptibility to consensus of a collective. In other words, diversity has a negative impact on susceptibility to consensus in small collectives but not so in large collectives.

As the future work, we intend to investigate how independence and decentralization play a role in the effectiveness of intelligent collectives. Furthermore, we also will take into consideration other representation of individual predictions such as interval values.

Acknowledgment. This research is partially supported by the projects DArDOS (TIN2015-65845-C3-1-R (MINECO/FEDER)) and SICOMORo-CM (S2013/ICE-3006) and and partially funded by Vietnam National University Ho Chi Minh City (VNU-HCM) under grant number C2018-26-09.

References

1. Adebambo, B.N., Bliss, B.: The value of crowdsourcing: Evidence from earnings forecasts. Working Paper (2015)
2. Arazy, O., Halfon, N., Malkinson, D.: Forecasting rain events - meteorological models or collective intelligence? In: Proceedings of EGU General Assembly Conference, vol. 17, pp. 15611–15614 (2015)
3. Armstrong, J.S.: Combining Forecasts. In: Armstrong, J.S. (ed.) Principles of Forecasting. International Series in Operations Research & Management Science, vol. 30. Springer, Boston (2001). https://doi.org/10.1007/978-0-306-47630-3_19
4. Bassamboo, A., Cui, R., Moreno, A.: The Wisdom of Crowds in Operations: Forecasting Using Prediction Markets (2015)
5. Campbell, K., Mínguez-Vera, A.: Gender diversity in the boardroom and firm financial performance. J. Bus. Ethics **83**, 435–451 (2008)
6. Gigone, D., Hastie, R.: Proper analysis of the accuracy of group judgments. Psychol. Bull. **121**, 149–167 (1997)
7. Hong, L., Page, S.E.: Groups of diverse problem solvers can outperform groups of high-ability problem solvers. Proc. Nat. Acad. Sci. U.S.A **101**, 16385–16389 (2004)
8. Kozierkiewicz-Hetmańska, A.: Analysis of susceptibility to the consensus for a few representations of collective knowledge. Int. J. Softw. Eng. Knowl. Eng. **24**, 759–775 (2014)
9. Lang, M., Bharadwaj, N., Di Benedetto, C.A.: How crowdsourcing improves prediction of market-oriented outcomes. J. Bus. Res. **69**, 4168–4176 (2016)
10. Larrick, R.P., Soll, J.B.: Intuitions about combining opinions: misappreciation of the averaging principle. Manag. Sci. **52**, 111–127 (2006)
11. Lorge, I., Fox, D., Davitz, J., Brenner, M.: A survey of studies contrasting the quality of group performance and individual performance. Psychol. Bull. **55**, 337–372 (1958)
12. Mann, R.P., Helbing, D.: Optimal incentives for collective intelligence. Proc. Nat. Acad. Sci. **114**, 5077–5082 (2017)
13. Nguyen, N.T.: Advanced Methods for Inconsistent Knowledge Management. Advanced Information and Knowledge Processing. Springer-Verlag, London (2008). https://doi.org/10.1007/978-1-84628-889-0
14. Nguyen, V.D., Nguyen, N.T.: An influence analysis of diversity and collective cardinality on collective performance. Inf. Sci. **430**, 487–503 (2018)
15. Nguyen, V.D., Nguyen, N.T.: An influence analysis of the inconsistency degree on the quality of collective knowledge for objective case. In: Nguyen, N.T., Trawiński, B., Fujita, H., Hong, T.-P. (eds.) ACIIDS 2016, Part I. LNCS (LNAI), vol. 9621, pp. 23–32. Springer, Heidelberg (2016). https://doi.org/10.1007/978-3-662-49381-6_3
16. Nguyen, V.D., Nguyen, N.T.: A method for improving the quality of collective knowledge. In: Nguyen, N.T., Trawiński, B., Kosala, R. (eds.) ACIIDS 2015, Part I. LNCS (LNAI), vol. 9011, pp. 75–84. Springer, Cham (2015). https://doi.org/10.1007/978-3-319-15702-3_8
17. Nguyen, V.D., Nguyen, N.T., Hwang, D.: An improvement of the two-stage consensus-based approach for determining the knowledge of a collective. In: Nguyen, N.-T., Manolopoulos, Y., Iliadis, L., Trawiński, B. (eds.) ICCCI 2016, Part I. LNCS (LNAI), vol. 9875, pp. 108–118. Springer, Cham (2016). https://doi.org/10.1007/978-3-319-45243-2_10
18. Nofer, M., Hinz, O.: Are crowds on the internet wiser than experts? the case of a stock prediction community. J. Bus. Econ. **84**, 303–338 (2014)
19. Page, S.E.: The Difference: How the Power of Diversity Creates Better Groups, Firms, Schools, and Societies. Princeton University Press, Princeton, NJ (2007)

20. Robert, L., Romero, D.M.: Crowd size, diversity and performance. In: Proceedings of the 33rd Annual ACM Conference on Human Factors in Computing Systems, pp. 1379–1382 (2015)
21. Surowiecki, J.: The Wisdom of Crowds. Doubleday/Anchor, New York (2005)
22. Wagner, C., Suh, A.: The wisdom of crowds: impact of collective size and expertise transfer on collective performance. In: Proceedings of 47th Hawaii International Conference on System Sciences, pp. 594–603 (2014)

The Increasing Bias of Non-uniform Collectives

Marcin Maleszka[(⊠)]

Faculty of Computer Science and Management, Wroclaw University of Science
and Technology, st. Wyspianskiego 27, 50-370 Wroclaw, Poland
marcin.maleszka@pwr.edu.pl

Abstract. In this paper we make initial study of the influence of initial
bias in a collective of agents on its knowledge or opinion, after taking
into account internal communication between agents. We provide details
about the model of collective that we use, with different levels of communication and different strategies utilized by agents to integrate messages
into their internal knowledge base. We then perform a simulation of such
collective, with introduced different number of biased agents. We observe
how these agents influence the overall knowledge of the collective over
time. The experiment shows that even a small percentage of biased agents
changes the views of the whole collective. We discuss the implications of
this result in possible practical applications.

Keywords: Collective intelligence · Collective knowledge
Knowledge integration

1 Introduction

There are many challenges present in the modern research in the area of collective
intelligence and more and more of them come to represent real world problems
that need automated solutions. One of those challenges is the modeling of groups
of people communicating about some issues and different situations that may
occur during this process. Collective knowledge integration is one such problem,
in which a group of agents work towards determining some common opinion,
or sometimes, an external observer tries to determine what can be derived from
knowledge of each individual in the group, when treated as a whole. In that
case it may be done by simple calculation of a common knowledge state (if it is
intermediate of others [16]) or summarization and inference based on the gathered knowledge. There are multiple methods solving this task as a mathematical
calculation, many of them derived from methods used to determine medians of
data structures [2] or situated in a centralized supervisor agent calculating this
for a whole multi-agent system [10].

The approach used in our research goes back to the underlying assumption
of those methods – that the calculation occurs immediately. We extend the integration process over time, making it closer to knowledge diffusion models in

© Springer Nature Switzerland AG 2018
N. T. Nguyen et al. (Eds.): ICCCI 2018, LNAI 11055, pp. 23–30, 2018.
https://doi.org/10.1007/978-3-319-98443-8_3

social network research (but as the aim of the model is different, our research is only marginally related to that area). We define the behavior of each member of the collective and their relations, then observe how the group behaves and how its opinions change over time. Thus we mostly model the communication occurring in the collective and its influence on the consensus of the whole group. During the development of the model it became similar to before-mentioned social network research, thus we further improve on that aspect to make it more similar to Twitter social network – specifying two different levels of communication between agents and importing the notion of sentiment inherent in Twitter communicates.

With the collective communication model we tackle the problem of inherent bias in the collective and how it changes over time. In this paper we investigate how large a portion of the collective needs to be biased (have opinion that is significantly different than the consensus of the collective and not be influenced by other members of the group) to over time change the collective knowledge or opinion of the whole group. This could be applied in general research into behavior of collectives and in areas such as so-called "fake news" analysis, in terms of their spread and influence.

This paper is organized as follows: in Sect. 2 we present other research relevant to one presented in this paper. Section 3 contains the detailed description of a collective model we use for our simulation. In Sect. 4 we describe the setup of the simulation and the results derived from it, while in Sect. 5 we present some concluding remarks and detail the work necessary for further enhancing the model and other observations on intelligent collectives that it may help to obtain.

2   Related Works

In this paper we discuss a model of a collective with internal communication occurring during the integration process between individual agents in the group, therefore we present some relevant works covering all the elements of the model.

Communication between agents in multi-agent systems is done by a whole spectrum of approaches between centralized and decentralized ones. The first group is used in systems such as traffic control [10] or power distribution [15]. In those, there is a group of agents that observe some situations or are actuators for some function; and a single supervisor or facilitator agent that gathers observations, makes decisions, or provides an interface for the human operator (e.g. presenting him pre-calculated possible decisions, so the user may decide the actual course of action). A single agent may oversee single traffic light, crossroads or power distribution node. They may share information between each other or be organized in a hierarchy (by geographical level or functionalities), where at each higher level the agents integrate data from lower level and forward it towards the top of the hierarchy [1].

The decentralized systems of agents on the other hand don't have the central supervisor. All agents with the same functions or sensors work towards the same goal using the same methods, in extreme cases with all agents in the multi-agent system being identical and only exchanging small snippets of information in their neighborhood [17]. The system may include some verification agents that might influence individual agents, but that do not make large scale operations in the system [5]. Hybrid centralized-decentralized systems have advantages of both types of systems, including reduced delays and increased robustness [9]. There may also be some additional structure of communication acting as a second layer of this, as shown in [14] to increase productivity in groups working towards the same purpose. This approach is most similar to the one we call asynchronous decentralized communication with underlying structure, that we use in our research.

Another area of computer science research we consider in our work are methods of data and knowledge integration, specifically the approaches to consensus calculation. This is done in multiple time-dependent consensus systems, e.g. the continuous-time approach that is used in network systems and autonomous robots [18] and applied to flocking, formation control, negotiations and more; and the finite-time approach that is considered in terms of multi-agent systems stability [4] (if all agents converge to the same consensus in finite time). These systems also cover centralized and decentralized approaches, e.g. the leader-follower and leaderless systems observed in [12].

The consensus of some data (knowledge) is also considered as a result of some mathematical function, e.g. based on graph theory [6]. More often the mathematical methods are based on research done by bio-mathematicians to determine medians of phylogenetic trees and the following works [2]; and often put under the common topic of consensus theory [8,16]. This methodology has been used over years and has been proven useful in solving conflicts, as well as being effective for knowledge inconsistency resolution (which may occur during knowledge integration). The general notion of consensus and postulates for consensus choice functions (also called knowledge functions) were defined as: reliability, unanimity, simplification, quasi-unanimity, consistency, Condorcet consistency, general consistency, proportion, 1-optimality and 2-optimality. Over time, multiple classes of consensus functions were analyzed in terms of these postulates, as well as criteria specific to more practical applications or data structures (logical, numeric, hierarchical, graph, ontological). In this paper we use some specific functions for determining consensus derived from those works. The main difference is that we consider knowledge instead of data.

3   Model of the Collective

In our research we are working with a model of a collective that is geared towards the representation of processes occurring in the collective during the process when it determines its overall knowledge (opinion). In most cases this could be named the integration process, but it also includes elements of knowledge diffusion and may be used towards this purpose (but it is not competitive with models in this area, as it is not geared towards this purpose). This section provides an overview of the version of the model used in this part of our research.

Each agent stores its knowledge in the form of declarative statements (or in the case of opinions, single issues) in the form "'A – not A"' (e.g. "'light bulbs require electricity – light bulbs do not require electricity"'). We consider a total of K_T such statements, and each agent has a vector of $K_i \in \{1, \ldots, K_T\}$ statements. Additionally, there is a function S_i, which for each agent assigns a numerical value that the agent associates with each statement, in range $[0, 10]$ (where 0 means neutral opinion towards statement, 10 means that the agents finds the issue important). We call this the sentiment function, as it corresponds to the notion of sentiment in Twitter social network analysis.

Agents may generate communicates about their knowledge. A single issue is selected with a distribution based on sentiment (statements with high sentiment occur more often, statements with neutral sentiment will never occur) and a communicate is a pair consisting of this statement and its sentiment value.

The collective is a set of agents with knowledge and sentiment defined as above. Additionally, agents have a list of other agents that they consider to be preferential receivers (modeling e.g. friends, followers). Periodically agents may send out communicates and continuously they monitor for any incoming messages. The communicates are formulated as previously described. Each discrete time moment there is a given chance that each agent will send out a message. In that case, he determines the number of receivers, either from the list of preferential receivers (with a given chance) or from the overall population (preferential receivers may be also randomly selected from the second group) – the selection is done with uniform distribution from a given group, with repetitions (to reflect real world situations when a person reads the same message twice without realizing).

When receiving the messages, the agents work to integrate it with their own knowledge, according to one of possible integration strategies. Each agent has its own strategy, selected at random, which in this model does not change over time. There are five main strategies available:

- Discard – in this strategy agents do not react to incoming messages, instead discarding all new knowledge.
- Substitute – in this strategy agents value all incoming knowledge as more important than their own. If they already had the received elements in their internal knowledge base, the new sentiment is used instead of previous one. If the element itself is new to agents knowledge, it is added to the knowledge base as another statement with the received sentiment.
- Extend – this strategy is similar to the previous one, but if the agent already knew some issue, their own sentiment is kept.
- Merge now – in this integration strategy the received knowledge immediately influences the knowledge and sentiment of the agent. If the received knowledge is new (previously unknown), the agent incorporates it, with the sentiment as received. If the received statements were already known, new sentiment is calculated as the average of the old one and the received one.
- Merge later – in this strategy the agents store incoming messages and do not immediately incorporate them. Instead, once a larger number of

communicates about the same issue is received, the new value of sentiment for this issue is calculated. There are several variants of this complex strategy depending on the method of calculating the new sentiment:

- Voting – the actual value of the sentiment in received messages is disregarded. The number of positive and negative opinions is calculated and the winning vote will be given maximum weight. In case of a tie, the sentiment is neutral (0).
- Average – the new sentiment is the average of own knowledge and received communicates.
- Simplified median – the sentiment values from communicates and own knowledge are ordered from smallest to largest, without repetitions, and the middle one is selected. In case of even number of opinions, one of two middle elements is chosen randomly.
- Median – the approach is similar to the previous one, but repetitions may occur.

The whole model consists of a whole network of such agents (collective) operating in the mentioned time moments. Additionally, the state of the collective and of individual agents is saved both in the initial and in the final state. If the model were to be considered in terms of social network knowledge diffusion one, it would be a predictive linear threshold model [7] or an evolutionary SIRS model [11]. In both cases it only incorporates elements of methodology from that area and is not geared towards that research.

4   Simulation

We use the model described in the previous section to observe how the initial bias of the collective influences its final consensus. By bias we understand an unbalanced number of agents favoring some issue (non-uniform distribution of sentiment with average different than neutral). The initialized collective is biased (a number of agents have maximum sentiment on some issue) and remains so (these agents have discard strategy and otherwise operate as defined in the model). We measure its state at the beginning and after a given number of iterations.

The simulation environment itself was created in JADE agent environment [3] and runs on a standard PC. There are a number of standard agents created according to the model and additional centralized agents required by JADE itself to function – these do not change the fact that the simulation itself is decentralized, but we use them to gather data about the initial state of the collective, as well as the final state before ending the simulation.

Upon initialization the agents function for some time to simulate a real collective discussing different issues. In initial runs we tuned the parameters of the system to the following, for best presentation and to better simulate real collectives (as shown in our previous research [13]):

- Total number of knowledge statements possible $K = 100$
- Maximum number of issues for each agent at initialization $K_A = 20$

- Maximum weight of an issue $W = 10$
- Total number of agents in simulation $N = 1000$
- Maximum number of preferred receivers for each agent $F = 10$
- Probability of communication in each tick $P_c = 20$
- Probability of communication to preferred receiver $P_f = 50$
- Number of receivers for each issue $R = 2$
- Size of message buffer for "Merge later" strategy $T = 10$
- Number of ticks, length of simulation $\tau = 1000$

Additionally, all agents have various strategies outlined in the model, with uniform distribution. Statistically, the discard agents in uniform distribution do not introduce the bias. For this, we add additional agents to the collective (the uniformly distributed agents and the biased ones sum up to N). Also, for implementation simplicity, we simulate negative statements with negative sentiment – this is not in conflict with the model and does not change the results of the simulation.

We conducted the simulation in 100 repetitions for each number of biased agents between 0 and 100 (up to 10% of the collective). We calculated the initial sentiment of the group on the discussed issue and the final sentiment (the sum of sentiments). Then we determined the global change and divided it by the number of agents to show the average increase of sentiment on the given issue for each individual agents. The results for the smaller number of biased agents

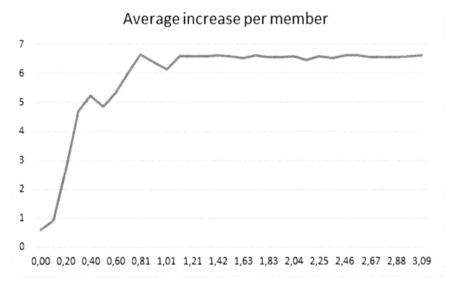

Fig. 1. The increase of average sentiment per member after the course of the simulation, with increasing participation of biased collective members in the initial phase. Sentiment is scaled from -10 to 10 (increase on graph is from 0 to 6.5). Number of biased members is in range of 0–3.1% of the total size of the collective.

are shown in Fig. 1. In latter phases, the final sentiment remains similar, but as the initial is larger (a larger number of agents are biased), the calculated average decreases. As this result was expected and not important for our research, we do not show these results in this paper.

In general, the obtained results show that even for small number of biased members of the collective (less than 1%), the impact of their influence is very visible. The average opinion of a collective member changes from neutral (from -0.1 to $+0.1$) to positive (around 6.5–6.6). Higher values are not attained due to influence of uniformly distributed discard agents and other aspects of communication. Higher numbers of biased members of collective are not required, as the change in collective opinion does not increase over some threshold. This result is especially important for practical applications in advertising and for determining the impact of so-called "fake news", as it shows that even small numbers of advertisement agents (social network accounts) may be enough to influence the whole collective.

5   Conclusions

This paper contains the description of a model of an agent collective with a focus on their communication as they work towards exchanging knowledge (opinion). We use it to observe how the initial bias of the collective influences the further evolution of its knowledge. We define this bias as an unbalanced number of agents favoring some issue, that is non-uniform distribution of sentiment with average different than neutral. In an agent simulation we observe how this initial bias leads to large changes in the overall collective. We have shown that for this type of collective that threshold is around 6.5–6.6 points increase of sentiment value (in a total range 0–10), but it may differ for other types of knowledge structures or other parameters of the collective.

One of our intended areas of future research is studying those other situations and the influence of the bias in them. We also intend to move the model towards an ontological representation of knowledge, as it should better represent knowledge of people. In that case the ontology would be a graph with two types of nodes (concepts, individuals) and relations represented by edges. Sentiment would describe any of those elements and a communicate of an agent would be any possible sub-graph of the whole.

Another area of research to consider are other approaches to internalization of knowledge by the agents. This will require extensive testing in cooperation with sociologists to better tune the model to real-world groups of users. More future work can be also done to move the model towards an influence maximization model of social network, as those have many important practical applications.

Acknowledgment. This research was co-financed by Polish Ministry of Science and Higher Education grant.

References

1. Abed-alguni, B.H., Chalup, S.K., Henskens, F.A., Paul, D.J.: A multi-agent cooperative reinforcement learning model using a hierarchy of consultants, tutors and workers. Vietnam J. Comput. Sci. **2**(4), 213–226 (2015)
2. Barthelemy, J.P., Janowitz, M.F.: A formal theory of consensus. SIAM J. Discrete Math. **4**, 305–322 (1991)
3. Bellifemine, F., Agostino, P., Giovanni, R.: JADE-A FIPA-compliant agent framework. In: Proceedings of PAAM, vol. 99, no. 97–108 (1999)
4. Bhat, S.P., Bernstein, D.S.: Finite-time stability of continuous autonomous systems. SIAM J. Control Optim. **38**(3), 751–766 (2000)
5. Chaimontree, S., Atkinson, K., Coenen, F.: A multi-agent based approach to clustering: harnessing the power of agents. In: Cao, L., Bazzan, A.L.C., Symeonidis, A.L., Gorodetsky, V.I., Weiss, G., Yu, P.S. (eds.) ADMI 2011. LNCS (LNAI), vol. 7103, pp. 16–29. Springer, Heidelberg (2012). https://doi.org/10.1007/978-3-642-27609-5_3
6. Chen, G., Lewis, F.L., Xie, L.: Finite-time distributed consensus via binary control protocols. Automatica **47**, 1962–1968 (2011)
7. Chen, H., Wang, Y.T.: Threshold-based heuristic algorithm for influence maximization. J. Comput. Res. Dev. **49**, 2181–2188 (2012)
8. Dubois, D., Liu, W., Ma, J., Prade, H.: The basic principles of uncertain information fusion. An organised review of merging rules in different representation frameworks. Inf. Fusion **32**, 12–39 (2016)
9. Hale, M.T., Nedic, A., Egerstedt, M.: Cloud-Based centralized/decentralized multi-agent optimization with communication delays. arXiv preprint arXiv:1508.06230 (2015)
10. Iscaro G., Nakamiti G.: A supervisor agent for urban traffic monitoring. In: IEEE International Multi-Disciplinary Conference on Cognitive Methods in Situation Awareness and Decision Support (CogSIMA), pp. 167–170. IEEE (2013)
11. Jin, Y., Wang, W., Xiao, S.: An sirs model with a nonlinear incidence rate. Chaos, Solitons Fractals **34**, 1482–1497 (2007)
12. Li, S., Dua, H., Lin, X.: Finite-time consensus algorithm for multi-agent systems with double-integrator dynamics. Automatica **47**, 1706–1712 (2011)
13. Maleszka, M.: Observing collective knowledge state during integration. Expert Syst. Appl. **42**(1), 332–340 (2015)
14. De Montjoye, Y.-A., Stopczynski, A., Shmueli, E., Pentland, A., Lehmann, S.: The strength of the strongest ties in collaborative problem solving. Sci. Rep. 4, 5277 (2014).
15. Nagata, T., Sasaki, H.: A multi-agent approach to power system restoration. IEEE Trans. Power Syst. **17**(2), 457–462 (2002)
16. Nguyen, N.T.: Advanced Methods for Inconsistent Knowledge Management. Springer, London (2007). https://doi.org/10.1007/978-1-84628-889-0
17. Peterson, C.K., Newman, A.J., Spall, J.C.: Simulation-based examination of the limits of performance for decentralized multi-agent surveillance and tracking of undersea targets. In: Signal Processing, Sensor/Information Fusion, and Target Recognition XXIII, vol. 9091, p. 90910F. International Society for Optics and Photonics (2014)
18. Ren, W., Beard, R.W., Atkins, E.M.: A survey of consensus problems in multi-agent coordination. In: Proceedings of the American Control Conference 2005, pp. 1859–1864. IEEE (2005)

Framework for Merging Probabilistic Knowledge Bases

Van Tham Nguyen[1,3(✉)], Ngoc Thanh Nguyen[2],
and Trong Hieu Tran[3]

[1] Faculty of Information Technology,
Nam Dinh University of Technology Education, Namdinh, Vietnam
thamnv.nute@gmail.com, 17028002@vnu.edu.vn
[2] Faculaty of Computer Science and Management,
Wroclaw University of Science and Technology, Wroclaw, Poland
Ngoc-Thanh.Nguyen@pwr.edu.pl
[3] Faculty of Information Technology,
VNU University of Engineering and Technology, Hanoi, Vietnam
hieutt@vnu.edu.vn

Abstract. Knowledge merging is of major concern in developing probabilistic expert systems. Each system provides a consistent probabilistic knowledge while the merged knowledge base is often inconsistent. Because of this reason, a wide range of approaches has been put forward to merge probabilistic knowledge bases. However, the input of the models is the set of possible probabilistic functions representing the original probabilistic knowledge bases. In this paper, we investigate a framework for merging probabilistic knowledge bases represented by the new form. To this aim, a process to merge probabilistic knowledge bases is introduced, several transformation methods for the representation of the original probabilistic knowledge base is presented, a set of merging operators is proposed, and several desirable logical properties are investigated and discussed.

Keywords: Probabilistic knowledge base · Knowledge merging
Merging operator

1 Introduction

In order to build and maintain the action of knowledge-based systems, it is necessary to develop appropriate methods. To ensure the interaction among systems, the knowledge provided by these systems has to be consistent. Knowledge merging could be understood as a process of creating a new single consistent knowledge from a set of knowledge bases belonged to different systems [1]. Therefore, knowledge merging is an important problem and its applications are many and diverse [2]. This is a difficult problem because merging process could yield the knowledge change requirement in order to ensure the consistency of merged result. It is therefore essential to consider these changes [3]. Knowledge merging could be observed in two respects: Merging knowledge bases which are not similar and merging a set of different representations for same knowledge base but different in representing degree [4]. Therefore, the first task in

© Springer Nature Switzerland AG 2018
N. T. Nguyen et al. (Eds.): ICCCI 2018, LNAI 11055, pp. 31–42, 2018.
https://doi.org/10.1007/978-3-319-98443-8_4

the knowledge merging process is to solve the inconsistency in knowledge bases. A class of basic inconsistency measures for probabilistic framework has been introduced in [5, 6]. In a probabilistic environment, some strategies have been developed to address the inconsistency of a knowledge base. The method of modification of probabilities is one of the most common approaches. This method has been applied in [7–10].

A general model of knowledge merging referring to its inconsistency in distribution aspect was presented in [1]. The knowledge merging problem, postulates for knowledge merging and algorithms for this process were also worked out and discussed in this work. There are some merging approaches presented in the propositional knowledge [11–13]. Those strategies often build a family of merging operators that based on distance functions and point out the logical relation between such operators and a set of desirable properties. There are many different methods to solve the possibilistic knowledge merging issue [14, 15]. The main idea of these methods is that merging process includes two stages: the separation and the combination. The merging results are usually consistent when using the propositional merging operators, while the merging results may be inconsistent if using the possibilistic merging operators.

The idea of probabilistic merging has been introduced and discussed in [16–19]. A common method for merging probabilistic knowledge bases in expert systems could be found in [16]. This method was based on the KL-projection means. A family of probabilistic merging operators was first introduced [17] and they were further studied in [18, 19]. These operators defined by convex Bregman divergences satisfied several logical properties in [19]. The probabilistic knowledge merging process is to build a joint probability distribution from an input set of low-dimensional distributions.

The main contribution of this paper is threefold. First, we propose a method to represent a consistent probabilistic knowledge base through a set of probability functions satisfied such base. Second, we introduce a family of operators which are used to merge probabilistic knowledge bases. Then, we consider several logical properties being desirable when merging probabilistic knowledge bases and prove the satisfaction of the proposed family of operators obeying these properties.

This paper is organized as follows. In Sect. 2 we review some necessary preliminaries about probability function, probabilistic constraint, the representation of knowledge bases in a probabilistic framework, and define a probabilistic knowledge base profile. Afterward, in Sect. 3 we will propose a common model to merge probabilistic knowledge bases into a consistent one. The way to determine probabilistic merging vector is also presented in this section. The desirable logical properties for merging operators are also drawn and discussed in Sect. 3. Conclusion and future work are brought out in the last Section.

2   Probabilistic Knowledge Base Representation

A sample space, denoted by $\mathcal{S}$, is a set including all possible outcomes of a statistical experiment. A finite set of events, denoted by $\hat{\mathcal{E}}$, $\hat{\mathcal{E}} = \{E_1, \ldots, E_n\}$ where $E_i \in \mathcal{S}, \forall 1 \leq i \leq n$. The intersection of F and G, denoted by $F \cap G$, is an event containing all elements that are common to F and G where $F, G \in \hat{\mathcal{E}}$; the negation of event F, denoted by $\neg F$, is abbreviated by $\overline{F}$. A complete conjunction of $\hat{\mathcal{E}}$, denoted by Θ, $\Theta = \tilde{E}_1 \cap \tilde{E}_2 \cap \ldots \cap \tilde{E}_n$ where $\tilde{E}_i = \{E_i, \overline{E}_i\}$. A set of all complete conjunctions of $\hat{\mathcal{E}}$, denoted by $\Delta\left(\hat{\mathcal{E}}\right), \Delta\left(\hat{\mathcal{E}}\right) = \{\Theta_1, \ldots, \Theta_{2^n}\}$. A complete conjunction $\Theta \in \Delta\left(\hat{\mathcal{E}}\right)$ satisfies an event F, denoted by $\Theta \vDash F$, iff F appears in Θ. Also, we said that $\Theta \vDash \{E_1, \ldots, E_n\}$ if $\Theta \vDash E_i, \forall 1 \leq i \leq n$. Let $SM(X) = \left\{\Theta \in \Delta\left(\hat{\mathcal{E}}\right) | \Theta \vDash X\right\}$, where X is an event or a set of events. Let $m = \left|\Delta\left(\hat{\mathcal{E}}\right)\right| = 2^n$ be the numbers of complete conjunctions of $\hat{\mathcal{E}}$. Let $\mathbb{R}_{\geq 0}$ be the set of non-negative real values including $+\infty$. Let $\mathbb{R}_{[0,1]}$ be the set of all real values from 0 to 1.

Definition 1. *Let* $\mathcal{P} : \Delta\left(\hat{\mathcal{E}}\right) \rightarrow \mathbb{R}_{[0,1]}$ *such as* $\sum_{i=1}^{m} \mathcal{P}(\Theta_i) = 1$, $\mathcal{P}$ *is called a probability function.*

Let $\widehat{\mathbf{P}}\left(\hat{\mathcal{E}}\right)$ be a set of all probability functions $\mathcal{P}$ over set of events $\hat{\mathcal{E}}$, defined by $\widehat{\mathbf{P}}\left(\hat{\mathcal{E}}\right) = \{\mathcal{P}(\Theta_1), \ldots, \mathcal{P}(\Theta_{2^n})\}$. With $F \in \hat{\mathcal{E}}$, $\mathcal{P}(F) = \sum_{\Theta \in \Delta(\hat{\mathcal{E}}):\Theta \vDash F} \mathcal{P}(\Theta)$. Let $\vec{\vartheta} = (\vartheta_1, \ldots, \vartheta_m)^T$ be a column vector, where an auxiliary variable ϑ_i corresponding to a Probability $\mathcal{P}(\Theta_i)$.

Definition 2. *Let* $F, G \in \hat{\mathcal{E}}$ *and* $\rho \in \mathbb{R}_{[0,1]}$. *A probabilistic constraint is an expression of the form* $c[\rho]$, *where* $c = (F|G)$.

Intuitively, a constraint $(F|G)[\rho]$ means that our knowledge in F given that G hold has probability value ρ. If F does not depend on G, $G \equiv \top$, we abbreviate $(F|\top)[\rho]$ by $(F)[\rho]$.

Definition 3. *A probabilistic knowledge base, denoted by* $\mathcal{K} = \langle \kappa_1, \ldots, \kappa_h \rangle$, *is a finite set of probabilistic constraints, where* $\kappa_i = c_i[\rho_i]$ *for all* $i = 1, \ldots h$.

Let $h = |\mathcal{K}|$ be the total number of constraints in $\mathcal{K}$. A probability function $\mathcal{P} \in \widehat{\mathbf{P}}\left(\hat{\mathcal{E}}\right)$ satisfies a probabilistic constraint $c[\rho]$, denoted by $\mathcal{P} \vDash c[\rho]$, iff $\mathcal{P}(FG) = \rho\mathcal{P}(G)$. A probability function $\mathcal{P}$ satisfies $\mathcal{K}$, denoted by $\mathcal{P} \vDash \mathcal{K}$, iff $\mathcal{P} \vDash \kappa \forall \kappa \in \mathcal{K}$. Let $SM(\mathcal{K}) = \left\{\mathcal{P} \in \widehat{\mathbf{P}}\left(\hat{\mathcal{E}}\right) | \mathcal{P} \vDash \mathcal{K}\right\}$ be the set of all probability function values satisfying $\mathcal{K}$. Knowledge base $\mathcal{K}_1$ is equivalent to $\mathcal{K}_2$, denoted by $\mathcal{K}_1 \equiv \mathcal{K}_2$, iff $SM(\mathcal{K}_1) = SM(\mathcal{K}_2)$. The set of constraints $\mathcal{K}_1 \cup \mathcal{K}_2$ corresponds to $SM(\mathcal{K}_1 \cup \mathcal{K}_2) = SM(\mathcal{K}_1) \cap SM(\mathcal{K}_2)$.

Definition 4. *A probabilistic knowledge base* $\mathcal{K}$ *is consistent, denoted by* $\mathcal{K} \nvDash \perp$, *iff* $SM(\mathcal{K}) = \emptyset$. *Otherwise,* $\mathcal{K}$ *is inconsistent, denoted by* $\mathcal{K} \vDash \perp$.

Definition 5. *A probabilistic knowledge base profile, denoted by* $\mathcal{B} = \{\mathcal{K}_1, \ldots, \mathcal{K}_n\}$, *is a finite set of probabilistic knowledge bases satisfying the following conditions:*

1. $\forall \mathcal{K}_i \in \mathcal{B} : \mathcal{K}_i \nvDash \bot$

2. $\forall \mathcal{K}_i, \mathcal{K}_j (i \neq j) :$
   $+ |\mathcal{K}_i| = |\mathcal{K}_j|$
   $+ \forall c_k[\rho_k] \in \mathcal{K}_i; c'_{k'}[\rho_{k'}] \in \mathcal{K}_j; k = k' : c_k = c'_{k'}.$

Example 1. Ministry of natural resources and environment makes a survey of pollution of the rivers. This survey was performed by two specialized departments.

- The first department obtains the results in which the probability that the river is polluted (denoted by R) is P(R) = 0.5; the probability that a sample of water tested detects pollution (denoted by W) is P(W) = 0.5; and the probability that a sample of water detects pollution, given that the river is polluted is P(W|R) = 1.0.
- The second department obtains the results in which the probability that the river is polluted is P(R) = 0.7; the probability that a sample of water tested detects pollution is P(W) = 0.5; and the probability that a sample of water detects pollution, given that the river is polluted is P(W|R) = 0.6.

According to the above results, we have the probabilistic knowledge bases as follows:

$$\mathcal{K}_1 = \langle (R)[0.5], (W)[0.5], (W|R)[1.0] \rangle; \mathcal{K}_2 = \langle (R)[0.7], (W)[0.5], (W|R)[0.6] \rangle.$$

We have $h = |\mathcal{K}_1| = 3, \hat{\mathcal{E}} = \{W, R\}$, and thus $\Delta\left(\hat{\mathcal{E}}\right) = \{W \cap R, W \cap \overline{R}, \overline{W} \cap R,$
$\overline{W} \cap \overline{R}\}$, m $= \left|\Delta\left(\hat{\mathcal{E}}\right)\right| = 4$, $\hat{P}\left(\hat{\mathcal{E}}\right) = \{\mathcal{P}(W \cap R), \mathcal{P}(W \cap \overline{R}), \mathcal{P}(\overline{W} \cap R), \mathcal{P}(\overline{W} \cap \overline{R})\}$.
Clearly, $\mathcal{K}_1 \nvDash \bot, \mathcal{K}_2 \nvDash \bot$. We have $\mathcal{B} = \{\mathcal{K}_1, \mathcal{K}_2\}$.

3   Model for Merging Probabilistic Knowledge Bases

In this work, the merging problem is defined in the context of priority as follows:

(1) **Input:** A probabilistic knowledge base profile $\mathcal{B} = \{\mathcal{K}_1, \ldots, \mathcal{K}_n\}$,
(2) **Output:** A consistent probabilistic knowledge base $\mathcal{K}$
(3) **Scope of problem:** Knowledge base is represented in a probabilistic framework.
(4) **Merging process:**

- Step 1: Computing $SM(\mathcal{K}_i)$ for all $i = 1, \ldots n$.
  - Finding characteristic matrix $A_{\mathcal{K}_i}$
  - Computing $SM(\mathcal{K}_i)$ by evaluating optimization problems with respect to with p-norm (p = 1, 2, ∞) or unnormalized optimization problem.
- Step 2: Computing the merging result $\Gamma(\mathcal{B})$ as a probabilistic knowledge base.

(5) **Result:** A consistent probabilistic knowledge base.

3.1 Desirable Properties for Probabilistic Merging Operators

In the rest later, we will put forward several properties of our family of probabilistic merging operators which are used to characterize merging operators. Let Γ be a function $\Gamma : \mathcal{B} \rightarrow \mathrm{SM}(\mathcal{B})$ that maps a probabilistic knowledge base profile onto a set of all probability function values satisfying $\mathcal{B}$. The following definition is similar to a set of logical properties which was stated in [1, 20].

Definition 6. *Let* $\mathcal{B} = \{\mathcal{K}_1, \ldots, \mathcal{K}_n\}$ *and,* $\mathcal{C} = \{\mathcal{K}'_1, \ldots, \mathcal{K}'_m\}$ *be probabilistic knowledge base profiles. Let* $\mathcal{K}$ *be a consistent probabilistic knowledge bases. Function* $\Gamma :$ $\mathcal{B} \rightarrow \mathrm{SM}(\mathcal{B})$ *is called a probabilistic merging operator iff the following properties hold:*

(CP) Consistency Principle. If $\bigcap_{i=1}^{n} \mathrm{SM}(\mathcal{K}_i) \neq \emptyset$ *then* $\Gamma(\mathcal{B}) = \bigcap_{i=1}^{n} \mathrm{SM}(\mathcal{K}_i)$

The property CP assures that if the input probabilistic knowledge base profile is consistent then the result of merging will be the union of all knowledge bases in it.

(IP) Ignorance Principle. $\forall \mathcal{K} = \emptyset: \Gamma(\mathcal{B} \cup \mathcal{K}) = \Gamma(\mathcal{B})$

The property IP state that the result of merging process should not be affected when an empty probabilistic knowledge base is added.

(EP) Equivalence Principle. If there exists a permutation μ *on {1,...,n} such that* $\mathrm{SM}(\mathcal{K}_i) = \mathrm{SM}\left(\mathcal{K}'_{\mu(i)}\right)$ *then* $\Gamma(\mathcal{B}) = \Gamma(\mathcal{C})$.

The property EP implies that result of merging process should not be contingent on the occurrence order of probabilistic knowledge bases.

(AP) Agreement. If $\Gamma(\mathcal{B}) \cap \Gamma(\mathcal{C}) \neq \emptyset$ *then* $\Gamma(\mathcal{B}) \cap \Gamma(\mathcal{C}) = \Gamma(\mathcal{B} \cup \mathcal{C})$.

This property states that the intersection of the merging results of two distinct sets of knowledge bases is not empty, it should be similar to the merging of all probabilistic knowledge bases in those two sets.

(DP) Disagreement Principle. If $\bigcap_{i=1}^{m} \mathrm{SM}\left(\mathcal{K}'_i\right) \neq \emptyset$ *and* $\mathrm{SM}(\Gamma(\mathcal{B})) \neq \mathrm{SM}(\Gamma(\mathcal{C}))$ *then* $\Gamma(\mathcal{B}) \cap \Gamma(\mathcal{C}) = \emptyset$.

This property states that if the merged probabilistic knowledge base $\Gamma(\mathcal{B})$ *is inconsistent with merged probabilistic knowledge base* $\Gamma(\mathcal{C})$ *of a distinct set of knowledge bases, the intersection of the merging results should be empty.*

(SDP) Strong Disagreement Principle. $\Gamma(\mathcal{B}) \cap \Gamma(\mathcal{C}) = \emptyset$.

This property states that the intersection of the merging results should be empty.

3.2 Model for Satisfied Probabilistic Functions

Definition 7. *Function* $\varphi(H, \Theta)$ *is called an indicator function whenever it is defined as follows:* $\varphi : \hat{\mathcal{E}} \times \Delta\left(\hat{\mathcal{E}}\right) \rightarrow \mathbb{R}_{[0,1]}$,

$$\varphi(H, \Theta) = \begin{cases} 1 & \text{if } \Theta \vDash H \\ 0 & \text{otherwise} \end{cases}$$

Definition 8. (*Characteristic Matrix*). $A_{\mathcal{K}}$ *is called a characteristic matrix of* $\mathcal{K}$ *whenever it is defined as follows:*

$$A_{\mathcal{K}} = (a_{ij}) \in \mathbb{R}^{h \times m}$$

where $a_{ij} = \varphi(F_i \cap G_i, \Theta_j)(1 - \rho_i) - \varphi(\overline{F}_i \cap G_i, \Theta_j)\rho_i$.

Intuitively, a_{ij} is either 1-ρ, -ρ or 0 dependent on whether $F_i \cap G_i$ satisfies Θ_j, $\overline{F}_i \cap G_i$ satisfies Θ_j, or not computable.

Proposition 1. *p-norm satisfied probabilistic vector of* $\mathcal{K}$, *noted* $\vec{\vartheta}^p_{\mathcal{K}}$, *corresponding to* $\vec{\vartheta}^*$ *of the optimal solution of the following problem:*

$$arg\,min_{\vec{\vartheta} \in \mathbb{R}^m} \left\| A_{\mathcal{K}}\vec{\vartheta} \right\|_p \tag{1}$$

$$subject\,to: \quad \sum_{i=1}^{m} \vartheta_i = 1; \vec{\vartheta} \geq \vec{0} \tag{2}$$

Proposition 1 is the result of Proposition 4 which was proposed and proved in [10].

Proposition 2. *1-norm satisfied probabilistic vector of* $\mathcal{K}$, *noted* $\vec{\vartheta}^1_{\mathcal{K}}$, *corresponding to* $\vec{\vartheta}^*$ *of the optimal solution of the following linear programming problem:*

$$arg \min_{(\vec{\vartheta}, \vec{\lambda}) \in \mathbb{R}^{m+h}} \sum_{i=1}^{h} \lambda_i \tag{3}$$

$$subject\,to: A_{\mathcal{K}}\vec{\vartheta} - \vec{\lambda} \leq \vec{0}; \ A_{\mathcal{K}}\vec{\vartheta} + \vec{\lambda} \geq \vec{0}; \ \sum_{i=1}^{m} \vartheta_i = 1; \vec{\vartheta} \geq \vec{0}; \ \vec{\lambda} \geq \vec{0} \tag{4}$$

Proposition 2 is the result of Proposition 5 which was proposed and proved in [10].

Proposition 3. ∞-*norm satisfied probabilistic vector of* $\mathcal{K}$, *noted* $\vec{\vartheta}^\infty_{\mathcal{K}}$, *corresponding to* $\vec{\vartheta}^*$ *of the optimal solution of the following linear programming problem:*

$$arg \min_{(\vec{\vartheta}, \lambda) \in \mathbb{R}^{m+1}} \lambda \tag{5}$$

$$subject\,to: A_{\mathcal{K}}\vec{\vartheta} - \vec{1}\lambda \leq \vec{0}; A_{\mathcal{K}}\vec{\vartheta} + \vec{1}\lambda \geq \vec{0}; \sum_{i=1}^{m} \vartheta_i = 1; \vec{\vartheta} \geq \vec{0}; \lambda \geq 0 \tag{6}$$

Proposition 3 is the result of Proposition 6 which was proposed and proved in [10].

Definition 9. $\overline{A}_{\mathcal{K}}$ *is called a diagonal double matrix of* $\mathcal{K}$ *whenever it is defined as follows:*

$$\overline{A}_{\mathcal{K}} = (\bar{a}_{ij}) \in \mathbb{R}^{h \times 2h}$$

where

$$\bar{a}_{ij} = \begin{cases} 1 & \text{if } i = j \text{ and } i, j = 1, \ldots, h \\ -1 & \text{if } i = j \text{ and } i, j = h+1, \ldots, 2h \\ 0 & \text{otherwise} \end{cases}$$

Let $\vec{\Delta}\left(\vec{\ell}, \vec{\zeta}\right) = (\ell_1, \ldots, \ell_h, \zeta_1, \ldots, \zeta_h)^T$ be a column vector.

Proposition 4. *Unnormalized satisfied probability vector of $\mathcal{K}$, noted $\vec{\vartheta}^u_{\mathcal{K}}$, corresponding to $\vec{\vartheta}^*$ of the optimal solution of the following linear programming problem:*

$$arg \min_{(\vec{\vartheta}, \vec{\Delta}) \in \mathbb{R}^{m \times 2h}} \sum_{i=1}^{h} (\ell_i + \zeta_i) \tag{7}$$

$$\text{subject to}: \quad \bar{A}_{\mathcal{K}}\vec{\Delta} \leq \vec{1} - \vec{\rho}; \; \bar{A}_{\mathcal{K}}\vec{\Delta} \geq -\vec{\rho}; \sum_{i=1}^{m} \vartheta_i = 1; \vec{\vartheta} \geq \vec{0} \tag{8}$$

$$(\rho_i + \ell_i - \zeta_i) \sum_{\Theta \in SM(F_i \cap G_i)} \vartheta_i - \sum_{\Theta \in SM(G_i)} \vartheta_i = 0, \forall i = 1, \ldots n \tag{9}$$

Proposition 1 is the result of Proposition 9 which was proposed and proved in [10].

Proposition 5. $SM(\mathcal{K}) = \vec{\vartheta}^1_{\mathcal{K}} = \vec{\vartheta}^p_{\mathcal{K}} = \vec{\vartheta}^\infty_{\mathcal{K}} = \vec{\vartheta}^u_{\mathcal{K}}$

Proof. Because $\mathcal{K}$ is consistent and satisfied probabilistic vectors $\vec{\vartheta}^1_{\mathcal{K}}, \vec{\vartheta}^p_{\mathcal{K}}, \vec{\vartheta}^\infty_{\mathcal{K}}, \vec{\vartheta}^u_{\mathcal{K}}$ represent the same consistent knowledge base, $SM(\mathcal{K}) = \vec{\vartheta}^1_{\mathcal{K}} = \vec{\vartheta}^p_{\mathcal{K}} = \vec{\vartheta}^\infty_{\mathcal{K}} = \vec{\vartheta}^u_{\mathcal{K}}$.

1-norm and ∞-norm satisfied probabilistic vector of $\mathcal{K}$ in Propositions 2 and Propositions 3 are the solution of the linear programming problems because the objective and constraint functions of problems are linear. Therefore, we use LPSolve (http://lpsolve.sourceforge.net) based on the Simplex algorithm proposed in [21] to compute $\vec{\vartheta}^1_{\mathcal{K}}, \vec{\vartheta}^\infty_{\mathcal{K}}$. However, the objective function of the problem in Propositions 1 and the constraint function of the problem in Propositions 4 are not linear, so we employ interior-point algorithm proposed in [22] and IBM-CPLEX (https://www-01.ibm.com/software/commerce/optimization/cplex-optimizer/) to compute $\vec{\vartheta}^2_{\mathcal{K}}, \vec{\vartheta}^u_{\mathcal{K}}$.

Example 2. Consider the knowledge bases from Example 1. We will compute $SM(\mathcal{K}_1)$, $SM(\mathcal{K}_2)$

- Finding characteristic matrix $A_{\mathcal{K}_1}$

$$A_{\mathcal{K}_1} = \begin{pmatrix} 0.5 & 0.5 & -0.5 & -0.5 \\ 0.5 & -0.5 & 0.5 & -0.5 \\ 0 & -1.0 & 0 & 0 \end{pmatrix}$$

- Computing $SM(\mathcal{K}_1)$.

We have $\vec{\lambda} = (\lambda_1, \lambda_2, \lambda_3)^T$, $\vec{\vartheta} = (\vartheta_1, \vartheta_2, \vartheta_3, \vartheta_4)^T$, $\vec{\rho} = (0.5, 0.5, 1.0)$

By Proposition 2, $\vec{\vartheta}^1_{\mathcal{K}_1}$ could be computed by the following linear programming program:

$$\arg\min\ (\lambda_1 + \lambda_2 + \lambda_3) \tag{10}$$

$$\text{subject to}:\quad A_{\mathcal{K}_1}\vec{\vartheta} - \vec{\lambda} \le \vec{0}; A_{\mathcal{K}_1}\vec{\vartheta} + \vec{\lambda} \ge \vec{0}; \vartheta_1 + \vartheta_2 + \vartheta_3 + \vartheta_4 = 1 \tag{11}$$

$$\vartheta_1 \ge 0;\ \vartheta_2 \ge 0; \vartheta_3 \ge 0; \vartheta_4 \ge 0; \lambda_1 \ge 0; \lambda_2 \ge 0; \lambda_3 \ge 0 \tag{12}$$

We have $\text{SM}(\mathcal{K}_1) = \{(0.5, 0, 0, 0.5)\}$. Similarly, $\text{SM}(\mathcal{K}_2) = \{(0.42, 0.28, 0.08, 0.22)\}$. We find that regardless of whether we apply the calculation of satisfied probability functions to represent a consistent knowledge base by evaluating optimization problems with respect to with 1-norm or 2-norm or ∞-norm, or unnormalized optimization problem, we obtain the same results.

3.3   The Families of Probabilistic Merging Operators

For any $x \in \widehat{P}\left(\hat{\mathcal{E}}\right), y \in \widehat{P}\left(\hat{\mathcal{E}}\right)$, it is said that x dominates y, denoted by $x \gg y$, if $x_i = 0$ implies $y_i = 0\ \forall 1 \le i \le m$. Let $I(x) = \{i : x_i \ne 0\}$.

Definition 10. *For a set of probability functions $\widehat{P}\left(\hat{\mathcal{E}}\right)$, a divergence distance function, is a map $d : \widehat{P}\left(\hat{\mathcal{E}}\right) \times \widehat{P}\left(\hat{\mathcal{E}}\right) \to \mathbb{R}_{\ge 0}$ satisfying following conditions:*

1. $d(x, y) = 0$ iff $x = y$
2. $d(x, y) = d(y, x)$
3. $d(x, y) + d(y, z) \ge d(x, z)$

where x,y, and z are probability functions.

Definition 11. *Function $d^{SQ}(y, x)$ is called a SQ divergence distance function whenever it is defined as follows: $d^{SQ} : \widehat{P}\left(\hat{\mathcal{E}}\right) \times \widehat{P}\left(\hat{\mathcal{E}}\right) \to \mathbb{R}_{\ge 0}$,*

$$d^{SQ}(y, x) = \sum_{i \in I(x)} (y_i - x_i)^2$$

Definition 12. *Function $d^{KL}(y, x)$ is called a KL divergence distance function whenever it is defined as follows: $d^{KL} : \widehat{P}\left(\hat{\mathcal{E}}\right) \times \widehat{P}\left(\hat{\mathcal{E}}\right) \to \mathbb{R}_{\ge 0}$,*

$$d^{KL}(y, x) = \begin{cases} \sum\limits_{i \in I(x)} y_i \log \frac{y_i}{x_i} & \text{if } x \gg y \\ +\infty & \text{otherwise} \end{cases}$$

(1) The probabilistic merging operator Γ^{SQ}

Proposition 6. *Probability merging vector of $\mathcal{B}$, noted $\vec{q}_{\mathcal{B}}$, is the optimal solution of the following problem:*

$$arg\,min_{\vec{q}_{\mathcal{B}} \in \mathbb{R}^n} \left(\sum_{i=1}^{n} d^{SQ}(q, x_i) : q \in \widehat{P}\left(\hat{\mathcal{E}}\right); x_i \in SM(\mathcal{K}_i); \mathcal{K}_i \in \mathcal{B} \right) \qquad (13)$$

$$subject\ to\quad \sum_{i=1}^{n} q_i = 1; \vec{q}_{\mathcal{B}} \geq \vec{0} \qquad (14)$$

Definition 13. *(SQ Merging Operator). Function $\Gamma^{SQ} : \mathcal{B} \rightarrow SM(\mathcal{B})$ is called SE Merging Operator whenever it is defined as follows: $\Gamma^{SQ}(\mathcal{B}) = \vec{q}_{\mathcal{B}}$.*

Proposition 7. *The merging operator Γ^{SQ} fulfills EP, CP, IP, DP and SDP.*

(2) The probabilistic merging operator Γ^{SE}

Proposition 8. *Probability merging vector of $\mathcal{B}$, noted $\vec{y}_{\mathcal{B}}$, is the optimal solution of the following problem:*

$$arg\,min_{\vec{y}_{\mathcal{B}} \in \mathbb{R}^n \mathbb{F}} \left(\sum_{i=1}^{n} d^{KL}(y, x_i) : y \in \widehat{P}\left(\hat{\mathcal{E}}\right); x_i \in SM(\mathcal{K}_i); \mathcal{K}_i \in \mathcal{B} \right) \qquad (15)$$

$$subject\ to\quad \sum_{i=1}^{n} y_i = 1; \vec{y}_{\mathcal{B}} \geq \vec{0} \qquad (16)$$

Definition 14. *(SE Merging Operator). Function $\Gamma^{SE} : \mathcal{B} \rightarrow SM(\mathcal{B})$ is called SE Merging Operator whenever it is defined as follows: $\Gamma^{SE}(\mathcal{B}) = \vec{y}_{\mathcal{B}}$.*

Proposition 9. *The merging operator Γ^{SE} fulfills EP, CP, IP.*

(3) The probabilistic merging operator Γ^{LE}

Proposition 10. *Probability merging vector of $\mathcal{B}$, noted $\vec{z}_{\mathcal{B}}$, is the optimal solution of the following problem:*

$$arg\,min_{\vec{z}_{\mathcal{B}} \in \mathbb{R}^n} \left(\sum_{i=1}^{n} d^{KL}(x_i, z) : z \in \widehat{P}\left(\hat{\mathcal{E}}\right); x_i \in SM(\mathcal{K}_i); \mathcal{K}_i \in \mathcal{B} \right) \qquad (17)$$

$$subject\ to\quad \sum_{i=1}^{n} z_i = 1; \vec{z}_{\mathcal{B}} \geq \vec{0} \qquad (18)$$

Definition 15. *(LE Merging Operator). Function $\Gamma^{LE} : \mathcal{B} \rightarrow SM(\mathcal{B})$ is called LE Merging Operator whenever it is defined as follows: $\Gamma^{LE}(\mathcal{B}) = \vec{z}_{\mathcal{B}}$.*
    We found that the objective functions of problems in Propositions 6, 8, 10 are not linear, so we could compute $\vec{q}_{\mathcal{B}}, \vec{y}_{\mathcal{B}}, \vec{z}_{\mathcal{B}}$ by using the method that is similar to one employed to compute $\vec{\vartheta}_{\mathcal{K}}^2, \vec{\vartheta}_{\mathcal{K}}^u$.

Proposition 11. *The merging operator* Γ^{LE} *fulfills EP, CP, IP and AP.*

Example 3. Consider the knowledge bases from Example 2. We will compute $\Gamma(\mathcal{B})$ as follows:

Step 1: We have $SM(\mathcal{K}_1) = \{(0.5, 0, 0, 0.5)\}$; $SM(\mathcal{K}_2) = \{(0.42, 0.28, 0.08, 0.22)\}$

Step 2: Computing the probabilistic knowledge base merging result $\Gamma(\mathcal{B})$

- $\vec{q}_{\mathcal{B}}$ could be computed by solving the following optimization program:

$$\arg\min(d^{SQ}(q, x_1) + d^{SQ}(q, x_2)) \tag{19}$$

$$\text{subject to} \quad q_1 + q_2 + q_3 + q_4 = 1; q_1 \geq 0, q_2 \geq 0, q_3 \geq 0, q_4 \geq 0 \tag{20}$$

where $d^{SQ}(q, x_1) = (q_1 - 0.5)^2 + (q_4 - 0.5)^2$

$$d^{SQ}(q, x_2) = (q_1 - 0.42)^2 + (q_2 - 0.28)^2 + (q_3 - 0.08)^2 + (q_4 - 0.22)^2$$

Therefore, we have $\Gamma^{SQ}(\mathcal{B}) = \vec{q}_{\mathcal{B}} = \{(0, 0.33, 0.33, 0.33)\}$

- $\vec{y}_{\mathcal{B}}$ could be computed by solving the following optimization program:

$$\arg\min(d^{KL}(y, x_1) + d^{KL}(y, x_2)) \tag{21}$$

$$\text{subject to} \quad y_1 + y_2 + y_3 + y_4 = 1; y_1 \geq 0, y_2 \geq 0, y_3 \geq 0, y_4 \geq 0 \tag{22}$$

where $d^{KL}(y, x_1) = y_1 \log\frac{y_1}{0.5} + y_4 \log\frac{y_4}{0.5}$

$$d^{KL}(y, x_2) = y_1 \log\frac{y_1}{0.42} + y_2 \log\frac{y_2}{0.28} + y_3 \log\frac{y_3}{0.08} + y_4 \log\frac{y_4}{0.22}$$

Therefore, we have $\Gamma^{SE}(\mathcal{B}) = \vec{z}_{\mathcal{B}} = \{(0, 0.33, 0.33, 0.33)\}$

- $\vec{z}_{\mathcal{B}}$ could be computed by solving the following optimization program:

$$\arg\min(d^{KL}(x_1, z) + d^{KL}(x_2, z)) \tag{23}$$

$$\text{subject to} \quad z_1 + z_2 + z_3 + z_4 = 1; z_1 \geq 0, z_1 \geq 0, z_1 \geq 0, z_1 \geq 0 \tag{24}$$

where $d^{KL}(x_1, z) = 0.5 \log\frac{0.5}{z_1} + 0.5 \log\frac{0.5}{z_4}$

$$d^{KL}(x_2, z) = 0.42 \log\frac{0.42}{z_1} + 0.28 \log\frac{0.28}{z_2} + 0.08 \log\frac{0.08}{z_3} + 0.22 \log\frac{0.22}{z_4}$$

Therefore, we have $\Gamma^{LE}(\mathcal{B}) = \vec{z}_{\mathcal{B}} = \{(0, 0.33, 0.33, 0.33)\}$.

By using the merging operator Γ^{SQ}, Γ^{SE} or Γ^{LE} to merge consistent probabilistic knowledge bases, we receive the same results. The new probabilistic knowledge base merged from $\mathcal{K}_1, \mathcal{K}_2$ is a consistent probabilistic knowledge base represented by a probability function satisfying desirable properties. We also find that the sum of the probability functions of the new probabilistic knowledge base equals 1, that is, satisfies

Definition 1 about the total of the probability of complete conjunctions over the set of events.

4 Conclusion

In this paper, we have investigated probabilistic merging operators based on divergence concepts and adapted them to new probabilistic merging framework by building distance functions between probabilities of constraints. We also introduced problems to calculate satisfied probabilistic functions of original base. Then, we gave postulates about logical relationships between the family of merging operators and desirable properties. However, the output of the proposed model is only the probabilistic functions representing newly merged knowledge base being not represented in the same form as the original knowledge bases. Furthermore, the input probabilistic knowledge bases are required to be in the same form. Therefore, in the future, we will go on investigating methods to ensure that the output of merged probabilistic knowledge base is similar to original bases in representation and allows the original knowledge bases to vary in form.

Acknowledgment. The authors would like to thank Professor Quang Thuy Ha and Knowledge Technology Lab, Faculty of Information Technology, VNU University of Engineering and Technology for expertise support.

References

1. Nguyen, N.T.: Advanced Methods for Inconsistent Knowledge Management. Advanced Information and Knowledge Processing, pp. 1–351. Springer, London (2008). https://doi.org/10.1007/978-1-84628-889-0
2. Bloch, I., et al.: Fusion: general concepts and characteristics. Int. J. Intell. Syst. **16**(10), 1107–1134 (2001)
3. Murray, K.S.: Learning as Knowledge Integration. Ph.D. thesis, The University of Texas at Austin (1995)
4. Reimer, U.: Knowledge integration for building organizational memories. In: Proceedings of the 11th Banff Knowledge Acquisition for Knowledge Based System Workshop, pp. 61–620 (1998)
5. Potyka, N., Thimm, M.: Probabilistic reasoning with inconsistent beliefs using inconsistency measures. In: International Joint Conference on Artificial Intelligence 2015 (IJCAI15), pp. 3156–3163. AAAI Press ©2015 (2015)
6. Nguyen, V.Y., Tran, T.H.: Inconsistency measures for probabilistic knowledge bases. In: Proceedings KSE 2017, pp. 148–153. IEEE Xplore (2017)
7. Potyka, N., Thimm, M.: Consolidation of probabilistic knowledge bases by inconsistency minimization. In: Proceedings ECAI, pp. 729–734. IOS Press (2014)
8. Potyka, N.: Solving Reasoning Problems for Probabilistic Conditional Logics with Consistent and Inconsistent Information. FernUniversitat, Hagen (2016)

9. Nguyen, V.T., Tran, T.H.: Solving inconsistencies in probabilistic knowledge bases via inconsistency measures. In: Nguyen, N.T., Hoang, D.H., Hong, T.-P., Pham, H., Trawiński, B. (eds.) ACIIDS 2018. LNCS (LNAI), vol. 10751, pp. 3–14. Springer, Cham (2018). https://doi.org/10.1007/978-3-319-75417-8_1

10. Nguyen, V.T., Nguyen, N.T., Tran, T.H., Nguyen, D.K.L.: Method for restoring consistency in probabilistic knowledge bases. J. Cybern. Syst. **49**, 1–22 (2018). Taylor & Francis

11. Konieczny, S., Pérez, R.P.: Merging with integrity constraints. In: Hunter, A., Parsons, S. (eds.) ECSQARU 1999. LNCS (LNAI), vol. 1638, pp. 233–244. Springer, Heidelberg (1999). https://doi.org/10.1007/3-540-48747-6_22

12. Lin, J., Mendelzon A.O.: Knowledge base merging by majority. In: Dynamic Worlds: From the Frame Problem to Knowledge Management (1999)

13. Konieczny, S., Lang, J., Marquis, P.: DA2 merging operators. Artif. Intell. **157**(1–2), 49–79 (2004)

14. Qi, G., Liu, W., Glass, D.H.: A split-combination method for merging inconsistent possibilistic knowledge bases. In: KR 2004, pp. 348–356. AAAI Press (2004)

15. Qi, G., Liu, W., Bell, D.A.: A revision-based approach to resolving conflicting information. In: UAI 2005, pp. 477–484. AUAI Press (2005)

16. Vomlel, J.: Methods of Probabilistic Knowledge Integration. Ph.D. thesis, Czech Technical University, Prague (1999)

17. Wilmers, G.: The social entropy process: axiomatising the aggregation of probabilistic beliefs. In: Probability, Uncertainty and Rationality (2010)

18. Adamcík, M., Wilmers, G.: The irrelevant information principle for collective probabilistic reasoning. Kybernetika **50**(2), 175–188 (2014)

19. Adamcik, M.: Collective Reasoning under Uncertainty and Inconsistency. Ph.D. thesis, University of Manchester, UK (2014)

20. Konieczny, S., Pérez, R.P.: Logic based merging. J. Philos. Log. **40**(2), 239–270 (2011)

21. Matousek, J., Gartner, B.: Understanding and Using Linear Programming. Universitext. Springer, Heidelberg (2007). https://doi.org/10.1007/978-3-540-30717-4

22. Boyd, S.P., Vandenberghe, L.: Convex Optimization. Cambridge, New York (2004)

Representation of Autoimmune Diseases with RDFS

Martina Husáková(✉)

Faculty of Informatics and Management, University of Hradec Králové,
Hradec Králové, Czech Republic
martina.husakova.2@uhk.cz

Abstract. Complex systems are systems consisting of many diverse and autonomous independent subsystems interacting with each other. Huge amount of interactions with many feedback loops complicate their investigation. Immune system is a typical complex system that attracts medical experts and also non-professionals especially because of its "ambient" nature and amazing complexity. Understandable information about immunity is required not only by the experts but also by the non-professionals. This paper is focused on development of the ontology providing fundamental facts about autoimmune diseases because these facts are not well-structured and presented for the end users on the web. The ontology should improve navigation among diverse pieces of information about these diseases and decrease information overloading.

Keywords: Navigation · Autoimmune disease · Ontology · RDFS
Protégé

1 Introduction

The immune system is complex, adaptive and self-organised system able to maintain homeostasis of a living organism. Homeostasis is based on intensive coordination, cooperation and communication between immune cells ensuring innate and adaptive regulatory processes. The immune system is not faultless. The autoimmunity is an exceptional state when the immune system attacks against own tissues and organs. Statistics show that 80 kinds of autoimmune disorders exist in 5–8% of the population [1]. Understandable information about autoimmune diseases is not required only by the experts but also by the non-professional audience. There is a huge amount of information about autoimmune diseases in the web space. It was found out that in many cases these pieces of information are not well-presented and structured for the final user, i.e. medical facts are not often interlinked with other medical facts, only the lists of keywords with short descriptions are provided, additional useful facts (web links to audio (video) files, pictures, synonyms or less known abbreviations) are missing or visualisation of facts is only based on the browsing of many pages. This paper investigates usefulness of ontologies for improvement navigation in the "space of autoimmune diseases" where the facts are easily interpretable by machines and humans. The Sect. 2 mentions selected applications of ontologies representing autoimmune diseases. The Sect. 3 describes development process of the ontology

© Springer Nature Switzerland AG 2018
N. T. Nguyen et al. (Eds.): ICCCI 2018, LNAI 11055, pp. 43–52, 2018.
https://doi.org/10.1007/978-3-319-98443-8_5

representing specific autoimmune diseases. The Sect. 4 proposes test cases where SPARQL-based queries are applied for receiving adequate feedbacks from the onto-logical repository. The Sect. 5 discusses proposed solution together with future directions and the Sect. 6 concludes the paper.

2   Representation of Autoimmune Diseases with Ontologies

At present, there is a huge amount of projects focusing on development, publishing, sharing, usage and maintaining of biomedical ontologies. Biomedical ontologies are mainly used for categorisation of health and medicine related terms, annotation, querying and inference. The paper mentions only a small fragment of structures trying to categorise various types of diseases, especially autoimmune diseases.

The Human Disease ontology (HDO) is a collaborative project of researchers at the Northwestern University and the University of Maryland School of Medicine [2]. The HDO is a knowledge base representing various inherited, developmental and acquired human diseases. The main aim of this ontology is to connect concepts about diseases with symptoms, findings, attributes of diseases and genes contributing to these diseases. The ontology-based knowledge structure is associated with particular biomedical data for enabling inference between heterogeneous datasets. The ontology also categorises immune system diseases. These ones are structured into the following categories: hematopoietic system diseases, hypersensitivity reaction diseases, lymphatic system diseases and primary immunodeficiency diseases. The ontology is mainly available in the OBO and the OWL format. It is available in the BioPortal (a community-based repository of more than 300 biomedical ontologies [3]) or in the Aber-OWL repository of OWL ontologies [4].

Experimental factor ontology (EFO) is the ontology supporting annotation, analysis and visualisation of data used by the European Molecular Biology Laboratory - European Bioinformatics Institute (EMBL-EBI) [5]. It covers types of cells, cell lines, chemical compounds, assay information and various aspects of diseases including autoimmune diseases. Some particular autoimmune diseases are categorised into more specific subclasses, e.g. multiple sclerosis is divided into the chronic progressive multiple sclerosis, neuromyelitis optica and relapsing-remitting multiple sclerosis. The EFO is also accessed in the BioPortal.

MeSH is a thesaurus of the National Library of Medicine in Bethesda (Maryland). It is used for indexing articles of biomedical journals for databases as MEDLINE/PubMed, cataloguing and searching for biomedical and health-related information [6]. The aim of the MeSH is to provide standardised keywords used for description of content of medicine-related documents. Thesaurus can be downloaded in various formats, e.g. XML or RDF. MeSH is not the ontology, but there is a reason to mention it because it represents particular autoimmune diseases (e.g. Addison disease, Hemotylic anemia, Rheumatoid arthritis, Diabetes mellitus (type 1), etc.). The autoimmune diseases are not categorised into specific subclasses, except the autoimmune diseases of the nervous system.

3   Ontological Modelling of Autoimmune Diseases

3.1   Motivation

Our initial research of autoimmune diseases with multi-agent-based modelling was initiated by the information and knowledge mapping of autoimmune disorders [7]. We consulted the autoimmunological concepts with the immunologist, but we also studied the autoimmune processes independently mainly by browsing the web sources. It was find out that web sources do not provide satisfying facts about autoimmune diseases, e.g. because relations between them are not presented and it is not possible to find out the context. Navigation between these facts is often based on browsing huge amount of web pages. An immunologist has strong background about autoimmune diseases, but a non-professional does not. The RDFS-based ontology is developed mainly for non-professionals which need to receive fundamental information about autoimmune diseases without information overloading. The ontology provides vocabulary of autoimmune concepts which should be also interpretable by machines for intelligent navigation.

3.2   Concept Mapping

Concept maps are non-formal and graph-based structures visualising relevant information and knowledge of particular application domain [8]. Concept maps are used for building of the ontology because they help to receive understandable and useful structures which are easily interpreted and extended by humans. They became valuable during consultations among immunologists and computer scientists preparing conceptual and computational models. Concept maps are used for facilitating understanding of complex immunological concepts by computer scientists. Concept maps are also chosen because of their graph-based nature in comparison to mind mapping. These ones are mainly based on the "star-like" structure which is not suitable if the complex models are developed.

Three concept maps are built in the VUE (Visual Understanding Environment) [9]. The first concept map (CMap-overview, see Fig. 1) proposes basic information about the autoimmunity. It explains what the autoimmune disorder is, risk factors influencing development of autoimmune diseases, typical attributes of the autoimmunity, possible treatment of autoimmune disorders, negative effects of autoimmunity, typical phases of autoimmune diseases, ways of the autoimmune diseases initiation).

Possible classifications of the autoimmune diseases are modelled by the second concept map (CMap-classifications). Generally speaking, classification of the autoimmune diseases is difficult because we do not know causal mechanisms for most of the autoimmune diseases and some of diseases also share some characteristics. Four types of classifications are proposed in [10] where the first one is the most frequently mentioned by scientists:

- classification according to the most affected area,
- classification according to the mechanisms of tissue damage,
- classification according to the type of an immune response,
- classification according to the hypersensitivity of reaction.

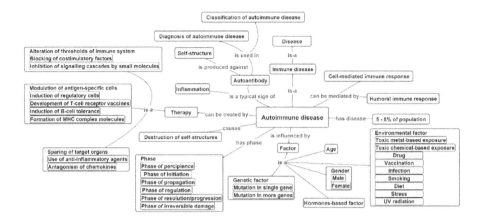

Fig. 1. General concept map about autoimmune diseases (CMap-overview)

The third concept map (CMap-attributes) visualises properties which are going to be observed for each autoimmune disease, see Fig. 2.

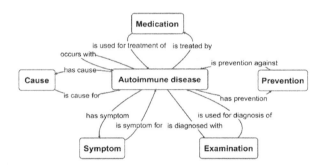

Fig. 2. General concept map about autoimmune diseases (CMap-attributes)

3.3   Ontology Development

Concept maps are used for development of the classes-based layer of the ontology, see Fig. 3. This layer is extended by the specific autoimmune diseases located in the individuals-based layer of the ontology, see Fig. 3. The ontology of autoimmune diseases is developed with the RDFS (Resource Description Framework Schema) model – W3C standard (2004) for development of formal vocabularies which are machine processable [11]. It is based on the RDF (Resource Description Framework) data model used for metadata representation and structuring data on the web where everything can be said about something in the form of the RDF statements [12, 20], see Fig. 3.

Concept maps-based layer

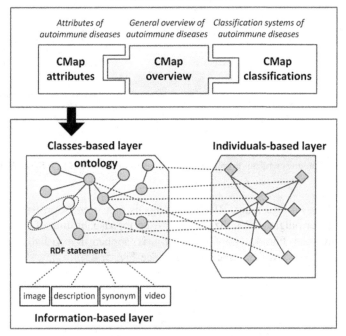

Fig. 3. Layers of the RDFS ontology

The following requirements are specified for the content of the RDFS ontology:

- categorisation of the autoimmune diseases and their characteristics,
- representation of the associations between the autoimmune diseases and their characteristics,
- representation of the relations between the autoimmune diseases which can occur together,
- description of the autoimmune diseases and not commonly known immunological terms,
- representation of the synonymous words (if they exist),
- web links to images, videos and other relevant information sources.

Classes represent general categories of autoimmune diseases according to the most affected area [10], see Fig. 4. Classes also represent attributes of these diseases, i.e. Cause, Symptom, Test, Prevention, Medication, Characteristic. Object properties represent relations between classes, i.e. hasCause, hasSymptom, isDiagnosedBy, hasPrevention, isTreatedBy, isCharacterisedBy and occurrsWith.

Particular autoimmune diseases, symptoms, medications, characteristics and causes are modelled as the individuals. Figure 5 depicts specific autoimmune disease called Addison's disease. Annotation properties are used for adding additional information of these individuals, i.e. rdfs:comment, rdfs:seeAlso (subproperties: seeAlsoexplanation, seeAlsovideo, seeAlsoImage) and a synonym. RDFS ontology is developed in the

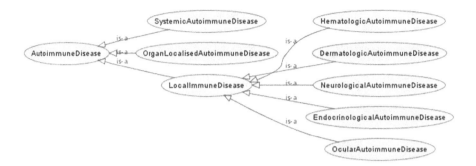

Fig. 4. Categories of autoimmune diseases represented in the RDFS ontology

ontological editor Protégé 5.2.0. RDFS ontology is consistent. Consistency is tested with the Debugger plugin. OWLViz plugin is used for visualisation of the ontological class hierarchy, see Fig. 4. OntoGraf plugin is applied for visualization of the autoimmune diseases and their characteristics, see Fig. 5.

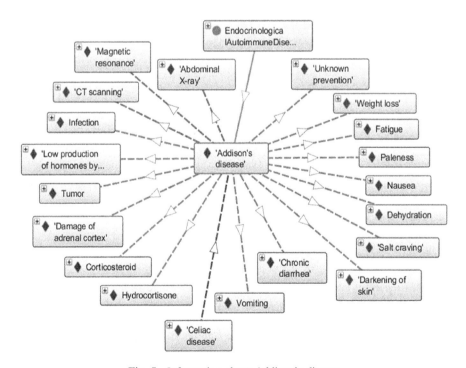

Fig. 5. Information about Addison's disease

The RDFS ontology contains 30 classes, 8 object properties, 8 annotation properties, 229 individuals, 620 logical axioms and 373 annotation assertions (additional facts about autoimmune concepts (e.g. comments (string), a synonym (string), explanations (web source), image (web source), video (web source)).

4   Querying the RDFS Ontology

The RDFS ontology represents 229 individuals and 620 logical axioms about them (including membership of an individual in a class). SPARQL (SPARQL Protocol and RDF Query Language) is a standardized query language mainly used for querying RDF (S) graphs [13, 19]. The following paragraph explains how SPARQL language is used for querying of the RDFS ontology. The following name spaces and prefixes are used during querying:

```
PREFIX oad: http://www.autoimmunediseases-ontology/
PREFIX rdfs: http://www.w3.org/2000/01/rdf-schema#
```

1. Which autoimmune disease has unknown cause and can be treated by any medication?

```
SELECT ?aDisease
WHERE {?aDisease oad:hasCause oad:UnknownCause.
?aDisease oad:isTreatedBy ?medication.}
```

Output (examples): Sjögren´s syndrome, Myasthenia gravis, Local scleroderma, Rheumatoid arthritis, Celias disease.

2. Which autoimmune diseases can be treated by Enbrel or Vitamin A?

```
SELECT ?aDisease
WHERE {
{?aDisease oad:isTreatedBy oad:Enbrel.}
UNION
{?aDisease oad:isTreatedBy oad:VitaminA.}}
```

Output: Sjogren´s syndrome, Psoriasis

3. How many symptoms exist in the ontology of autoimmune diseases?

```
SELECT (COUNT (?symptom) AS ?CountOfSymptoms)
WHERE {
?subclass rdfs:subClassOf oad:Symptom.
?symptom a ?subclass.}
```

Output: "50"^^xsd:integer

4. Which autoimmune disease can occur with different autoimmune disease?

```
SELECT ?aDiseaseA ?aDiseaseB
WHERE {
?aDiseaseA oad:occursWith ?aDiseaseB.}
ORDER BY ?aDiseaseA
```

Output (examples): Celias disease and rheumatoid arthritis; Uveitis and Psoriasis; Uveitis and Sarcoidosis.

5. Which autoimmune diseases can be diagnosed by biopsy method?

```
SELECT ?aDisease
WHERE {?aDisease oad:isDiagnosedBy oad:Biopsy.}
ORDER BY ?aDisease
```

Output (examples): Psoriasis, Sjögren´s syndrome, Sarcoidosis

5   Discussion and Future Directions

Presented ontology of autoimmune diseases represents fundamental facts about 17 specific autoimmune diseases. This collection is not too large because approximately 80 autoimmune disorders are registered. It is inevitable to include more autoimmune diseases into the ontology for provision of complete view on the "world" of this kind of disease.

One of the key questions can be: "Why the formal ontology is used for modeling of autoimmune diseases?" "Is there any other approach how to represent them?" If we want to access specific information on the web, we use HTML language for web pages development in most of cases. In this case, each web page would correspond with a particular autoimmune disease. The problem is that pieces of information and knowledge about autoimmune diseases are represented on the web pages together with the inner structure of the HTML document (e.g. semantic or non-semantics tags). Separation of the specific information and knowledge from the HTML structure can ease management of facts about particular application domain. Non-semantic databases can be used for storing information about specific concepts but if we want to represent semantics of these concepts, it is problematic. Non-semantic databases (e.g. relational databases) are not designed for this purpose. Formal ontologies can formally represent specific concepts with semantics of these concepts. Semantics representation is crucial for machine interpretation and understanding. Formal ontology can be also used for inference of the new facts which can be hidden in the ontology. The formal ontologies are developed mainly for machines, but we would like to use tools or frameworks for accessing their inner ontological structure. There exist various ways how to access the ontologies in the web space, e.g.:

- OWLDoc [14]: OWLDoc is a plugin accessible in the Protégé tool. It takes the whole ontological structure and generates HTML pages for components of the ontology (especially for axioms, classes, properties and individuals).
- LODE [15]: Live OWL Documentation Environment is a web service that automatically extracts components of the ontology and renders them in the HTML web page.
- OntoDia [16]: OntoDia is a web application for ontology data visualisation. Diagrams are generated from the ontological data and can be shared with others.
- jOWL [17]: jOWL is a jQuery plugin for navigating and visualising RDFS and OWL ontologies.

The RDFLib is an open source Python library for working with RDF files [18]. This library is going to be used for development of a web application where the ontology of autoimmune diseases will play the role of the repository of medical information and a final user will be able to navigate, to browse, to visualise of the ontological structure together with the SPARQL querying.

6   Conclusion

The paper presents steps of the RDFS ontology development focusing on the autoimmune diseases. Concept mapping is used for receiving of a global view on the autoimmune diseases in a graph-based form. Collection of concept maps is used for development of this formal ontology. The main reason for development of this ontology is that information and knowledge about autoimmune diseases is found in various places of the web and a user has to spend significant effort on finding specific information about this type of disease, i.e. a lot of web pages has to browse for receiving answers on questions. The main idea behind the RDFS ontology is to provide a web application that will use this ontology and provide fundamental information about autoimmune diseases in one place where semantics and relations between specific immunological concepts will be provided together with asking questions that relate to autoimmunity. The basic structure of the ontology is provided, but more autoimmune diseases have to be represented. Protégé editor is used for RDFS ontology development, but this solution is not friendly for all users. The RDFLib - the Python-based library is going to be used for web application development that will be available in public.

Acknowledgements. The support of the FIM UHK Specific Research Project "Socio-economic models and autonomous systems" is gratefully acknowledged. The author would like to thank Thomas Nachazel for article formatting and depicting specific figures.

References

1. Delves, P.J., et al.: Roitt's Essential Immunology. 12 edn., 560 p. Wiley-Blackwell (2011). ISBN 978-1405196833
2. Schriml, L.M., et al.: Disease ontology: a backbone for disease semantic integration. J. Nucleic Acids Res. **40**(database issue) (2012). https://doi.org/10.1093/nar/gkr972. http://www.ncbi.nlm.nih.gov/pmc/articles/PMC3245088/
3. Salvadores, M., et al.: BioPortal as a dataset of linked biomedical ontologies and terminologies in RDF. J. Semant. Web **4**(3), 277–284 (2014)
4. Hoehndorf, R., et al.: Aber-OWL: a framework for ontology-based data access in biology. BMC Bioinf. **16**(1), 13 (2014). http://arxiv.org/pdf/1407.6812v1.pdf
5. Malone, J., et al.: Modeling sample variables with an experimental factor ontology. Bioinformatics **26**(8), 1112–1118 (2010)
6. U.S. National Library of Medicine: Fact Sheet: Medical Subject Headings (MeSH). https://www.nlm.nih.gov/pubs/factsheets/mesh.html. Accessed 20 Jan 2018
7. Cimler, R., Husáková, M., Koláčková, M.: Exploration of autoimmune diseases using multi-agent systems. In: Nguyen, N.-T., Manolopoulos, Y., Iliadis, L., Trawiński, B. (eds.) ICCCI 2016. LNCS (LNAI), vol. 9876, pp. 282–291. Springer, Cham (2016). https://doi.org/10.1007/978-3-319-45246-3_27
8. Novak, J.: Concept maps and Vee diagrams: two metacognitive tools to facilitate meaningful learning. Instr. Sci. **19**, 29–52 (1990)
9. VUE: VUE. http://vue.tufts.edu/. Accessed 23 Jan 2018
10. Murphy, K.P., et al.: Janeway's Immunobiology. Garland Science. 8th edn., p. 888 (2008). ISBN 978-0815342434
11. W3C: RDF Schema 1.1 – W3C Recommendation 25, February 2014. (Brickley, D., Guha, R.V. (eds.)). https://www.w3.org/TR/2014/REC-rdf-schema-20140225/. Accessed 10 Mar 2018
12. W3C: RDF 1.1 Primer – W3C Working Group Note 24 June 2014 (Schreiber, G. (eds.) VU University Amsterdam, Raimond, Y. BBC). https://www.w3.org/TR/2014/NOTE-rdf11-primer-20140624/. Accessed 10 Mar 2018
13. W3C: SPARQL 1.1 Query Language (Harris, S. (eds.) Garlik, a part of Experian Seasborne, A., The Apache Software Foundation). https://www.w3.org/TR/2013/REC-sparql11-query-20130321/. Accessed 10 Mar 2018
14. Drummond, N., Horridge, M., Redmond, T.: OWLDoc. https://protegewiki.stanford.edu/wiki/OWLDoc. Accessed 10 Mar 2018
15. Peroni, S., Shotton, D., Vitali, F.: The live OWL documentation environment: a tool for the automatic generation of ontology documentation. In: ten Teije, A., et al. (eds.) EKAW 2012. LNCS (LNAI), vol. 7603, pp. 398–412. Springer, Heidelberg (2012). https://doi.org/10.1007/978-3-642-33876-2_35. http://speroni.web.cs.unibo.it/publications/peroni-2012-live-documentation-environment.pdf
16. Ontodia: Ontodia – a platform to build web applications for exploration and visualization of graph data. http://ontodia.org/. Accessed 10 Mar 2018
17. jOWL. http://jowl.ontologyonline.org/. Accessed 10 Mar 2018
18. RDFLib 4.2.2. https://rdflib.readthedocs.io/en/stable/. Accessed 10 Mar 2018
19. Pinczel, B., Nagy, D., Kiss, A.: The pros and cons of RDF structure indexes. Ann. Uni. Sci Bp. Sect. Comp. **42**, 283–296 (2014)
20. Rácz, G., Gombos, G., Kiss, A.: Visualization of semantic data based on selected predicates. In: Nguyen, N.T. (ed.) Transactions on Computational Collective Intelligence XIV. LNCS, vol. 8615, pp. 180–195. Springer, Heidelberg (2014). https://doi.org/10.1007/978-3-662-44509-9_9

EVENTSKG: A Knowledge Graph Representation for Top-Prestigious Computer Science Events Metadata

Said Fathalla[1,2]([✉]) and Christoph Lange[1,3]

[1] Smart Data Analytics (SDA), University of Bonn, Bonn, Germany
{fathalla,langec}@cs.uni-bonn.de
[2] Faculty of Science, University of Alexandria, Alexandria, Egypt
[3] Fraunhofer IAIS, Sankt Augustin, Germany

Abstract. Digitization has made the preparation of manuscripts as well as the organization of scientific events considerably easier and efficient. In addition, data about scientific events is increasingly published on the Web, albeit often as raw dumps in unstructured formats, immolating its semantics and relationships to other data and thus restricting the reusability of the data for, e.g., subsequent analyses. Therefore, there is a great demand to represent this data in a semantic representation using Semantic Web technologies. In this paper, we present the EVENTSKG dataset to offer a comprehensive semantic descriptions of scientific events of six computer science communities for 40 top-prestigious event series over the last five decades. We created a new, publicly available and improved release of the EVENTSKG dataset as a unified knowledge graph based on our Scientific Events Ontology (SEO). It is of primary interest to event organizers, as it helps them to assess the progress of their event over time and compare it to competing events. Furthermore, it helps potential authors looking for venues to publish their work. We shed light on these events by analyzing the EVENTSKG data.

Keywords: Scientific events dataset · Scholarly communication
Linked data · Semantic Web · Metadata analysis · Knowledge graph

1 Introduction

The exponential growth of Web data places an excessive pressure on researchers who are working on scholarly communication to assess, analyze, and organize this huge amount of data produced every day [13]. This paper introduces the EVENTSKG dataset, a new release of our previously presented EVENTS dataset [7], in the form of knowledge graph (KG), containing 60% additional event series belonging to six CS communities. A notable feature of the new release is the use of our Scientific Events Ontology (SEO)[1] as a reference ontology for event metadata modeling and link related data that was not previously,

---

[1] http://sda.tech/SEOontology/Documentation/SEO.html.

© Springer Nature Switzerland AG 2018
N. T. Nguyen et al. (Eds.): ICCCI 2018, LNAI 11055, pp. 53–63, 2018.
https://doi.org/10.1007/978-3-319-98443-8_6

i.e. in EVENTS, linked. EVENTSKG is a knowledge graph containing metadata of top-40 prestigious events series as a *unified graph* rather than individual RDF dumps for each event series in the previous release. The benefits of publishing EVENTSKG as linked data are:

- *Data linking*: establish links between dataset elements so that machines can explore related information,
- *Semantic querying*: query the dataset using the SPARQL query language,
- *Data enrichment*: inference engines could be used to infer implicit knowledge which does not explicitly exist,
- *Data validation*: semantically validate data against inconsistencies.

Events are linked by research fields, hosting country, and publishers. For instance, EVENTSKG is able to answer competency questions such as:

- What are the events related to *"Computational Intelligence"* with an acceptance rate less than 20% and proceedings published by *"Springer"*?,
- Which countries have hosted most of the events related to *"Semantic web"* over the last 20 years?
- Which of the six CS communities has attracted growing interest (in terms of the number of submissions) in the last 10 years?
- Which of the six CS communities has a growing production (in terms of the number of accepted papers) in the last 10 years?

The main goal of EVENTSKG is to facilitate the analysis of events metadata, by enabling them to be queried using semantic query languages, such as SPARQL. A key research question that motivates our work is: *What is the effect of digitization on scholarly communication in computer science events?* In particular, we address the following questions: *(a) What is the orientation of submissions and corresponding acceptance rates of prestigious events in computer science? (b) How did the number of publications of a CS sub-community fluctuate? (c) Did the date of prestigious events changes from year to year? (e) Which countries host most events in different CS communities?* In terms of events' impact, we address the following questions: *(a) What are the high-impact events of computer science? (b) How are the high-impact events currently ranked in the available ranking services?* By analyzing the dataset content, we gain some insights to answer these questions. Exploratory data analysis is performed aiming at exploring some facts and figures about CS events over the last five decades. Top-40 prestigious event series have been identified based on several criteria (see subsection 5.1). These event series fall into six CS communities[2]: information systems (IS), security and privacy (SEC), artificial intelligence (AI), computer systems organization (CSO), software and its engineering (SE) and web (WWW).

We believe that EVENTSKG dataset closes an important gap in analyzing the progress of a particular event series and CS community, using prestigious event series in the community, in terms of submissions and publications over a long-term period. Furthermore, it have a momentous influence on the research community, in particular:

---

[2] Using ACM Computing Classification System: https://dl.acm.org/ccs/ccs.cfm.

(a) *event chairs* – to asses the progress/impact of the event,
(b) *potential authors* – to find out events with high impact to submit their work,
(c) *proceedings publishers* – to trace the impact of their events.

The rest of the paper is structured as follows: Sect. 2 presents a brief review of the related work. Section 3 introduces the SEO ontology. Section 4 presents the main characteristics of EVENTSKG. Section 5 explains the curation process of creating and evolving the dataset. Section 6 discusses the results of analyzing the EVENTSKG data. Section 7 concludes and outlines our future work.

2   Related Work

Scholarly communication and data publishing have received much attention in the past decade [2,7,9,10,12,14]. The first considerable work to provide a comprehensive semantic description of scientific events metadata is the Semantic Web Conference (SWC) ontology [15]. The Semantic Web Dog Food (SWDF) dataset and its successor *ScholarlyData* are among the pioneers of datasets of comprehensive scholarly communication metadata [14]. A first attempt to create a dataset of prestigious events in five computer science communities is represented by our own EVENTS dataset [7]. It covers historical information about 25 top-prestigious events, describing each of them with 15 metadata attributes. The main shortcoming of this dataset is that it published as individual RDF dumps rather than one knowledge graph, by which it loses the potential links between dataset elements. Biryukov and Dong [4] analyzed a sets of top-ranked conferences in different Computer Science Communities and compared them in terms of publication growth rate, population stability and collaboration trends using DBLP. In a different work, we analyzed the evolution of key characteristics of CS conferences over a period of 30 years, including frequency, geographic distribution, and submission and acceptance numbers [9]. Hiemstra et al. [12] analyzed the SIGIR community in terms of authors' countries, number of papers per year for each country and co-authorship. Yan and Lee [19] proposed two measures for ranking academic venues by defining the goodness of a venue. Vasilescu et al. [18] presented a dataset of eleven prestigious software engineering conferences, such as ICSE and ASE, containing accepted papers along with their authors, programme committee members and the number of submissions each year. Agarwal et al. [1] analyzed the bibliometric metadata of seven ACM conferences in information retrieval, data mining, and digital libraries.

Despite these continuous efforts, none of these previous works was done on top of a unified KG. What additionally distinguishes our work from the related work mentioned above is that our analysis is based on a comprehensive list of metrics, considering quality in terms of event-related metadata in six CS communities and the dataset is published as a unified knowledge graph of all events.

3   Scientific Events Ontology

It is considered a good practice to reuse vocabularies from well-known ontologies wherever possible in order to facilitate ontology development and to lower the barrier for third party ontology-aware services reuse one's Linked Data. In this section, we introduce the scientific events ontology (SEO) [8], a reference ontology for modeling data about scientific events such as conferences, symposiums, and workshops. SEO reuses several well-designed ontologies, such as SWC[3], FOAF, SIOC, Dublin Core and SWRC (Semantic Web for Research Communities), and defines some of its own vocabulary as discussed later in this section. The vocabulary of SEO is defined in a new namespace prefixed *seo*, e.g., as *seo:EventTrack* and *seo:Symposium*. All namespace prefixes are used according to prefix.cc[4]. Several classes have been used such as *ConferenceSeries*, *AcademicEvent* and NonAcademicEvent and data properties such as acronym, startDate and endDate, and object properties such as *hasLocation*, *hasTopic*, and *isProceedingsOf*. Namely, *Topic* for describing a research topic or scientific area of an event (e.g., Database systems), *OrganisedEvent* for describing events related to an academic or non-academic event and *Role* for describing the roles held by people involved in an event. We define inverse relations, e.g., as *seo:isTrackOf* is the inverse relation of *seo:hasTrack* and *seo:isSponsorOf* is the inverse relation of *seo:hasSponsor*. Thus, if an event E *seo:hasTrack* T, then it can be inferred that T *seo:isTrackOf* E. Also, some symmetric relations are defined, such as *seo:colocatedWith*. Such definitions allow to reveal implicit information and increase the coherence and thus the value of event metadata [11].

4   Characteristics of the EVENTSKG Dataset

EVENTSKG covers three types of events since 1969: conferences, workshops, and symposia[5]. It contains metadata of 1048 editions of 40 events series with 15 attributes each. It is available in four different formats: RDF/XML, Turtle, CSV, and JSON-LD. The number of submissions and publications of each event involves all tracks' submissions and publications. There are several challenges to pursuing the maintenance of EVENTSKG for the future and keeping it sustainable; here is how we address them:

- *Availability:* EVENTSKG is publicly available online under a persistent URL (PURL): http://purl.org/events_ds. It is subjected to the Creative Commons Attribution license.
- *Extensibility:* There are three dimensions to extend the dataset to meet future requirements: (a) increase the number of events in each community, (b) cover more CS communities and (c) add more event properties such as hosting organization, registration fees and event sponsors.
- *Validation:* we perform two types of validation: *syntactic* and *semantic* validation. We syntactically validate EVENTSKG to conform with the W3C RDF standards using the online RDF validation service[6] and semantically validate it using Protégé reasoners.

---

[3] http://data.semanticweb.org/ns/swc/swc_2009-05-09.html.

[4] namespace look-up tool for RDF developers: http://prefix.cc/.

[5] Symposium is a small scale conference, with a smaller number of participants.

[6] https://www.w3.org/RDF/Validator/.

– *Documentation:* the documentation of the dataset has been checked using the W3C Markup Validation Service[7] and is available online on the dataset web page[8].

A concrete use case for querying EVENTSKG is supporting the research community in taking decisions on what event to submit their work to, or whether to accept invitations for being a chair or PC member. For example, finding all events related to *Artificial Intelligence*, which took place in the USA along with their acceptance rate; this requires joins between Field/Topic and event entities:

```
SELECT ?event ?eventTitle ?acc ?topic WHERE {
   ?event       rdfs:label        ?eventTitle .
   ?event       seo:country       http://dbpedia.org/resource/United_States .
   ?event       seo:field         ?topic
   ?event       seo:acceptanceRate ?acc .
   ?topic       rdfs:label        "Artificial Intelligence" .}
```

5 Data Curation

While collecting data from different sources, several problems have been encountered, such as duplicate data, incomplete data, incorrect data, and the change of event title over time. Therefore, a data curation process has been carried out comprising data acquisition, preprocessing, augmentation, Linked Data generation, data enrichment, and publication. *Data Curation* refers to the activities related to data organization, integration, annotation, and publication collected from various sources [17]. The curation of EVENTSKG dataset is an incremental process involving (see Fig. 1): Data acquisition and completion, Data Preprocessing, Data Augmentation, Linked Data Generation, Linked Data Enrichment and Data Publication.

5.1 Data Acquisition and Completion

First, top prestigious events have been identified based on several criteria, such as events ranking services (e.g., CORE, Qualis and GII rankings[9]) and Google h5-index. Second, we have collected the metadata of these events from different data sources, such as IEEE Xplore, ACM DL, DBLP, OpenResearch.org and events' official websites. Data has been collected in either structured or unstructured format to be exposed as Linked Data.

5.2 Data Preprocessing

To prepare the raw data for the further steps, we carried out four preprocessing processes: *data integration, data cleansing, data structure transformation* and *event name unification.* In the following we briefly describe the main steps of preprocessing the dataset:

---

[7] https://validator.w3.org/.

[8] http://sda.tech/EVENTS-Dataset/EVENTS.html.

[9] http://www.core.edu.au/, http://qualis.ic.ufmt.br/, https://goo.gl/3kDfFB.

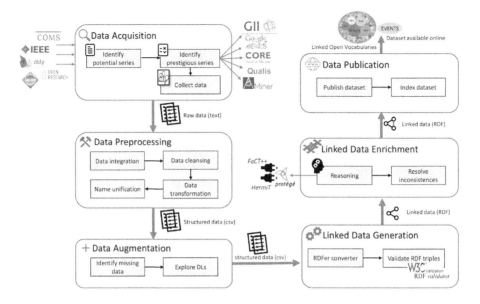

Fig. 1. Data curation of EVENTSKG.

- *Data integration:* involves integrating collected data from disparate sources into a unified view,
- *Data cleansing:* Data integration might result in data redundancy, therefore, data cleansing is crucial. This process involves eliminating redundancy, and identifying and mending unsound data,
- *Data structure transformation:* involves transforming cleaned data to a structured format, such as CSV, RDF/XML and JSON-LD,
- *Event name unification:* involves the unification of the names of all editions of an event series which changed their name. The most recent name has been selected because it is the current name of the event.

5.3   Data Augmentation

The objective of the data augmentation process is to add new events to the dataset and fill in missing data. To achieve this objective, we periodically explore online digital libraries for the missing information. The output of this process is structured data in CSV format.

5.4   Linked Data Generation

The adoption of the Linked Data best practices has led to the enrichment of data published on the Web by connecting data from diverse domains such as scholarly communication, people, digital libraries, and medical data [6]. The objective of the Linked Data Generation process here is to generate linked data from unlinked data in the CSV format. We developed RDFer, a custom Java tool to convert data from CSV format to linked data (RDF/XML syntax). Therefore, the next step is to validate (i.e. syntactic validation) the produced data using a standard validation tool (e.g., the W3C RDF online validation service). A sample of the output of RDFer, in RDF/XML syntax, for ICDE in 2017 can be found on the EVENTSKG documentation page.

5.5 Linked Data Enrichment

The Linked Data enrichment (LDE) process is important in order to discover the interlinking relationships between RDF triples by using inference engines, i.e., reasoners. The input of LDE is the RDF triples produced by the Linked Data generation and the output is a set of *consistent* RDF triples, including the newly discovered relationships, where available. Semantic inference can be used to improve the quality of data integration in a dataset by discovering new relationships, detecting possible inconsistencies and inferring logical consequences from a set of asserted facts or axioms in an ontology. To enrich and validate (i.e. semantics validation) RDF data, generated from the previous process, we use two reasoners integrated in Protégé, FaCT++ and HermiT[10], which support three types of reasoning: (1) detecting inconsistencies, (2) identifying subsumption relationships, and (3) instance classification [16]. Detecting inconsistencies is a crucial step in LDE because inconsistency results in false semantic understanding and knowledge representation. We resolve detected inconsistencies and run the reasoner again to ensure that no other inconsistencies arise.

5.6 Data Publication

The goal of Linked Data publishing is to enable humans and machines to share structured data on the Web. EVENTSKG is published according to the Linked Data community best practices [3,5] and registered in a GitHub repository (https://github.com/saidfathalla/EVENTS-Dataset). The final step is to index the dataset in a public data portal (e.g., DataHub), which is the fastest way for individuals and teams to find, share and publish high-quality data online. Events is published on DataHub at https://datahub.ckan.io/dataset/eventskg.

6 Dataset Content Analysis

Dataset content has been analyzed to answer the research questions presented in Sect. 1. Table 1 provides the results obtained from the preliminary analysis of the dataset[11]. The following metrics are used for the analysis:

---

[10] https://github.com/ethz-asl/libfactplusplus,   https://github.com/phillord/hermit-reasoner.

[11] This table is an extension to the table on the scientometric profile of events series in the EVENTS dataset as presented by Fathalla et al. [7].

Submissions and publications: we see a clear upward trend in the number of submitted and accepted papers during the whole time span, while, roughly speaking, the acceptance rate remains the same.

Time distribution: we observe that prestigious events usually take place around the same month each year (i.e. usual month in Table 1). This helps potential authors to expect when the event will take place next year, which helps for submission schedule organization. *Usual month* refers to the number of times an event has occurred in a specific month. Namely, CVPR conference has been held 26 times (out of 28) in June and POPL has been held 41 times (out of 45) since 1973 in January.

Acceptance rate: Avg. AR of an event series refers to the average of the acceptance rate of all editions. In the last five decades, we observe that the *Avg. AR* for all event series falls between 15% to 30%, except for FOGA, COLT, and IJCAR. Roughly speaking, the largest acceptance rate is the one of FOGA of 59%, while PERCOM has the smallest one of 15%.

H5-index: CVPR has the largest h5-index of 158, while FOGA has the smallest one of 9. Among all considered CS communities, SEC has the largest average h5-index of 58.16, while CSO has the smallest one of 40.2.

Geographical distribution: we analyze the geographical distribution of each event series by recording the country which hosted the most editions of the series (*Most freq. country*). The remarkable observation to emerge from this analysis is that US hosted most editions of events in all communities and most editions of all

Table 1. Scientometric profile of newly added events to EVENTSKG. N is the number of editions in 2018.

Acronym	Comm.	CORE 2018	GII	Q	h5	N	Avg. AR	Most freq. Country	Usual Month	Usual Month Freq.	Since	Publisher
IJCAR	AI	A*	A	B1	45	10	0.41	UK	Jul	4	2001	ACM
COLT		A*	A+	A2	33	31	0.49	US	Jun	11	1988	PMLR
KR		A*	A+	A2	26	16	0.28	US	Apr	4	1989	AAAI
ISMAR		A*	A	A2	26	21	0.24	US	Oct	10	1999	IEEE
VR		A	A-	A2	17	25	0.26	US	Mar	24	1993	IEEE
FOGA		A*	A-	B3	9	14	0.59	US	Jan	7	1990	ACM
PODS	IS	A*	A+	A1	26	37	0.24	US	Jun	17	1982	ACM
POPL	SE	A*	A++	A1	46	45	0.20	US	Jan	41	1973	ACM
OOPSLA		A*	A++	A1	37	33	0.22	US	Oct	26	1986	ACM
OSDI	SEC	A*	A+	A1	39	13	0.16	US	Oct	7	1994	USENIX
TheWeb	WWW	A*	A++	A1	75	23	0.17	US	May	12	1989	TheWeb
WSDM		A*	A+	B1	54	11	0.18	US	Feb	10	2008	ACM
ISWC		A	A+	A1	40	21	0.24	US	Oct	12	1997	Springer
ESWC		A	A	A1	40	15	0.25	Greece	May	9	2004	Springer
ICWS		A	A	A1	26	25	0.21	US	Jun	6	1995	IEEE

SE events. France comes second, having hosted most editions of PKDD and EuroCrypt.

Publishers: we observe that several events series organizers publish the proceedings of their events on their own digital library, such as AAAI, VLDB, and TheWeb. On the other hand, ACM publishes the proceedings of most events, and IEEE comes next.

History: IJCAI is the oldest series since it has been established in 1969 (i.e. 50 editions), while RecSys is the most recent one since it has been established in 2007 (i.e. 12 editions).

7   Conclusions and Future Work

We present a new release of the EVENTS dataset, called EVENTSKG, as a unified Knowledge graph of top-40 prestigious events based on the SEO ontology. To the best of our knowledge, this is the first time a dataset is published as a knowledge graph of metadata of prestigious events in IS, SEC, AI, CSO, SE, and WWW. EVENTSKG closes an important gap in analyzing the progress of a CS community in terms of submissions and publications and it is of primary interest to steering committees, proceedings publishers and prospective authors. The most striking findings to emerge from analyzing EVENTSKG content is that:

- Among all considered CS communities, SEC has the largest average h5-index, while CSO has the smallest one,
- The number of submissions has kept growing over the last five decades, while, roughly speaking, the acceptance rate has remained the same. The reason may be the digitization of scholarly communication.
- The average acceptance rate for all events, since the first edition, falls into the range 15% to 31%,
- US leads by far, having hosted most editions of events in all communities and most editions of all SE events,
- ACM publishes most of the proceedings of the event, and IEEE comes next.

These findings highlight the usefulness of EVENTSKG for events organizers as well as CS researchers.

To further our research, we are planning to add more events from other CS communities such as computer vision, data management and computational learning and elaborate on the set of features that could be used to efficiently compare events in the same community, such as acceptance rate, h-index, and organizers' reputation, defined, e.g., in terms of their h-index and i10-index. Furthermore, we are planning to perform more complex semantic data analysis by querying EVENTSKG using auto-generated SPARQL queries from user selections using a web service.

Acknowledgments. Said Fathalla would like to acknowledge the Ministry of Higher Education (MoHE) of Egypt for providing a scholarship to conduct this study.

References

1. Agarwal, S., Mittal, N., Sureka, A.: A glance at seven acm sigweb series of conferences. In: ACM SIGWEB Newsletter, p. 5 (2016) (Summer)
2. Barbosa, S.D.J., Silveira, M.S., Gasparini, I.: What publications metadata tell us about the evolution of a scientific community: the case of the Brazilian human-computer interaction conference series. Scientometrics **110**(1), 275–300 (2017)
3. Berrueta, D., Phipps, J., Miles, A., Baker, T., Swick, R.: Best practice recipes for publishing RDF vocabularies. In: Working draft, W3C (2008). http://www.w3.org/TR/swbp-vocab-pub/
4. Biryukov, M., Dong, C.: Analysis of computer science communities based on DBLP. In: Lalmas, M., Jose, J., Rauber, A., Sebastiani, F., Frommholz, I. (eds.) ECDL 2010. LNCS, vol. 6273, pp. 228–235. Springer, Heidelberg (2010). https://doi.org/10.1007/978-3-642-15464-5_24
5. Bizer, C., Cyganiak, R., Heath, T., et al.: How to publish linked data on the web (2007). www4.wiwiss.fu-berlin.de/bizer/pub/LinkedDataTutorial/
6. Bizer, C., Heath, T., Berners-Lee, T.: Linked data-the story so far. Int. J. Semant. Web Inf. Syst. **5**(3), 1–22 (2009)
7. Fathalla, S., Lange, C.: EVENTS: a dataset on the history of top-prestigious events in five computer science communities. In: International Workshop on Semantic, Analytics, Visualization. Springer, Heidelberg (2018). (in Press)
8. Fathalla, S., Vahdati, S., Auer, S., Lange, C.: The scientific events ontology of the OpenResearch.org curation platform (2018)
9. Fathalla, S., Vahdati, S., Lange, C., Auer, S.: Analysing scholarly communication metadata of computer science events. In: Kamps, J., Tsakonas, G., Manolopoulos, Y., Iliadis, L., Karydis, I. (eds.) TPDL 2017. LNCS, vol. 10450, pp. 342–354. Springer, Cham (2017). https://doi.org/10.1007/978-3-319-67008-9_27
10. Fathalla, S., Vahdati, S., Auer, S., Lange, C.: Towards a knowledge graph representing research findings by semantifying survey articles. In: Kamps, J., Tsakonas, G., Manolopoulos, Y., Iliadis, L., Karydis, I. (eds.) TPDL 2017. LNCS, vol. 10450, pp. 315–327. Springer, Cham (2017). https://doi.org/10.1007/978-3-319-67008-9_25
11. Gangemi, A.: Ontology design patterns for semantic web content. In: Gil, Y., Motta, E., Benjamins, V.R., Musen, M.A. (eds.) ISWC 2005. LNCS, vol. 3729, pp. 262–276. Springer, Heidelberg (2005). https://doi.org/10.1007/11574620_21
12. Hiemstra, D., Hauff, C., De Jong, F., Kraaij, W.: SIGIR's 30th anniversary: an analysis of trends in IR research and the topology of its community. In: ACM SIGIR Forum. vol. 41. no. 2. ACM. pp. 18–24 (2007)
13. McAfee, A., Brynjolfsson, E., Davenport, T.H., Patil, D., Barton, D.: Big data: the management revolution. Harvard Bus. Rev. **90**(10), 60–68 (2012)
14. Möller, K., Heath, T., Handschuh, S., Domingue, J.: Recipes for semantic web dog food — the ESWC and ISWC metadata projects. In: Aberer, K., et al. (eds.) ASWC/ISWC -2007. LNCS, vol. 4825, pp. 802–815. Springer, Heidelberg (2007). https://doi.org/10.1007/978-3-540-76298-0_58
15. Nuzzolese, A.G., Gentile, A.L., Presutti, V., Gangemi, A.: Semantic web conference ontology - a refactoring solution. In: Sack, H., Rizzo, G., Steinmetz, N., Mladenić, D., Auer, S., Lange, C. (eds.) ESWC 2016. LNCS, vol. 9989, pp. 84–87. Springer, Cham (2016). https://doi.org/10.1007/978-3-319-47602-5_18
16. Rubin, D.L., Knublauch, H., Fergerson, R.W., Dameron, O., Musen, M.A.: Protegeowl: creating ontology-driven reasoning applications with the web ontology language. In: AMIA Annual Symposium Proceedings, vol. 2005. American Medical Informatics Association, p. 1179 (2005)

17. Sabharwal, A.: Digital curation in the digital humanities: Preserving and promoting archival and special collections. Chandos Publishing, Cambridge (2015)
18. Vasilescu, B., Serebrenik, A., Mens, T.: A historical dataset of software engineering conferences. In: Proceedings of the 10th Working Conference on Mining Software Repositories, pp. 373–376. IEEE Press (2013)
19. Yan, S., Lee, D.: Toward alternative measures for ranking venues: a case of database research community. In: Proceedings of the 7th ACM/IEEE-CS joint conference on Digital libraries, pp. 235–244. ACM (2007)

Assessing the Performance of a New Semantic Similarity Measure Designed for Schema Matching for Mediation Systems

Aola Yousfi[(✉)], Moulay Hafid Elyazidi, and Ahmed Zellou

SPM Team, ENSIAS, Mohammed V University, Rabat, Morocco
aola.yousfi@gmail.com, elyazidi.hafid@gmail.com,
zellou_ahmed@hotmail.com

Abstract. Rising adoption of mediation systems leading to plenty of room for improvement. As a matter of fact, schema matching is one of the biggest pressing challenges facing the improvement of mediation systems. The purpose of this research paper is to take aim at schema matching for mediation systems with a semantic similarity measure, which is highly expected to outperform the most well-known semantic similarity measures out there. Using WordNet information, this research study introduced a new semantic similarity measure (as well as a pre-matching strategy) and compared its performance against all the following measures: Resnik's measure, Jiang and Conrath's distance, Lin's measure, and Nababteh's measure. The results indicated that the new measure provides much better results than the aforementioned measures. This paper definitely solves the problem regarding schema matching for mediation systems, and even though it is still in an early stage, the new measure will benefit both researchers and organizations.

Keywords: Semantic similarity measures
Information content-based measures · Schema matching · Mediation systems
Virtual data integration · WordNet

1 Introduction

In the near future, data will be generated by every single object on Earth, data sources will be spread to every place on Earth where favorable conditions occur, and considerable amounts of data updates will take place per second. It is reasonable to believe that, by then, data integration will get even harder than ever before.

Data integration is a term that is clearly used to refer to the gathering of different data sources. In this way, the user might access all data via a unified interface. Thanks to that, the user doesn't have any more to send his query to each existing data source separately. Under those circumstances, the importance of data integration is undeniable. Due to this fact, many data integration approaches have been defined among which virtual approach, data warehousing approach, redevelopment approach and SOA approach. In this paper, special attention will be devoted to virtual data integration approach in general, and semantic matching in particular.

© Springer Nature Switzerland AG 2018
N. T. Nguyen et al. (Eds.): ICCCI 2018, LNAI 11055, pp. 64–74, 2018.
https://doi.org/10.1007/978-3-319-98443-8_7

Virtual data integration approach aims at integrating virtually multiple, heterogeneous, autonomous, distributed and scalable data sources by means of a mediator incorporated between data sources and user applications [1] (see Fig. 1). Semantic matching's goal is to find semantic correspondences between elements of the mediated schema, also known as global schema, and elements of the sources schemas, also known as local schema [2]. The results are used later to describe how elements of the global schema are related to elements of the local schema, also known as schema mapping [2]. Many research studies have addressed schema matching and mapping still to this day. Indeed, plenty of semantic matchers are out there, among which COMA, Cupid, and S-Match [3]. And there are plenty of mapping approaches as well, for instance, Global As View (GAV), Local As View (LAV), and Global Local As View (GLAV) [4]. Semantic matchers and mapping approaches keep on coming and outdoing existing matchers and mapping approaches, respectively, and we're left wondering, "What's next? What solution can potentially last for a long period of time?" Every powerful mediation system requires a schema matching approach as well as a mapping approach, every last one, it's not going to work with one without the other. To our knowledge, no previous research works have addressed the problem of schema matching and mapping as a unit, we decided to take matters into our own hands and come up with both a matching approach and a mapping approach that can be used together.

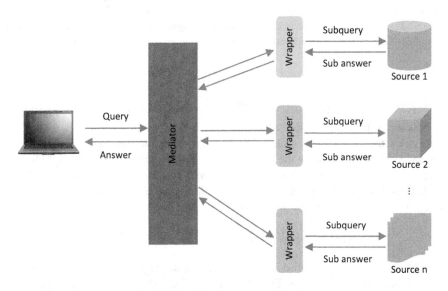

Fig. 1. Mediation system architecture

This paper is devoted to schema matching for mediation systems. As for the mapping, we'll shortly reserve a whole paper to discuss it. Our plan for schema matching is to turn the idea of comparing two elements semantically into computing the value of a mathematical equation. This could be a big step forward for mediation systems.

The rest of this paper is organized as follows. The second section provides a general overview of some semantic similarity measures. The third section introduces our semantic similarity measure. The fourth section assesses the performance of our measure. The fifth section presents an example. The sixth section summarizes this paper and suggests new perspectives in order to open the door to all-important improvements.

2   Related Work

According to [5], semantic similarity measures are grouped into four categories: edge-based measures (also known as path-based measures), information content-based measures, feature-based measures, and hybrid-based measures. But since information content-based measures are the high-performance ones [5], we decided to only direct our attention to some information content-based measures. There are basically three classical information content-based measures we need to present and a recent one we choose to present as well, so we can use them all later to judge the performance of our measure. WordNet is a lexical database for the English language created by a research team at Princeton University [6]. It groups words into sets of synonyms called synsets, which are interlinked by means of semantic relationships, for instance, is-a relationship which connects a hyponym to a hypernym [6]. And it's commonly used by semantic similarity measures. Indeed, the following measures all use WordNet.

Resnik's measure computes the Information Content (IC) of the Least Common Subsumer (LCS) of two compared concepts denoted by C_1 and C_2 as follows [7]:

$$Sim_{Resnik}(C_1, C_2) = IC(LCS(C_1, C_2)) \tag{1}$$

Where $IC(C) = -\log(p(C))$, $p(C) = \frac{frequency(C)}{N}$ the probability of a given concept C, and N refers to the total number of nouns.

Resnik's measure's main issue is the following: any pair of concepts having the same LCS will definitely have the same semantic similarity value [5]. Luckily, Jiang & Conrath (J&C) and Lin found out a way to overcome Resnik's measure's problem [8, 9]. In addition to the IC of the LCS, both J&C and Lin consider the IC of each concept [8, 9]. J&C define the distance between two concepts as follows [8]:

$$Dist_{J\&C}(C_1, C_2) = IC(C_1) + IC(C_2) - 2 * IC(LCS(C_1, C_2)) \tag{2}$$

It differs from similarity measures in a way that the higher it gets, the less similar the two compared concepts are. Typically, given J&C's distance, one can revert it to serve as a similarity measure and vice versa. Note that in this paper, we are going to use the similarity measure.

Lin describes the semantic similarity between two concepts as follows [9]:

$$Sim_{Lin}(C_1, C_2) = \frac{2 * IC(LCS(C_1, C_2))}{IC(C_1) + IC(C_2)} \tag{3}$$

Lin's measure's main problem is the following: if the IC of LCS, C_1 or C_2 is equal to 0, then the semantic similarity value is equal to 0 as well [10].

In order to deal with Lin's measure's problem, Nababteh (N.) suggests to divide 2 times the IC of the LCS of two compared concepts by the sum of the IC of the direct hypernym of the first concept and the IC of the direct hypernym of the second concept [10].

$$Sim_N.(c_1, c_2) = \frac{2 * IC(LCS(C_1, C_2))}{IC(P(C_1)) + IC(P(C_2))} \tag{4}$$

For the time being, the aforementioned semantic similarity measures are quite successful, they remain, however, some issues that require more attention, otherwise they will come home to roost every time the corresponding measure is applied to a new mediation system. Indeed, schema matching for mediation systems is quite different and isn't necessarily between two concepts that belong to WordNet. Truth is, none of the above measures were designed especially for virtual data integration systems, the thing is, if we want to make important improvements, then we have to come up with a semantic similarity measure that fits mediation systems requirements. Furthermore, according to [7–10], the aforementioned measures might not provide the right match all the time since when compared to human similarity judgments on Miller and Charles' (M&C) benchmark dataset the results were not promising.

3 Pairing Global Schema Elements with Local Schema Elements

As said earlier, this paper's main objective is to take aim at schema matching for mediation systems with a new semantic similarity measure. But before we do that, we need to get schema elements prepared for our measure.

3.1 Schema Elements Ready for Pre-matching, Preparing for Schema Matching

With the objective to determine the semantic similarity value between two schema elements, the use of a pre-matching strategy is mandatory, since both global schema elements and local schema elements might involve expressions that don't necessarily belong to WordNet. Indeed, let's take for example an XML schema, an element might be a word, a concatenation of words or a mix of letters, digits, hyphens, underscores or periods. The pre-matching strategy we came up with goes like this:

- Step 1: Turn complex elements into simple elements.

Every XML schema element that belongs to a complex element is turned into a union of its element and its complex element, every last one.

- Step 2: Transform non-WordNet elements into sets of WordNet concepts.

Extract all WordNet concepts that a non-WordNet element might have. In this way, a non-WordNet element is from now on a set of WordNet concepts.

- Step 3: Seek the right sense.

We might face two cases for which the measure we have to use, in fact, should be different (5). Indeed, an XML schema element denoted by W_0 (W_0 might have more than one sense) can either be a WordNet concept or, a set of WordNet concepts. Either way, we based both measures on the Word Sense Disambiguation (WSD) approach (we refer the reader to [11] for a thorough description of the WSD approach) to determine the most accurate sense of W_0 in the given context [11]. WSD uses the context of W_0 (i.e. elements surrounding W_0 in the schema), denoted by W_k. The sense of W_0 retained for later use is the one with the highest score value.

$$\begin{cases} Sense_{W_0} = \max_{1 \le j \le c_0} Score_j = \max_{1 \le j \le c_0} \sum_{k=1}^{|W_k|} \sum_{l=1}^{c_k} relatedness(s_{0,j}, s_{k,l}), \ if \ card(W_0)=1 \\ Sense_{W_0} = \max_{\substack{1 \le i \le |W_0| \\ 1 \le j \le c_{0,i}}} Score_{i,j} = \max_{\substack{1 \le i \le |W_0| \\ 1 \le j \le c_{0,i}}} \sum_{k=1}^{|W_k|} \sum_{l=1}^{c_k} relatedness(s_{i,j}, s_{k,l}), \ otherwise \end{cases} \quad (5)$$

Where:

- $s_{0,j}$, $s_{k,l}$ and $s_{i,j}$ refer to the jth sense of W_0, the lth sense of the kth element of W_k, and the jth sense of the ith element of W_0, respectively.
- c_0, c_k and $c_{0,i}$ refer to the total number of senses of W_0, the total number of senses of the kth element of W_k, and the total number of senses of the ith element of W_0, respectively.

With our three-step pre-matching strategy in place, both elements of the mediated schema and elements of the local schema are in a state of preparedness for the next step, namely semantic matching.

3.2 Semantic Matching with a Semantic Similarity Measure

After an in-depth analysis of WordNet, we noticed that the hypernyms along with the direct hyponyms of a given concept C contribute to different degrees to the definition of C. Hence, we decided to include both the compared concept's hypernyms and direct hyponyms in our semantic similarity measure in a way that comparing two concepts semantically, C_L (element of the local schema) and C_G (element of the global schema), would be equivalent to comparing $\{C_L, P_{C_L}, H_{C_L}\}$ with $\{C_G, P_{C_G}, H_{C_G}\}$ i.e. compare semantically C_L to C_G (6), C_L to P_{C_G} (7), C_L to H_{C_G} (8), P_{C_L} to C_G (9), P_{C_L} to P_{C_G} (10), P_{C_L} to H_{C_G} (11), H_{C_L} to C_G (12), H_{C_L} to P_{C_G} (13), and H_{C_L} to H_{C_G} (14) (P_{C_L} and P_{C_G} refer to C_L's hypernyms and C_G's hypernyms, respectively. Note that $P_{C_G} \cap P_{C_L} = \emptyset$; we only consider non-shared hypernyms. H_{C_L} and H_{C_G} refer to C_L's direct hyponyms and C_G's direct hyponyms, respectively.). The above comparisons calculated by the following equations (Note that S_T and Sy_T refer to the sense and the synset of a given concept T (T is a concept of WordNet), respectively.):

$$SM_1(C_L, C_G) = card(S_{C_L} \cap S_{C_G}) + card(S_{C_L} \cap (C_G \cup Sy_{C_G})) + card(S_{C_G} \cap (C_L \cup Sy_{C_L})) \quad (6)$$

$$SM_2(C_L, P_{C_G}) = \sum_{i=1}^{|P_{C_G}|} card\left(S_{C_L} \cap S_{P_{C_{G_i}}}\right) + card\left(S_{C_L} \cap \left(P_{C_{G_i}} \cup Sy_{P_{C_{G_i}}}\right)\right)$$
$$+ card\left(S_{P_{C_{G_i}}} \cap (C_L \cup Sy_{C_L})\right) \tag{7}$$

$$SM_3(C_L, H_{C_G}) = \sum_{i=1}^{|H_{C_G}|} card\left(S_{C_L} \cap S_{H_{C_{G_i}}}\right) + card\left(S_{C_L} \cap \left(H_{C_{G_i}} \cup Sy_{H_{C_{G_i}}}\right)\right)$$
$$+ card\left(S_{H_{C_{G_i}}} \cap (C_L \cup Sy_{C_L})\right) \tag{8}$$

$$SM_4(P_{C_L}, C_G) = \sum_{i=1}^{|P_{C_L}|} card\left(S_{P_{C_{L_i}}} \cap S_{C_G}\right) + card\left(S_{P_{C_{L_i}}} \cap (C_G \cup Sy_{C_G})\right)$$
$$+ card\left(S_{C_G} \cap \left(P_{C_{L_i}} \cup Sy_{P_{C_{L_i}}}\right)\right) \tag{9}$$

$$SM_5(P_{C_L}, P_{C_G}) = \sum_{i=1}^{|P_{C_L}|} \sum_{j=1}^{|P_{C_G}|} card\left(S_{P_{C_{L_i}}} \cap S_{P_{C_{G_j}}}\right) + card\left(S_{P_{C_{L_i}}} \cap \left(P_{C_{G_j}} \cup Sy_{P_{C_{G_j}}}\right)\right)$$
$$+ card\left(\left(P_{C_{L_i}} \cup Sy_{P_{C_{L_i}}}\right) \cap S_{P_{C_{G_j}}}\right) \tag{10}$$

$$SM_6(P_{C_L}, H_{C_G}) = \sum_{i=1}^{|P_{C_L}|} \sum_{j=1}^{|H_{C_G}|} card\left(S_{P_{C_{L_i}}} \cap S_{H_{C_{G_j}}}\right) + card\left(S_{P_{C_{L_i}}} \cap \left(H_{C_{G_j}} \cup Sy_{H_{C_{G_j}}}\right)\right)$$
$$+ card\left(\left(P_{C_{L_i}} \cup Sy_{P_{C_{L_i}}}\right) \cap S_{H_{C_{G_j}}}\right) \tag{11}$$

$$SM_7(H_{C_L}, C_G) = \sum_{i=1}^{|H_{C_L}|} card\left(S_{H_{C_{L_i}}} \cap S_{C_G}\right) + card\left(S_{H_{C_{L_i}}} \cap (C_G \cup Sy_{C_G})\right)$$
$$+ card\left(S_{C_G} \cap \left(H_{C_{L_i}} \cup Sy_{H_{C_{L_i}}}\right)\right) \tag{12}$$

$$SM_8(H_{C_L}, P_{C_G}) = \sum_{i=1}^{|H_{C_L}|} \sum_{j=1}^{|P_{C_G}|} card\left(S_{H_{C_{L_i}}} \cap S_{P_{C_{G_j}}}\right) + card\left(S_{H_{C_{L_i}}} \cap \left(P_{C_{G_j}} \cup Sy_{P_{C_{G_j}}}\right)\right)$$
$$+ card\left(\left(H_{C_{L_i}} \cup Sy_{H_{C_{L_i}}}\right) \cap S_{P_{C_{G_j}}}\right) \tag{13}$$

$$SM_9(H_{C_L}, H_{C_G}) = \sum_{i=1}^{|H_{C_L}|} \sum_{j=1}^{|H_{C_G}|} card\left(S_{H_{C_{L_i}}} \cap S_{H_{C_{G_j}}}\right) + card\left(S_{H_{C_{L_i}}} \cap \left(H_{C_{G_j}} \cup Sy_{H_{C_{G_j}}}\right)\right)$$
$$+ card\left(\left(H_{C_{L_i}} \cup Sy_{H_{C_{L_i}}}\right) \cap S_{H_{C_{G_j}}}\right) \tag{14}$$

When conducting some experiments including those presented in Sect. 4, we want it to calculate the degree of involvement of each one of $SM_{1 \leq i \leq 9}$ to the semantic similarity between C_L and C_G, so we introduced two parameters, $\alpha, \beta \in [0, 1]$, and then we computed the correlation coefficient between human similarity judgments as stated in M&C's experiment and our measure for all possible combinations ($[\alpha * \sum_{i=1}^{9} SM_i = \beta * \sum_{i=1}^{9} SM_i = \sum_{i=1}^{9} SM_I$, (where $\alpha = \beta = 1$)], $[\alpha * SM_1 = 0,1*SM_1$ and $\beta * \sum_{i=2}^{9} SM_i = 0,9* \sum_{i=2}^{9} SM_i]$, $[\alpha * SM_2 = 0,1*SM_2$ and $\beta * \sum_{\substack{i=1 \\ i \neq 2}}^{9} SM_i = 0,9* \sum_{\substack{i=1 \\ i \neq 2}}^{9} SM_i]$ etc.). We found that the highest correlation coefficient is the one with the following combination $0.8 * (SM_1 + SM_5 + SM_9) + 0.2 * \sum_{\substack{i=2 \\ i \neq 5}}^{8} SM_i$.

When comparing C_L and C_G semantically, we might face one of two cases: either C_L and C_G are each a WordNet concept and we have to use (15) or, at least one of the two, C_L and C_G, is a set of WordNet concepts and we should use (16).

$$\begin{cases} Sim_{AoLa/1-1}(C_G, C_L) = 1, \text{ if } C_L, C_G \text{ are synonyms or one is a direct hyponym of the other} \\[2ex] Sim_{AoLa/1-1}(C_G, C_L) = 0, \text{ if } [0.8*(SM_1 + SM_5 + SM_9) + 0.2*\sum_{\substack{i=2 \\ i \neq 5}}^{8} SM_i] e^{\frac{\sum_{\substack{i=1 \\ SM_i \neq 0}}^{9} 1}{9}} \leq 1 \\[3ex] Sim_{AoLa/1-1}(C_G, C_L) = \dfrac{[0.8*(SM_1 + SM_5 + SM_9) + 0.2*\sum_{\substack{i=2 \\ i \neq 5}}^{8} SM_i] e^{\frac{\sum_{\substack{i=1 \\ SM_i \neq 0}}^{9} 1}{9}} - 1}{[0.8*(SM_1 + SM_5 + SM_9) + 0.2*\sum_{\substack{i=2 \\ i \neq 5}}^{8} SM_i] e^{\frac{\sum_{\substack{i=1 \\ SM_i \neq 0}}^{9} 1}{9}} + 1}, \text{ otherwise} \end{cases}$$

(15)

$$\begin{cases} Sim_{AoLa}(C_G, C_L) = \frac{1}{card(C_G)} \sum_{i=1}^{card(C_G)} max(a_{i,j})_{1 \leq j \leq card(C_L)}, & \text{if } card(C_G) \leq card(C_L) \\ Sim_{AoLa}(C_G, C_L) = \frac{1}{card(C_L)} \sum_{j=1}^{card(C_L)} max(a_{i,j})_{1 \leq i \leq card(C_G)}, & \text{otherwise} \end{cases}$$

(16)

Where $A = (a_{i,j})_{\substack{1 \leq i \leq card(C_G) \\ 1 \leq j \leq card(C_L)}}$ is a $card(C_G) \times card(C_L)$ matrix whose individual items are defined as follows: $a_{i,j} = Sim_{AoLa/1-1}(C_{G_i}, C_{L_j})$.

We define the threshold value 0.8 of our semantic similarity measure beyond which schema element pairs (i.e. pairs of elements of the mediated schema and elements of the local schema) are retained.

Note that our measure doesn't depend on the language itself, that is to say our measure can be used with any language having WordNet.

4 Evaluation of the New Semantic Similarity Measure

We use M&C's benchmark dataset [7] to assess the performance of our measure. Table 1 presents semantic similarity values by word pair when each of the following measures was applied: Resink's measure [7], J&C's measure [8], Lin's measure [9], N.'s measure [10], and our measure (AoLa). As for the second column, it presents human similarity judgments as stated in M&C's experiment [7].

Table 1. Semantic similarity values by word pair

Word pair	M&C	Resnik	J&C	Lin	N.	AoLa
Automobile/Car	0.98	0.9962	1	1	1	1
Journey/Voyage	0.96	0.9907	0.9165	0.8277	0.857335	1
Gem/Jewel	0.96	1	1	0.2434	0.31453	1
Boy/Lad	0.94	0.9971	0.8613	0.6433	1	1
Coast/Shore	0.925	0.9994	0.9567	0.96	1	1

(*continued*)

Table 1. (*continued*)

Word pair	M&C	Resnik	J&C	Lin	N.	AoLa
Asylum/Madhouse	0.9025	1	0.9379	0.769	0.879	1
Magician/Wizard	0.875	0.9999	1	0.1958	0.28158	1
Midday/Noon	0.855	0.9998	1	1	1	1
Furnace/Stove	0.7775	0.6951	0.593	0.2294	0.26674	0.79
Food/Fruit	0.77	0.9689	0.7925	0.0956	0.103839	0.98
Bird/Cock	0.7625	0.9984	0.8767	0.7881	0.930014	1
Bird/Crane	0.7425	0.9984	0.815	0	0.850943	0.95
Implement/Tool	0.7375	0.9852	0.977	0.914	1	1
Brother/Monk	0.705	0.8722	0.6656	0	1	1
Crane/Implement	0.42	0.8722	0.6526	0	0.513459	0.73
Brother/Lad	0.415	0.8693	0.6775	0.24	0.29735	0.62
Car/Journey	0.29	0	0.5883	0	0	0
Monk/Oracle	0.275	0.8722	0.6203	0.1828	0.191595	0.75
Food/Rooster	0.2225	0.5036	0.5885	0.0762	0.095302	0.66
Coast/Hill	0.2175	0.9867	0.8487	0.127	0.19414	0.49
Forest/Graveyard	0.21	0	0.484	0.1119	0.1706	0.61
Monk/Slave	0.1375	0.8722	0.6962	0.2011	0.34281	0.25
Coast/Forest	0.105	0	0.5179	0	0	0.2
Lad/Wizard	0.105	0.8722	0.6905	0.2241	0.34155	0.2
Cord/Smile	0.0325	0.8044	0.5845	0	0	0.11
Glass/Magician	0.0275	0.5036	0.5699	0.0663	0.09335	0
Rooster/Voyage	0.02	0	0.4168	0	0	0
Noon/String	0.02	0	0.4329	0	0	0

In order to compare the performance of our measure to the aforementioned measures, we consider two factors: correlation coefficient and Mean Square Error (MSE). Table 2 presents the correlation coefficients between human similarity judgments and all the previously mentioned measures. It presents the MSE of each measure as well.

Table 2. Correlation coefficients and MSE

Measure	Correlation coefficient	MSE
Resnik	0.6671	0.1373
J&C	0.8363	0.1018
Lin	0.6852	0.1188
N.	0.7654	0.0699
AoLa	0.9102	0.0453

The graph representing the correlation between human similarity judgments as stated in M&C's experiment and our measure is presented in Fig. 2.

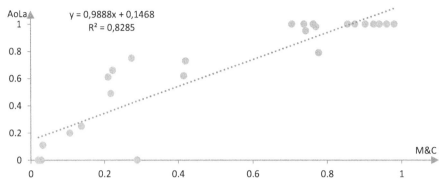

Fig. 2. Graph of AoLa vs. M&C

The results indicate that the correlation coefficient of our measure is +0.9102, which is the highest coefficient compared to the other measures in this experiment. The findings also indicate that the MSE of our measure is 0.0453, which is the smallest MSE compared to the other measures in our experiment. These values show that our measure outperforms all the others.

5 Experimentation and Discussion

With our new measure in place, schema matching for mediation systems will no longer be an issue. Let's take for example two excerpts, E_G and E_L, from a global schema and a local schema, respectively. Suppose that both schemas are written in XML.

- Excerpt from a global schema (E_G)

```
<xs:element name="employee">
 <xs:complexType>
  <xs:sequence>
   <xs:element name="familyName" type="xs:string"/>
   <xs:element name="firstName" type="xs:string"/>
   <xs:element name="age" type="xs:integer"/>
  </xs:sequence>
 </xs:complexType>
</xs:element>
```

- Excerpt from a local schema (E_L)

```
<xs:element name="firstName_employee" type="xs:string"/>
<xs:element name="lastName_employee" type="xs:string"/>
<xs:element name="birthDate_employee" type="xs:date"/>
<xs:element name="academicDegree_employee"/>
 <xs:simpleType>
  <xs:restriction base="xs:string"/>
   <xs:enumeration value="Bachelor"/>
   <xs:enumeration value="Master"/>
   <xs:enumeration value="Doctorate"/>
  </xs:restriction/>
 </xs:simpleType>
</xs:element/>
```

When the aforementioned pre-matching strategy (we refer the reader to Sect. 3.1 for more details) was applied to E_G and E_L, we got the following sets: $set_{E_{G_1}}$ ={family name#n#1, employee#n#1}, $set_{E_{G_2}}$ ={first name#n#1, employee#n#1}, $set_{E_{G_3}}$ = {age#n#1, employee#n#1} and $set_{E_{L_1}}$ = {first name#n#1, employee#n#1}, $set_{E_{L_2}}$ = {last name# n# 1, employee# n#1}, $set_{E_{L_3}}$ = {birth#n#1, date#n#1, employee#n#1}, $set_{E_{L_4}}$ = {academic degree#n#1, employee#n#1} respectively. Word#n#k denotes the kth sense of a given noun of WordNet.

Table 3 presents semantic similarity values between $set_{E_{G_{1 \leq i \leq 3}}}$ and $set_{E_{L_{1 \leq i \leq 4}}}$ as well as schema element pairs retained (denoted by bold table cells).

We only take into consideration pairs whose semantic similarity values are the highest and greater than 0.8 as well.

Table 3. Schema matching results

	$set_{E_{G_1}}$	$set_{E_{G_2}}$	$set_{E_{G_3}}$
$set_{E_{L_1}}$	0.83	**1**	0.66
$set_{E_{L_2}}$	**1**	0.83	0.67
$set_{E_{L_3}}$	0.68	0.67	**0.84**
$set_{E_{L_4}}$	0.59	0.58	0.5

6   Conclusion and Perspectives

The objective of this paper was to take aim at schema matching for mediation systems with a new semantic similarity measure. The findings determined a strong positive correlation, 0.9102 (which is the highest correlation coefficient compared to Resink's measure, J&C's distance, Lin's measure, and N.'s measure), between our measure and human similarity judgments on M&C's benchmark dataset. The experimental results also indicated a weak MSE, 0.0453 (which is the smallest MSE compared to the previously mentioned measures), of our measure. These values provide evidence of the usefulness of our measure, along with some indications of what makes our measure much better than the aforementioned measures.

We intend to use our semantic similarity measure to create initially a schema matching-mapping tool for mediation systems and then, subsequently, to extend this tool gradually, in order to fit other data integration approaches systems requirements. This tool is believed to be all-important to both researchers and organizations.

References

1. Samuel, J.: Feeding a Data Warehouse with Data Coming from Web Services. A Mediation Approach for the DaWeS Prototype. Ph.D. thesis supervised by Toumani, F. and Rey, C. Blaise Pascal University, France (2014)
2. Bellahsene, Z., Bonifati, A., Duchateau, F., Velegrakis, Y.: On evaluating schema matching and mapping. In: Bellahsene, Z., Bonifati, A., Rahm, E. (eds.) Schema Matching and Mapping. Data Centric-Systems and Applications, pp. 253–291. Springer, Heidelberg (2011). https://doi.org/10.1007/978-3-642-16518-4_9
3. Sutanta, E., Wardoyo, R., Mustofa, K., Winarko, E.: Survey: models and prototypes of schema matching. Int. J. Electr. Comput. Eng. **6**(3), 1011–1022 (2016)
4. Lenzerini, M.: Data integration: a theoretical perspective. In: Proceedings of the Twenty-First ACM SIGMOD-SIGACT-SIGART Symposium on Principles of Database Systems, pp. 233–246. ACM, USA (2002)
5. Meng, L., Huang, R., Gu, J.: A review of semantic similarity measures in WordNet. Int. J. Hybrid Inf. Technol. **6**(1), 1–12 (2013)
6. Miller, G.A.: WordNet: a lexical database for English. Commun. ACM **38**(11), 39–41 (1995)
7. Resnik, P.: Using information content to evaluate semantic similarity in a taxonomy. In: Proceedings of the 14th International Joint Conference on Artificial Intelligence, vol. 1, pp. 448–453. Canada (1995)
8. Jiang, J.J., Conrath, D.W.: Semantic similarity based on corpus statistics and lexical taxonomy. In: Proceedings of International Conference Research in Computational Linguistics (ROCLING X), pp. 19–33. Taiwan (1997)
9. Lin, D.: An information-theoretic definition of similarity. In: Proceedings of the 15th ICML, pp. 296–304. USA (1998)
10. Mohammed, N., Mohammed, D.: New modified semantic similarity measure based on information content approach. Int. J. Comput. Sci. Network Secur. **17**(3), 73–76 (2017)
11. Banerjee, S., Pedersen, T.: Extended gloss overlaps as a measure of semantic relatedness. In: Proceedings of the 18th International Joint Conference on Artificial Intelligence, pp. 805–810. Mexico (2003)

The Impact of Data Dispersion
on the Accuracy of the Data Warehouse
Federation's Response

Rafał Kern[✉]

Faculty of Computer Science and Management, Wroclaw University of Science
and Technology, Wybrzeze Wyspianskiego 27, 50-370 Wroclaw, Poland
rafal.kern@pwr.edu.pl

Abstract. This article is a preliminary attempt to verify how the values of the metrics that characterize data warehouses affect the accuracy of the data warehouse federation's response. The federation can be build almost always, but sometimes the heterogeneity of data stored in the component data warehouse is too big to properly handle the user's request. If that happens, the effort put on the integration is a waste. Some work was done, but the federation does not give accurate response. In this paper some dependencies between the accuracy of federation's response and data warehouse metrics are discussed.

Keywords: Data warehouses · Federation · Data integration

1 Introduction

At the times when the speed of data collection exceeds the speed of data analysis and some social websites gathers hundreds of TB of data a day, the data analysis challenges are getting more and more crucial. There is no space for delays. Some of these delays are inevitable because they rely very much on the human skills. Some of them can be supported by decision support systems. These systems, to work properly, have to be filled with good quality test sets. It can find any relations or rules, that will be later used for making decision suggestions.

At this time millions of IT systems exists, are further developed and maintained. Many systems are legacy, but still are useful and their maintenance is still cheaper than investment in new solution. Database systems are not an exception. Their specialized type: data warehouses are widely used especially in large organizations. Federated structures allow to use existing infrastructure and subsystems to deliver working solution cheaper and faster than building new one from beginning [4,17].

Federation is a set of components with unified external interface that allows accessing data stored in each component in transparent way. It works on assumption that each component has reliable data about their area of interest. The federation structure described previously in [5,6] must have the ability to

© Springer Nature Switzerland AG 2018
N. T. Nguyen et al. (Eds.): ICCCI 2018, LNAI 11055, pp. 75–84, 2018.
https://doi.org/10.1007/978-3-319-98443-8_8

balance the partial responses to provide reliable, integrated response. It depends on many factors: e.g query type, integration method, transformation methods etc. However, some topics might be discussed regardless mentioned factors. The goal of this paper is to examine factors that may suggest if this task is possible to be done.

The paper is organized as follows: Sect. 2 contains analysis of related works, Sect. 3 contains necessary definitions and notions. In Sect. 4 an overview and some results details are described. Summary and some propositions for future works are presented in Sect. 5.

2   Related Works

Data integration process is a natural consequence of distributed data source heterogeneity and the need of transparent access to their data. The initial user query must be decomposed to sub-queries that are processable on these distributed data warehouses. Many papers describe federation as a good solution for data processing problems. It work on the logical level, where the coordinator layer [1] is responsible for maintenance of the structure of federation. Sometimes it may be handled by global sql views [12]. In the area of medical treatment it has been also successfully used [13,14].

This paper is a continuation of a topic described in papers [5–7,15]. The procedures of sources schemas integration, user queries decomposition and source data warehouses responses integration were implemented and used for the experiment mentioned in introduction. Comparing to results received in previous experiments, several additional metrics propositions and distance functions were added. The responses integration procedure requires the context of data in each component. This metadata is collected both during schema integration and ETL processes [4].

A good model of Data Integration System was described in [8]. In (like in [1]) authors suggested that some semantic data might be very useful in responses integration. In [2] authors mentioned, that the heterogeneity may occur on two levels: schema or instance level. Operations on corrupted, inconsistent, incomplete or outdated data may make the problem even worse [9,11]. The last example fortunately does not exists in the field of data warehouses. No data is outdated. It only needs a well defined time dimension. Additionally, in most cases the data warehouse works on numerical data. This fact significantly narrows the area of possible inconsistency situations. Conflicted data may be moved into common, canonical model [3]. A good review of methods for inconsistency solving was presented in [10].

During a design process, especially in query execution, many factors should be taken into account, e.g. [16]:

– complexity of different tasks
– computational capabilities of central site - the federation layer
– shared data among local data warehouses

A decision if given data warehouse is worth to be integrated with federation should be taken very carefully. The proposed method does not need to rebuilt the whole canonical, federated data model. However, still it is a complex task, that should be supervised by an domain expert. Therefore, any additional information, that could e.g. tell the experts that this process is pointless because the responses will be inaccurate, are extremely valuable.

3   Basic Notions

In this section some basic definition are presented:

Definition 1 Data warehouse federation. *Data warehouse federation is defined as a tuple:*

$$\hat{F} = (F, H, U, q, l) \tag{1}$$

where:

F - *federation schema*
H - *set of data warehouses*
U - *user/external interface*
q - *query decomposition procedure*
l - *responses integration procedure*

The end-user interface U contains defined query language L_Q.

Definition 2 Data warehouse federation schema. *Data warehouse federation schema is defined as a tuple:*

$$F = \{D_0, D_1, ..., D_n\} \tag{2}$$

where:

D_0 - *facts table schema*
D_j - *dimensions tables schemas for $j \in [1, n]$. Between dimension table D_j and facts table there occurs a relationship of type $1 - \infty$ which means, that one row from dimension table may be associated with more than one row from facts table, but one facts table is associated with exactly one row from given dimension table.*

Each attribute from federation facts table or dimension table must have its representative in at least one source data warehouse.

Each of data warehouse associated with the federation may be describe by factors listed and described below:

– **Power** - ratio between number of facts in given data warehouse and the whole federation.

$$p = \frac{Card(H_i)}{Card(F)} \tag{3}$$

where:

$p \in\, <0,1>\; Card(H_i)$ - facts number in data warehouse H_i
$Card(F)$ - facts number in federation

- **Coverage** - the coverage of federation schema by schema of a given data warehouse.

$$c = \frac{AC(H_i) + MC(H_i)}{AC(F) + MC(F)} \tag{4}$$

where:

$AC(H_i)$ - number of attributes in dimensions of data warehouse H_i
$MC(H_i)$ - number of measures in H_i
$AC(F)$ - dimensions attributes number in federation
$MC(F)$ - measures number in federation

Moreover, some additional factors that described the entire federation were formulated:

1. **Average power**
$$P_a = median(p_1, p_2, ..., p_x) \tag{5}$$

   where:
   $p_i \in\, <0,1>$ is a power factor for i-th data warehouse in given federation
   $i \in (1...x)$
   median is standard, statistical function that gives median from set

2. **Power deviation**
$$P_d = stdev(p_1, p_2, ..., p_x) \tag{6}$$

   where:
   $p_i \in\, <0,1>$ is a power factor for i-th data warehouse in given federation
   $i \in (1...x)$
   stdev means standard deviation from set

3. **Average coverage**
$$C_a = median(c_1, c_2, ..., c_x) \tag{7}$$

   where:
   $p_i \in\, <0,1>$ is a coverage factor for i-th data warehouse in given federation
   $i \in (1...x)$
   median is standard, statistical function that gives median from set

4. **Coverage deviation**
$$C_d = stdev(c_1, c_2, ..., c_x) \tag{8}$$

   where:
   $p_i \in\, <0,1>$ is a coverage factor for i-th data warehouse in given federation
   $i \in (1...x)$
   stdev means standard deviation from set

5. **Attributes domain dispersion**

$$Disp_a = \frac{a_{dev}}{a_{median}} \qquad (9)$$

where:

a_{dev} - standard deviation of attributes values

a_{median} - median of attributes values

Some more factors are described in [15], but were not tested during the experiment described in next section. The last four factor's goals is to characterize the existing and hypothetical federations in order to estimate its future profitability.

4 Experiment

4.1 Overview

This paper investigates possible relations between the lack of precision of federations response and factors presented in Sect. 3. The lack of precision is defined as a distance:

$$dist(R_p, R) = \frac{\sum\limits_{i=1}^{n}(\frac{\sum\limits_{j=1}^{s} diff(R_p^{ij} - R^{ij})}{s})}{n} \qquad (10)$$

where:

R_p - ideal warehouse response

R - federation response

n - rows number in pattern data warehouse response

R_p^{ij} - value of j element in i row pattern data warehouse response

R^{ij} - value of j element in i row federation response

s - number of columns in single row of response.

There was a lot of difficulties with obtaining very large set of real-world data which are quite similar (from semantic point of view). This condition is very important because when the structures of source data warehouses are not similar enough, the schemas integration step will be very simple. That is why the next steps would be trivial. These factors entailed the decision about using randomly generated test datasets.

Figure 1 presents the general idea of the experiment. Each experiment consists of two main phases:

1. Preparation phase:
   (a) Generate source data warehouses according to following parameters:
       i. Number of source data warehouses: 10
       ii. Number of dimensions: N(10,5) - normal distribution, mean = 10, standard deviation = 5.
       iii. Number of dimensions attributes: N(6,4)

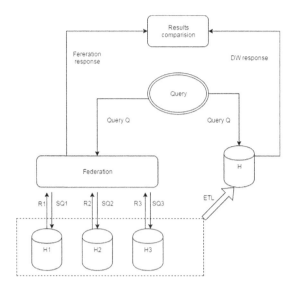

Fig. 1. Experiments overview

    iv. Attributes names dictionary size: 44

    v. Number of measures: N(3,3)

    vi. Measures names dictionary size: 10

(b) Generate data warehouses instances and fill them with data

(c) Calculate data warehouse factors

(d) Integrate source data warehouses schemas and generate global schema

(e) Generate pattern data warehouse on the federation schema basis

(f) Load data from source data warehouses to pattern data warehouse

2. Execution phase:

(a) Send query to federation

(b) Send query to pattern data warehouse

(c) Compare results from federation and pattern data warehouse.

(d) Save results, input data parameters and factors in database for further analysis.

    Factors described in Sect. 3 are handled in two different strategies. Attributes domain dispersion is calculated from data gathered directly from experiment input. Average power, power deviation, average coverage and coverage deviation are calculated from data generated during experiment preparation phase. Each data warehouse is characterized by coverage and power. When the federation layer decomposes given query, sends partial queries to data warehouses that are able to process the query properly and generate partial response. For example $P_a = median(p_1, p_3, p_5)$, where p_1, p_3, p_5 are power factors of data warehouses that processed the partial queries. Other factors are calculated analogously.

4.2 Results

Several responses types were described in [7]. During this experiment one of them - the 'rank' was checked. All factors described in Sect. 3 were applied.

A linear regression model based on Pearson's linear correlation coefficient was used. It presumes the existence of a linear relation between the studied variables. The summary of the obtained results is presented in Table 1. The strength of the linear relation between the tested variables in two cases was significant. The confidence level was set to 0.95. In cases 1, 2 and 4 the strength of linear relations was very poor. For the average coverage the correlation coefficient r = −0.4412 (Fig. 2).

Table 1. Experiments results

	Factor name	Pearson correlation - r
1	Average power	0.0029
2	Power deviation	−0.1146
3	Average coverage	−0.4112
4	Coverage deviation	−0.1756
5	Attributes domain dispersion	0.4471

– standard error of estimation: $S_E = 0.1662481058$
– the result of t-test: $p - value = 0.0001669357301$

For the attributes domain dispersion the correlation coefficient r = 0.4471 (Fig. 3)

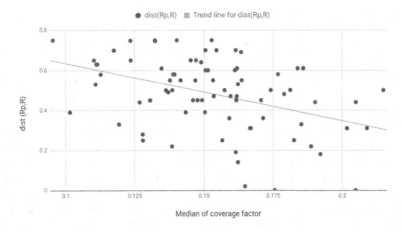

Fig. 2. Pearson's linear correlation coefficient between the median of coverage factor and dist(Rp, R)

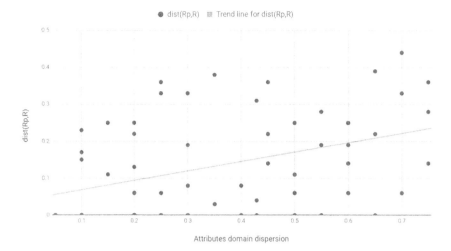

Fig. 3. Pearson's linear correlation coefficient between the attributes' domain dispersion and dist(Rp,R)

– standard error of estimation: $S_E = 0.1870702258$
– the result of t-test: $p - value = 0.00026921325$

The result for average coverage can be understood as: the more data warehouses have in common, the more possible is that the federation query will be better balanced on the sources. The result for attributes domain dispersion means that if the attributes domain is not too wide, the distance between federation response and the pattern data warehouse is relatively small. The "wideness" should be treated relatively to number of rows in dimensions and facts table.

5   Conclusions and Future Work

This paper is a preliminary review of possible estimator towards data warehouse usefulness. It continues development of model described initially in [5] and further developed in [6,7,15]. Five new factors were proposed and all of them were tested on dedicated experimental environment. 308 test iterations were executed. Two factors: *Average coverage* and *Attributes domain dispersion*, for given boundaries, have promising results. The early results give clues that some of these factors may be considered as a part of much more complicated and sophisticated model, that could support the integration process. It could estimate the usefulness of the new schema that might be considered as possible extension of the component data warehouses set. It is quite obvious that in real business scenarios hardly ever attributes' values have the same distribution in all dimensions. Therefore the dispersion metric itself is too trivial to work outside experimental environment, but combined with other factors or calculated for all corresponding dimensions could be very useful for real-life cases.

Continuously developed experimental environment is ready to examine additional query types and use different integration methods. As next steps some more types of queries will be tested. It will be also investigated how the influence of each factor compensates others. For example: increasing coverage should increase the precision of response. On the other hand, when the domain dispersion raises, the precision of response falls down. Finally, presented model will be tested with data coming from real world scenarios.

References

1. Berger, S., Schrefl, M.: From federated databases to a federated data warehouse system. In: Proceedings of the 41st Annual Hawaii International Conference on System Sciences, HICSS 2008, pp. 394–404 (2008)
2. Dong, X.L., Berti-Equille, L., Srivastava, D.: Data fusion: resolving conflicts from multiple sources. In: Wang, J., Xiong, H., Ishikawa, Y., Xu, J., Zhou, J. (eds.) WAIM 2013. LNCS, vol. 7923, pp. 64–76. Springer, Heidelberg (2013). https://doi.org/10.1007/978-3-642-38562-9_7
3. Fan, W., Lu, H., Madnick, S.E., Cheung, D.: Discovering and reconciling value conflicts for numerical data integration. Inf. Syst. **26**(8), 635–656 (2001)
4. Jindal, R., Acharya, A.: Federated Data Warehouse Architecture. White paper, Wipro Technologies (2003)
5. Kern, R., Ryk, K., Nguyen, N.T.: A framework for building logical schema and query decomposition in data warehouse federations. In: Jędrzejowicz, P., Nguyen, N.T., Hoang, K. (eds.) ICCCI 2011, Part I. LNCS (LNAI), vol. 6922, pp. 612–622. Springer, Heidelberg (2011). https://doi.org/10.1007/978-3-642-23935-9_60
6. Kern, R., Stolarczyk, T., Nguyen, N.T.: A formal framework for query decomposition and knowledge integration in data warehouse federations. Expert Syst. Appl. **40**, 2592–2606 (2013)
7. Kern, R., Dobrowolski, G., Nguyen, N.T.: A method for response integration in federated data warehouses. In: Camacho, D., Kim, S.-W., Trawiński, B. (eds.) New Trends in Computational Collective Intelligence. SCI, vol. 572, pp. 63–73. Springer, Cham (2015). https://doi.org/10.1007/978-3-319-10774-5_6
8. Lenzerini, M.: Data integration: a theoretical perspective. In: Proceedings of the Twentyfirst ACM SIGMOD-SIGACT-SIGART Symposium on Principles of Database Systems, PODS 2002, pp. 233–246 (2002)
9. Motro, A., Anokhin, P.: Fusionplex: Resolution of data inconsistencies in the integration of heterogeneous information sources. Inf. Fusion **7**(2), 176–196 (2006)
10. Nguyen, N.T.: Advanced Methods for Inconsistent Knowledge Management. Advanced Information and Knowledge Processing. Springer, New York (2007). https://doi.org/10.1007/978-1-84628-889-0
11. Vo, T.N.C., Nguyen, H.P., Vo, T.N.T.: Making kernel-based vector quantization robust and effective for incomplete educational data clustering. Vietnam J. Comput. Sci. **3**(2), 93–102 (2016)
12. Schneider, M.: Integrated vision of federated data warehouses CEUR-WS. In: Proceedings of International Workshop on Data Integration and Semantic Web (DisWeb 2006), vol. 238, pp. 336–347 (2006)
13. Marcos, M., Maldonado, J.A., Martinez-Salvador, B., Bosc, D., Robles, M.: Interoperability of clinical decision-support systems and electronic health records using archetypes: a case study in clinical trial eligibility. J. Biomed. Inf. **46**(4), 676 (2013)

14. Banek, M., Tjoa, A.M., Stolba, N.: Integrating different grain levels in a medical data warehouse federation. In: Tjoa, A.M., Trujillo, J. (eds.) DaWaK 2006. LNCS, vol. 4081, pp. 185–194. Springer, Heidelberg (2006). https://doi.org/10.1007/11823728_18

15. Kern, R.: Data warehouses federation as a single data warehouse. In: Nguyen, N.-T., Manolopoulos, Y., Iliadis, L., Trawiński, B. (eds.) ICCCI 2016, Part I. LNCS (LNAI), vol. 9875, pp. 356–366. Springer, Cham (2016). https://doi.org/10.1007/978-3-319-45243-2_33

16. Getta, J.R., Handoko: On transformation of query scheduling strategies in distributed and heterogeneous database systems. In: Nguyen, N., Trawiński, B., Kosala, R. (eds.) Intelligent Information and Database Systems. Lecture Notes in Computer Science, vol. 9011. Springer, Cham (2015). https://doi.org/10.1007/978-3-319-15702-3_14

17. Waddington, R.: An Architected Approach to Integrated Information. White paper, Kalido (2004)

Social Network Analysis

Physical Activity Contagion and Homophily in an Adaptive Social Network Model

Marit van Dijk and Jan Treur[(✉)]

Behavioural Informatics Group, Vrije Universiteit Amsterdam,
Amsterdam, The Netherlands
maritvdijk@gmail.com, j.treur@vu.nl

Abstract. Regular physical activity contributes to higher levels of well-being, healthy aging and prevention of several chronic diseases such as depression. To establish or change behaviours concerning physical activity, social contagion may play a role. The aim of this study was to model the contagion of physical activity based on empirical Twitter data and to assess the role of homophily within this contagion. To model the contagion of physical activity, an adaptive temporal-causal network model was designed, and accordingly, the parameters of the model were tuned using empirical data obtained from Twitter. Two variants of the adaptive temporal-causal network model were created, in which one calculated the weights of the connections between the nodes based on follow relations on Twitter, while in the other the connection weights were modulated by the homophily principle. The results indicate that within the considered social network of already active persons homophily does not play an important role in the physical activity behaviour.

1   Introduction

A sufficient amount of physical activity is important to stay healthy (Warburton et al. 2006). Regular physical activity contributes to higher levels of well-being (Penedo and Dahn 2005), healthy aging (Warburton et al. 2006) and prevention of several chronic diseases such as depression (Warburton et al. 2006). However, one third of the global adults population and four-fifths of the adolescents do not reach the recommended amount of physical activity according to the public health guidelines (Hallal et al. 2012). Behavioural changes towards a more (physically) active lifestyle are therefore highly recommended. One theory to describe changes in behaviour, beliefs and attitudes is the social contagion phenomenon (Marsden 1998). Social contagion is defined as 'the spread of ideas, attitudes, or behaviour patterns in a group through imitation and conformity' (Colman 2009). A more specific definition of social contagion is the 'spread of affect, attitude or behaviour from Person A ("the initiator") to person B (the "recipient") where the recipient does not perceive an intentional influence attempt on the part of the initiator' (Levy and Nail 1993). Thus, to establish or change behaviours concerning physical activity social contagion may play a role.

Furthermore, it seems that individuals interact more with similar than with dissimilar others, which is called the principle of homophily (McPherson et al. 2001). As contagion depends on bonds and interaction, contagion between persons will be

© Springer Nature Switzerland AG 2018
N. T. Nguyen et al. (Eds.): ICCCI 2018, LNAI 11055, pp. 87–98, 2018.
https://doi.org/10.1007/978-3-319-98443-8_9

influenced by the extent of homophily between the persons. For example, a significant reduction of physical activity is found among smokers and cannabis users (Pearson et al. 2006), relating to the phenomenon of homophily. Simpkins et al. (2013) investigated the interconnections among BMI and physical activity, and friendship groups as noted in bioecological and homophily theories. They found that friendships were more likely when adolescents were similar to one another in BMI and physical activity.

To assess friendships in a large social network, the data of online social platforms such as Twitter or Facebook can be used. Further, the contagion principle can easily be assessed in such online social networks. Online social networks are therefore suitable to develop and verify a social network model of contagion of certain behaviours. With the right social network model, it is possible to intervene in the most influential nodes of the network to stimulate physical activity behaviours. However, in order to do this in the right way, more insights are needed into the modelling of physical activity and the role of contagion and homophily in social networks.

The aim of the current study was to model the contagion of physical activity in a social network of already active persons based on empirical Twitter data and to assess the role of homophily for such a type of network.

2 Method Used

To develop a social network model for the contagion of physical activity, based on the Network-Oriented Modeling approach described in (Treur 2016) a generic adaptive temporal-causal network model was created, including a number of parameters, and accordingly these parameters were tuned using empirical data obtained from Twitter. Data from Twitter were chosen because they were easy accessible and allow one-way connections. The designed adaptive temporal-causal network model consist of nodes (for certain states of persons) and connections between them. The changes in states are influenced by the states of the connected nodes and their connection weights. Furthermore, a speed factor determines the timing of state changes and a combination function determines how the influences of multiple impacts were aggregated and regulated. The parameters of the model were tuned in a way that the model approximates the real world, i.e., the empirical data.

The empirical data was collected from a social network within Twitter. Data of the tweets during eight consecutive weeks of all the persons in the a certain social network was collected via Python. According to the pre-defined metrics, scores were calculated per person per week. These scores were based on the amount of tweets containing terms about physical activity. Subsequently, the scores were used to obtain numerical data for the states over time. Using this empirical data set, the temporal-causal network model was tuned. The initial states of the model were set similar to the first (calculated) states in the empirical data. For the connections between the states, two different variants of the network model were created. One in which connections were only based on the follow relations on Twitter, and another in which the connections were influenced by the amount of homophily between the persons. In both variants, parameter tuning was used to assess the best speed factors and combination function parameter values. The best parameters are the ones with which the empirical states and the modelled states are the most similar. This accuracy is also used to assess whether the addition of homophily increases or decreases the accuracy of the model. The data was collected using Python while data processing and parameter tuning was completed using Matlab. The visualization software Gephi was used to visualize the information gathered from Twitter to create the social network models.

For the empirical data, we first searched for a proper starting point, in this case a person on Twitter to create a social network around. The person had to be relatively active on Twitter and its tweets should contain some terms to describe the physical activity. We chose for these characteristics to be able to model the contagion of physical activity to other persons in this network. Before gathering data from all followers of this 'central' person, the decision was made to exclude all followers that had more than 5000 friends. The reason for this was the fact that Twitter accounts that have more than 5000 friends were likely to be companies or famous people and the focus of our project was on 'regular' people. After one week, the social network of the same person was created again following the same procedure. This was done to see if more connections between the 'nodes' of the original network are created. If this would stroke with our calculated connection values based on homophily, this could be used to verify the model concerning homophily.

To determine the states values for the empirical data, first the scores per person per week were calculated. The calculation of the scores was based on the amount of tweets about physical activity they posted in this week. The metrics that was used for this is shown in Table 1. The total score in a certain time interval (week) was determined by the sum of scores per tweet following the described metrics. The maximal score per tweet was 0.3 and when none of the words in the metrics was used, the score was 0. After this, the obtained scores were used to calculate the state of each person according to:

$$X(t+1) = X(t) * 0.9 + (X(t) * (1 - X(t))) * (score(t)/(scoremax(t) - 0.1))$$

in which X is the state, t is time in weeks, *scoremax* is the maximal score. A 53×8 matrix was created, containing the states of 53 persons over 8 weeks, that was used later to tune the parameters of the temporal causal network model.

3 The Adaptive Temporal-Causal Network Model

First the Network-Oriented Modeling approach used to model the adaptive agent is briefly explained. As discussed in detail in (Treur 2016, Chap. 2) this approach is based on temporal-causal network models which can be represented at two levels: by a conceptual representation and by a numerical representation. These model representations can be used not only to display graphical network pictures, but also for numerical simulation. Furthermore, they can be analyzed mathematically and validated by comparing their simulation results to empirical data. They usually include a number of parameters for domain, person, or social context-specific characteristics. A conceptual representation of a temporal-causal network model in the first place involves representing in a declarative manner states and connections between them that represent (causal) impacts of states on each other, as assumed to hold for the application domain addressed. The states are assumed to have (activation) levels that vary over time. In reality not all causal relations are equally strong, so some notion of *strength of a connection* is used. Furthermore, when more than one causal relation affects a state, some way to *aggregate multiple causal impacts* on a state is used. Moreover, a notion of *speed of change* of a state is used for timing of processes. These three notions are covered by elements in the Network-Oriented Modelling approach based on temporal-causal networks, and form the defining part of a conceptual representation of a specific temporal-causal network model:

- **Strength of a connection $\omega_{X,Y}$**
  Each connection from a state X to a state Y has a *connection weight value* $\omega_{X,Y}$ representing the strength of the connection, often between 0 and 1, but sometimes also below 0 (negative effect) or above 1.
- **Combining multiple impacts on a state $c_Y(..)$**
  For each state (a reference to) a *combination function* $c_Y(..)$ is chosen to combine the causal impacts of other states on state Y.
- **Speed of change of a state η_Y**
  For each state Y a *speed factor* η_Y is used to represent how fast a state is changing upon causal impact.

Figure 1 shows a small example social network and shows the way nodes and their states, connections, weights and combination functions are modelled. Combination functions can have different forms, as there are many different approaches possible to address the issue of combining multiple impacts. The applicability of a specific combination rule for this may depend much on the type of application addressed, and even on the type of states within an application. Therefore the Network-Oriented Modelling approach based on temporal-causal networks incorporates for each state, as a kind of

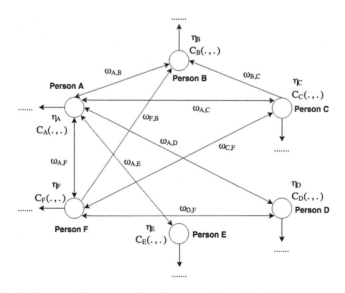

Fig. 1. Conceptual representation of a temporal-causal social network model

label or parameter, a way to specify how multiple causal impacts on this state are aggregated. For this aggregation a number of standard combination functions are available as options and a number of desirable properties of such combination functions have been identified (see Treur 2016, Chap. 2, Sects. 2.6 and 2.7. The above three concepts (connection weight, speed factor, combination function) can be considered as parameters or labels representing characteristics in a network model.

For the modelling of physical activity level and contagion of this within a social network, a temporal-causal network was created. In this network, the nodes stand for states of persons in the social network. In the temporal-causal network model, the states are calculated based on the other, connected states and the connection weights. The initial values for the states can be set equal to the first states of empirical data from a domain addressed. The way in which these calculations take place is based on difference equations that can be obtained from the conceptual representation as follows. A conceptual representation of any temporal-causal network model can be transformed in a systematic or even automated standard manner into an equivalent numerical representation of the model as follows (Treur 2016, Chap. 2):

- at each time point t each state Y in the model has a real number value in the interval [0, 1], denoted by $Y(t)$
- at each time point t each state X connected to state Y has an impact on Y defined as $\mathbf{impact}_{X,Y}(t) = \omega_{X,Y} X(t)$ where $\omega_{X,Y}$ is the weight of the connection from X to Y
- The *aggregated impact* of multiple states X_i on Y at t is determined using a *combination function* $\mathbf{c}_Y(..)$:

$$\mathbf{aggimpact}_Y(t) = \mathbf{c}_Y(\mathbf{impact}_{X_1,Y}(t), \ldots, \mathbf{impact}_{X_k,Y}(t))$$
$$= \mathbf{c}_Y(\omega_{X_1,Y} X_1(t), \ldots, \omega_{X_k,Y} X_k(t))$$

where X_i are the states with connections to state Y
- The effect of $\mathbf{aggimpact}_Y(t)$ on Y is exerted over time gradually, depending on speed factor η_Y:

$$Y(t + \Delta t) = Y(t) + \eta_Y[\mathbf{aggimpact}_Y(t) - Y(t)]\Delta t$$

or $$\mathbf{d}Y(t)/\mathbf{d}t = \eta_Y[\mathbf{aggimpact}Y(t) - Y(t)]$$

- Thus, the following *difference* and *differential equation* for Y are obtained:

$$Y(t + \Delta t) = Y(t) + \eta_Y[\mathbf{c}_Y(\omega_{X_1,Y} X_1(t), \ldots, \omega_{X_k,Y} X_k(t)) - Y(t)]\Delta t$$
$$\mathbf{d}Y(t)/\mathbf{d}t = \eta_Y[\mathbf{c}_Y(\omega_{X_1,Y} X_1(t), \ldots, \omega_{X_k,Y} X_k(t)) - Y(t)]$$

To obtain an adaptive temporal-causal network model, also the connection weights are determined in a dynamic manner; in that case they are modelled as if they are states, with their own combination function and speed factor. The determination of the connection weights is different for both variants of the model. In the first variant, the connection weights were static and for the application to the domain their values were 1 if person A follows person B, otherwise 0. In the second variant of the model, the connection weights were dynamic and based on homophily between persons. When A and B are similar in state value, the connection weight will increase (with 1 as a maximal value). If they are not similar it will decrease (with 0 as a minimal value). When the minimal value 0 or maximal value 1 is reached, no further change will occur. In other words, the more similar the levels of two states, the stronger their connection will become. Some of the connections initially get assigned a value 1 (for the application domain related to follow relations), and then will stay static. The follow relations were based on the social network at 13-12-2017 (Sect. 4). It is assumed that follow relations did not change over the eight weeks from which the empirical data was attained.

The combination functions were also different for both variants. In the first variant, for the combination function for the states a scaled sum function $\textbf{sum}_\lambda(...)$ was used, in the second variant the sum function $\textbf{sum}(...)$:

$$\textbf{sum}(V_1, \ldots, V_k) = V_1 + \ldots + V_k$$
$$\textbf{sum}_\lambda(V_1, \ldots, V_k) = (V_1 + \ldots + V_k)/\lambda \quad \text{with } \lambda > 0$$

In the second variant the combination function that was used to model the homophily principle was the logistic homophily combination function; see (Treur 2016, p. 308):

$$\textbf{c}_{A,B}(V_1, V_2, W) = W + \alpha W(1 - W) * (0.5 - 1/(1 + e^{-\sigma(|V_1 - V_2| - \tau)}))$$

where V_1 and V_2 stand for the two state values of the connected states and W for the (current) value of the connection weight. Parameter α is an amplification parameter, σ is the steepness and τ is the threshold for connection adaptation. When $|V_1 - V_2|$ is below τ, increase of W takes place, when it is above τ decrease. Note that indeed always $\textbf{c}_{A,B}(V_1, V_2, 0) = 0$ and $\textbf{c}_{A,B}(V_1, V_2, 1) = 1$; therefore connections that are set at 1 initially will stay 1, and for 0 the same.

4   Acquisition of the Empirical Data from the Application Domain

To apply the designed adaptive network model to the application domain, data were collected about the considered real world social network. The twitter account that was used for data collection is 'Erwintrailrun' (https://twitter.com/Erwintrailrun). The first social network was acquired on 13th of December 2017 and consisted of 53 persons; see Fig. 2, left hand side.

This network contains 250 connections in which 52 were from other persons to the chosen central person. The second social network, acquired on 20th of December 2017, consisted of 125 persons and 2242 connections. However, the number of connections between the 53 persons that were included in the model was again 250. This second network is shown in Fig. 2, right hand side. These findings suggest that there are no differences between the connections of the 'original' network after one week. The sudden increase of the social network around the central person is therefore solely based on newly created follow relations, and no new relations between the persons

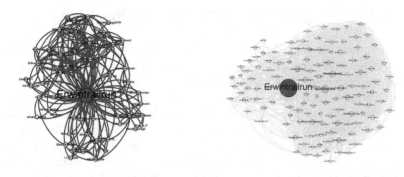

Fig. 2. Social networks acquired on 13-12-2017 and 20-12-2017, respectively

included in the first network. This can thus not be used for the homophily connection values. Based on the calculated scores from the tweets in eight consecutive weeks from 25th of November until 13th of December 2017, the empirical data points for the state values were determined, as shown in Table 1.

Table 1. Metrics used to determine the empirical data points for the state values

Tokenization	Points per tweet	Argumentation
Fit/sport/bewegen	0.3	General terms about (stimulating) an active lifestyle
Workout/exercise	0.2	General terms about an active moment
Gym/runn/hardlop/cycling/wiel-rennen/bootcamp/ wandel/fiets	0.1	Specific terms about (interest in) an active lifestyle or moment

Figure 3 displays the scores (left hand side) and state values (right hand side) of the persons over the weeks.

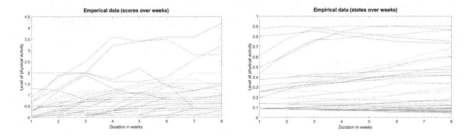

Fig. 3. Empirical data with scores and state values over weeks, respectively

5   Tuning the Adaptive Network Model to the Empirical Data

To explore in how far the designed adaptive network model is able to simulate the processes in the chosen application domain, the model had to be tuned to the empirical data. This has been done using a parameter tuning tool: the Optimization Toolbox, an inbuilt Matlab application that provides solvers for optimization problems. To find the best parameter values we used two solver algorithms: Simulated Annealing and Pattern Search. After running the algorithm until the end, the calculated values for which the mean squared error was the lowest, was used for the model. The (root) mean squared error was calculated as: $\sqrt{(SSR/n)}$ in which SSR is the Sum of the Squared Residuals between the state values of the empirical data and the modelled data, and n is the number of data points used, which is the number of nodes in the network multiplied by the number of time points used for each state.

Model variant 1: Connections static, based on follow-relations. For two parameters, speed factor and scaling factor, the values were optimized. First the values for speed factors were optimized. After manual optimization, the 'best' values were filled in as starting points and then different algorithms were used. After Pattern Search the best result was obtained. The corresponding mean squared error was 0.1026. The found values were the new speed factor values that were used from here. The second parameter for which the values were optimized was the scaling factor of the scaled sum combination function. For this, the same procedure was used. In the end, the best results were obtained using Simulated Annealing. This resulted in a final root mean squared error of 0.0951. The results are shown in Fig. 4. In Fig. 6 the results are plotted against the empirical data for both variants.

Model variant 2: Connections based on homophily. For four parameters (threshold, steepness and two speed factors), the values were optimized. First the values for the speed factors of the states were optimized. Again, the Pattern Search algorithm resulted in the smallest root of mean squared error. The error after filling in the found values was 0.1023. Accordingly, the values for threshold and amplification parameters were optimized. It was assumed that these values were the same for every connection, and thus 2809 times the same value. The best values were obtained after Simulated Annealing and corresponded to a steepness of 50 and a threshold of 50. At last the values for the speed factors of the connections were optimized. The best values were attained after the Pattern Search algorithm, which resulted in a root of mean squared error of 0.1015.

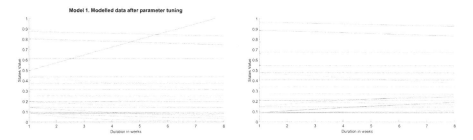

Fig. 4. Simulation of states by the tuned model variants 1 and 2, respectively

Next, using the parameter values found model variant 2 was used to make predictions for after two years. The results can be found in Fig. 5.

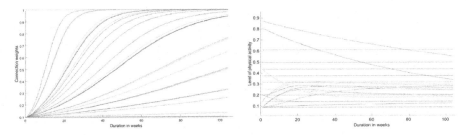

Fig. 5. Predicted connection weights (left) and activity states (right) over two years

6   Verification of the Model by Mathematical Analysis

To verify the behaviour and to test correctness of the models, a mathematical analysis was used. For verification, analysis of the stationary points was performed to determine the accuracy of the models in these points. At time t state variable Y has a stationary if $\mathbf{d}Y(t)/\mathbf{d}t = 0$. For a temporal-causal network there is a simple criterion for this:

$$\text{state } Y \text{ has a stationary point at } t \Leftrightarrow \mathbf{aggimpact}_Y(t) = Y(t)$$

$$\Leftrightarrow \mathbf{c}_Y(\omega_{X_1,Y}X_1(t), \ldots, \omega_{X_k,Y}X_k(t)) = Y(t)$$

where X_i are the states with connections to state Y.

Stationary points of the simulated data were identified by careful inspection of the data and a point was considered stationary if the state values between two consecutive time intervals differed less than 0.0001. For this, Δt was set on 0.01 and a duration of 20 weeks was modelled. After determining the stationary points, the aggregated impact

Table 2. Outcomes of the verification by mathematical analysis

Model variant 1								
Person	15	18	20	29	36	37	48	51
Time point (in weeks)	0.50	1.09	19.95	1.27	19.98	19.91	1.23	19.97
State value	0.0900	0.1968	0.2263	0.1417	0.0194	0.0055	0.0900	0.0564
Aggregated impact	0.0885	0.2263	0.2082	0.1382	0.0206	0.0083	0.0885	0.0360
Absolute deviation	0.0015	0.0295	0.0181	0.0035	0.0012	0.0028	0.0015	0.0204
Model variant 2								
Person	5	10	13	16	20	37	43	50
Time point (in weeks)	19.98	20.00	19.98	12.27	20.00	19.99	3.51	2.11
State value	0.1349	0.1354	0.2404	0.2546	0.2691	0.2010	0.0900	0.1417
Aggregated impact	0.1671	0.1493	0.2068	0.2456	0.1716	0.1716	0.1409	0.1491
Absolute deviation	0.0322	0.0139	0.0336	0.0090	0.0975	0.0294	0.0509	0.0074

of the model was compared to the state value at this time point. In this way, absolute deviations are calculated and the model can be verified. The results are satisfactory and can be found in Table 2. The average deviation is 0.01 and 0.03 for model 1 and 2 respectively.

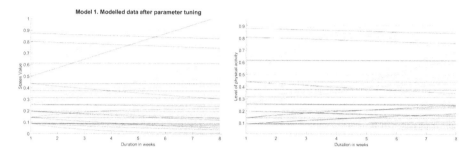

Fig. 6. Modelled data (blue) and empirical data (red) in one figure for variants 1 and 2 respectively (Color figure online)

7   Discussion

The aim of the current study was to model the contagion of physical activity based on empirical Twitter data and to assess the role of homophily within this contagion. The accuracy of the model compared to the empirical data is reasonable: around 0.1. However, when the homophily principle was added, the accuracy of the model was almost the same (it slightly decreased) indicating that within the considered social network for the contagion of physical activity homophily does not play an important role. In the current study, Twitter was used to verify the model about physical activity contagion. However, the use of social media involves some assumptions. The amount of physical activity, for example, was determined based on the tweets about physical activity terms. Nevertheless, whether or not people actually tweet about (all) their physical activity is hard to verify. Besides, some people may tweet about others' physical activity when they are, for example, watching a match or reading about sports in the newspaper. For events like this, the data and model were not corrected. However, Paul and Dredze (2011) analysed user messages of social media for measures of public health and reported quantitative correlations with public health data and evaluations of model output, and concluded that Twitter is applicable to attain public health data.

To our knowledge this was the first study that modelled the contagion and homophily of physical activity behaviour based on Twitter data. However, some studies investigated contagion and homophily of physical activity within other social networks. In these studies, the amount of physical activity was determined by special electronic equipment (Aral and Nicolaides 2017; Shen et al. 2016) or questionnaires (Flatt et al. 2012; Pearson et al. 2006). In contrast to the results of the present study, these studies found indications for a role of homophily. Flatt et al. (2012) investigated the effects of homophily on the health behaviours of a social network of older adults. They reported strong effects for homophily, this effect was the strongest for those who smoked and were physically inactive. Pearson et al. (2006) investigated sporting activity and substance use among adolescents and concluded the same. However, they found that homophily prevailed among tobacco and alcohol users, with a significant reduction of sporting activity among smokers. Shen et al. (2016) also used homophily to model physical activity in social networks and reported that 'social influences can propagate physical activity in health social networks at both user-level and community-level'.

Overall, homophily may play a role in the propagations of physical activity among social networks. However, it seems that the influence of homophily on the amount of physical activity especially works in making people more inactive. This was also concluded by Aral and Nicolaides (2017) who found that exercise is socially contagious and that its contagiousness varies with the relative activity of friends and gender relationships between friends. They reported that less active runners influence more active runners, while the reverse was not found. This effect is also seen to some extent in our model. After two years, the overall physical activity seems to be decreased. However, in general the effects are small. This may be due to the chosen social network, which generally involved people who already were physically active. Furthermore, Twitter may be not the most suitable means to measure physical activity.

Although the results are promising, some limitations should be mentioned. First of all, we did not take a random person to build our network from. To get proper results, we chose for a person that tweeted regularly and used physical activity terms. This was done in order to have at least one factor that was more than average physically active on the Twitter medium. Second, the social network that was used was restricted to 53 persons which may be described as a relatively small network. Third, we did not correct for the amount of tweets people post in general. This may underestimate the state values of people that do not tweet that much. The last limitation is that the translation between actual physical activity and 'tweeted' physical activity was not made.

To conclude, the current model extends our knowledge concerning the contagion of physical activity and the role of homophily within this. Physical activity behaviours can be modelled using an online social platform. The results indicate that not in all social networks homophily modulates physical activity behaviour, and in particular not within the social network consisting of already active persons considered here. To assess the exact mechanism in which contagion plays a role in physical activity behaviours, more knowledge about the difference between online and 'real' physical activity and a broader social network is required. Future research should analyse a larger network, without a highly active central person and eventually include also other online platforms. When the correlation between actual physical activity and online physical activity is assessed, the outcomes will add more insights.

Acknowledgements. We thank Jeroen Borst for the inspiring discussions and his support by creating the conceptual representation of the network models, and Fawad Taj for his support for the Matlab software environment used.

References

Aral, S., Nicolaides, C.: Exercise contagion in a global social network. Nat. Commun. **8**, 14753 (2017)

Colman, A.M.: A Dictionary of Psychology 3rd edn. Oxford University Press, Oxford (2009)

Flatt, M.J.D., Agimi, M.Y., Albert, S.M.: Homophily and health behavior in social networks of older adults. Fam. Community Health **35**(4), 312–321 (2012)

Hallal, P.C., Andersen, L.B., Bull, F.C., Guthold, R., Haskell, W., Ekelund, U., Lancet Physical Activity Series Working Group: Global physical activity levels: surveillance progress, pitfalls, and prospects. The lancet **380**(9838), 247–257 (2012)

Levy, D.A., Nail, P.R.: Contagion: a theoretical and empirical review and reconceptualization. Genet. Soc. Gener. Psychol. Monogr. **119**(2), 233–284 (1993)

Marsden, P.: Memetics and social contagion: two sides of the same coin. J. Memet. Evol. Models Inf. Transm. **2**(2), 171–185 (1998)

McPherson, M., Smith-Lovin, L., Cook, J.M.: Birds of a feather: homophily in social networks. Ann. Rev. Sociol. **27**(1), 415–444 (2001)

Paul, M.J., Dredze, M.: You are what you tweet: analyzing Twitter for public health. Icwsm **20**, 265–272 (2011)

Pearson, M., Steglich, C., Snijders, T.: Homophily and assimilation among sport-active adolescent substance users. Connections **27**(1), 47–63 (2006)

Penedo, F.J., Dahn, J.R.: Exercise and well-being: a review of mental and physical health benefits associated with physical activity. Curr. Opin. Psychiatry **18**(2), 189–193 (2005)

Shen, Y., Phan, N., Xiao, X., Jin, R., Sun, J., Piniewski, B., Kil, D., Dou, D.: Dynamic socialized gaussian process models for human behavior prediction in a health social network. Knowl. Inf. Syst. **49**(2), 455–479 (2016)

Simpkins, S.D., Schaefer, D.R., Price, C.D., Vest, A.E.: Adolescent friendships, BMI, and physical activity: untangling selection and influence through longitudinal social network analysis. J. Res. Adolesc. **23**(3), 537–549 (2013)

Treur, J.: Network-Oriented Modeling: Addressing Complexity of Cognitive, Affective and Social Interactions. Springer, Heidelberg (2016). https://doi.org/10.1007/978-3-319-45213-5

Warburton, D.E., Nicol, C.W., Bredin, S.S.: Health benefits of physical activity: the evidence. Can. Med. Assoc. J. **174**(6), 801–809 (2006)

e-School: Design and Implementation of Web Based Teaching Institution for Enhancing E-Learning Experiences

Md. Shohel Rana[✉], Touhid Bhuiyan, and A. K. M. Zaidi Satter

Daffodil International University, Dhaka, Bangladesh
rana.swe@diu.edu.bd,
{t.bhuiyan,zsatter}@daffodilvarsity.edu.bd

Abstract. Numerous technological improvements have found in the academic setting which removes the binding of educators and students from time and space. Day by day, the rate of drop-out is increasing by separation of them from learning. The main effort of our project is to fulfill a mission allowing individuals to learn or educate without physically attending. In this paper, we build an innovative web-based application enabling the teachers and students with numerous educational exercises using computer/smart devices. Using this application teacher, students and parents can collaborate on a single podium, while teachers can counsel with students in a real-time and share the performance and actions with parents as well as administrators. A method to modification of our traditional education system but not the replacement of teaching, it's only the enhancement idea for teaching helps to learn easily and fill up their liking.

Keywords: Technology · Innovation · Education · Time and space
Educators and learners · Whiteboard · Counseling

1 Introduction

E-LEARNING, an asynchronous and synchronous communication for the resolution of creating and performing knowledge can be defined electronically [1]. The technological foundation of e-learning is internet and associated communication technology like PC, cell phone, iPad etc. The main goal of E-learning is to allow people to learn anytime from anywhere using of ethernet technology including training, teaching, learning; in one word is the delivery of information and supervision from experts remotely [2].

The term "e-learning" first introduced in October of 1999 as a system where an individual has some information needed for him/her to make effective communication, collaboration, and training or learning. The expression precisely referred to learning anything through the Internet or any other interactive electronic media sources. According to the e-Learning fundamentals website it can also be treated as "online learning" and/or "distance learning" where the students have freedom to learn lessons and complete assignments without physically attending in the classroom [3].

At present, it has been thought that new innovative technologies can decisively be benefitted in education. Students can expand their skills, pieces of knowledge, world's

© Springer Nature Switzerland AG 2018
N. T. Nguyen et al. (Eds.): ICCCI 2018, LNAI 11055, pp. 99–108, 2018.
https://doi.org/10.1007/978-3-319-98443-8_10

acuity under their parents monitoring with the interaction of new media. Since the traditional education system cannot be replaced or removed, every person can be equipped with basic knowledge in technology during the time of technological development and reduce of distance. It can be used as a medium to reach a precise goal line through the use of the Internet with the help of web technology.

Having numerous ways to present e-learning, we describe two: (i) Synchronous and (ii) Asynchronous. Synchronous way encompasses the interaction of learners/students with tutor through the use of the web in real time. For example – Virtual classrooms (real classroom in online). Asynchronous way allows the learners to accomplish the training or learning without communicating the instructor real time. For example – Khan Academy, 10 min School, YouTube, etc. It's like a self-help basis 24/7 h information booth offering learners the study materials available whenever they require.

There are numerous e-learning programs, applications, software already developed around the world (e.g. Google Classroom, Moodle, Canvas, etc.) where most of them are not suitable to meet the full needs of e-learning system [4–7]. In this paper, we design and implement a real-time web-based system for teaching institution to enhance the e-learning experiences to fulfill the maximum needs of an online education system (see Fig. 1). We develop the system using a model called PST (Parents-Students-Teachers) which already proposed our previous article described in the proposed methodology section. Our proposed system provides a solid basis to alteration of the traditional learning system but not the replacement of teaching. Our proposed application enables the teachers to involve students using different educational exercises like assignments, quizzes, etc. Teachers are able to give performance points to students and share students' performance and action with guardians. Teachers can counsel with students about numerous topics as one-to-one or one-to-many (two-way conversations) at any time anywhere in real-time. Students are also able to study as group from different location at any time. A smartboard concept has also been added to help both teachers and students during Counselling, Lecture delivery in Classroom, Group Study, and related works to replace or minimize the use of papers, pencils, pens.

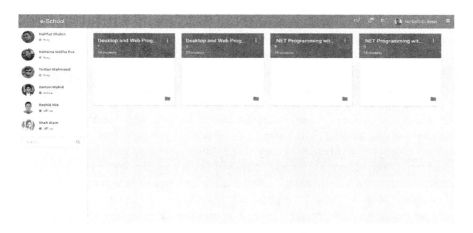

Fig. 1. A Snapshot of our proposed system 'e-School' home page for teacher

The remainder of the paper is organized as follows: Sect. 2 describes the literature review, Sect. 3 describes the overview of application including types of users, architecture, application features, Sect. 4 presents proposed methodology including data flow diagram, collaboration diagram, system design, Sect. 5 describes system evaluation including objective of study, design of study, requirements and data collections, respondents, and result and finally Sect. 6 gives conclusions and future work.

2 Literature Review

To develop our web-based e-learning system firstly we have visited numerous teaching institutions to survey that whether the application is suitable for them or not. Secondly, we accept suggestions act as the key factor from numerous teaching institutions to make our web-based e-learning system cheaper and user-friendly. Thirdly we read numerous research papers, articles, visited websites related to our works that describe web-based e-learning.

In the current method of education, the presence of technology driven device like PC, Laptop in the classroom is habitually symbolized as the momentous modernization. The use of chalkboards was introduced in Prussian classrooms in the late 18th century (Konrad 2007) while the bodily presence of classroom has not changed. It has been observed that the arrival of computers, tablets, and the Internet which led to the re-thinking of many traditional teaching practices as the opportunity for improvement of education style (The Economist 2013) [8]. It does not mean that other innovative ideas are less operative where numerous ideas or techniques are considered as innovation in e-learning system including Project-based learning, study in group, use of games for learning like math game.

A meta-study online education applied by the US Department of Education (2009) proves that the online courses can play important role to improve student performances. 16 experimentations of them and only online and live lecture by the same instructor are compared. It has been observed that for the same lecture, the students had higher achievement and higher satisfaction who attended a 45 min e-learning lecture than the lecture where the students attended physically 45 min in the traditional classroom, Zhang et al. (2006) [9].

3 Design and Implementation

3.1 Overview of Application

Business Owners. The business partners who own to provide the application through their organization. For example, a university can provide the application through their system to the teachers and students to automate the way of teaching and learning. Similarly, coaching centers, high schools, and different online schools can be the business owners. After registering the business with the application, the business owners can add, modify and control to the courses, and the teachers, students and parents.

End Users. Basically, those who directly use the application and get benefitted. Teachers, students, and parents are the key end users. However, there are some other individuals like admins are also the end users. Teacher creates the classroom for a specific subject and the students enroll the classroom with the generated token given by the teachers.

3.2 Architecture

We proposed a model called PST (Parent-Student-Teacher) in our previous paper [10]. According to the PST model (see Fig. 2), students and educators are no longer bound by time and space. Our proposed we-based application enables teachers to create, organize and share course curricula, lesson plans, and classroom materials into the web portal to make available for the students. The web-based application built using the PST model empowers the teachers to share information, knowledge, experiences, collect feedback from students to give act opinions. A monitoring system has been added to track and record of student's performance that can be shared with the guardians by connecting teachers, students and parents to collaborate on a single podium. A smart board is added as an invention where teachers can provide lectures and save it to the file as class note that can be shared into the common portal of student and teacher.

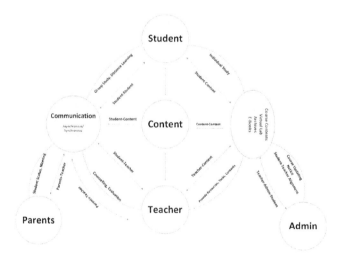

Fig. 2. A proposed PST model

3.3 Tools and Technology

3-tiered architecture has been applied to design our application. We have used JetBrains PHP Storm IDE by IntelliJ for the development of the PHP web application using one of the PHP framework called Laravel. The web-based application is developed using PST model by PHP, MySQL, WebRTC, Ratchet (Web Sockets for PHP), XMPP Server. PHP is used for programming language, MySQL is for database,

WebRTC is responsible for real-time communication (e.g. Audio/Video call), XMPP (Extensive Messaging Presence Protocol) responsible for students and teachers' real-time presence, and Ratchet Server is for broadcasting lectures, presentation, group chat, counseling, etc. on real-time.

4 Methodology

4.1 Collaboration Diagram

A collaboration diagram (see Fig. 3), represents a birds-eye glimpse of the overall progress of our system in real-time, meanwhile, representing the characters, functionalities, and actions of different actors. The diagram is actually considered as the

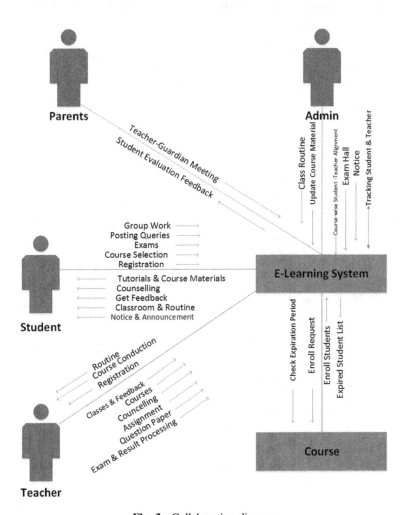

Fig. 3. Collaboration diagram

origin of information used to shaping class responsibilities and interfaces. In our system, teacher, students, and parents are assigned with high priority to perform essential roles such as course conduction, notice board, consultation with guardians, course contents update into the portal, announcement, etc. The admin controls in central as the interaction happens in a very close manner [10].

4.2   Data Flow Diagram

A significant part of the several parts of our system called Data Flow Diagram (see Fig. 4). Throughout our PST model, DFD illuminates the graphical representation of the flow of data. It also represents the processes of information sharing within the system. The processes include tutoring, counseling, exam, course update, teacher-guardian collaboration etc. The internal objects include Teacher, Parents, Student and Admin, where Course-Resources, Result, Routine, Progress, Exam etc. [10].

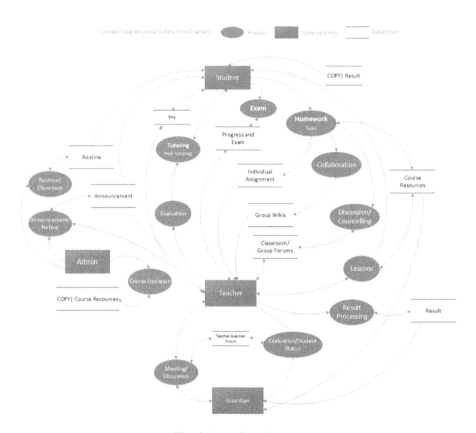

Fig. 4. Data-flow diagram

4.3   Entity Relationship Diagram

A graphical representation of entities and their relationships to each other is called entity-relationship (ER) diagram. In the below (see Fig. 5), that represents the relationship among numerous entities of the database which had been used in our system. Broadly, how the data of several entities throughout the application is getting stored, used and retrieved when it's needed, hence the transaction of data. In this system, admin is responsible for administrative tasks. Teachers can monitor the system and can proved essential materials while student's acts as learner and treated as regular user of the system [10].

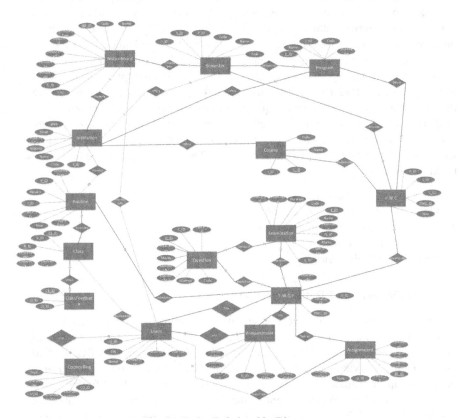

Fig. 5. Entity Relationship Diagram

5   System Evaluation

We have conducted an online survey to evaluate our system. The survey report has been explained through this section.

5.1 Study Objective

According to our education system, collection of user's standpoint are the major objectives of this survey regarding the impact of our system. Principally, two main sub-topics are prioritized to obtain the information in this survey.

- Impact of PST model in the education system and its acceptance.
- The elementary envisioned features of users in Bangladesh.

5.2 Study Design

In order to take full advantage of the data collection and anonymity of participants, we have chosen an online based methodology. And the survey we have designed by Google Online Form to streamline the processing including 20 questions. The questions are developed based on the key issues in the academic and experimental knowledge. The questionnaire also follows the standard using well-defined utensil of a survey [11]. The main goal of this survey is to obtain the acceptance and effect of e-learning using the PST model. The survey questions have designed in such a way so that it's simple, clear and precise queries to collect the forthright information with predetermined responses about the users' opinion. The study title, organization, and the aim of the survey has given in the introductory statement hence the questionnaire would be consistent and clear to the participants. The following steps we have followed to do this survey.

- Objective of this survey
- Identification of the group of participants
- Writing the questionnaire and organization
- Administration of the survey
- Understanding the outcome

5.3 Requirement and Data Collection

In Bangladesh, the users of Facebook are higher in terms of percentage. We used this opportunity to get a speedy response to our survey. We shared the survey (link: https://docs.google.com/forms/d/e/1FAIpQLScSurQ77gLfKsvfT1YneK3x0CRGDlEsQiiFG G0OqXDpjIcpYA/viewform?fbzx=8269331306802716000) through numerous university group and timeline. However, we did send email to the different university students and teachers. We also went to the different university research laboratory to conduct the survey. The time frame of our survey was from October 2017 to January 2018. The survey wasn't for specific people, rather we welcomed everybody who was interested and related to doing so. Due to having the time limitation, we obtained a limited number of respondents from this survey.

5.4 Respondents

Through the survey 315 respondents are observed from all over different universities of Bangladesh and different organizations related to education (see Fig. 6). The participants are not just students and teachers. Rather various designated participants whoever familiar with distance learning, contributed in the study. Both male and female online users have participated varies from different age groups.

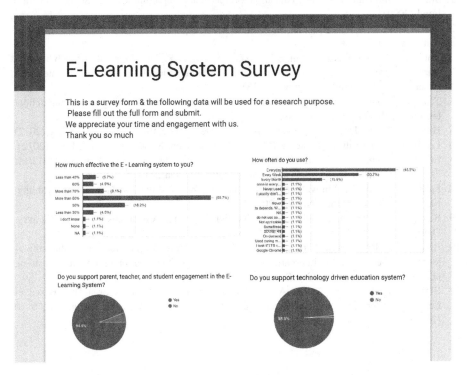

Fig. 6. A Snapshot of survey of our proposed e-learning application 'e-School'

6 Conclusion

Evaluating a system especially an e-learning platform is always challenging. However, this e-school, as per the conduction of the survey and analyzed result, shows ways to add values in the teaching institution. In this paper, this e-school developed using the PST model of technology driven education system. It allows instructors to occupy more times with those students whom are facing numerous difficulties while promising the effectiveness. It supports to improve the classroom efficiency and consents one-on-one attention in bigger classroom. It allows the guardians and teachers to maintain communication about classroom events as the students, parents, and teachers all are connected through the web. A teacher can create a specific classroom, homework and activities blog regarding education related panels and to distribute the token to the students to get enrolled in. The e-school unlocks the parent-student-teacher real-time communication and consultation. The documents, conversation, and overall all activities can be archived for future use.

References

1. Smith, K.: Innovation in public education: problems and opportunities. NewSchools Venture Fund (2009)
2. Garrison, D.R.: E-Learning in the 21st Century, pp. 1–15. Taylor & Francis, United Kingdom (2011)
3. Bhuiyan, T., Yousuf, K.B., Urmi, S.K., Nahar, A., Ali, N.Y.: Development of a web based e-learning system for teaching institution. In: 3rd International Conference on Intelligent Computational Systems, Singapore (2013)
4. Teaching with technology. https://federation.edu.au/staff/learning-andteaching/clipp/teaching-with-technology. Accessed 17 Jan 2018
5. Hill, P.: Online educational delivery models: A Descriptive View, Educes e review (2012)
6. Lewin, T.: Universities Reshaping Education on the Web. New York Times, USA (2012)
7. Education Technology and Mobile Learning. http://www.educatorstechnology.com/2012/11/7-awesomecollaborative-whiteboard.html. Accessed 17 Jan 2018
8. Falch, T., Mang, C.: Innovations in education for better skills and higher employability, EENEE Analytical Report No. 23 (2015)
9. Beyond the Classroom: A New Digital Education for Young Australians in the 21st Century. Digital Education, Australia (2012)
10. Bhuiyan, T., Rana, M.S., Sarker, K., Wahid, Z.: A proposed PST model for enhancing e-learning experiences. In: International Conference on Digital Technology in Education, pp. 44–48. ACM, Taiwan (2017)
11. Bhuiyan, T., Khan, S., Nahar, A.: Evaluation of the Effectiveness of a Web-Based e-Learning System for Tertiary Educational Institution. Lect. Notes Softw. Eng. 2(1), 6–10 (2014)

Categorizing Air Quality Information Flow on Twitter Using Deep Learning Tools

Brigitte Juanals[1,2] and Jean-Luc Minel[3(✉)]

[1] Centre Norbert Elias, Aix Marseille University, Marseille - CNRS - EHESS,
Marseille, France
[2] UMI IGlobes, CNRS - University of Arizona, Tucson, USA
[3] MoDyCo, CNRS - Université Paris Nanterre, Nanterre, France
jean-luc.minel@parisnanterre.fr

Abstract. Environmental health is an emerging and hotly debated topic that covers several fields of study such as pollution in urban or rural environments and the consequences of these changes on health populations. In this field of intersectorial forces, the complexity of stakeholders' logics is realized in the production, use and communication of data and information on air quality. The Twitter platform is a "partial public space" that can throw light on the different types of stakeholders involved, the information and issues discussed and the dynamics of articulation between these different aspects. A methodology aiming at describing and representing, on the one hand, the modes of circulation and distribution of message flows on this social media and, on the other hand, the content exchanged between stakeholders, is presented. To achieve this, we developed a classifier based on Deep Learning approaches in order to categorize messages from scratch. The conceptual and instrumented methodology presented is part of a broader interdisciplinary methodology, based on quantitative and qualitative methods, for the study of communication in environmental health.

Keywords: Environmental health · Air quality
Instrumented methodology · Circulation of information · Mediation
Social network · Twitter

1 Introduction

Environmental health is an emerging and hotly debated topic that covers several fields of study such as pollution in urban or rural environments and the consequences of these changes on health populations. Its progressive conceptualization reveals divergent positions that may include "an extremely wide range of environmental factors related to the notion of human well-being". The environmental factors analyzed in the work of the French School of Advanced Studies in Public Health (EHESP) fall into four broad thematic dimensions relating to

© Springer Nature Switzerland AG 2018
N. T. Nguyen et al. (Eds.): ICCCI 2018, LNAI 11055, pp. 109–118, 2018.
https://doi.org/10.1007/978-3-319-98443-8_11

polluted sites and soils, water quality and air quality and habitat; Among these factors, we focused on air quality, which was the subject of many alerts in major cities at the end of 2016 and which is becoming a national concern with the regular peaks of fine particles matter in urban areas.

From a societal perspective, air is one of the "commons" [15] that are described as natural or material resources that can be extended to information and knowledge commons [16]. Pooling and sharing these resources raises issues about their modes of governance. Air is a common good that is subject to conflicts of interest. As a result, data and information on air quality become political issues that provoke debate, even clashes, concerning the modes of production and dissemination of these data. In particular, the modes of production and interpretation of these data call for the specification of technical criteria in the distribution and control domains. The specification of measurements is a political, industrial and environmental issue. It is linked to the scientific validity of the selection of pollutants and pollution thresholds that are used. These open data make sense only in the light of standards used to interpret them and to take decisions at European, national and regional levels.

The production and dissemination of these data are regulated. They are part of European regulations and national or regional public policies, interdisciplinary cross-knowledge, militant positions as well as industrial and commercial logics. In France, environmental health (with which air quality is associated) is part of the State prerogatives; it is therefore governed by public health policies.

The European Directive on Ambient Air Quality - Clean Air for Europe - was adopted on 14 April 2008; it merged the main Directive (96/62/EC), the first three specific directives (99/30/EC, 2000/69/EC, 2002/3/EC) and the Council of Europe decision concerning data exchange (97/101/EC). These legal constraints were gradually applied in France by a 1991 decree then by laws: the so-called Barnier law (1995) and the law on air and the rational use of energy (LAURE Act) of December 30, 1996.

In France, the Ministry of Environment and Sustainable Development no longer centralizes actions in the field of education, awareness raising, production and dissemination of information. Communication initiatives are now carried out at the regional scale defined by the administrative organization in the context of decentralization. In this perspective, the LAURE act stipulates that the State entrusts the monitoring of air quality to approved Non-governmental organizations (NGO) in charge of this surveillance. The result is the distribution of some 40 NGOs on the national territory. These organizations are "monitoring networks" characterized by a hybrid status. They bring together companies, public organizations and NGOs [1]. These NGOs are now linked by their status and their missions to the public authorities while maintaining their vocation of integrating the demands of civil society in their initiatives. These developments led the Air Quality Monitoring Associations (AASQA) to build a national federation, ATMO France, in order to become a lobby able to negotiate with national and local authorities on the subject of air quality.

In the US, the Clean Air Act (CAA) is the comprehensive federal law that regulates air emissions from stationary and mobile sources. "Among other things, this law authorizes the Environmental Protection Agency (EPA) to establish National Ambient Air Quality Standards (NAAQS) to protect public health and public welfare and to regulate emissions of hazardous air pollutants."[1]. In June 2014, EPA proposed The Clean Power Plan, an Obama administration policy aimed at combating anthropogenic climate change, but in March 28, 2017 Donald Trump, signed an executive order mandating the EPA to review the plan and following his announcement on June 1, 2017, United States withdrawal from the Paris Agreement.

In this legal and regulatory framework, the setting up of a public debate on the implementation of a policy to monitor air quality opens up political questions about the production and the circulation of information on air quality in the public space. In this field of intersectorial forces, the complexity of stakeholders' logics is realized in the production, use and communication of data and information on air quality. The Twitter platform is a "partial public space" that can throw light on the different types of stakeholders involved, the information and issues discussed and the dynamics of articulation between these different aspects.

The purpose of our work is to conceive, by relying jointly on methods anchored in social sciences and digital humanities, a representation of the modes of circulation and distribution of message flows on Twitter about air quality in relation with stakeholders and the content exchanged.

The outline of this paper is the following. First, in Sect. 2, we present the litterature review and then specify the contribution of the proposed methodology to tackle this field. In Sect. 3, we will present the specificities and the contribution of our approach, our methodology to collect and analyze flows of tweets and the results from classification analysis. Finally, we conclude in Sect. 4.

2   Literature Review and Methodology

In this section, we begin with a review of the literature, and then we present the interdisciplinary and instrumented methodology we developed to conduct the analysis of messages.

2.1   Literature Review

As mentioned in [6], tweet analysis has led to a large number of studies in many domains such as ideology prediction in Information Sciences [5], spam detection [18], dialog analysis in Linguistics [2], and natural disaster anticipation in Emergency [17], while work in Social Sciences and Digital Humanities has developed tweet classifications. However, few studies aim at classifying tweets according to communication classes. They mostly rely on small reference sets analyzed by experts in Information Communication (InfoCom) rather than by Twitter users. Two exceptions worth mentioning are first the work presented in Lovejoy

---

[1] https://www.epa.gov/.

and Saxton [12] in which the authors (Twitter users) analyze the global behavior of nonprofit organizations on Twitter based on three communication classes: Information, Community and Action classes. The second work [11] compares the use of the Chinese Weibo service during a 2013 smog emergency with data on Twitter concerning a weather event in North America. The data collected were hand-coded and the authors discussed the implications of their results for agencies designing social media campaigns to inform and motivate the public affected by those events, which is one of the goals of the agency we are working with.

Recently, several studies on tweet classification have been carried out in NLP. Basically, these analyses aim at categorizing open-domain tweets using a reasonable amount of manually classified data and either small sets of specific classes (e.g. positive versus negative classes in sentiment analysis) or larger sets of generic classes (e.g. News, Events and Memes classes in topic filtering). The advantage of NLP approaches is that they can automatically classify large corpora of tweets. Until recently, the most commonly used models were supervised learning, Support Vector Machine (SVM), Random Forest, Gradient Boosting Machine and Naive Bayes (NB). In supervised machine learning, features are extracted from tweets and metadata and then vectorized as training examples to build models.

2.2   An Instrumented Methodology Using Convolutional Neural Networks

The main drawback of shallow supervised machine learning approaches presented above, is that they require a very time consuming step to identify linguistic or semiotic features and raise issues about the relevance of these features. Recently, new approaches based on Deep Learning techniques and especially on convolutional neural networks (convets), which no longer require researchers to identify features, have been proposed. A second advantage of convets is that they obtain better results in terms of accuracy than shallow machine learning systems [8,13]. It is for these reasons that we developed a classifier based on convets. The proposed methodology, based on convets, aims to describe and analyze the informational and communicational dynamics at work on the Twitter platform. It explores the circulation patterns of message flows and exchanges, apprehended as a dynamic process, as well as the relationships which are established between different stakeholders. Our aim is to question the forms of engagement, participation and relationships between organizations and audiences by analyzing the flow of their messages. We consider that the field of environmental health and the socio-technical device Twitter contribute to configuring the relations and the interactions between the participants. In this perspective, hashtags on Twitter may be seen as meeting points of different categories of stakeholders interested in the same themes while being anchored in different spheres - the environment and health, politics, the media, industry, the economy, etc.

The architecture of our classifier is composed of several layers as shown on Fig. 1. The first layer is a pre-trained word embedding as proposed by [14] with a kernel that matches the 5 words used as neighbors. A word-embedding is a distributed representation where each word is mapped to a fixed-sized vector of continuous values. The benefit of this approach is that different words with a similar meaning will have a similar representation. A fixed-vector size of 100 was chosen. The following layers, a Conv1D with 200 filters and a MaxPooling1D are based on the works reported in [4,8]. The back-end of the model is a standard Multilayer Perceptron layer to interpret the convets features. The output layer uses a softmax activation function to output a probability for each of the three classes affected at the tweet processed (see Sect. 3.2). Finally, only the class with the highest probability is kept.

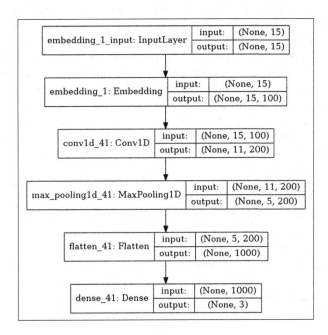

Fig. 1. Architecture of the convets classifier

We would like to point out several important features: first, although this classifier is not a very deep convets, the results are satisfactory (see Sect. 3.2) and moreover it does not require significant computing resources; second, the different parameters of the architecture were determined after several trials and errors; and finally, building such a network does not require advanced programming, so it is within the reach of an interdisciplinary team.

We implemented our methodology by building a workflow based on the open access tools Gephi[2] (graph visualization), Neo4j[3] (graph mining), Scikit-learn[4]

---

[2] https://gephi.org/.
[3] https://neo4j.com/.
[4] http://scikit-learn.org/stable/.

and Keras library (deep learning)[5]. We also developed some scripts python to manage the interoperability between all these different tools.

3 Analyzing Tweets About Air Quality

Data Acquisition. The data acquisition stage consisted in harvesting tweets with the following hashtags :the hashtag #Air and one other hashtag among the following list :#pollution, #santé(health), #qualité(quality) or #environnement(environment). Twitter maintains an Application Programming Interface (API) that returns approximately 200 features about a tweet. We developed a Python script, based on the Twarc module proposed by Ed Summers (http://github.com/docnow/twarc) using the search option of Twitter API.

Our analysis focuses on Twitter messages (called tweets) sent by accounts of organizations and non-institutional stakeholders. A first step was to build a terminology to describe the objects studied according to three dimensions: the message, the stakeholders, and the forms of stakeholder participation.

Concerning messages, we will call a message sent by a twitter account an 'original tweet' and an original message sent by an account different from the issuing account a 'retweet'. The current Twitter API gives access to the original tweet (and its sending account) of a retweet. The generic term tweet includes 'original tweet' and 'retweet'.

Regarding stakeholder qualification, we distinguished Twitter accounts, accounts managed by institutions (called 'organizational account'), and accounts managed by individuals (called 'private account'). This distinction was carried out by human analysts based on the description field filled out by the account holder. When this field was unfilled, the private account category was assigned.

In this paper, we limit the analysis to French tweets, by using the "lang" features in Tweeter API, sent between the first of November 2017 and the 17th of March 2018. This period of time is considered as a proof of concept and we intent to use the classifier to process all the tweets that will be sent during the year 2018.

The main figures are the following: 3027 tweets of which 30% of original tweets and 70% of retweets sent by 902 participants (405 organizational accounts, 497 private accounts). More specifically: 41% of organizational accounts and only 25% of private accounts produced original tweets. Participation for private accounts was largely limited to the action of retweeting (75% of tweets) the messages sent by the institutional partners. A peak during the week 46 in 2017 is the consequence of the COP22 event (World Climate Conference) which was the major event of the observed period of time.

3.1 Analyzing the Modes of Involvement and Interaction Between Organizational and Private Accounts

Categorical analysis relying on automatic classification is a relevant processing method to characterize the semantics of the messages as it makes possible to

---

[5] https://keras.io/.

analyze the modes of engagement of the stakeholders on Twitter. Automatic classification implies a prior human classification (supervised machine learning). Taking into account the size of the corpus of tweets, a human analysis would still have been possible but first, as mentioned in [10], the inter-coder minimum reliability is usually around 0.74, and secondly, we intend to process in real time all the tweets that will be sent during the year 2018.

These two reasons argue in favour of developing an automatic classification.

In order to build a classifier, a classification analysis of the contents of a sample of 350 randomly chosen original tweets was carried out in two stages. First, a team composed of one linguist and one researcher in communication analyzed the sample of tweets to determine the classes in which to categorize the tweets. Based from three classes were identified ("informative", "promoting", "humorous" (in the sense of emotional or expressive).

"Informative" tweets were those providing information concerning technical aspects of air quality or specific mitigation efforts.

If an advertisement, generally associated with a link on a Web site was present, the tweet was categorized as "promoting".

"Humorous" tweets contained the expression of emotions (such as worry, anger, dread) or political comments. It must be noted that there were no tweets containing insults. Consequently, 350 tweets were annotated by hand by the same two experts according to the categories defined in the previous step.

In the second stage, a classifier, based on convets (see Sect. 2.2), was trained on the annotated samples. It must be pointed out that none modification of the raw text of tweets was carried out, that is to say, we did not use stop words list or stemming transformation. A simple tokenizer from the Keras library was used to compute embedded vectors associated with tokens. The training loss decreases with every epoch, and the training accuracy increases with every epoch; to prevent overfitting, the training was stopped after ten epochs.

Results of the evaluation using the standard cross-validation 10-fold test [9] gave an accuracy of 0.97 in line with the state of the arts [8].

The classifier was applied to the corpus of all tweets to categorize them. Figure 2 show the findings. There is a main difference between institutional and individuals accounts concerning the "humorous" class. Private accounts sent twice as many "Humorous" tweets as institutional accounts. Similarly, institutional accounts sent much more "Promoting" tweets.

3.2 Analyzing Networks of Accounts

Observing the flow of information through the circulation of messages involves looking at the modes of stakeholder participation. They are materialized in information-communication practices. To answer this question, the analysis is based on the classification of the accounts.

In order to characterize Twitter accounts we used some of attributes proposed by [7]. The attribute "relayed" is assigned to an account if at least one of its tweets was retweeted or quoted. The "relaying" attribute is assigned to an account if the account retweeted or quoted at least one tweet. The "mentioned"

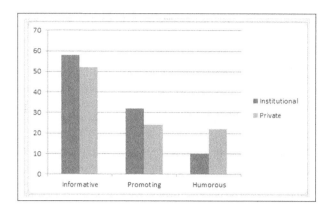

Fig. 2. Comparing tweets classification

attribute is assigned to an account if its Twitter account name was mentioned at least once in a tweet. And finally, the attribute "passing" is assigned to an account if it was both "relayed" and is "relaying"; this attribute is computed as the product of the "relayed" and "relaying score".

As pointed out by [7], the value of this index is not significant in itself; it simply provides a means of comparing accounts. There are several ways to compute the importance of a node in a network [3]. Network theory proposes the notion of degree and betweenness centrality. The degree is the sum of the indegree and the outdegree. But from our point of view, an influencer must be both. Vertices with high betweenness have control over information passing between others vertices. A vertex can have quite a low degree, be connected to others that have a low degree and still have high betweenness. Vertices in roles like this are sometimes referred to as brokers because they play the role of a bridge. But from our point of view, they are not necessarily influencers. For all these reasons, we preferred to use the passing score proposed by [7], which is the product of the indegree and the outdegree. The hypothesis is that an account with a high passing score is a key influential user who actively participates in the circulation of information. We indentified six top accounts and a remarkable point is the non-correlation between their passing score and their number of followers. The Kendall-correlation computed on these six passing accounts is 0.04761, i.e. is close to zero.

The list of the top six passing accounts illustrates our description of the stakeholders (see Sect. 1) who have a hybrid status, both public and private. The analysis of these passing accounts makes it possible to identify some of their characteristics. These are all accounts of organizations with the exception of one influencer.

@Ambassad Air is the official account of an operation run by the city of Rennes, located in the West of France, and the House of Consumption and the Environment (in connection with the other associative actors and citizens) mobilizing inhabitants about the air quality in Rennes. @Anne Hidalgo is the official account of the mayor of the city of Paris who is also the chairman of @C40Cities and @AIMFrancophones. Since her election in 2014, she has promoted a very strong policy about air quality. The number of her followers (1 375 410), twenty times higher than the second passing account (in followers number), attests her notoriety. @ARS Paca is the official account ot the Regional Health Agency of the South of France. It is an autonomous public institution at the administrative and financial level. It ensures coordination between the services of the State and the Health insurance by grouping several structures dealing with public health and the organization of care'. @INERIS is the official account of the French National Institute of Industrial Environment and Risks. @RennesVilleMetropole is the official account of the city of Rennes, located in the West of France. It must be pointed out that the city of Rennes is also involved in the @Amabassad Air account. @ATMOFRANCE is the official account of the network of associations authorized to monitor air quality. French law has entrusted this network (see Sect. 2.1) with the responsibility for implementing air quality monitoring and public information. @Charlotte Marchandis is the private account ("My tweets are only my commitment") of an elected member of Rennes City Council, chairman of @VillesSanteOMS who ran for the French presidential election in 2017.

It is remarkable that the number of followers of these passing accounts extends over a scale of 1 to 3300 and that these key influential accounts do not share their communities of accounts.

4   Conclusions

The construction of an instrumented methodology for the analysis of the flow of messages about air quality on the Twitter platform reveals the complexity of this object of study, which led us to choose a multi-dimensional approach to be able to apprehend it. This research is part of a broader framework of ongoing work on the evolution of environmental communication in the public space.

We developed a classifier based on convolutional neural networks to categorize the flow of tweets. This classifier will be running throughout 2018 year in order to feed a National agency in charge of informing inhabitants and changing their behavior.

Acknowledgments. This study is partially funded by iGlobes (UMI 3157).

References

1. Boutaric, F.: Les réseaux de la qualité de l'air : des associations stratèges de l'action publique. Développement durable et territoires, pp. 1–14 (2007)
2. Boyd, D., Golder, S., Lotan, G.: Tweet, tweet, retweet conversational aspects of retweeting on twitter. In: 43rd Hawaii International Conference on System Sciences, HICSS, pp. 1–10 (2010)
3. Brandes, U., Erlebach, T.: Network Analysis. Springer, Heidelberg (2005). https://doi.org/10.1007/b106453

4. Brownlee, J.: Deep Learning for Natural Language Processing. Machine Learning Mystery, Vermont, Australia (2017)

5. Djemili, S., Longhi, J., Marinica, C., Kotzinos, D., Sarfat, G.E.: What does twitter have to say about ideology? In: NLP 4 CMC: Natural Language Processing for Computer-Mediated Communication, pp. 16–25 (2014)

6. Foucault, N., Courtin, A.: Automatic classification of tweets for analyzing communication behavior of museums. In: LREC 2016, pp. 3006–3013 (2016)

7. Juanals, B., Minel, J.L.: Information flow on digital social networks during a cultural event: Methodology and analysis of the "european night of museums 2016" on twitter. In: Proceedings of the 8th International Conference on Social Media and Society, pp. 13:1–13:10. #SMSociety17, ACM, New York (2017). http://doi.acm.org/10.1145/3097286.3097299

8. Kim, Y.: Convolutional neural networks for sentence classification. CoRR abs/1408.5882 (2014). http://arxiv.org/abs/1408.5882

9. Kohavi, R.: A study of cross-validation and bootstrap for accuracy estimation and model selection, pp. 1137–1143. Morgan Kaufmann (1995)

10. Lachlan, K., Spence, P., Lin, X., Del Greco, M.: Screaming into the wind: examining the volume and content of tweets associated with hurricane sandy. Commun. Stud. **65**(5), 500–518 (2014)

11. Lin, X., Lachlan, K., Spence, R.: Exploring extreme events on social media: a comparison of user reposting retweeting behaviors on twitter and weibo. Comput. Hum. Behav. **65**, 576–581 (2016)

12. Lovejoy, K., Saxton, G.D.: Information, community, and action: how nonprofit organizations use social media. J. Comput.-Mediat. Commun. **17**(3), 337–353 (2012)

13. Manning, C.D.: Computational linguistics and deep learning. Comput. Linguist. **41**(4), 701–707 (2015). https://doi.org/10.1162/COLI_a_00239

14. Mikolov, T., Chen, K., Corrado, G., Dean, J.: Efficient estimation of word representations in vector space. CoRR abs/1301.3781 (2013). http://arxiv.org/abs/1301.3781

15. Ostrom, E.: The institutional analysis and development framework and the commons. Cornell Law Rev. **95**, 807–816 (2010)

16. Ostrom, E., Hess, C.: Understanding Knowledge As a Commons: From Theory to Practice. MIT Press, Cambridge (2007)

17. Sakaki, T., Okazaki, M., Matsuo, Y.: Tweet analysis for real-time event detection and earthquake reporting system development. IEEE Trans. Knowl. Data Eng. **25**(4), 919–931 (2013)

18. Yamasaki, S.: A trust rating method for information providers over the social web service: a pragmatic protocol for trust among information explorers, and provider in formation. In: 11th Annual International Symposium on Applications and the Internet (SAINT 2011), pp. 578–582 (2011)

A Computational Network Model
for the Effects of Certain Types of Dementia
on Social Functioning

Charlotte Commu[1], Jan Treur[1(✉)], Annemieke Dols[2,3],
and Yolande A. L. Pijnenburg[2,3]

[1] Behavioural Informatics Group, Vrije Universiteit,
Amsterdam, The Netherlands
c.a.commu@student.vu.nl, j.treur@vu.nl
[2] Ouderenpsychiatrie, GGZ inGeest, Amsterdam, The Netherlands
A.Dols@ggzingeest.nl, YAL.Pijnenburg@vumc.nl
[3] Alzheimer Centre, VUmc, Amsterdam, The Netherlands

Abstract. This paper introduces a temporal-causal network model that describes the recognition of emotions shown by others. The model can show both normal functioning and cases of dysfunctioning, such as can be the case for persons with certain types of dementia. Simulations have been performed to test the model in both these types of behaviours. A mathematical analysis was done which gave evidence that the model as implemented does what it is meant to do. The model can be applied to obtain a virtual patient model to study the way in which recognition of emotions can deviate for certain types of persons.

1 Introduction

Computational methods are more and more used to get more insight in human functioning and dysfunctioning. By designing a human-like computational model for certain normal functioning of certain mental and/or social processes, it can be found out what alterations make the model show dysfunctional behavior, and verify how that relates to the empirical literature. Such a computational model can be a basis for a so-called virtual patient model. An important source of knowledge for the design of a human-like computational model can be found in the fields of Cognitive and Social Neuroscience, and in what is encountered in the practice of medical clinics. The work reported in this paper results from a cooperation between researchers in AI and in medical practice.

The focus here is on social functioning and dysfunctioning resulting from a certain type of dementia, in particular the behavioral variant of frontotemporal dementia (bvFTD); see (Piguet et al. 2011). As will be explained in Sect. 2 in more detail, one of the problems encountered is difficulty in recognizing emotions of others, in particular the negative ones, even while emotion contagion still can function properly.

To model such human processes in a way that is justifiable from a neuroscientific perspective, knowledge of the underlying mechanisms in the brain is needed. Usually dynamics and cyclic connections play an important role in such mechanisms, and

© Springer Nature Switzerland AG 2018
N. T. Nguyen et al. (Eds.): ICCCI 2018, LNAI 11055, pp. 119–133, 2018.
https://doi.org/10.1007/978-3-319-98443-8_12

therefore a modeling approach is needed that can handle such cyclic dynamic processes well. The Network-Oriented Modeling approach based on temporal-causal networks used here is indeed able to do so (Treur 2016b; 2018).

In the paper, first in Sect. 2 background knowledge is described on the processes addressed. In Sect. 3 the temporal-causal network model is introduced. Section 4 describes simulation experiments for the addressed case, both for normal functioning and for dysfunctioning. Finally, Sect. 5 is a discussion.

2 Neuropsychiatric Background

The behavioral variant of frontotemporal dementia (bvFTD) is a neurodegenerative disorder associated with progressive degeneration of the frontal lobes, anterior temporal lobes, or both (Piguet et al. 2011). The disease is a leading cause of early-onset dementia and the third most common form of dementia across all age groups (Ratnavalli et al. 2002). Alterations in social cognition represent the earliest and core symptoms of bvFTD resulting in emotional disengagement and socially inappropriate responses or activities (Ibanez and Manes 2012; Kumfor and Hodges 2017). As part of their impaired social cognition, bvFTD patients are often unconcerned about their relatives, unable to adjust to their environment, lacking usual social inhibitions, and unable to recognize and attribute mental states to self and others. Consequently, dissolution of social attachment can be profound and the implications on patients' life and their relatives are far-reaching (Diehl-Schmid et al. 2007; Riedijk et al. 2006). In this paper, the following case is used as an illustration.

A 55 year old man who was recently diagnosed with bvFTD visited our outpatient clinic with his wife. While explaining the difficulties she met in the home situation, she started crying. The patient followed the conversation, and at this point he looked at her, his own eyes got watery, but he looked dazzled. Upon the question how he thought his wife was feeling, he answered that his wife was probably feeling happy. On the Ekman 60 faces test, he scored 43 out of 60 items correctly, which is below the cutoff of 46. His subscores were: Anger 8/10, Disgust 9/10, Anxiousness 8/10, Happiness 8/10, Sadness 5/10, Surprise 5/10.

This case illustrates that in bvFTD there may be a dissociation between emotion contagion and facial emotion recognition, in this case in particular the recognition of sadness.

Studies on social cognition in bvFTD have shown that facial emotion recognition is severely disturbed, with the exception of happiness. In particular, impaired recognition of negative emotions such as anger and disgust have been reported (Gossink et al. 2018). Applying the animal model of empathy of Frans de Waal, emotional contagiousness is the most inner layer, present from early evolution in most vertebrate animals (de Waal 2009). Following the hypothesis that empathy in humans, and more specific in bvFTD, will exhibit a 'Recapitulati in reverse', the outlayers of the Russian Doll, symbolic for more advanced evolutionary social cognitive abilities will be lost first and the inner layer of emotion contagion will be preserved and even more prominent in advanced dementia: 'Heightened emotional contagion in mild cognitive impairment and Alzheimer's disease is associated with temporal lobe degeneration', by (Sturm 2013), as is illustrated in our case.

3   The Temporal-Causal Network Model

This section describes the temporal-causal network model for interpretation of emotions. The model describes how interpretation of emotions takes place, with a focus on recognizing emotions showed by others. Patients with frontotemporal dementia (bvFTD) show emotional disengagement and social responses or activities that are not suitable. In particular, this model focuses on the part where people with bvFTD are unable to recognize and attribute emotional states to self and others. This can lead to the effect that emotions are misinterpreted or even not recognized at all. The model can both show how the process of recognizing and attributing emotional states works regularly and when it is affected by bvFTD.

A conceptual representation of a temporal-causal network model represents in a declarative manner states and connections between them that indicate (causal) impacts of states on each other, as assumed to hold for the application domain addressed. The states have (activation) levels that vary over time. The following three notions are main elements of a conceptual representation of a temporal-causal network model:

Connection weight $\omega_{X,Y}$. Each connection from a state X to a state Y has a *connection weight value* $\omega_{X,Y}$ representing the strength of the connection, between -1 and 1.
Combination function $c_Y(..)$. For each state a *combination function* $c_Y(..)$ to aggregate the causal impacts of other states on state Y.
Speed factor η_Y. For each state Y a *speed factor* η_Y to represent how fast a state is changing upon causal impact.

The conceptual and numerical representation of the model introduced here will be presented in this section. The model was designed by integrating a number of theories some of which were discussed in Sect. 2, and also elements from Damasio (1994; 1999; 2018)'s view on emotions and feelings, and Iacoboni (2009) on mirror neurons and social contagion.

The developed model shows the difficulties that persons with bvFTD can have regarding recognition of emotions. Not only the recognition of emotions of others is included, but also the experience of own emotional feelings which includes mirror links from observed emotions. Figure 1 gives an overview of the conceptual representation of the model. The following notations are used for the state names:

ws	world state	ss sensor state	srs sensory representation state
bs	belief state	ps preparation state	cs control state es execution state

For each state a label LP_n refers to the corresponding numerical representation of the update equation of the state, as described below. An overview of the states, their connections and weights can be found in Table 1. States or weights with subscript h or s correspond to the emotional feelings happy or sad. An example is ss_h meaning the

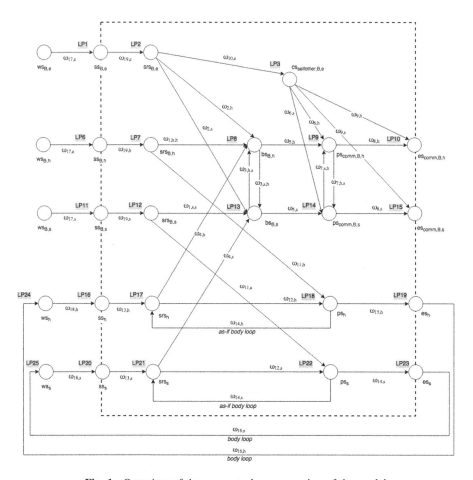

Fig. 1. Overview of the conceptual representation of the model

sensor state for the own emotional response for happy (sensing the own body state, for example, the own smile). States indicated by a B correspond to the observation of emotional expression(s) of another person B. For example, $srs_{B,h}$ means the sensory representation state of B having a happy face. Finally, subscript e is used to indicate if someone is showing any emotion. Therefore, $ws_{B,e}$ means the world state of person B showing an emotion, for example, an emotional face. Overall, the upper part (the first three causal pathways) are used for recognizing the emotional state of someone else (person B).

The lower part (the other two causal pathways) are used to model feeling the own emotions using body loops and as-if body loops as described by Damasio (1994; 1999; 2018). The model presented here incorporates parts of the model described in (Treur 2016b, Ch. 9). The part that is included from this model are the bottom two cycles of states with the body loops affecting the body state x of a person, representing own emotional feeling according to the theory of Damasio (1994; 1999; 2018). In this

Table 1. Overview of the connections, their weights, and their explanations; see also Fig. 1

From state	To state	Weight	Connection	LP	Explanation
$ws_{B,e}$	$ss_{B,e}$	$\omega_{17,e}$	sensing e of B	LP1	Sensing body state e (emotional) of person B
$ss_{B,e}$	$srs_{B,e}$	$\omega_{19,e}$	representing e of B	LP2	Representing the stimulus: B showing emotional
$srs_{B,e}$	$cs_{selfother,B,e}$	$\omega_{10,e}$	monitoring e of B	LP3	Control state for self-other distinction from represented emotion of person B
$ws_{B,h}$	$ss_{B,h}$	$\omega_{17,h}$	sensing h of B	LP6	Sensing body state h (happy) of person B
$ss_{B,h}$	$srs_{B,h}$	$\omega_{19,h}$	representing h of B	LP7	Representing the stimulus of B showing happy
$srs_{B,e}$ $srs_{B,h}$ srs_h $bs_{B,s}$	$bs_{B,h}$	$\omega_{2,h}$ $\omega_{1,h,h}$ $\omega_{4,h}$ $\omega_{2,s,h}$	interpreting e of B interpreting h of B interpreting own h suppressing belief of s	LP8	Believing that B is feeling *happy* (h) - from showing emotional by B - from emotion h showed by B - from own emotional feeling h - decreases by belief state for emotion s
$cs_{selfother,B,e}$ $bs_{B,h}$ $ps_{comm,B,s}$	$ps_{comm,B,h}$	$\omega_{6,h}$ $\omega_{5,h}$ $\omega_{7,s,h}$	controlling communication believing h of B suppressing preparation state s of B	LP9	Preparing for body state h: communicating that B feels happy: - controlled by self-other distinction - from believing B has emotion h - suppressed by preparation state that B has emotion s
$cs_{selfother,B,e}$ $ps_{comm,B,h}$	$es_{comm,B,h}$	$\omega_{9,h}$ $\omega_{8,h}$	controlling communication executing response	LP10	Expressing communication of body state h of B (communicating that B feels happy) - controlled by self-other distinction - from preparation state for h
$ws_{B,s}$	$ss_{B,s}$	$\omega_{17,s}$	sensing s of B	LP11	Sensing body state s (sad) of person B
$ss_{B,s}$	$srs_{B,s}$	$\omega_{19,s}$	representing s of B	LP12	Representing the stimulus of B showing sad
$srs_{B,e}$ $srs_{B,n}$ $srs_{B,s}$ srs_s $bs_{B,h}$	$bs_{B,s}$	$\omega_{2,s}$ $\omega_{1,s,s}$ $\omega_{4,s}$ $\omega_{2,h,s}$	interpreting e of B interpreting s of B interpreting own s suppressing belief of h	LP13	Believing that B is feeling *sad* (s) - from showing emotional by B - from emotion s showed by B - from own emotional feeling s - decreases by belief state for emotion h

(continued)

Table 1. (*continued*)

From state	To state	Weight	Connection	LP	Explanation
$cs_{selfother,B,e}$ $bs_{B,s}$ $ps_{comm,B,h}$	$ps_{comm,B,s}$	$\omega_{6,s}$ $\omega_{5,s,h}$ $\omega_{7,h,s}$	controlling communication believing s of B suppressing preparation state h of B	LP14	Preparing for body state s: communicating that B feels sad - controlled by self-other distinction - from believing B has emotion s - suppressed by preparation state that B has emotion h
$cs_{selfother,B,s}$ $ps_{comm,B,s}$	$es_{comm,B,s}$	$\omega_{9,s}$ $\omega_{8,s}$	controlling communication executing response	LP15	Expressing communication of body state s of B (communicating that B feels sad) - controlled by self-other distinction - from preparation state for s
ws_h	ss_h	$\omega_{18,h}$	sensing own h	LP16	Sensing body state h (happy) for feeling happy
ss_h ps_h	srs_h	$\omega_{13,h}$ $\omega_{14,h}$	representing h of B predicting h	LP17	Representing a *body map* for h: emotion h felt (own feeling of happy) - from sensing own body state h - via as-if body loop for body state h
srs_h $srs_{B,h}$	ps_h	$\omega_{12,h}$ $\omega_{11,h}$	amplifying mirroring h of B to own emotional feeling	LP18	Preparing for body state h: emotional response h (own feeling h) - via emotion integration from own emotion - via mirroring of emotion that B shows
ps_h	es_h	$\omega_{15,h}$	Executing emotional response	LP19	Expressing emotional response of h
ws_s	ss_s	$\omega_{18,s}$	sensing own s	LP20	Sensing body state s (sad), own feeling of sad
ss_s ps_s	srs_s	$\omega_{13,s}$ $\omega_{14,s}$	representing s of B predicting s	LP21	Representing a *body map* for s: emotion s felt (own feeling of sad) - from sensing own body state s - via as-if body loop for body state s

(*continued*)

Table 1. (*continued*)

From state	To state	Weight	Connection	LP	Explanation
srs_s $srs_{B,s}$	ps_s	$\omega_{12,s}$ $\omega_{11,s}$	amplifying mirroring s of B to own emotional feeling	LP22	Preparing for body state s: emotional response s (own feeling s) - via emotion integration from own emotion - via mirroring of emotion that B shows
ps_s	es_s	$\omega_{15,s}$	Executing emotional response	LP23	Expressing emotional response of s
es_h	ws_h	$\omega_{16,h}$	Effectuating h	LP24	Effectuating actual body state
es_s	ws_s	$\omega_{16,s}$	Effectuating s	LP25	Effectuating actual body state

model body state x can be either s (sad) or h (happy) corresponding to the emotion. This emotion can also be expressed by another person B. Therefore, the communication of, for example, body state h (happy) to B expresses that the person *self* knows that B feels h (happy). The connections from $srs_{B,h}$ and $srs_{B,s}$ to ps_h and ps_s, respectively, provide mirroring functionality to the preparation states, following Iacoboni (2009). These connections make the person feel what the other person expresses.

Most connection weights have a positive value between 0 and 1 according to the strength of the effect they have on consecutive states. However, suppressing effects are modeled by using a negative weight. A few of those negative weights occur in the model. The connection weights with a negative value are $\omega_{3,h,s}$, $\omega_{3,s,h}$, $\omega_{7,h,s}$, $\omega_{7,s,h}$, $\omega_{3,h}$, and $\omega_{3,s}$.

A conceptual representation of the temporal-causal network model can be transformed in a systematic manner into a numerical representation of the model (Treur, 2016b, Ch 2):

- at each time point t each state X connected to state Y has an *impact* on Y defined as **impact**$_{X,Y}(t)$ = $\omega_{X,Y} X(t)$ where $\omega_{X,Y}$ is the weight of the connection from X to Y
- Based on the combination function $c_Y(\ldots)$ the *aggregated impact* of multiple states X_i on Y at t is: **aggimpact**$_Y(t)$ = c_Y(**impact**$_{X1,Y}(t)$, ..., **impact**$_{Xk,Y}(t)$)

$$= c_Y(\omega_{X_1,Y} X_1(t), \ldots, \omega_{X_k,Y} X_k(t))$$

where X_i are the states with connections to state Y
- Using the speed factor η_Y the effect of **aggimpact**$_Y(t)$ on Y is exerted *over time gradually*: $Y(t + \Delta t) = Y(t) + \eta_Y$ [**aggimpact**$_Y(t)$ − $Y(t)$] Δt or

$$dY(t)/dt = \eta_Y[\mathbf{aggimpact Y(t)} - Y(t)]$$

- Thus, the following *difference* and *differential equation* for Y are obtained:

$$Y(t + \Delta t) = Y(t) + \eta_Y \left[c_Y(\omega_{X_1,Y} X_1(t), \ldots, \omega_{X_k,Y} X_k(t)) - Y(t) \right] \Delta t$$

$$dY(t)/dt = \eta_Y \left[c_Y(\omega_{X_1,Y} X_1(t), \ldots, \omega_{X_k,Y} X_k(t)) - Y(t) \right]$$

The states related to LP1, LP2, LP3, LP6, LP7, LP11, LP12, LP16, LP19, LP20, LP23, LP24, and LP25 make use of the identity combination function $c(V) = \textbf{id}(V) = V$. Those for LP8, LP9, LP13, LP14, LP17, LP18, LP21, and LP22 make use of the scaled sum combination function, which is represented numerically by:

$$c(V_1, \ldots, V_k) = \textbf{ssum}_\lambda(V_1, \ldots, V_k) = (V_1 + \ldots + V_k)/\lambda$$

where λ is the scaling factor. Finally, states related to LP10 and LP15 make use of a logistic function to get a binary all-or-nothing effect of these communications.

$$c(V_1, \ldots, V_k) = \textbf{alogistic}_{\sigma,\tau}(V_1, \ldots, V_k)$$
$$= \left[(1/(1 + e^{-\sigma(V_1 + \cdots + V_k - \tau)})) - 1/(1 + e^{\sigma\tau}) \right](1 + e^{-\sigma\tau})$$

4 Simulation Experiments

To explore the behaviour of the designed temporal-causal network model, two scenarios were simulated in Matlab. The first scenario describes the case of a normal person who would recognize emotions of others. In this case, it is expected that when person B shows an emotion, the person will correctly communicate this emotion at the communication states $es_{comm,B,h}$ or $es_{comm,B,s}$. Also, the own feeling of that specific observed emotion will be activated through mirror neurons. The second scenario describes the specific case in which a person has difficulties recognizing the right emotions due to bvFTD. It is expected that when person B shows the emotion sad, this emotion will be wrongly interpreted as happy as the case explained in Sect. 2. Therefore, the communication states will yield activations that differ from the ones in the first scenario, although through the mirroring system still contagion takes place through which the sadness is felt.

The weights for the connection strengths ω_k are for most connections set to 1; the exceptions are shown in Table 2 lower part. For $\omega_{7,h,s}$ and $\omega_{7,s,h}$ a value of -0.2 has been chosen, since the preparation states for communication that either it is a sad emotion that person B is showing or a happy emotion normally will not have a high activation level at the same time. In this way, negative weights will cause suppression between the states if one of them is activated. Similarly, for weights $\omega_{3,h,s}$ and $\omega_{3,s,h}$ a value of -0.05 has been chosen, to express that the belief states for either believing person B shows a happy emotion or a sad emotion usually will not have high activations at the same time. Note that the values 0.7 and 0.05 for $\omega_{2,h}$ and $\omega_{2,s}$, respectively, indicate that when no specific emotion is recognized, usually an emotional face is more believed to indicate happiness than sadness.

Table 2. Settings for the scaling factors used and connection weights deviating from 1

	LP8	LP9	LP13	LP14	LP17	LP18	LP21	LP22				
λ	2.2	1.3	1.55	1.3	2	2	2	2				
$\omega_{2,h}$	$\omega_{2,s}$	$\omega_{4,h}$	$\omega_{4,s}$	$\omega_{3,s,h}$	$\omega_{3,h,s}$	$\omega_{7,h,s}$	$\omega_{7,s,h}$	$\omega_{6,s}$	$\omega_{6,h}$	$\omega_{9,s}$	$\omega_{9,h}$
0.7	0.05	0.5	0.5	-0.05	-0.05	-0.2	-0.2	0.3	0.3	0.25	0.25

The simulations have been performed with speed factor $\eta = 0.5$ for all states, $\Delta t = 0.5$, and the scaling factors as displayed in Table 2 upper part. Since LP10 and LP15 make use of a logistic function, they have a threshold and steepness. Both states use a logistic function with steepness 200 and threshold 0.5. On the figures, time can be seen on the horizontal axis of the figures and the activation levels of the states are on the vertical axis.

The graphs in Figs. 2 and 3 display the results of the simulations that have been performed, for Scenario 1 and Scenario 2, respectively. The upper and lower graphs show a part of the results, to get a better view at them. The graph in Fig. 2 highlights a few of the states which show the results of the simulation. It can be seen that the states for a person showing emotional ($ws_{B,e}$) and for a person showing a sad face ($ws_{B,s}$) are highly activated at the start (blue and orange striped lines). Naturally, the sensor states and sensory representation states are becoming active as well ($srs_{B,e}$ and $srs_{B,s}$) which can be seen by the yellow and black striped lines. Those states for the representation of the happy face ($srs_{B,h}$) stay low, visible by the pink striped line. Furthermore, it can be seen that the believe state for recognizing a happy face ($bs_{B,h}$) shows some activation (purple line). This is caused by the fact that the state for recognizing an emotional face is high, but when it gets clear to the person that the emotion is about a sad emotion the feeling that it might be a happy emotion is quickly reduced and it can be seen that the communication state for a happy emotion ($es_{comm,B,s}$) stays low (dark blue line). In the end, the person communicates that a sad face has been observed ($es_{comm,B,s}$, red line). Also, the mirror neuron system for the own sad feeling becomes active, showing that emotion contagion takes place for the observed sadness. This can be seen by the activation of es_h which is in the emotion contagion cycle. When performing the simulation with the activation of a happy face instead of a sad face at the start, similar results are shown (with the activation of communication a happy face instead of a sad face). Therefore, this simulation shows what is expected of how someone without any condition that affects these processes would interpret an emotion.

For the second scenario, the settings of four weights have been changed. The weights for $\omega_{1,h,h}$, $\omega_{1,s,s}$, $\omega_{4,h}$, and $\omega_{4,s}$ have been set to a connection weight of 0.05. This has been done because in the second scenario the case of a person with bvFTD is simulated, which means that those links are damaged and therefore have a low weight. Figure 3 displays the result of the second scenario. In the graph, it can be seen that the external states for showing an emotional face ($ws_{B,e}$) and showing a sad face ($ws_{B,s}$) are high from the start, and are kept high, to simulate their presence.

These are the black and pink lines on top of the graph. However, due to the damaged links, the communication state for saying that a person shows a sad face ($es_{comm,B,s}$, dark blue line) is not activated in the end, while still by emotion contagion

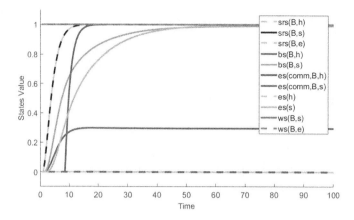

Fig. 2. Simulation results for Scenario 1: normal functioning (Color figure online)

the own sad feeling develops (es$_s$, orange line). This can be seen as the dark blue dotted line at the bottom of the graph stays low throughout the entire simulation, which implies no communication of an observed sad feeling while the orange line indicates the own sad feeling to be active. As can be seen, in contrast the communication state for saying that a person shows a happy face (es$_{comm,B,h}$) does get activated while the person never showed a happy face (red line), and no contagion of happiness took place. This can be explained from the fact that the person does recognize that there is an emotion visible (activation of srs$_{B,e}$, yellow line). However, the interpretation of the specific kind of emotion is disrupted. Therefore, the simulation shows the specific case that has been observed in patients: how damaged links can cause someone with bvFTD to misinterpret emotions.

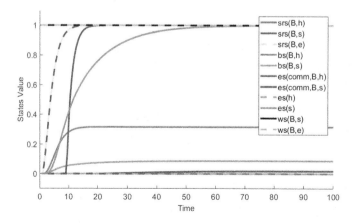

Fig. 3. Simulation results for Scenario 2: the case with bvFTD (Color figure online)

5 Verification of the Network Model by Mathematical Analysis

Dedicated methods have been developed for temporal-causal network models to verify whether an implemented model shows behaviour as expected; see (Treur 2016a, 2016b, Ch. 12). In this section equilibria of the designed model are addressed. By Mathematical Analysis their values are found and by comparing them to simulated values the model is verified. Stationary points and equilibria are defined as follows.

A state Y in a temporal-causal network model has a *stationary point* at t if $dY(t)/dt = 0$. A temporal-causal network model is in an *equilibrium state* at t if all states have a stationary point at t. In that case the above equations $dY(t)/dt = 0$ for all states Y are called the *equilibrium equations*. These are general notions, for temporal network models the following simple criterion was obtained in terms of the basic elements defining the network, in particular, the states Y, connection weights $\omega_{X,Y}$ and the combination functions $c_Y(..)$; see (Treur 2016a; 2016b, Ch. 12).

Criterion for Stationary Points and Equilibria in a Temporal-Causal Network Model

A state Y in an adaptive temporal-causal network model with nonzero speed factor has a stationary point at t if and only if

$$c_Y(\omega_{X_1,Y}(t)X_1(t), \ldots, \omega_{X_k,Y}(t)X_k(t)) = Y(t)$$

where $X_1, \ldots, X_k$ are the states with outgoing connections to Y.

A temporal-causal network model is in an equilibrium state at t if and only if for all states with nonzero speed factor the above criterion holds at t.

Equilibrium equations for an identity function **id(.)** or scaled sum combination function **ssum**$_\lambda$**(..)** are

$$\mathbf{id}(\omega_{X,Y}X(t)) = \omega_{X,Y}X(t) = Y(t)$$
$$c_Y(\omega_{X_1,Y}X_1(t), \ldots, \omega_{X_k,Y}X_k(t)) = (\omega_{X_1,Y}X_1(t) + \ldots + \omega_{X_k,Y}X_k(t))/\lambda_Y = Y(t)$$

So, they are linear equations in the state values involved with connection weights and scaling factors as coefficients:

$$\omega_{X,Y}X(t) = Y(t)$$
$$\omega_{X_1,Y}(t)X_1(t) + \ldots + \omega_{X_k,Y}(t)X_k(t) = \lambda_Y Y(t)$$

In the presented model the scaling factors have been set as the sum of the positive weights of the incoming connections; therefore all coefficients are built from connection weights. Using this, the following equilibrium equations for the states were obtained for the presented network model here; to simplify the notation the reference to t has been left out, and underlining is used to indicate that this concerns equilibrium state values, not state names. Here the connection weights are named as shown in Table 1, and A_1 to A_3 are constants.

$$\underline{srs}_{B,h} = A_1 \qquad \underline{srs}_{B,s} = A_2 \qquad \underline{srs}_{B,e} = A_3$$

$$(\omega_{1,X,X} + \omega_{1,Y,X} + \omega_{2,X} + \omega_{4,X}) \, \underline{bs}_{B,X} = \omega_{1,X,X} \, \underline{srs}_{B,X} + \omega_{1,Y,X} \, \underline{srs}_{B,Y} + \omega_{2,X} \, \underline{srs}_{B,e} + \omega_{4,X}$$
$$\underline{srs}_X + \omega_{3,Y,X} \, \underline{bs}_{B,Y}$$

$$(\omega_{5,X} + \omega_{6,X}) \, \underline{ps}_{comm,B,X} = \omega_{5,X} \, \underline{bs}_{B,X} + \omega_{6,X} \, \underline{cs}_{selfother,B,e} + \omega_{7,Y,X} \, \underline{ps}_{comm,B,Y}$$

$$(\omega_{8,X} + \omega_{9,X}) \, \underline{es}_{comm,B,X} = \omega_{8,X} \, \underline{ps}_{comm,B,X} + \omega_{9,X} \, \underline{cs}_{selfother,B,e}$$

$$\underline{cs}_{selfother,B,e} = \omega_{10,e} \, \underline{srs}_{B,e}$$

$$(\omega_{11,X} + \omega_{12,X}) \, \underline{ps}_X = \omega_{11,X} \, \underline{srs}_{B,X} + \omega_{12,X} \, \underline{srs}_X$$

$$(\omega_{13,X} + \omega_{14,X}) \, \underline{srs}_X = \omega_{13,X} \, \underline{ss}_X + \omega_{14,X} \, \underline{ps}_X$$

$$\underline{es}_X = \omega_{15,X} \, \underline{ps}_X$$

$$\underline{ss}_X = \omega_{16,X} \, \underline{es}_X$$

Note that in the above equations in the equilibrium state values, variable names X and Y are used that have multiple instances for h (happy) and s (sad). If these equilibrium state values are instantiated and renamed as shown in Table 3, 19 linear equations in X_1 to X_{19} are obtained with coefficients based on the connection weights and the constants A_1 to A_3.

Table 3. State names used in the equilibrium equations

X_1	X_2	X_3	X_4	X_5	X_6	X_7	X_8	X_9
$srs_{B,h}$	$srs_{B,s}$	$srs_{B,e}$	srs_h	srs_s	$bs_{B,h}$	$bs_{B,s}$	$ps_{comm,B,h}$	$ps_{comm,B,s}$
X_{10}	X_{11}	X_{12}	X_{13}	X_{14}	X_{15}	X_{16}	X_{17}	X_{18}
$es_{comm,B,h}$	$es_{comm,B,s}$	$cs_{selfother,B,e}$	ps_h	ps_s	es_h	es_s	ss_h	ss_s

These 19 linear equations can be solved symbolically, for example using the WIMS Linear Solver (see WIMS 2018), thereby obtaining complex algebraic expressions for the equilibrium values, linear in the constants A_1 to A_3 with as coefficients rational (broken) functions in terms of the connection weights. For verification all connection weights have been set as the simulation shown and Table 2. For these connection weight values, the following solution was found in terms of A_1 to A_3:

$X_1 = A_1$ $X_2 = A_2$ $X_3 = A_3$ $X_4 = A_1$ $X_5 = A_2$

$X_6 = 0.3176815847395451\, A_3$ $- 0.02201027146001467\, A_2$ $+ 0.682318415260455\, A_1$

$X_7 = 0.02201027146001311\, A_3$ $+ 0.9684519442406457\, A_2$ $- 0.02201027146001467\, A_1$

$X_8 = 0.4476266702238825\, A_3$ $- 0.134729540452211\, A_2$ $+ 0.5402521176549057\, A_1$

$X_9 = 0.1788345672424897\, A_3$ $+ 0.7656906556392985\, A_2$ $- 0.1000466884546121\, A_1$

$X_{10} = 0.558101336179106\, A_3$ $- 0.1077836323617688\, A_2$ $+ 0.4322016941239245\, A_1$

$X_{11} = 0.3430676537939918\, A_3$ $+ 0.6125525245114387\, A_2$ $- 0.08003735076368973\, A_1$

$X_{12} = A_3$ $X_{13} = A_1$ $X_{14} = A_2$ $X_{15} = A_1$ $X_{16} = A_2$ $X_{17} = A_1$ $X_{18} = A_2$

For the above connection weight values and values $A_1 = 1$, $A_2 = 0$, and $A_3 = 1$, the solution was found shown in the third and sixth row of Table 4.

Table 4. Results of the mathematical analysis

X_1	X_2	X_3	X_4	X_5	X_6	X_7	X_8	X_9
$srs_{B,h}$	$srs_{B,s}$	$srs_{B,e}$	srs_h	srs_s	$bs_{B,h}$	$bs_{B,s}$	$ps_{comm,B,h}$	$ps_{comm,B,s}$
1	0	1	1	0	1	0	0.987879	0.078788
X_{10}	X_{11}	X_{12}	X_{13}	X_{14}	X_{15}	X_{16}	X_{17}	X_{18}
$es_{comm,B,h}$	$es_{comm,B,s}$	$cs_{selfother,B,e}$	ps_h	ps_s	es_h	es_s	ss_h	ss_s
0.990303	0.263030	1	1	0	1	0	1	0

A logistic function with steepness 200 and threshold 0.625 applied to the communication execution states X_{10} and X_{11} (multiplied by the scaling factor 1.25 to undo the scaling) provides $X_{10} = 1$, and $X_{11} = 2.613\ 10^{-21}$. Similarly, for other values of A_1 to A_3, the equilibrium values have been found. For example, for $A_1 = 0$, $A_2 = 0$, $A_3 = 1$, it was found $X_6 = 0.3176815847395451$, $X_7 = 0.02201027146001467$, $X_8 = 0.4476$ 266702238826, $X_9 = 0.1788345672424908$, $X_{10} = 0.5581013361791062$, $X_{11} = 0.343$ 0676537939928 (a logistic function with steepness 200 and threshold 0.625 applied to the communication execution states X_{10} and X_{11} multiplied by the scaling factor 1.25 to undo the scaling provides $X_{10} = 1$, and $X_{11} = 0$), and for $A_1 = 0$, $A_2 = 1$, $A_3 = 1$, $X_6 = 0.2956713132795305$, $X_7 = 0.9904622157006603$, $X_8 = 0.3128971297716712$, $X_9 = 0.9445252228817893$, $X_{10} = 0.4503177038173369$, $X_{11} = 0.9556201783054313$ (a logistic function with steepness 200 and threshold 0.625 applied to the communication execution states X_{10} and X_{11} multiplied by the scaling factor 1.25 to undo the scaling provides $X_{10} = 1$, and $X_{11} = 0$). All these values have been checked with the values of the simulation scenarios and were found very accurate (deviations less than 0.001). This provides evidence that the implemented model does what is expected.

6   Discussion

This paper introduces a temporal-causal network model that describes the interpretation of emotions showed by others. The model can also show cases of when the interpretation of emotions is incorrect, such as can be the case of persons with bvFTD; this is based on the assumption that it is at least observed that there is an emotional face, although the specific type of emotion is not recognized correctly. Several simulations have been performed to test the model in both these behaviours. In the presented scenario for a person with bvFTD it was shown how an observed sad face led to contagion of sadness by the mirror system in a correct way, but at the same time the emotional face was nevertheless not recognized as sad, but instead as happy. A mathematical analysis was done confirming the simulation outcomes; this gave evidence that the model as implemented does what it is meant to do.

The model can be applied as the basis for human-like virtual agents, for example, to obtain a virtual patient model to study the way in which recognition of emotions can deviate for certain types of persons. Also, to study how to potentially enhance the recognition of emotions when damaged. In further research real data can be used to test the model in more detail. Furthermore, more scenarios or cases could be simulated to analyse more and different outcomes of the model. An example is the addition of specific therapies to enhance the damaged links and therefore see if this can have a positive effect on interpreting emotions for people that have difficulties with recognizing as a result of bvFTD. In future extensions of the model also more emotions than sad and happy can be addressed.

References

Damasio, A.R.: Descartes' Error: Emotion, Reason, and the Human Brain. Quill Publishing, New York (1994)

Damasio, A.R.: The Feeling of What Happens: Body and Emotion in the Making of Consciousness. Harcourt Incorporated, New York (1999)

Damasio, A.R.: The Strange Order of Things: Life, Feeling, and the Making of Cultures. Knopf Doubleday Publishing Group (2018)

de Waal, F.B.M.: The Age of Empathy. Random House, New York (2009)

Diehl-Schmid, J., Pohl, C., Ruprecht, C., Wagenpfeil, S., Foerstl, H., Kurz, A.: The Ekman 60 faces test as a diagnostic instrument in frontotemporal dementia. Arch. Clin. Neuropsychol. **22**(4), 459–464 (2007). https://doi.org/10.1016/j.acn.2007.01.024

Gossink, F., Schouws, S., Krudop, W., Scheltens, P., Stek, M., Pijnenburg, Y., Dols, A.: Social Cognition Differentiates Behavioral Variant Frontotemporal Dementia From Other Neurodegenerative Diseases and Psychiatric Disorders. Am. J. Geriatr. Psychiatry **26**(5), 569–579 (2018). https://doi.org/10.1016/j.jagp.2017.12.008

Iacoboni, M.: Mirroring People: The New Science of How We Connect with Others. Farrar, Straus and Giroux, New York (2009)

Ibanez, A., Manes, F.: Contextual social cognition and the behavioral variant of fronto-temporal dementia. Neurology **78**(17), 1354–1362 (2012). https://doi.org/10.1212/wnl.0b013e318 2535d0c

Kumfor, F., Hodges, J.R.: Social cognition in frontotemporal dementia proceedings: special lecture neuropsychology association of Japan 40th annual meeting. Jpn. J. Neuropsychol. **33** (1), 9–24 (2017). https://doi.org/10.20584/neuropsychology.33.1_9

Piguet, O., Hornberger, M., Mioshi, E., Hodges, J.R.: Behavioural-variant frontotemporal dementia: diagnosis, clinical staging, and management. Lancet Neurol. **10**(2), 162–172 (2011). https://doi.org/10.1016/s1474-4422(10)70299-4

Ratnavalli, E., Brayne, C., Dawson, K., Hodges, J.R.: The prevalence of frontotemporal dementia. Neurology **58**(11), 1615–1621 (2002). https://doi.org/10.1212/wnl.58.11.1615

Riedijk, S.R., De Vugt, M.E., Duivenvoorden, H.J., et al.: Caregiver burden, health-related quality of life and coping in dementia caregivers: a comparison of frontotemporal dementia and Alzheimer's disease. Dement. Geriatr. Cogn. Disord. **22**(5–6), 405–412 (2006). https://doi.org/10.1159/000095750

Sturm, V.E., Yokoyama, J.S., Seeley, W.W., Kramer, J.H., Miller, B.L., Katherine, P., Rankin, K.P.: Heightened emotional contagion in mild cognitive impairment and Alzheimer's disease is associated with temporal lobe degeneration. PNAS **110**(24), 9944–9949 (2013). http://www.pnas.org/cgi/doi/10.1073/pnas.1301119110

Treur, J.: Verification of temporal-causal network models by mathematical analysis. Vietnam J. Comput. Sci. **3**, 207–221 (2016a)

Treur, J.: Network-Oriented Modeling: Addressing Complexity of Cognitive, Affective and Social Interactions. Springer, Cham (2016b). https://doi.org/10.1007/978-3-319-45213-5

Treur, J.: The ins and outs of network-oriented modeling: from biological networks and mental networks to social networks and beyond. Transactions on Computational Collective Intelligence, to appear. Springer, AG (2018)

WIMS: Web Interactive Multipurpose Server; Linear Solver (2018). https://wims.unice.fr/wims/wims.cgi?session=K06C12840B.2&+lang=nl&+module=tool%2Flinear%2Flinsolver.en

Homophily Independent Cascade Diffusion Model Based on Textual Information

Thi Kim Thoa Ho[1,2(✉)], Quang Vu Bui[1,3(✉)], and Marc Bui[1]

[1] CHArt Laboratory EA 4004, EPHE, PSL Research University, Paris, France
{thi-kim-thoa.ho,quang-vu.bui}@etu.ephe.psl.eu
[2] University of Education, Hue University, Hue, Vietnam
[3] University of Sciences, Hue University, Hue, Vietnam

Abstract. In this research, we proposed homophily independent cascade model based on textual information, namely *Textual-Homo-IC*. This model based on standard independent cascade model; however, we exploited the aspect of infected probability estimation relied on homophily. Particularly, homophily is measured based on textual content by utilizing topic modeling. The process of propagation takes place on agent's network where each agent represents a node. In addition to expressing the *Textual-Homo-IC* model on the static network, we also revealed it on dynamic agent's network where there is not only transformation of the structure but also the node's properties during the spreading process. We conducted experiments on two collected data sets from NIPS and a social network platform-Twitter and have attained satisfactory results.

Keywords: Social network · Dynamic network · Diffusion
Independent cascade model · Agent-Based Model
Latent Dirichlet Allocation · Author-Topic Model

1 Introduction

In recent years, research on the process of information diffusion through social networks has attracted the attention of researchers with applications in various fields including computer science, economy, and biology. Information propagation has been extensively researched in networks, with the objective of observing the information spreading among objects when they are connected with each other. Recently, there are numerous diffusion models which have been proposed including linear threshold (LT), independent threshold (IT) model [7], independent cascade (IC) model [6] and so on. The IC model has been used extensively since it is the simplest cascade model and is successful at explaining diffusion phenomena in social networks [6]. In IC, each edge is associated with a probability of infection independently which is usually assigned by a uniform distribution [9–11]. Nevertheless, perhaps in the fact that the infected probability from one

© Springer Nature Switzerland AG 2018
N. T. Nguyen et al. (Eds.): ICCCI 2018, LNAI 11055, pp. 134–145, 2018.
https://doi.org/10.1007/978-3-319-98443-8_13

object to another depends on similarity or homophily among them, for instance, the probability that two scientists in the common field incorporate to write a paper is higher in comparison with the different field. In another instance, a user *A* on Twitter is easy to *'follow'* user *B* when *A* have common interests with *B*. Therefore, in this study, we discover the aspect of infected probability estimation based on similarity or homophily.

Homophily is the tendency of individuals to associate with similar others [13,15]. There are two principal approaches to measure homophily including the first one based on a single characteristic and the combination of multiple features for the second. For the first approach, homophily is classified into two types including status homophily and value homophily in which the former refers to the similarity in socio-demographic traits, such as race, age, gender, etc while the similarity in internal states for the later, such as opinions, attitudes, and beliefs [13,15]. Besides, Laniado et al. analyzed the presence of gender homophily in relationships on the Tuenti Spanish social network [12]. On other hands, with the second approach, Aiello et al. [1] discovered homophily from the context of tags of social networks including Flickr, Last.fm, and aNobii. Additionally, Cardoso et al. [4] explored homophily from hashtags on Twitter. However, in general, these methods have not exploited the textual information related to users yet while it contains significant information for similarity analysis, for instance, based on content of papers, we can define whether the authors research in the same narrow subject or not, or we can determine which are common interests between two users on Twitter based on their tweets. For that reason, we propose a method of homophily measurement based on textual content. A fundamental technology for text mining is *Vector Space Model* (VSM) [18] where each document is represented by word-frequency vector. Nevertheless, two principal drawbacks of VSM are the high dimensionality as a result of the high number of unique terms in text corpora and insufficient to capture all semantics. Therefore, topic modeling was proposed to solve these issues. Recently, there are dissimilar methods of topic modeling which include *Latent Dirichlet Allocation* (LDA) [14], *Author-Topic Model* (ATM) [17], etc. In this study, we chose LDA and ATM to estimate topic's probability distribution of users.

In this study, we propose an expanded model of independent cascade model, namely *Textual-Homo-IC*. In *Textual-Homo-IC*, the infected probability from an active object to inactive another is estimated based on their similarity or homophily. Particularly, homophily is measured from textual information by utilizing topic modeling. The spreading process is performed on agent's network where each node is represented by an agent. *Textual-Homo-IC* is demonstrated on static agent's network and dynamic agent's network in which the network structure and characteristics of agents have remained during propagation process for the former while there is a variation for the later. Some experiments were implemented on co-author network and Twitter with the combination of two methods LDA and ATM for estimating topic's distribution of users and two distance measurements Hellinger distance and Jensen-Shannon distance for measuring homophily. On the static networks, the results demonstrated that the effectiveness of *Textual-Homo-IC* outperforms in comparison with random diffusion. Additionally, our results also illustrated the fluctuation of active number for the diffusion process on a dynamic network instead of attaining and remaining stable state on a static network.

The structure of our paper is organized as follows: Sect. 2 reviews preliminaries; the *Textual-Homo-IC* models are proposed in Sect. 3; Sect. 4 illustrates experiments; results and evaluation are demonstrated in Sect. 5; finally we conclude our work in Sect. 6.

2   Preliminaries

2.1   Independent Cascade Model (IC)

We assume a network $G = (V, E, P)$, where $P : V \times V \rightarrow [0, 1]$ is probability function. $P(u, v)$ is the probability of node u infecting node v. The diffusion process occurs in discrete time steps t. The set of active nodes at step t is considered as I_t. At each step t where I^{newest} is the set of the newly activated nodes at step $t - 1$, each $u \in I^{newest}$ infects the inactive neighbors $v \in \eta^{out}(u)$ with a probability $P(u, v)$.

2.2   Agent-Based Model

An agent-based model (ABM) is a class of computational models for simulating the actions and interactions of autonomous agents. ABM has been utilized in numerous fields, for instance, biology, ecology, and social science [16]. ABM contains three principal elements including agents, their environment, and interactive mechanisms among agents. The essence of an ABM is the transformation of agent's properties which result from the interactions process.

2.3   Topic Modeling

Latent Dirichlet Allocation (LDA). Latent Dirichlet Allocation (LDA) [14] is a generative statistical model of a corpus. In LDA, each document may be taken into account as a combination of multiple topics and each topic is demonstrated by a probability distribution of words.

Author-Topic Model (ATM). Author-Topic model (ATM) [17] is a generative model for documents that expands LDA to incorporate author's information. Each author is associated with a mixture of topics where topics are multinomial distributions over words. The words in a collaborative paper are assumed to be the result of a mixture of the authors' topics.

Update Process of LDA and ATM. LDA and ATM can be updated with additional documents after training has been finished. This update procedure is executed by Expectation Maximization (EM)-iterating over new corpus until the topics converge. This process is equal to the online training of Hoffman [8]. There are already several available packages for topic modeling including *topicmodels* or *lda* in R, or *Gensim*[1] in Python. In this study, we chose Gensim for training and updating the topic modeling.

3 Homophily Independent Cascade Model Based on Textual Information (Textual-Homo-IC)

In this section, we proposed an expanded model of independent cascade diffusion model, namely *Textual-Homo-IC*. This diffusion model is demonstrated in detail on both static and dynamic network. We present in the following, the latter steps in more details.

3.1 Agent's Network

In this study, the network that we take into account for spreading process is agent's network $G(V, E)$ in which V is set of agents represented for nodes and E is set of edges among nodes. Agents are heterogeneous with three principal properties including *ID*, *Neighbors* and *TP-Dis* (topic's probability distribution). LDA and ATM can be utilized to estimate *TP-Dis* of users.

To demonstrate the dynamic of agent's network, we exploit not only the structure of network but also agent's properties. Firstly, the structure of the network can be transformed with the appearance of new agents or new connections. Moreover, topic's distribution of agents can fluctuate since agents own more text information through the interactive process. The problem is to how to update the transformation of topic's distribution of users after a time period based on existing topic modeling. In LDA, to make an estimate of topic's distribution of users, we consider each user correspond to each document. Therefore, we can not utilize update mechanism of LDA to update topic's distribution of users when users have more documents in interaction. Instead of unusable update mechanism of LDA, we can make use of ATM to estimate topic's distribution of users and simultaneously update mechanism to update user's topic's distribution since each author can own various documents.

3.2 Homophily Measure

In this study, we estimate homophily between two agents based on their topic's probability distribution. If we consider a probability distribution as a vector, we can choose some distances measures related to the vector distance such as Euclidean distance, Cosine Similarity, Jaccard Coefficient, etc. However, experimental results in our previous work [2] demonstrated that it is better if we choose distances measures related to the probability distribution such as KullbackLeibler Divergence, Jensen-Shannon divergence, Hellinger distance, etc.

---

[1] https://pypi.python.org/pypi/gensim.

Algorithm 1. Random-IC on static agent's network

Require: agent's network G=(V, E), I_0: seed set
1: **procedure** RANDOM-IC-STATIC-NETWORK(G, I_0)
2:     $t \leftarrow 0, I^{total} \leftarrow I_0, I^{newest} \leftarrow I_0$
3:     **while** infection occur **do**
4:         $t \leftarrow t + 1; I_t \leftarrow \emptyset$
5:         **for** u $\in I^{newest}$ **do**
6:             $I_t(u) \leftarrow \{v^{inactive} \in \eta^{out}(u), p <= q\}; p, q \sim U(0, 1)$
7:             $I_t \leftarrow I_t \bigcup I_t(u)$
8:         $I^{total} \leftarrow I^{total} \bigcup I_t; I^{newest} \leftarrow I_t$
9:     **return** I^{total} ▷ Output

In this study, we chose Hellinger distance and Jensen-Shannon divergence to measure distance. Let probability distributions on k topics $P = (p_1, p_2, ..., p_k)$ and $Q = (q_1, q_2, ..., q_k)$ correspond to user u and v.

Hellinger distance:

$$d_H(P, Q) = \frac{1}{\sqrt{2}} \sqrt{\sum_{i=1}^{k} (\sqrt{p_i} - \sqrt{q_i})^2} \tag{1}$$

Jensen-Shannon distance:

$$d_{JS}(P, Q) = \frac{1}{2} \sum_{i=1}^{k} p_i ln \frac{2p_i}{p_i + q_i} + \frac{1}{2} \sum_{i=1}^{k} q_i ln \frac{2q_i}{p_i + q_i} \tag{2}$$

Homophily:
$$\mathbf{Homo(u, v)} = \mathbf{1 - d(P, Q)} \tag{3}$$

3.3   Random-IC on Static Agent's Network

In this section, we illustrate IC model on a static network in which infected probability based on uniform distribution, namely *Random-IC*. This model plays as a benchmark model for comparing performance with *Textual-Homo-IC* model that we will propose at Sect. 3.4. At each step t where I^{newest} is the set of the newly active nodes at time $t - 1$, each $u \in I^{newest}$ infects the inactive neighbors $v \in \eta^{out}(u)$ with a probability $P(u, v)$ randomly. The propagation continues until no more infection can occur (see Algorithm 1).

3.4   Textual-Homo-IC on Static Agent's Network

Propagation mechanism of *Textual-Homo-IC* on static agent's network is similar to *Random-IC*, but the difference is that each active agent $u \in I^{newest}$ infects the inactive neighbors $v \in \eta^{out}(u)$ with a probability $P(u, v)$ equal *Homophily*(u, v) instead of a random probability (see Algorithm 2).

Algorithm 2. Textual-Homo-IC on static agent's network

Require: agent's network G=(V, E), I_0: seed set
1: **procedure** TEXTUAL-HOMO-IC-STATIC-NETWORK(G, I_0)
2:    $t \leftarrow 0, I^{total} \leftarrow I_0, I^{newest} \leftarrow I_0$
3:    **while** infection occur **do**
4:       $t \leftarrow t + 1; I_t \leftarrow \emptyset$
5:       **for** u $\in I^{newest}$ **do**
6:          $I_t(u) \leftarrow \{v^{inactive} \in \eta^{out}(u), p <= Homo(u,v)\}; p \sim U(0,1)$
7:          $I_t \leftarrow I_t \bigcup I_t(u)$
8:       $I^{total} \leftarrow I^{total} \bigcup I_t; I^{newest} \leftarrow I_t$
9:    **return** I^{total} ▷ Output

3.5   Textual-Homo-IC on Dynamic Agent's Network

Although IC model on the dynamic network has been researched in [5,19], the dynamic concept of a network has only been considered under the structure transformation while the activated probability from an active node to inactive another is always fixed during spreading process. Therefore, we propose *Textual-Homo-IC* model on a dynamic agent's network in which not only discover the variation of network's structure but also agent's topics distribution. It can be said that infected probability among agents can change over time because of their homophily transformation.

There is the resemblance in the propagation mechanism of *Textual-Homo-IC* on the dynamic network in comparison with the static network; however, in spreading process at step $t \in C$, agent's network G will be updated as shown in the Sect. 3.1 (see Algorithm 3).

Algorithm 3. Textual-Homo-IC on dynamic agent's network

Require: agent's network $G = (V, E)$; I_0: seed set
Require: $C = \{k_1, k_2, ..., k_n\}$, at step k_i G is updated; n: number steps of diffusion
1: **procedure** TEXTUAL-HOMO-IC-DYNAMIC-NETWORK(G, I_0, C)
2:    $t \leftarrow 0, I^{total} = I_0, I^{newest} = I_0$
3:    **while** $t < n$ **do**   ▷ ($n > max\{C\}$)
4:       $t \leftarrow t + 1; I_t = \emptyset$
5:       **if** $t \in C$ **then:**
6:          **Update G**; $I^{newest} = I^{total}$
7:       **loop for** u $\in I^{newest}$
8:          **Begin**
9:             $I_t(u) \leftarrow \{v^{inactive} \in \eta^{out}(u), p <= Homo(u,v)\}; p \sim U(0,1)$
10:            $I_t \leftarrow I_t \bigcup I_t(u)$
11:         **End**
12:      $I^{total} = I^{total} \bigcup I_t; I^{newest} = I_t$
13:   **return** I^{total} ▷ Output

4   Experiments

4.1   Data Collection

The proposed *Textual-Homo-IC* models have been tested on a well-known social network platform-Twitter and co-author network. For Twitter network, we have aimed to 1524 users in which links are *"follow"* relations. We crawled 100 tweets for each user and textual data stretched from 2011 to April 2018. For co-author network, we have targeted authors who have participated in Neural Information Processing Systems Conference (NIPS) from 2000 to 2012. The dataset contains 1740 papers which are contributed by 2479 scientists.

4.2   Setup

Firstly, we defined the number of topic for the whole corpus based on the Harmonic mean of Log-Likelihood (HLK) [3]. We calculated HLK with the number of topics in the range [10, 200] with sequence 10. We realized that the best number of topics is in the range [40, 100] for Twitter network (Fig. 1a) and [50, 90] (Fig. 2a) for co-author network. Therefore, we ran HLK again with sequence 1 and obtained the best is 69 for Twitter network (Fig. 1b) and 67 for co-author network (Fig. 2b).

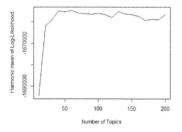
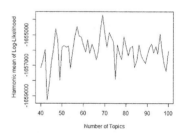

(a) ♮ topics in range [10, 200], seq 10   (b) ♮ topics in range [40, 100], seq 1

Fig. 1. Log-likelihood for Twitter network

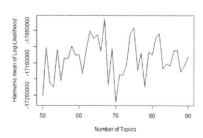

(a) ♮ topics in range [10, 200], seq 10   (b) ♮ topics in range [50, 90], seq 1

Fig. 2. Log-likelihood for co-author network

Textual-Homo-IC diffusion is implemented on the static Twitter network and co-author network. Agent's networks which are constructed as shown in Sect. 3.1 in which *"follow"* relation for Twitter network and *"co-author"* relation for co-author network. For each network, we implemented four experiments of *Textual-Homo-IC* with combination two methods of estimating topic's distribution (LDA and ATM) and two kinds of distance measurements (Hellinger distance and Jensen-Shanon distance). Besides, we also conducted *Random-IC* as a benchmark to compare the performance with *Textual-Homo-IC*.

To simulate *Textual-Homo-IC* on dynamic agent's network, we conducted experiments on the dynamic Twitter network and co-author network. For co-author network, we collected textual data between 2000 to 2009 for train corpus and estimating the author's topic distribution using ATM. An agent's network is formed with *"co-author"* relation. In another hand, for Twitter network, textual data is gathered from 2011 to January 2018 for train corpus. Unfortunately, it is impossible to get the exact date that a user starts to follow another on Twitter, including in the API or Twitter's web interface. This leads to the inability to express the fluctuations in network structure with *"follow"* relation. Therefore, we took into account agent's network with *"major topic"* relation (R_{MTP}) which will appear when two agents are interested in the common topic with a probability greater than threshold p_0. In this study, we considered R_{MTP} with $p_0 = 0.1$.

The diffusion process on dynamic network starts as soon as an agent's network is formed. For each kind of distance measurement, we implemented four experiments in which the first one is the propagation on agent's network without dynamic. The last investigations are that after every 5, 10 and 15 steps of diffusion agent's network will fluctuate once follow mechanism presented in Sect. 3.1. Agent's networks will be updated in 3 times corresponding to each month from February to April 2018 for Twitter network and each year from 2010 to 2012 for co-author network.

4.3   Model Evaluation

To evaluate the performance of diffusion models, we can use the number of active nodes or the active percentage which are standard metrics in information diffusion field [5,19]. In this research, we utilise the active number to evaluate the performance of spreading models. We compare the performance of proposed *Textual-Homo-IC* diffusion model with baseline model (*Random-IC*).

5   Results and Evaluation

5.1   Compare Textual-Homo-IC and Random-IC Diffusion

The results of *Textual-Homo-IC* on static agent's networks are shown in Fig. 3. For both networks, we can see that the active number of *Textual-Homo-IC* is always greater than *Random-IC* in four cases which are the combination of two

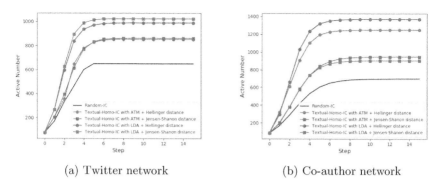

(a) Twitter network   (b) Co-author network

Fig. 3. Textual-Homo-IC diffusion on static networks

methods of topic modeling and two distance measurements. Firstly, in Twitter network (Fig. 3a), the number of active agents reaches approximately 650 for *Random-IC* diffusion while *Textual-Homo-IC* attains about 862 for both cases where ATM combine with two distances. Particularly, *Textual-Homo-IC* that incorporate LDA with Hellinger distance and Jensen-Shanon distance obtain the higher number of active agents in comparison with cases utilizing ATM, about 989 and 1023 active agents respectively. On the other hand, in co-author network (Fig. 3b), the number of active agents reaches approximately 700 for *Random-IC* diffusion while *Textual-Homo-IC* attains about 903 for the case using LDA combined with Jensen-Shanon distance. Besides, 946 active agents are reached by collaborating ATM and Jensen-Shanon distance. In addition, *Textual-Homo-IC* with ATM and Hellinger distance obtains approximately 1248 active agents while the highest number belongs to *Textual-Homo-IC* with LDA and Hellinger distance, around 1369 active agents. In summary, we can conclude that *Textual-Homo-IC* diffusion outperforms in comparison with *Random-IC*.

5.2   Textual-Homo-IC Diffusion on Dynamic Agent's Network

Results are shown in Figs. 4 and 5 which illustrate *Textual-Homo-IC* diffusion on the dynamic Twitter network and co-author network respectively. For Twitter network, there is only one agent that be activated from the seed set on the static network for both cases of distance measurements. The reason is the number of connection with R_{MTP} in the initial stage is too low to diffusion. However, if there is the network's transformation in the next 3 stages with the arrival of many new connections, there is a significant increase in the active number. For co-author network, *Textual-Homo-IC* on a static network reaches a steady state from the 12th step and 7th step onwards for using Hellinger and Jensen-Shanon distance respectively. However, if there is the network's fluctuation in the next 3 stages, the active number increase significantly. In short, we can conclude that the propagation process without dynamic of network reached and maintained the steady state while there is a significant transformation in the active number if agent's network has fluctuation in the diffusion process.

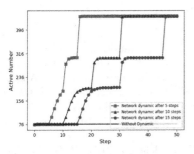

(a) ATM and Hellinger distance   (b) ATM and Jensen-Shanon distance

Fig. 4. Textual-Homo-IC diffusion on dynamic Twitter network

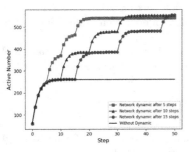

(a) ATM and Hellinger distance   (b) ATM and Jensen-Shanon distance

Fig. 5. Textual-Homo-IC diffusion on dynamic co-author network

6   Conclusion

In this study, we proposed an expansion model of independent cascade diffusion model, namely *Textual-Homo-IC*. In *Textual-Homo-IC*, we estimated the probability of an active user infecting inactive another based on homophily which is measured based on textual information by utilizing topic modeling. *Textual-Homo-IC* has been revealed details on both static and dynamic agent's network. Experimental results demonstrated that the effectiveness of *Textual-Homo-IC* on the static network outperforms *Random-IC*. In addition, experiments also illustrated the fluctuation of the active number on dynamic agent's network instead of obtaining and remaining a steady state in a static network. In future works, we will conduct experiments on other large-scale networks and compare proposed model with more other baseline models.

References

1. Aiello, L.M., Barrat, A., Schifanella, R., Cattuto, C., Markines, B., Menczer, F.: Friendship prediction and homophily in social media. TWEB **6**(2), 91–93 (2012)
2. Bui, Q.V., Sayadi, K., Amor, S.B., Bui, M.: Combining Latent Dirichlet Allocation and K-means for documents clustering: effect of probabilistic based distance measures. In: Nguyen, N.T., Tojo, S., Nguyen, L.M., Trawiński, B. (eds.) ACIIDS 2017. LNCS (LNAI), vol. 10191, pp. 248–257. Springer, Cham (2017). https://doi.org/10.1007/978-3-319-54472-4_24
3. Buntine, W.: Estimating likelihoods for topic models. In: Zhou, Z.-H., Washio, T. (eds.) ACML 2009. LNCS (LNAI), vol. 5828, pp. 51–64. Springer, Heidelberg (2009). https://doi.org/10.1007/978-3-642-05224-8_6
4. Cardoso, F.M., Meloni, S., Santanchè, A., Moreno, Y.: Topical homophily in online social systems. CoRR abs/1707.06525 (2017)
5. Gayraud, N.T., Pitoura, E., Tsaparas, P.: Diffusion maximization in evolving social networks. In: Proceedings of the 2015 ACM on Conference on Online Social Networks, USA, pp. 125–135 (2015)
6. Goldenberg, J., Libai, B., Muller, E.: Talk of the network: a complex systems look at the underlying process of word-of-mouth. Mark. Lett. **12**(3), 211–223 (2001)
7. Granovetter, M.: Threshold models of collective behavior. Am. J. Sociol. **83**(6), 1420–1443 (1978)
8. Hoffman, M.D., Blei, D.M., Wang, C., Paisley, J.W.: Stochastic Variational Inference. CoRR abs/1206.7051 (2012)
9. Kempe, D., Kleinberg, J., Tardos, É.: Maximizing the spread of influence through a social network. In: Proceedings of the Ninth ACM SIGKDD International Conference on Knowledge Discovery and Data Mining, USA, pp. 137–146 (2003)
10. Kempe, D., Kleinberg, J., Tardos, É.: Influential nodes in a diffusion model for social networks. In: Caires, L., Italiano, G.F., Monteiro, L., Palamidessi, C., Yung, M. (eds.) ICALP 2005. LNCS, vol. 3580, pp. 1127–1138. Springer, Heidelberg (2005). https://doi.org/10.1007/11523468_91
11. Kimura, M., Saito, K.: Tractable models for information diffusion in social networks. In: Fürnkranz, J., Scheffer, T., Spiliopoulou, M. (eds.) PKDD 2006. LNCS (LNAI), vol. 4213, pp. 259–271. Springer, Heidelberg (2006). https://doi.org/10.1007/11871637_27
12. Laniado, D., Volkovich, Y., Kappler, K., Kaltenbrunner, A.: Gender homophily in online dyadic and triadic relationships. EPJ Data Sci. **5**(1), 19 (2016)
13. Lazarsfeld, P.F., Merton, R.K.: Friendship as a social process: a substantive and methodological analysis. Freedom Control Mod. Soc. **18**(1), 18–66 (1954)
14. Blei, D.M., Ng, A.Y., Jordan, M.: Latent Dirichlet Allocation. J. Mach. Learn. Res. **3**, 601–608 (2001)
15. McPherson, M., Smith-Lovin, L., Cook, J.M.: Birds of a feather: homophily in social networks. Ann. Rev. Sociol. **27**(1), 415–444 (2001)
16. Niazi, M., Hussain, A.: Agent-based computing from multi-agent systems to agent-based models: a visual survey. Scientometrics **89**(2), 479 (2011)
17. Rosen-Zvi, M., Griffiths, T.L., Steyvers, M., Smyth, P.: The author-topic model for authors and documents. In: UAI 2004, Proceedings of the 20th Conference in Uncertainty in Artificial Intelligence, Canada, pp. 487–494 (2004)

18. Salton, G., Buckley, C.: Term-weighting approaches in automatic text retrieval. Inf. Process. Manag. **24**(5), 513–523 (1988)
19. Zhuang, H., Sun, Y., Tang, J., Zhang, J., Sun, X.: Influence maximization in dynamic social networks. In: 2013 IEEE 13th International Conference on Data Mining, pp. 1313–1318, December 2013

A Semi-automated Security Advisory System to Resist Cyber-Attack in Social Networks

Samar Muslah Albladi[✉] and George R. S. Weir

University of Strathclyde, Glasgow G1 1XH, UK
{samar.albladi,george.weir}@strath.ac.uk

Abstract. Social networking sites often witness various types of social engineering (SE) attacks. Yet, limited research has addressed the most severe types of social engineering in social networks (SNs). The present study investigates the extent to which people respond differently to different types of attack in a social network context and how we can segment users based on their vulnerability. In turn, this leads to the prospect of a personalised security advisory system. 316 participants have completed an online-questionnaire that includes a scenario-based experiment. The study result reveals that people respond to cyber-attacks differently based on their demographics. Furthermore, people's competence, social network experience, and their limited connections with strangers in social networks can decrease their likelihood of falling victim to some types of attacks more than others.

Keywords: Advisory system · Social engineering · Social networks

1 Introduction

Individuals and organisations are becoming increasingly dependent on working with computers, accessing the World Wide Web and, more importantly, sharing data through virtual communication. This makes cyber-security one of today's greatest issues. Protecting people and organisations from being targeted by cybercriminals is becoming a priority for industry and academia [1]. This is due to the huge potential damage that could be associated with losing valuable data and documents in such attacks.

Previous research focuses on identifying factors that influence people's vulnerability to cyber-attack [2] as the human has been characterised as the weakest link in information security research. When investigating human behaviour toward online threats, it is important to focus on the interaction between the individual's attributes, their current context, and the message persuasion tactic [3]. Most previous studies that have considered persuasion tactics in social engineering exploits, have focused on phishing as the common type of cyber-attack while limited research has investigated other types, such as malware, or clickjacking. Figure 1 shows that 37% of participants in the current study fell victim to a phishing scam attack that asked them to validate their Facebook account using a phishing link, while only 28% fell victim to a phishing attack that asked them to register their information to enter a prize draw. Consequently, the present study argues that people's vulnerabilities change depending upon the type of cyber-attack and our

© Springer Nature Switzerland AG 2018
N. T. Nguyen et al. (Eds.): ICCCI 2018, LNAI 11055, pp. 146–156, 2018.
https://doi.org/10.1007/978-3-319-98443-8_14

investigation addresses the human characteristics associated with victimisation for a range of cyber-attacks which, in turn, facilitates the design of a semi-automated advisory system that relies on the idea of people segmentation and targeting. Segmentation, Targeting, and Positioning (STP) strategic approach is a well-known model that has been extensively applied to modern marketing research [4]. According to this model, there are three main processes to segment people in order to deliver them an effective and 'focused-to-need' messages. We have adopted this approach to design a security advisory system based on SNs users' characteristic and associated threat vulnerability.

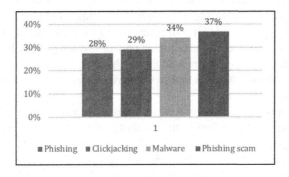

Fig. 1. Victim percentages

The material presented here is organised as follows. Section 2 provides a brief literature review. Section 3 describes the study methodology while Sect. 4 presents the results of the analysis. Discussion of the results is provided in Sect. 5. An outline approach to a semi-automated user advisory system is proposed in Sect. 6. Finally, the last section offers conclusions from the study.

2   Literature Review

Criminals in social communication channels use advanced methods to access sensitive information to help them increase the success rate of their attacks and this so-called social engineering attacks. In this type of attack, the aim is often not to target systems, but rather individual's users or organizations. In order to protect against this type of cyber-attack, it is necessary to investigate and understand the reasons why people are not able to detect these attempts to penetrate their data and devices.

It is important to investigate the main entities that encapsulate and contribute to the success of social engineering-based attacks in order to understand why people get easily deceived by this kind of attack. Krombholz et al. [5] proposed a taxonomy of social engineering sophisticated attacks in the virtual communication networks. The taxonomy comprised three main entities that have been argued to form every social engineering attack, the operator of the attack, the type of the attack, and the attack channel. The attack can be originated by either a person which reflected a limited number of victims such as spear phishing [6] or by a malicious software which usually targeted a considerable huge number of users such as the cross-site scripting attack in SN [7].

In the virtual environment of social networks, there has been limited research available to help explain why people are easily deceived by SE attacks. An investigation of people's social network habits and their relation to people's vulnerability to SN phishing attacks revealed that the willingness of raising the number of connected friends as well as maintaining frequent use of the network have a high impact on user behaviour [8]. Yet, another study [9] went further by investigating whether the impact of social network usage on people's likely victimisation differs among different social network platforms and found a statistically negative relationship between frequent usage of multipurpose dominant SN such as Facebook and victimization. This means that high frequency of using Facebook does not lead to an increase in the likelihood of victimisation. Furthermore, another study [10] found that connecting with a large number of profiles on Facebook would lead to a non-controllable online network which ultimately increases individuals' vulnerability. Perceptual-related factors have also been identified as affecting vulnerability to cyber-attacks, as a high-level of risk propensity could cause people to fall victim to cyber-attacks [9].

One of the proposed solutions to deal with enduring online threats is to understand the victim's background and examine their reaction by conducting a real attack, such as the case of sending phishing emails to a particular group of users [11, 12]. In contrast, due to ethical considerations, the majority of studies [13–15] used scenario-based experiments to examine people's vulnerabilities. Among many identified characteristics that are believed to predict potential victims [2, 16], demographics are the most controversial variables. Moreover, most of the earlier mentioned studies are focused on one type of attack although criminals have several ways to perform cyber-attacks. This has indicated the need for further investigation of people's vulnerability to different types of cyber-attack and the need to explore which groups of users are more vulnerable to specific types of cyber-attack in a social network context. Identifying the characteristics of most vulnerable individuals for a particular type of attack could help designing an advisory system to push awareness messages to vulnerable individuals. We expect that designing such security advisory system based upon observed user behaviour and characteristics could reduce people susceptibility to different types of cyber-attacks in SNs.

3   Methodology

In order to design our advisory system, we first collect the user data. An online questionnaire has been designed as an assessment tool to examine participants' perception and behaviour toward different threats in a social network context.

An invitation email was sent to faculty staff in two universities asking them to distribute the online-questionnaire among their students and staff. Participants were presented with the online-questionnaire which has 3 parts. The first part asked about demographics. The second part includes questions that measure study constructs such as the three scales used to measure the user's competence to deal with cyber-attacks [17].

The final part includes the scenario-based experiment as participants were presented with four images of Facebook posts, each post includes a type of cyber-attack such as phishing for sensitive information (Attack 1), clickjacking with an executable file

(Attack 2), malware attack (Attack 3), and a phishing scam that impersonates a legitimate organization (Attack 4). These four cyber-attacks have been chosen from the most prominent cyber-attacks that occur in social networks [18]. Participants were asked to indicate their response to these attacks, as if they encountered them in their real accounts, by rating a number of statements such as "I would click on this button to read the file" using a 5-point Likert-scale from 1 "Strongly Disagree" to 5 "Strongly Agree".

After that, we examine whether the collected user data could help us designing a semi-automated advisory system that classifies participants into different vulnerability segments in order to target their needs by providing personalised awareness messages.

4 Results

We have tested which group of people are vulnerable to each type of cyber-attack in the scenario-based experiment, based upon their rating response to the different statements. Table 1 describes the mean from the five-point Likert-scale and its corresponding vulnerability level.

Table 1. Description of the scale mean

Mean	Likert scale	Vulnerability Level
1.00–1.79	Strongly Disagree	Low vulnerable
1.80–2.59	Disagree	
2.60–3.39	Neither Agree nor Disagree	Moderately vulnerable
3.40–4.19	Agree	High vulnerable
4.20–5.00	Strongly Agree	

Demographics Differences

To examine whether user demographics have an impact on user susceptibility to SE victimization, every demographic has been tested individually to identify which group of people are more vulnerable to a certain type of attack. Figure 2 shows that female participants are more vulnerable than male participants to all considered cyber-attacks.

Generally, younger adults are less vulnerable to cyber-attacks than older adults (as appears in Fig. 3). Surprisingly, in the phishing and malware attacks, the oldest group (45–55) was most vulnerable (phishing = 2.60, malware = 2.80) while the mid-aged group (35–44) was the least susceptible to these attacks (phishing = 1.71, malware = 1.64).

The analysis of different groups with various education levels and their response to the four types of SE attacks revealed that master's degree holders are more vulnerable to clickjacking than to other types of cyber-attack (m = 2.14). While high school and bachelor's degree holders are more vulnerable to the phishing scam that impersonates a legitimate SN provider (with a mean of 2.10, and 2.31 respectively).

Users with a technical education background were shown to be less vulnerable to cyber-attacks. In contrast, Business School students are more vulnerable to the phishing attack that offers a prize than other attacks, while Humanities and Arts students are more vulnerable to the malware attack. Medical and Science students are more vulnerable to the phishing scam that impersonates a Facebook technical support message.

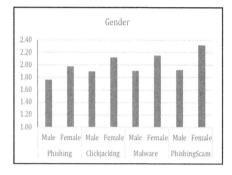

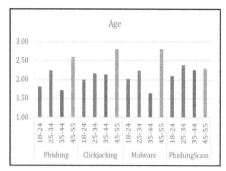

Fig. 2. Gender comparisons of vulnerability to SE   **Fig. 3.** Age comparisons of vulnerability to SE

Prevention Factors

In order to investigate if user characteristics can prevent user's vulnerability to specific types of attacks, three factors have been chosen (user's competence, Social network experience, low connections with strangers) to consider whether their prevention effect is similar across the four types of attacks. Multiple regression tests have been conducted to test the impact of these three variables on preventing users from falling victim to cyber-attacks. These factors are proved to decrease people's vulnerability to the four considered cyber-attacks, when combined together in our study. This section will present the result of their impact on the four types of attacks as shown in Fig. 4.

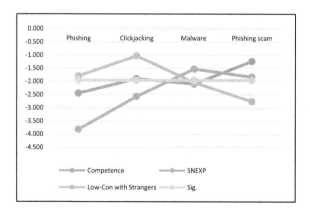

Fig. 4. Regression analysis results

User's Competence: When analysing the impact of users' competence on decreasing users' susceptibility to social engineering victimization, the result in Fig. 4 shows that measuring users' competence could identify less vulnerable individuals who can correctly detect phishing attack that offers a prize (t value = −2.447, P < 0.05) and also detectors of malware attack (t value $= -2.098$, $P < 0.05$). While competence could not prevent participants from falling victim to clickjacking and phishing scam attacks, as these relationships appear to be not significant (t value > -1.96).

Social Network Experience: Regression analysis of the impact of social network experience on decreasing individuals' response to different kinds of cyber-attack indicated that among the four types of cyber-attacks, phishing attack that offers a prize (t value $= -3.816$, $P < 0.05$) and clickjacking (t value $= -2.573$, $P < 0.05$) are attacks that experienced social network users seem to have the ability to deal with and detect. It is also worth noting that there is a negative impact of social network experience on the other two cyber-attacks, however, this effect is still considered weak and not significant.

Low Connections with Strangers in SN: People with limited connections to strangers are less vulnerable to malware attack (t value $= -2.049$, $P < 0.05$) as well as to the phishing scam that impersonates a legitimate organization (t value $= -2.759$, $P < 0.05$). The result also shows that such low connections decrease users' vulnerability to phishing and clickjacking, although these relationships are not strong enough to be significant.

5   Discussion

Some studies found no major variance between male and female in regards to response to email phishing [19]. Other studies have found women to be less susceptible to email phishing than men [20, 21]. Yet, female users have been repeatedly indicated as the weakest gender in detecting the risk in different cyber-attacks contexts such as email [14] or Social Network [15]. Our study also found that women are more vulnerable than men in all four types of attack. Furthermore, younger adults have been seen as a reckless group when dealing with risky emails, as stated by previous studies [14]. Yet, our study context is different and younger adults have shown their competency in detecting social engineering attacks in social networks. This might be because their awareness and experience with social network settings and environment surpasses their knowledge of email environments and associated risks.

Benson et al. [22] state that students are less likely to fall for cybercrimes in social networks when compared with non-students. However, their study did not distinguish between the education levels of participants. Our study found that master's degree holders are more vulnerable to clickjacking attacks. This might be due to the fact that educated people usually seek new information even if there is risk associated with it. Further, a recent study [23] found that business students are more likely to open email phishing than humanities students. That study argues that this might be because business students are accustomed to a competitive environment and try to show their commitment by quick response to university emails.

The current study indicated three factors that can protect people from being deceived on social network websites. The individual competence level to deal with cyber-crime can be measured based upon three dimensions as proposed by [17], i.e., security awareness, privacy awareness, and self-efficacy. The result shows that this measure can significantly predict the individual's ability to detect phishing and malware attacks while decreasing the individual's vulnerability to clickjacking and scam attacks. Perception of self-ability to control the content shared on social network websites is considered a predictor of detection ability of social network threats [9].

A recent study investigating social engineering attacks in Facebook [15] found that the time elapsed since joining Facebook can be a significant predictor of susceptibility to victimisation. The more time has elapsed, the less vulnerable is the person. This accords with our findings as the years the individual spent using the network has been used as a measure of social network experience which also appeared to increase the individual awareness of the risk associated with using the network. Experienced users are more familiar with phishing and clickjacking, and thereby, easily detect them. Despite the fact that the relationship between SN experience and vulnerability to the other two attacks (malware, scam) are not considered statistically significant, experience with the network has decreased the likelihood of victimisation.

Previous studies claim that having a large network size is positively associated with online vulnerability [8, 10] as larger network's size included strangers as well as friends. Our study found that low connections with strangers can protect people from being deceived in social network specifically when encountering malware and phishing scam attacks. Limited connection with strangers is also decreasing the individual vulnerability to phishing and clickjacking attacks.

Our user study results provided an insight on the possibility to segment SN users based on their characteristics and vulnerabilities as a basis for a semi-automated security advisory system that responds to individual user vulnerabilities.

6   The Architecture of a Security Advisory System

Our research on determining user vulnerabilities affords a basis for profiling users according to their weakness in respect of particular threats. In turn, this provides a means to design a personalised advisory system that sends awareness posts to target individual

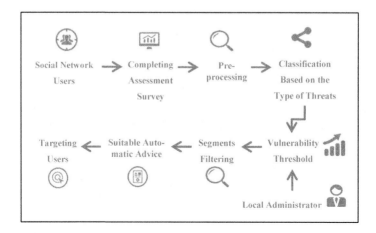

Fig. 5. Architecture of semi-automated advisory system

user needs. For example, if the characteristics of the user are similar to those who are vulnerable to clickjacking, the advisory system might send awareness posts to the user and advise him/her on how to deal with this type of threat. The architecture of our proposed semi-automated advisory system is shown in Fig. 5. A brief description of each component is given below.

Social Networks Users. A message must be sent to social networks users who want to register and benefit from the advisory system; *Completing assessment survey.* Any new user should start by completing a start-up survey that helps us assess participants' behaviour and perception in online social networks. This assessment survey result will profile the user in the most suitable segment later on to receive advice that suits the particular user needs. *Pre-processing.* The collected data will go through different screening and analysis tests such as construct reliability and validity tests. *User classification.* The segmentation process can be based on two different machine learning approaches: supervised or unsupervised [24]. Using unsupervised techniques such as clustering might be not suitable in our system as it requires no prior knowledge and clusters users based on patterns of unlabelled data. We aim to group users based on their vulnerability to different cyber-attacks. Therefore, supervised techniques such as classification are more appropriate to our goal, where the classes are predefined and the users grouped based on determined criteria. Thus, users will be classified into different groups based on the result of the scenario-based experiment in the assessment survey. Every segment should include users who shared similar characteristics that were found to increase vulnerability to a particular type of threat. For example, based on users' response to the phishing attack in the scenario-based experiment, users may be grouped into at least three segments: high, moderate, and low vulnerability. However, as we have considered user characteristics in the classification process, we might have more than one high vulnerable segment to a particular type of attack. For example, age and gender are among the factors that are included in the classification process, so it is possible to have two high vulnerable segments, e.g., one segment includes young-adult males and the other includes mid-aged females. This variation in the segmentation process can help us provide more individualised awareness messages.

Vulnerability Threshold. The local administrator can determine the threshold and the priority for each type of attack. For example, in our study we found that phishing scam is the most effective attack. Therefore, the threshold for this type of attack may be set to 3 which means high, moderate and low vulnerable segments will receive awareness advice on this type of threat. While the severity of malware attack is considered average in our study, we might set its threshold to 2, meaning that malware-related advice will be sent to the high and moderate rated vulnerability segments. Both phishing and clickjacking thresholds may be set to 1, meaning that only high vulnerable segments will receive advice for these two types of attack. Of course, a single user could be vulnerable to different types of attack and assigned to more than one segment. Therefore, the priority of the received type of advice is also determined by the attack's vulnerability threshold as assigned by the local administrator.

Segment Filtering. In this step, segments are filtered based on threat thresholds. For each type of attack, only segments in the threshold vulnerability level will be addressed. For instance, only segments with high vulnerability to phishing and clickjacking attacks may be considered. While according to the threshold of phishing scam, high, moderate, and low vulnerable segments may be taken into account. *Suitable automatic advice.* Different user segments are vulnerable to different threats and require advice that is tuned to their needs. With this in mind, each of the identified threats has a set of recommendations that would help individuals to avoid falling victim to a particular threat. *Targeting users.* Each segment of users will receive automatic advice that aims to sensitise them to threats to which they are more vulnerable, while each single user can receive more than one package of advice, based on attack priorities, that he/she is vulnerable to.

7   Conclusion

We are investigating why people easily fall victim to cyber-attacks in various online channels and whether vulnerabilities differ across cyber-attack categories in the context of social networks. The present study indicates that people respond differently to different types of cyber-attacks. A phishing attack that pretended to be from an authorized and legitimate organization (Facebook) is the most successful attack in our study with 37% of participants falling victim.

Female participants were found to be more vulnerable to social engineering victimisation than male participants. Younger and mid-aged adults show high detection ability compared to other age groups. Education is found to influence people's capability, as users with technical majors were found to be competent to detect cyber-attacks. Furthermore, the study result demonstrates that users' competence level, their experience with social networks, and low connections with strangers in the network play an important role in preventing people from falling victims to certain types of cyber-attacks.

The proposed semi-automated advisory system should help to address the problem of human vulnerabilities and weakness in detecting social engineering attacks. Assessing social network users and grouping them based on their behaviour and vulnerabilities is essential in order to focus relevant advice that meets users' needs. This is considered cost and time effective as users are only presented with insight on relevant threats. Furthermore, integrating individuals' needs as well as administrator's knowledge of existing threats, could avoid the overhead and inconvenience of sending blanket advice to all social network users.

References

1. Gupta, B.B., Arachchilage, N.A.G., Psannis, K.E.: Defending against phishing attacks: taxonomy of methods, current issues and future directions. Telecommun. Syst. **67**(2), 247–267 (2018)
2. Albladi, S.M., Weir, G.R.S.: User characteristics that influence judgment of social engineering attacks in social networks. Hum. Centric Comput. Inf. Sci. **8**(1), 5 (2018)

3. Williams, E.J., Beardmore, A., Joinson, A.N.: Individual differences in susceptibility to online influence: a theoretical review. Comput. Hum. Behav. **72**, 412–421 (2017)
4. Andaleeb, S.S.: Market segmentation, targeting and positioning. In: Strategic Marketing Management in Asia, pp. 179–207. Emerald Group Publishing Limited (2016)
5. Krombholz, K., Hobel, H., Huber, M., Weippl, E.: Advanced social engineering attacks. J. Inf. Secur. Appl. **22**, 113–122 (2015)
6. Bullee, J.-W., Montoya, L., Junger, M., Hartel, P.: Spear phishing in organisations explained. Inf. Comput. Secur. **25**(5), 593–613 (2017)
7. Rathore, S., Sharma, P.K., Park, J.H.: XSSClassifier: an efficient XSS attack detection approach based on machine learning classifier on SNSs. J. Inf. Process. Syst. **13**(4), 1014–1028 (2017)
8. Vishwanath, A.: Habitual Facebook use and its impact on getting deceived on social media. J. Comput. Commun. **20**(1), 83–98 (2015)
9. Saridakis, G., Benson, V., Ezingeard, J.N., Tennakoon, H.: Individual information security, user behaviour and cyber victimisation: an empirical study of social networking users. Technol. Forecast. Soc. Change **102**, 320–330 (2016)
10. Buglass, S.L., Binder, J.F., Betts, L.R., Underwood, J.D.M.: When 'friends' collide: social heterogeneity and user vulnerability on social network sites. Comput. Hum. Behav. **54**, 62–72 (2016)
11. Alseadoon, I., Othman, M.F.I., Chan, T.: What is the influence of users' characteristics on their ability to detect phishing emails? In: Sulaiman, H.A., Othman, M.A., Othman, M.F.I., Rahim, Y.A., Pee, N.C. (eds.) Advanced Computer and Communication Engineering Technology. LNEE, vol. 315, pp. 949–962. Springer, Cham (2015). https://doi.org/10.1007/978-3-319-07674-4_89
12. Vishwanath, A., Harrison, B., Ng, Y.J.: Suspicion, cognition, and automaticity model of phishing susceptibility. Commun. Res. (2016)
13. Iuga, C., Nurse, J.R.C., Erola, A.: Baiting the hook: factors impacting susceptibility to phishing attacks. Hum. Centric Comput. Inf. Sci. **6**(1), 8 (2016)
14. Sheng, S., Holbrook, M., Kumaraguru, P., Cranor, L.F., Downs, J.: Who falls for phish? A demographic analysis of phishing susceptibility and effectiveness of interventions. In: Proceedings of the 28th International Conference on Human Factors in Computing Systems, CHI 2010, pp. 373–382 (2010)
15. Algarni, A., Xu, Y., Chan, T.: An empirical study on the susceptibility to social engineering in social networking sites: the case of Facebook. Eur. J. Inf. Syst. **26**(6), 661–687 (2017)
16. Albladi, S., Weir, G.R.S.: Vulnerability to social engineering in social networks: a proposed user-centric framework. In: IEEE International Conference on Cybercrime and Computer Forensic (ICCCF), pp. 1–6 (2016)
17. Albladi, S.M., Weir, G.R.S.: Competence measure in social networks. In: IEEE International Carnahan Conference on Security Technology (ICCST), pp. 1–6 (2017)
18. Gao, H., Hu, J., Huang, T., Wang, J., Chen, Y.: Security issues in online social networks. IEEE Internet Comput. **15**(4), 56–63 (2011)
19. Kumaraguru, P., Cranshaw, J., Acquisti, A., Cranor, L., Hong, J., Blair, M.A., Pham, T.: School of phish: a real-world evaluation of anti-phishing training. In: Proceedings of the 5th Symposium on Usable Privacy and Security, SOUPS 2009, p. 1 (2009)
20. Flores, W., Holm, H., Svensson, G., Ericsson, G.: Using phishing experiments and scenario-based surveys to understand security behaviours in practice. Inf. Manag. Comput. Secur. **22**(4), 393–406 (2014)

21. Mohebzada, J., El Zarka, A., Bhojani, A., Darwish, A.: Phishing in a university community: two large scale phishing experiments. In: International Conference on Innovations in Information Technology (IIT), pp. 249–254 (2012)
22. Benson, V., Saridakis, G., Tennakoon, H.: Purpose of social networking use and victimisation: are there any differences between university students and those not in HE? Comput. Hum. Behav. **51**, 867–872 (2015)
23. Goel, S., Williams, K., Dincelli, E.: Got phished: internet security and human vulnerability. J. Assoc. Inf. Syst. **18**(1), 22–44 (2017)
24. Jordan, M.I., Mitchell, T.M.: Machine learning: trends, perspectives, and prospects. Science **349**(6245), 255–260 (2015)

The Role of Mapping Curve
in Swarm-Like Opinion Formation

Tomasz M. Gwizdałła[(✉)]

Faculty of Physics and Applied Informatics, University of Łódź,
Pomorska 149/153, 90-236 Łódź, Poland
tomgwizd@uni.lodz.pl

Abstract. Recently [1] we have proposed the scheme of performing the
opinion formation simulation based on popular global optimization mechanism - the Particle Swarm Optimization. The basic idea was to use the
interaction between two potential directions of agents' heading: those
forced by the global opinion and those forced by the opinion of neigh-
bors/colleagues. In the proposed paper some enhancement of the proposed model is shown. We assume that, when performing the binary
PSO-like update of system, we use the generalized version of logistic
function. The results are promising in the sense that the introduced
change increases explicitly the number of possible solutions.

1 Introduction

The first ideas concerning the mathematical description of human groups and
especially the relations between the individuals in group and its influence on
some final features being the result of complex processes inside the group arose
as early as in the 50'th of twentieth century. The revolution in the analysis of
social processes by means of computational methods came however mainly from
the physicists who found a lot of similarities between the formulation of problems
in sociology and some notions well known in different physical areas. The question whether "do people behave like atoms" (possibly "like magnetic moments")
although started by the Clifford's model [2] with the Glauber dynamics for some
spatial structures and has been later especially extensively studied at the turn of
centuries. Here especially several models have to be mentioned. Galam [3] proposed some analytical formulas for the determination of opinion changes. Very
simple but easy extensible Voter's model (see e.g. [4]) adopted a mechanism of
spreading the opinion into a selected neighbor. Sznajds in their approach [5]
proposed a deterministic method of outward opinion dissemination. This model
has also been studied and considerably extended by Stauffer [6] who presented
own models and reviews (see e.g. [7]). The dynamical models on the continuous
space were proposed by Hegselmann [8] and Deffuant [9].

The paper is the continuation of the former papers devoted to the study of
opinion formation process [1,10,11]. The main effort was concentrated on the
study of influence of different topologies, different sample sizes and different

© Springer Nature Switzerland AG 2018
N. T. Nguyen et al. (Eds.): ICCCI 2018, LNAI 11055, pp. 157–166, 2018.
https://doi.org/10.1007/978-3-319-98443-8_15

update techniques on the characteristics obtained in the process of simulation of social effects and the possible interpretation of observed results. It could be mentioned that very similar techniques can be used in studying the problem in various contexts. On the one hand we can look for some formation of set of opinions like in our works, on the other hand one can try to use them e.g. to discover the consensus and therefore find some solutions to a number of different problems (see e.g. [12–14]). In our study however, the main interest is related to the observation of election. We believe that the elections are the most directly observed, best evidenced and regularly performed event where the people's opinion is formulated and revealed.

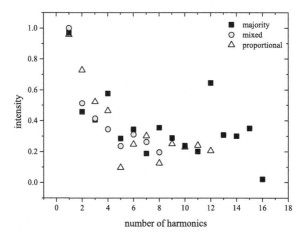

Fig. 1. The Fourier transform of election results for several countries with different electoral systems.

Additionally, in the last two of papers mentioned above [1, 11] we use another approach for visualization of results. We try to find some periodic characteristics by performing the Fourier transform of results. The effect of this trick is shown in Fig. 1 which is the same picture as Fig. 1 of paper [1].

The idea of using the Fourier transform is to try to find some basic pattern which could be characteristic for particular countries (particular societies). Maybe this remark can be generalized for e.g. same electoral systems. Thus we selected several countries with different systems and try to combine the data. In the presented Figure we have the data for three systems. They are: First Past The Post (United Kingdom and United States), mixed majority proportional system from Germany and proportional (d'Hondt) system from Denmark. The data for particular parties were transformed by FT and then averaged over different parties. We tried certainly to standardise the time differences, for example the elections in UK are every four years and in US every two year (House of Representatives).

The technique proposed for the update of opinion system is the Particle Swarm Optimization (PSO). The source of this method is the observation of the

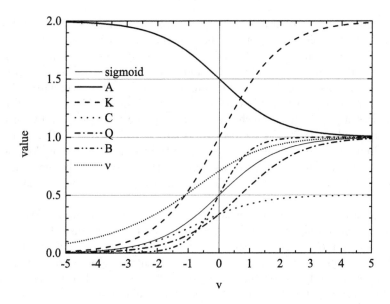

Fig. 2. The shape of generalized logistic function for some changed values of parameters (see text).

behavior of herds of animals, especially flocks of birds. We can generally show two types of such behavior, both of them found analogues in computational models. The first one is a conclusion of looking for aggregate motion of the flock and was described by Reynolds [15]. The second one, which lead directly to PSO as the optimization method is based on the complex behavior where single individuals/agents use the common knowledge as well as the information of their best memorized solutions to find the global optimum of the problem. This has been originally proposed by Kennedy and Eberhart [16] and now is considered as one of the most popular global optimization metaheuristics.

In the paper [1] we showed some features of the typical PSO technique used for binary variables. Different approaches are here mainly the successors of the other Kennedy and Eberhart proposition [17]. The crucial role in this approach is given to the logistic function which determines the selection of the direction of "velocity" which corresponds to the direction of change of binary state. In our current approach we pay more attention to the general logistic function which can significantly modify the features of the solution obtained by our model.

2 Model

In the presented paper we use the approach similar to the study of magnetic problems in the CA-like solutions. It means that generally three properties of the system have to be defined. There are: the set of states, the topology and the update functionality.

Considering the set of states we use the typical two-state system well known from the basic models of magnetism, like e.g. the famous Ising model. The opinions are, like in the model mentioned described simply as belonging to the set $\{-1, 1\}$ or equivalently $\{0, 1\}$. The selection among the above sets was especially important when using the so-called Glauber update which is originated directly from physics simulations. In the current approach its choice is not essential.

As it was shown earlier, the choice of topology can be the crucial factor influencing the results [10,11]. In the presented paper, mainly due to the limitation of the submission, we are going only to show the results for the topology which seems to best reproduce the relations between the individuals in the communities. i.e. the Barabasi-Albert network, well known for its usage in small world problems.

The main interest of the paper is related to the update function. We limited here our approach only to the PSO-like scheme and we don't use the mentioned earlier Glauber model. That is the reason why we do not pay attention to the strict formulation of set of opinions. In the presented paper we do not need to define the energy-related formula, like we made it earlier. The typical formula for Particle Swarm Optimization is to define the velocity as:

$$v_{ij}(t) = \chi \left(\omega v_{ij}(t-1) + c_1 * rnd() * (x_{global,j}(t) - x_{ij}(t)) \right.$$
$$\left. + c_2 * rnd() * (x_{ij}^{best}(t) - x_{ij}(t)) \right) \qquad (1)$$

and then to find the new position of member of the swarm:

$$x_{ij}(t+1) = x_{ij}(t) + v_{ij}(t). \qquad (2)$$

The crucial idea of formula (1) is to enforce the individuals/agents to follow the two directions. The first one is the, currently, best solution obtained during the simulation and remembered by a swarm $x_{global,j}(t)$, the second one is the best position remembered by the agent by himself $x_{ij}^{best}(t)$. By the cooperation and competition between this factors, enriched by some inertia ω and constriction χ we believe the system is able to find the global optimum. In our considerations needed for opinion formation we have to change two features. Firstly, we have to adapt the system in such a way that it will not lead the agent to the optimum but to some point determined by its neighborhood. That is why the formula (1) is replaced by the one where other centers of attraction are used.

$$v_{ij}(t) = \omega_{inert} v_{ij}(t-1) + c_{all} * rnd() * (x_{all,j}(t) - x_{ij}(t))$$
$$+ c_{group} * rnd() * (x_{ij}^{group}(t) - x_{ij}(t)) \qquad (3)$$

In the above proposition, the first term corresponds to the tendency to move to the average position of all agents in the sample $x_{all,j}(t)$ and its weighted strength is described by the coefficient c_{all}. The second term describes the heading into the average position of all agents' neighbors. As the neighbors we understand all agents that have direct links with the considered one in the Barabasi-Albert network.

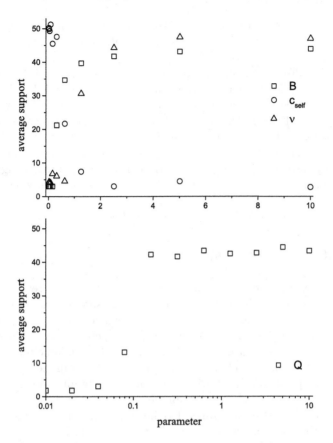

Fig. 3. The dependence of average support for particular opinion on the selected parameter. As it is described in the text, the values of variant parameter are distributed evenly in the $[0.01, 10.0]$ interval.

The second problem to be solved is the mapping of velocity (3) onto the one of only two possible states. Usually, as it was in seminal Kennedy paper [17] as well as in our earlier paper [1] the sigmoid function was used in order to determine the probability of reaching the particular state:

$$sig(v) = \frac{1}{1 + e^{-v}} \qquad (4)$$

$$x_{ij}(t+1) = \begin{cases} -1, & if\ rnd \geq sig(v_{ij}(t)) \\ 1, & if\ rnd < sig(v_{ij}(t)) \end{cases} \qquad (5)$$

In the current paper we are going to propose to use the generalized logistic function [18]:

$$sig(v) = A + \frac{K - A}{(C + Qe^{-Bv})^{1/\nu}}. \qquad (6)$$

The idea of generalisation can be better observed when commenting Fig. 2 where some curves obtained for different modifications of parameter in formula (6) is shown.

Every generalized sigmoid is described by the tuple of 6 parameters $\{A, K, C, B, Q, \nu\}$. The basic one corresponds to the tuple $\{0, 1, 1, 1, 1, 1\}$. In the Fig. 2 we plotted the basic sigmoid and 6 others where the parameter described in legend equals 2. It seems that especially 3 curves (for C, Q, ν changed) can be of special interest. That is due to its asymmetric character what can help us to describe the situation when some opinion can be somehow privileged.

We modify also the meaning of constriction factor by introducing the c_{self} coefficient which limits the possibility of opinion change for every agent.

$$prob_{change} = c_{self} * prob_{change} \qquad (7)$$

3   Results and Conclusions

It is certainly impossible to present here the comprehensive review of results obtained for a lot of settings which determine the simulation process. We are going just to present some selection of results which seems most interesting. Finally, it turned out that the best selection of parameters which are necessary to present results consists 3 parameters of generalized logistic function and 4 basic parameters of PSO-like model. That is why the sets of parameters for particular simulations are presented in the form of tuples containing seven parameters $\{Q, B, \nu, c_{all}, c_{group}, \omega, c_{self}\}$. Several numbers contained in the set of parameters below could seem a little bit strange looking like irrational numbers. It comes from the fact that we performed a series of test calculations for different values of all parameters spread evenly in the logarithmic scale in the period $[0.01, 10]$. Some effects turned out to be especially well visible for intermediate values of parameters.

The presentation of results we start from the plot which is well known from some sociophysical observations. The very general problem is to find the variable which enables the system to present different average or final properties in some interval of this variable. As an example we can show e.g. the result shown in Fig. 4 of [19]. The authors study there the averaged state dependence on the so-called level of nonconformity for q-voter model. Authors call their "state" as the public opinion and relate it to the physical notion of magnetization. In our model we can manipulate with several parameters and the results of this manipulation are shown in Fig. 3. As it can be seen, by changing the selected parameter with all other ones from the set constant we can produce every ordering of states from the unanimity up to balance. In the language of physics we can say that we can obtain ferromagnetic as well as paramagnetic state with our model.

An interesting observation is that, by using the generalized logistic function, we can produce the wanted dependence also for c_{self} which is the parameter of simulation model and not of the introduced function. One can also notice that the behavior is different for different parameters. While increasing B or c_{self} we

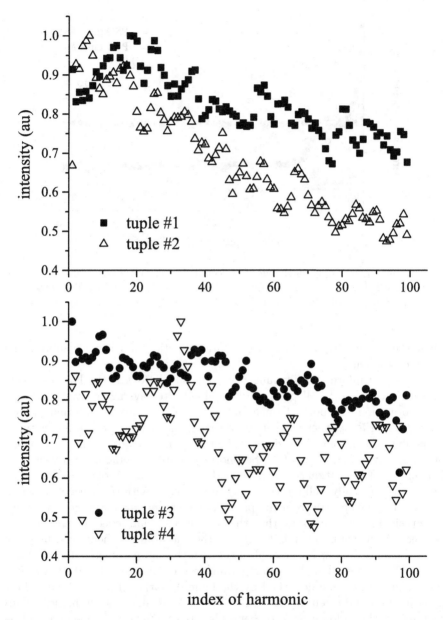

Fig. 4. The values of intensities of Fourier harmonics for selected parameters: tuple#1: {5.01187, 1, 0.1, 10, 0.01, 1, 0.01}, tuple#2: {10, 1, 10, 10, 0.01, 1, 0.01}, tuple#3: {0.0794328, 1, 0.1, 10, 0.01, 1, 0.01}, tuple#4: {10, 0.01, 0.01, 0.1, 0.75, 0.95, 0.25}.

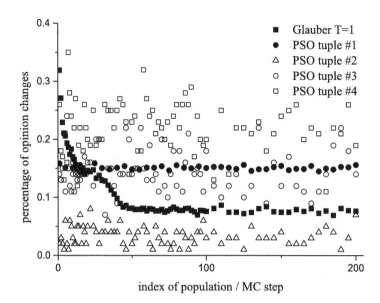

Fig. 5. The number of opinion changes in selected single run: tuple#1: {0.1, 0.75, 0.95, 0.25}, tuple#2: {1, 10, 0.1, 0.1, 0.3, 0.2, 0.25}, tuple#3: {1, 0.01, 0.01, 0.1, 0.75, 0.95, 0.25}, tuple#4: {10, 10, 10, 0.1, 2, 0.75, 0.5}.

change the average state from ferro- to paramagnetic, changing ν we obtain the reverse dependence.

The crucial information coming from our calculations is the Fourier transform, shown in Fig. 4. All points are obtained after averaging over 100 runs. This information is important since taking it into account we can formulate two observations. Firstly, there is visible linear dependence of average intensity on the number of harmonic. The slope of the spectrum can vary in the wide range, approximately $(-1.7, 5.0) \times 10^{-3}$. Secondly, even after averaging over relatively high number of simulation runs, the dependencies are characterized by large deviations. It is especially well noticeable when compared with the spectra shown in Fig. 2 of [1]. The spectra from the previous paper were also averaged and their shape is significantly smoother then the current ones. This effect is strongest for the spectrum which is the modification of the one from the mentioned figure of previous paper. Tuple#4 is created in the way that all parameters for PSO-like scheme (3) are tested with various parameters of function (6). It shows that every single run can be described by the large deviations from the line of best approximation and often higher harmonics can have significantly higher values of intensities that a lower ones. It seems also that there very detailed and subtle effects can determine the dependence of type of spectrum on the parameters of the update function.

Figure 5 contains the number of opinion changes during the single simulation run. In order to make possible the comparison with the earlier results shown in Fig. 4 of paper [1]. As compared to the previous characteristics, we can emphasize that we do not observe the stabilization phase. The number of changes starts to oscillate around some average value already from the first MC step. As it can be seen from the picture, the average number of changes can vary for different parameters, so can describe different societies with them.

The paper is the continuation of the initial one devoted to the presentation of a new technique prepared for the simulation of the opinion formation mechanism. When considering such models we have to think about several features. We must have the possibility to cover a lot of different social groups characterized with different stability or volatility. We have to find some characteristic which would be appropriate to describe the behavior of this groups in time. It seems that the PSO-like scheme can fill a lot of such requirements. The addition of generalized logistic function enhances significantly the capabilities of the model.

Certainly, the results obtained for the single set of election results, like it is presented in Fig. 1 is very hard when using the averaged values from a series of simulations. We think however that the presented approach is a promising idea.

References

1. Gwizdałła, T.M.: The swarm-like update scheme for opinion formation. In: Nguyen, N.T., Papadopoulos, G.A., Jędrzejowicz, P., Trawiński, B., Vossen, G. (eds.) ICCCI 2017. LNCS (LNAI), vol. 10449, pp. 66–75. Springer, Cham (2017). https://doi.org/10.1007/978-3-319-67077-5_7
2. Clifford, P., Sudbury, A.: A model for spatial conflict. Biometrika **60**, 581–588 (1973)
3. Galam, S.: Minority opinion spreading in random geometry. Eur. Phys. J. B **25**, 403–406 (2002)
4. Liggett, T.M.: Interacting Particle Systems. Springer, Heidelberg (1985). https://doi.org/10.1007/b138374
5. Sznajd-Weron, K., Sznajd, J.: Opinion evolution in closed community. Int. J. Mod. Phys. C **11**, 1157 (2000)
6. Stauffer, D., Sousa, A.O.: Generalization to square lattice of Sznajd sociophysics. Int. J. Mod. Phys. C **11**, 1239 (2000)
7. Fortunato, S., Stauffer, D.: Computer simulations of opinions and their reactions to extreme events. In: Albeverio, S., Jentsch, V., Kantz, H. (eds.) Extreme Events in Nature and Society, pp. 233–257. Springer, Heidelberg (2006). https://doi.org/10.1007/3-540-28611-X_11
8. Hegselmann, R., Krause, U.: Opinion dynamics and bounded confidence: models, analysis and simulation. J. Artif. Soc. Soc. Simul. **5**, 1–24 (2002)
9. Deffuant, G., Neau, D., Amblard, F., Weisbuch, G.: Mixing beliefs among interacting agents. Adv. Complex Syst. **03**, 87–98 (2000)
10. Gwizdałła, T.M.: Gallagher index for sociophysical models. Physica A **387**, 2937–2751 (2008)
11. Gwizdałła, T.M.: The influence of cellular automaton topology on the opinion formation. In: Malyshkin, V. (ed.) PaCT 2015. LNCS, vol. 9251, pp. 179–190. Springer, Cham (2015). https://doi.org/10.1007/978-3-319-21909-7_17
12. Lei, C., Ruan, J.: A particle swarm optimization-based algorithm for finding gapped motifs. BioData Min. **3**, 9 (2010)

13. Hristoskova, A., Boeva, V., Tsiporkova, E.: A formal concept analysis approach to consensus clustering of multi-experiment expression data. BMC Bioinform. **15**, 151 (2014)
14. Pan, F., Zhang, Q., Liu, J., Li, W., Gao, Q.: Consensus analysis for a class of stochastic PSO algorithm. Appl. Soft Comput. **23**, 567–578 (2014)
15. Reynolds, C.W.: Flocks, herds and schools: a distributed behavioral model. SIG-GRAPH Comput. Graph. **21**, 25–34 (1987)
16. Kennedy, J., Eberhart, R.: Particle swarm optimization. In: Grefenstette, J.J. (ed.) Proceedings of the IEEE International Conference on Neural Networks, Perth, pp. 1942–1948. IEEE Service Center, Piscataway (1995)
17. Kennedy, J., Eberhart, R.C.: A discrete binary version of the particle swarm algorithm. In: IEEE International Conference on Systems, Man, and Cybernetics. 'Computational Cybernetics and Simulation', vol. 5, pp. 4104–4108 (1997)
18. Richards, F.J.: A flexible growth function for empirical use. J. Exp. Bot. **10**, 290–301 (1959)
19. Radosz, W., Mielnik-Pyszczorski, A., Brzezińska, M., Sznajd-Weron, K.: Q-voter model with nonconformity in freely forming groups: does the size distribution matter? Phys. Rev. E **95**, 062302 (2017)

Popularity and Geospatial Spread of Trends on Twitter: A Middle Eastern Case Study

Nabeel Albishry[1,3], Tom Crick[2(✉)], Tesleem Fagade[1], and Theo Tryfonas[1]

[1] Faculty of Engineering, University of Bristol, Bristol, UK
{n.albishry,tesleem.fagade,theo.tryfonas}@bristol.ac.uk
[2] Department of Computer Science, Swansea University, Swansea, UK
thomas.crick@swansea.ac.uk
[3] Faculty of Computing and IT, King Abdulaziz University, Jeddah, Saudi Arabia
nalbishry@kau.edu.sa

Abstract. Thousands of topics trend on Twitter across the world every day, making it increasingly challenging to provide real-time analysis of current issues, topics and themes being discussed across various locations and jurisdictions. There is thus a demand for simple and extensible approaches to provide deeper insight into these trends and how they propagate across locales. This paper represents one of the first studies to look at geospatial spread of trends on Twitter, presenting various techniques to provide increased understanding of how trends on social networks can spread across various regions and nations. It is based on a year-long data collection ($N = 2{,}307{,}163$) and analysis between 2016–2017 of seven Middle Eastern countries (Bahrain, Egypt, Kuwait, Lebanon, Qatar, Saudi Arabia, and the United Arab Emirates). Using this year-long dataset, the project investigates the popularity and geospatial spread of trends, focusing on trend information but not processing individual topics, with the findings showing that likelihood of trends spreading to other locales is to a large extent influenced by the place in which it first appeared.

Keywords: Trends · Topic spread · Popularity · Network graphs
Twitter

1 Introduction

With the huge daily volume of generated content on Twitter – c.500 million tweets per day – trending topics serve as valuable sources of real-time information highlighting what is going on in the world, or in specific locations. Apart from the "official" trend lists provided by the platform (on the website or through

N. Albishry—This work has been supported by a doctoral research scholarship for Nabeel Albishry from King Abdulaziz University, Kingdom of Saudi Arabia.

N. T. Nguyen et al. (Eds.): ICCCI 2018, LNAI 11055, pp. 167–177, 2018.
https://doi.org/10.1007/978-3-319-98443-8_16

API endpoints), generating insight from trends and topics detection has been receiving increasing attention from across a variety of big social data-driven research domains, with varying results; in health for example, monitoring and analysis of trending topics on social media has been adopted to measure emerging public health issues, such as the spread of influenza [1,2]. Furthermore, across marketing and business domains, topic detection and classification are valuable approaches in extracting knowledge and insight on public opinions from posts on social media [3–5], including analysing voting intentions and political view of users [6]. With the increasing popularity and use of social networks across a wide range of domains, the impact of trends on public opinion and perceptions has transformed social media campaigns and public relations strategy. This has made trends a valuable target for manipulation [7], stuffing [8], spamming [9,10], and hijacking [11]. Interestingly, deeper analysis of trend hijacking cases suggests that increasing social media engagement may not always be beneficial for public relations strategies [12].

A common approach in analysing Twitter trends is through clustering and classification of trending topics based on content [13–16]. The study in [17] presented a content-independent method to model trends progression through the dynamics of users interactions; other studies have also attempted to provide real-time classification or detection of trends [18,19]. Many studies have dealt with trending topics, in terms of topic detection, for example. To the best of our knowledge, no study has investigated how trends spread spatially, and this study aims to provides methods for evaluating trending topics in regard to specific locales and how they spread from one place to another, without the need to collect tweeting activity of individual topics. The approach presented in this study provides a framework for understanding how topics capture interest of public across places and how likely it is for a topic to trend in other places based on its origin.

2   Methodology

The study is based on generating graph structures and conducting analyses of their properties, a widely-used approach (although we recognise there are other approaches and methodologies [20]). The graph construction approach presented in this paper was adopted as a simple way of capturing trends and related places, as well as temporal relationship between them. Furthermore, it can be used to produce simple visual representations of those data. The analysis approach involves constructing two graphs; the *temporal base graph* that captures the structure of trend raw data, and the *weighted aggregated graph* which is generated from the base graph to further explore its structure to provide additional insights. Figure 1 illustrates these graphs; the nodes and direction orientation of edges are the same in both graphs. Thus, nodes with zero indegree identify places, while trend nodes feature zero outdegree.

The *temporal base graph* is a directed graph that consists of three trend entities: *place*, *trend* and *timestamp*. Nodes represent place and trends, and edges

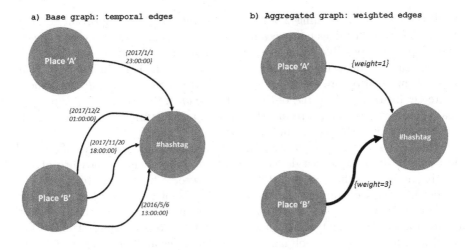

Fig. 1. Graphs constructed during analyses

are labelled with timestamps to indicate the time at which the trend appeared in a location. This graph is used to examine temporal properties, such as spread. The *weighted aggregated graph* is a graph that combines temporal edges between two nodes (in the base graph) into a single weighted edge. The feature of weighted edges in this graph is used to measure the popularity of trends, repetition rate, participation of countries, and the volume of the engagement.

2.1 Locales

Seven Middle Eastern countries were selected for this study: Bahrain, Egypt, Kuwait, Lebanon, Qatar, Saudi Arabia, and the United Arab Emirates (UAE). The selection includes countries with relatively large population (e.g. Egypt: 97,553,000) and relatively small populations (e.g. Bahrain: 1,493,000) [21]. Kuwait is reported to have the most active daily users on Twitter [22]; as of March 2016, Saudi Arabia and Egypt generated 33% and 20% of the tweets in the Middle Eastern region. Bahrain is the most balanced location in terms of gender breakdown of active users. Interestingly, between March 2014 and March 2016, Lebanon was the only location in the Middle Eastern states that has not seen growth in active users, while UAE increased by 60%. The Gulf Cooperation Council countries – Bahrain, Kuwait, Qatar, UAE, and Saudi Arabia – were reported to have the highest penetration rates [22].

2.2 Data Collection

Trending topic lists in the seven countries were monitored for the year between October 2016 and October 2017. Every hour, trending lists were collected through the Twitter REST API, which resulted in 7,948 hour's worth of records for all the countries, totalling 2,307,163 trend records. It is important to note that the Twitter API does not necessarily provide trends data for every request; for example, it is possible to receive no information for tweet volume. For each location, the list of available trending topic is returned. From this list, four pieces of information are extracted from each trend record: *woeid*: the Yahoo! Where On Earth ID (WOEID) of the location; *name*: text of trending topic (e.g '#Call_For_Action'); *as_of*: recorded timestamp of the trend; *tweet_volume*: volume of tweets over the past 24 h, if available.

While the Twitter API returns a list of trending topics for a specific *woeid* location, the tweet volumes do not provide a comprehensive measure of the tweeting activity in that location. Rather, the tweet volume refers to the overall number of tweets containing the trend, regardless of their location. Although the Twitter documentation[1] does not provide the necessary detail on this, it was apparent after observing trends that showed up in various locations. Trends were found with the same tweet volume across all locations and, hence, participation volume of each location was not possible to be accurately measured. Therefore, the context of this study does not include any reference to this volume entity.

3   Results

Observation of the weighted graph provided an overall evaluation of activity for trends and places. In total there were 76,266 distinct trends that trended 2,307,163 times across all locations; this suggests that trends may appear repeatedly over time. The overall repetition ratio in the dataset was 97%, and ranged from 80% to 98% for individual locations, with Saudi Arabia scoring lowest and Qatar scoring highest rate. *Indegree*, *outdegree*, and *edges* were used to conduct subsequent results, with further explanation to follow in the relevant sections.

3.1   Commonality and Popularity of Trends

The node indegree indicates the number of locations at which the trend showed (*commonality*), and the weighted indegree is used to measure the total number of times a trend showed (*popularity*). Therefore, grouping trends based on indegree has revealed 7 indegree groups, as shown in Table 1. Also, weighted indegree was used to analyse activity in these groups. Although 83% of trends have appeared in one location only, their total weighted indegrees was 40%; in other words, there were less common trends amongst locations, but their popularity was higher than isolated trends[2] – this implies that trends showing across multiple locations does not necessarily imply the prominence or importance of activity or topic.

---

[1] https://developer.twitter.com/en/docs/trends/trends-for-location/api-reference/get-trends-place.

[2] Isolated trends are those that have trended in one place, i.e. their indegree equal 1.

3.2 Location Participation

The node outdegree reflects how many unique trends a location is connected to (*diversity*), and weighted outdegree measures the ability of the location to generate trends (*activity*). The outdegree descriptive statistics, presented in Table 2, shows that Saudi Arabia came at the top of the list, with 42% of outgoing edges and weighing 20% of the total weight of the graph. Closeness in the table shows how close a location node is to all other trend nodes; it shows that Saudi Arabia has connections to 56% of trends in the graph. Nevertheless, Saudi Arabia was found lowest in terms of maximum edge weight, mean and standard deviation; Qatar was found to have the reverse values. This can be interpreted as the trends activity in Saudi Arabia was more diverse in total, but more consistent. In contrast, Qatar is connected to a limited number of trends with more focused activity. Also, Qatar's outdegree is just 60% of Bahrain's, although its weighted degree was 1.9 higher.

Table 1. Trends indegree groups

Indegree	No. trends	Ratio	Total W.	W. Ratio	Max.	Mean	Std
1	62,959	0.826	936,959	0.406	2,146	14.88	43.26
2	7,338	0.096	335,073	0.145	1,957	45.66	77.45
3	2,840	0.037	220,805	0.096	1,842	77.75	96.29
4	1,538	0.020	216,524	0.094	2,797	140.78	201.03
5	850	0.011	184,127	0.080	3,604	216.62	297.36
6	463	0.006	192,968	0.084	3,998	416.78	581.76
7	278	0.004	220,707	0.096	5,367	793.91	994.66

Table 2. Location outdegree descriptive statistics

Location	Outdegree	Out. ratio	W. Out.	W. Ratio	Closeness	Max.	Mean	Std
Bahrain	7,424	0.07	133,069	0.06	0.10	1,949	17.92	78.48
Egypt	14,282	0.14	383,830	0.17	0.19	1,408	26.88	58.68
Kuwait	13,891	0.14	397,960	0.17	0.18	1,400	28.65	51.67
Lebanon	6,044	0.06	294,761	0.13	0.08	2,146	48.77	133.64
Qatar	4,484	0.04	248,003	0.11	0.06	2,173	55.31	146.79
Saudi Arabia	42,767	0.42	468,081	0.20	0.56	1,175	10.95	17.43
UAE	12,389	0.12	381,459	0.17	0.16	1,655	30.79	70.75

3.3 Edges Properties

Edge weights in the graph were utilised to evaluate location activity in indegree groups. Overall, most of location activity went to common trends. Although Saudi Arabia was the highest in terms of total activity, the majority of its activity (61.26%) was identified as isolated trends. Moreover, observing originating locations for isolated trends shows that 30% of inbound edges came from Saudi Arabia, as shown in Fig. 3. Egypt contributed the most in 2 and 3 indegree trend groups, UAE in 4, 5 and 6 indegree trends, and for 7 indegree trends most of in edges originated from the Lebanon.

3.4   Common Trends

Based on results so far, following is further exploration of common trends amongst countries. In the context of this study, trend is regarded common between two places, if it appeared in the trend list of both places. While this network graph excludes weight of edges taken from weighted country-trend graph, edge weights here reflect number of common trends between two linked countries. Accordingly, undirected graph was constructed as shown in Fig. 2. Although, it was found with lowest tendency to relate to multi-indegree trends previously, Saudi Arabia was found with highest degree in this graph. This confirms that low weighted outdegree in common trends of a place does not necessarily imply low number of shared trends.

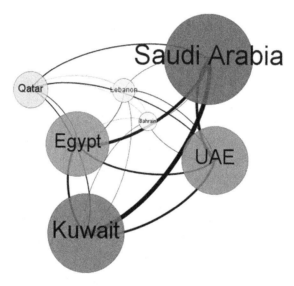

Fig. 2. Common trends graph

3.5   Temporal Spread and Reach

As shown in Table 1, 60% of weighted indegree was associated with common trends. To further examine temporal changes on those trends, timestamps on in-edges of trend nodes in the temporal graph were observed. Those timestamps were used to measure temporal order of locations for trend, as shown in Table 3. For instance, about 42% of first appearance of trends was in Saudi Arabia, while 36% of 7th trend appearance was in Bahrain.

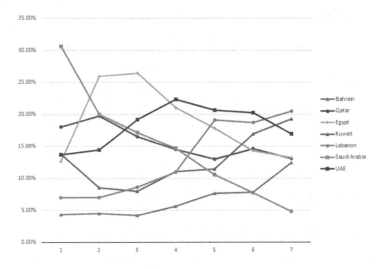

Fig. 3. Weighted contribution of countries toward trends indegree groups

Table 3. Distribution of temporal orders of location for multi-indegree trends

Location	1st	2nd	3rd	4th	5th	6th	7th
Bahrain	2.3	3.0	3.5	6.3	14.9	26.0	36.0
Egypt	16.3	13.2	21.3	22.8	16.8	11.7	2.5
Kuwait	16.5	29.5	23.8	15.1	13.2	11.2	8.3
Lebanon	5.7	5.4	6.2	7.0	8.7	11.3	23.7
Qatar	4.1	4.9	8.9	16.1	25.1	25.6	12.6
Saudi Arabia	42.2	28.7	10.2	7.1	7.3	8.1	14.4
UAE	12.9	15.2	26.0	25.4	14.0	5.9	2.5
Total	100	100	100	100	100	100	100

A similar observation was made on the outdegree measures, with timestamps on out-edges of place nodes in the temporal graph observed; the results are presented in Table 4. The highest portion of activities in Saudi Arabia, Egypt and Lebanon made them 1st locations for trends to appear in. However, Bahrain, UAE, Qatar, and Kuwait were more active with trends that have appeared previously.

Additionally, the reach of trend was measured to examine how many other locations a trend is likely to reach based on the location in which it first appeared. Therefore, edges and related nodes relating to the first column in Table 3 were used. The results presented in Table 5 show that 62.3% of trends that first appeared in Kuwait have also appeared in exactly one more location, and 5.1% of those that first appeared in Qatar have also appeared in six more locations.

Table 4. Distribution of appearance orders of locations

Order	Bahrain	Saudi	Egypt	UAE	Lebanon	Qatar	Kuwait
1st	18.4	53.6	34.6	26.9	32.2	19.1	26.3
2nd	24.2	36.4	28.0	31.9	30.5	22.9	47.1
3rd	12.8	5.8	20.3	24.4	15.8	18.6	17.1
4th	12.1	2.1	11.4	12.5	9.3	17.6	5.7
5th	14.5	1.1	4.3	3.5	5.9	13.9	2.5
6th	11.8	0.6	1.4	0.7	3.6	6.6	1.0
7th	6.1	0.4	0.1	0.1	2.8	1.2	0.3
Total	100	100	100	100	100	100	100

Table 5. Further reach of trends per locations

Reach	Bahrain	Egypt	UAE	Lebanon	Qatar	Kuwait	Saudi
1	54.8	59.2	54.2	50.1	41.2	62.3	53.1
2	19.3	22.7	21.2	21.3	24.3	18.9	21.7
3	8.3	9.8	11.1	10.4	12.6	10.1	13.2
4	8.3	4.4	6.4	8.1	12.0	4.1	7.2
5	4.7	2.7	4.3	6.5	4.7	2.6	3.3
6	4.7	1.2	2.8	3.7	5.1	2.1	1.6
Total	100	100	100	100	100	100	100

Finally, Table 6 shows the origin of common trends grouped by their indegree (connected places). As can be seen, 31.7% of trends that appeared in seven locations have originated from Saudi Arabia, and 5% from Bahrain. Nevertheless, Bahrain was better in terms of further reach.

Table 6. Origin of trends per reached locations

Location	2	3	4	5	6	7
Bahrain	2.2	2.0	1.6	2.9	3.0	5.0
Egypt	17.5	17.3	13.8	11.3	12.5	9.4
Kuwait	18.7	14.6	14.4	10.5	12.3	16.5
Lebanon	5.2	5.7	5.1	7.2	10.6	10.1
Qatar	3.1	4.7	4.5	7.8	5.6	10.0
Saudi Arabia	40.7	42.9	48.1	47.4	40.2	31.7
UAE	12.7	12.8	12.4	12.9	15.8	17.3
Total	100	100	100	100	100	100

4   Discussion and Conclusions

From the results presented in this first study, we can see that isolated trends were found to be most common across all countries, although the study includes countries with a high proportion of active users and high tweet generation rate, such as Saudi Arabia and Egypt [22]. As previously mentioned, the number of trends returned by the Twitter API does not accurately reflect the true activity of the location. Low trending topics may indicate low consensus on these discussed topics and does not necessarily reflect tweeting activity. Also, the number of trending topics is very likely to include repeated ones, and therefore a high number of trends does not necessarily imply more new topics. Furthermore, the number of trends was not found to correlate with the tendency of location to participate in common trends. For example, Saudi Arabia was found to be to connected to 56% of trends, however 61% them were isolated trends i.e. trends that only appeared in Saudi Arabia. Meanwhile, most of Qatar's trends (73.6%) were common ones, although it had edges with 6% of trends; this indicates that the activity of certain location is more focused on internal issues and concerns.

Also, the further reach of trends (i.e. appearing in other locations) was observed for each location. Although a specific location may do well in reaching other locations, the number of trends it generates may affect the total reach. For example, Qatar was highest in reaching other locations, however it was the 5th in being the origin of trends that reach all locations. This was certainly clear in the case of Saudi Arabia: its scores in reaching other locations were not comparable to its scores in being the origin of common trends, as shown in Tables 5 and 6.

In summary, this study has presented an new approach to analysing Twitter trends data using graphs and their properties. It has reinforced the utility of graph construction approaches to capture raw trends data, resulting in the temporal base graph. Then, we demonstrated how aggregated weighted graph can be generated from the base graph. The temporal graph was used to measure temporal properties such as spread and reach; the weighted graph was used to measures overall activities, such as commonality and popularity of trends, and diversity and activity of locations. This approach demonstrates how trends data can be used to evaluate topics and location activity without the need to crawl individual topics, a burdensome process. Furthermore, it shows how to measure spread of trends and reach based on their historical records as well as the originating location. This approach can be extended and applied to identify trends of other important features; for example, to extract high-spread trends, or how likely it is for a trend to reach a specific location from another one.

References

1. Lazer, D., Kennedy, R., King, G., Vespignani, A.: The parable of google flu: traps in big data analysis. Science **343**(6176), 1203–1205 (2014)
2. Parker, J., Yates, A., Goharian, N., Frieder, O.: Health-related hypothesis generation using social media data. Soc. Netw. Anal. Min. **5**(7), 7 (2015)

3. Blamey, B., Crick, T., Oatley, G.: 'The First Day of Summer': parsing tempo-ral expressions with distributed semantics. In: Bramer, M., Petridis, M. (eds.) SGAI 2013, pp. 389–402. Springer, Cham (2013). https://doi.org/10.1007/978-3-319-02621-3_29

4. Bello, G., Menéndez, H., Okazaki, S., Camacho, D.: Extracting collective trends from Twitter using social-based data mining. In: Bădică, C., Nguyen, N.T., Brezovan, M. (eds.) ICCCI 2013. LNCS (LNAI), vol. 8083, pp. 622–630. Springer, Heidelberg (2013). https://doi.org/10.1007/978-3-642-40495-5_62

5. Albishry, N., Crick, T., Tryfonas, T., Fagade, T.: An evaluation of performance and competition in customer services on Twitter: a UK telecoms case study. In: Com-panion of The Web Conference 2018. Social Sensing and Enterprise Intelligence: Towards a Smart Enterprise Transformation (2018)

6. Fang, A., Ounis, I., Habel, P., Macdonald, C., Limsopatham, N.: Topic-centric classification of Twitter user's political orientation. In: Proceedings of the 38th International ACM SIGIR Conference on Research & Development in Information Retrieval, pp. 791–794 (2015)

7. Zhang, Y., Ruan, X., Wang, H., Wang, H., He, S.: Twitter trends manipulation: a first look inside the security of Twitter trending. IEEE Trans. Inf. Forensics Secur. **12**(1), 144–156 (2017)

8. Irani, D., Webb, S., Pu, C., Drive, F., Gsrc, B.: Study of trend-stuffing on Twit-ter through text classification. In: Proceedings of the 7th Annual Collaboration, Electronic Messaging, AntiAbuse and Spam Conference (CEAS 2010) (2010)

9. Sedhai, S., Sun, A.: HSpam14: a collection of 14 million tweets for hashtag-oriented spam research. In: Proceedings of the 38th International ACM SIGIR Conference on Research & Development in Information Retrieval, pp. 223–232 (2015)

10. Chu, Z., Widjaja, I., Wang, H.: Detecting social spam campaigns on Twitter. In: Bao, F., Samarati, P., Zhou, J. (eds.) ACNS 2012. LNCS, vol. 7341, pp. 455–472. Springer, Heidelberg (2012). https://doi.org/10.1007/978-3-642-31284-7_27

11. VanDam, C., Tan, P.N.: Detecting hashtag hijacking from Twitter. In: Proceedings of the 8th ACM Conference on Web Science (WebSci 2016), pp. 370–371 (2016)

12. Sanderson, J., Barnes, K., Williamson, C., Kian, E.T.: 'How could anyone have pre-dicted that #AskJameis would go horribly wrong?' Public relations, social media, and hashtag hijacking. Public Relat. Rev. **42**(1), 31–37 (2016)

13. Zubiaga, A., Spina, D., Fresno, V., Martínez, R.: Classifying trending topics: a typology of conversation triggers on Twitter. In: Proceedings of the 20th ACM International Conference on Information and Knowledge Management, pp. 461–2464 (2011)

14. Benhardus, J., Kalita, J.: Streaming trend detection in Twitter. Int. J. Web Based Communities **9**(1), 122–139 (2013)

15. Ferragina, P., Piccinno, F., Santoro, R.: On analyzing hashtags in Twitter. In: Proceedings of the 9th International AAAI Conference on Web and Social Media (ICWSM 2015), pp. 110–119 (2015)

16. Albishry, N., Crick, T., Tryfonas, T.: *"Come Together!"*: interactions of language networks and multilingual communities on Twitter. In: Nguyen, N.T., Papadopou-los, G.A., Jędrzejowicz, P., Trawiński, B., Vossen, G. (eds.) ICCCI 2017, Part II. LNCS (LNAI), vol. 10449, pp. 469–478. Springer, Cham (2017). https://doi.org/10.1007/978-3-319-67077-5_45

17. ten Thij, M., Bhulai, S.: Modelling trend progression through an extension of the Polya Urn process. In: Wierzbicki, A., Brandes, U., Schweitzer, F., Pedreschi, D. (eds.) NetSci-X 2016. LNCS, vol. 9564, pp. 57–67. Springer, Cham (2016). https://doi.org/10.1007/978-3-319-28361-6_5

18. Mathioudakis, M., Koudas, N.: TwitterMonitor: trend detection over the Twitter stream. In: Proceedings of the 2010 ACM SIGMOD International Conference on Management of Data, pp. 1155–1158 (2010)
19. Zubiaga, A., Spina, D., Martínez, R., Fresno, V.: Real-time classification of Twitter trends. J. Assoc. Int. Sci. Tech. **66**(3), 462–473 (2015)
20. Tufekci, Z.: Big questions for social media big data: representativeness, validity and other methodological pitfalls. In: Proceedings of the 8th International AAAI Conference on Weblogs and Social Media (ICWSM 2014) (2014)
21. United Nations: World Population Prospects 2017. Technical report, Department of Economic and Social Affairs, Population Division (2017)
22. Salem, F.: Social media and the internet of things towards data-driven policymaking in the Arab world: potential, limits and concerns. Technical report, The Arab Social Media Report, MBR School of Government, Dubai, February 2017

On the Emergence of Segregation in Society: Network-Oriented Analysis of the Effect of Evolving Friendships

Christianne Kappert, Rosalyn Rus, and Jan Treur[(✉)]

Behavioural Informatics Group,
Vrije Universiteit Amsterdam, Amsterdam, Netherlands
christianne.kappert@hotmail.com,
rosalynrus@gmail.com, j.treur@vu.nl

Abstract. Segregation is a widely-observed phenomenon through history, different cultures and around the world. This paper addresses how in a network of immigrants segregation emerges from friendship homophily. The simulation results show that homophily results in clusters of lower and higher local language use. A mathematical analysis provides a more in depth understanding of the phenomena observed in the simulations.

1 Introduction

Segregation is the process or act of setting something or someone apart from others. It is a widely-observed phenomenon through history, different cultures and around the world. It seems to emerge easily everywhere, based on a variety of features, ranging from gender to disabilities, from age to religion. Race and ethnicity are however the strongest in dividing a network into groups [9]. Examples thereof are the segregation of Jews, not only in the Second World War but also during the Medieval times, the Apartheid in the 20th century and a more recent example is linked to immigration. Since the Second World War more than 65 million people had to flee their country, flee from famine, war or climate change [15]. After settling down in a new country, immigrants come into contact with the local population, a different culture and new language. Acculturation, the cultural change that results from these encounters, and the adaptation (outcome as a result of acculturation) thereof, is influenced by numerous factors such as age, gender, cultural distance and personality traits [2]. Studying racial segregation is important as it has big influences on society as a whole. The work environment is impacted through segregation as it increases employment and income inequality [12]. School achievements are affected as well, since segregation decreases academic performance [10]. Segregation however does not only have negative effects. An example is how segregation impacts substance use. Studies show that the less immigrants integrate with locals (low homophily) they consume fewer alcohol and cannabis [8].

An important phenomenon that influences the emergence of segregation is homophily, the principle that connections develop at a higher rate between similar people than dissimilar people [9]. Inter-ethnic friendships (caused by friendship homophily) has even become an important benchmark to reduce racial segregation [1]. When it comes to

© Springer Nature Switzerland AG 2018
N. T. Nguyen et al. (Eds.): ICCCI 2018, LNAI 11055, pp. 178–191, 2018.
https://doi.org/10.1007/978-3-319-98443-8_17

immigrants, the level of the new language usage can vary a lot. A longitudinal study of the language use of immigrants in Australia showed that the level of their language use increases over 2 years. The part of the population that was proficient in their new language use increased from 51.52% to 66.46% within 3 years [3]. Furthermore, language use plays a crucial role in creating social connections by new immigrants; lower usage of the new language is related to higher friendship homophily [4, 13].

This paper addresses how segregation emerges, using friendship homophily in a network of immigrants, as well as a mixed network of immigrants and local people. Friendship homophily is predicted by the use of the local language by immigrants as these are negatively correlated [13].

An adaptive temporal-causal network was designed by which a simulation experiments were performed. Moreover by Mathematical Analysis some more in depth understanding was obtained of the observed processes. The paper is structured as follows. Section 2 describes both the conceptual and numerical representation of the adaptive temporal-causal network model that was designed. In Sect. 3 the simulation scenarios are described. as well as the assumptions for the different scenarios. Section 4 presents the results of the simulations, and Sect. 5 contributes a Mathematical Analysis. Section 6 is a final discussion.

2   Network-Oriented Modeling of Evolving Friendships

The effect of the new (local) language usage on segregation has been studied by means of the Network-Oriented Modeling approach based on adaptive temporal-causal networks described in [14], including social contagion and an adaptive homophily effect. Causal modeling, causal reasoning and causal simulation have a long tradition in AI; e.g., [6, 7, 11]. The Network-Oriented Modeling approach based on temporal-causal networks described in [14] can be viewed on the one hand as part of this causal modeling tradition, and on the other hand from the perspective on mental states and their causal relations in Philosophy of Mind (e.g., [5]). It is a widely usable generic dynamic AI modeling approach that distinguishes itself by incorporating a dynamic and adaptive temporal perspective, both on states and on causal relations. This dynamic perspective enables modeling of cyclic and adaptive networks, and also of timing of causal effects. This enables modelling by adaptive causal networks for connected mental states and for evolving social interaction.

Temporal-causal network models can be represented by a conceptual representation and by an equivalent numerical representation. They usually include a number of parameters for domain, person, or social context-specific characteristics. A conceptual representation of a temporal-causal network model in the first place involves representing in a declarative manner states and connections between them that represent (causal) impacts of states on each other, as assumed to hold for the application domain addressed. The states are assumed to have (activation) levels that vary over time. In reality, not all causal relations are equally strong, so some notion of strength of a connection is used. Furthermore, when more than one causal relation affects a state, some way to aggregate multiple causal impacts on a state is used. Moreover, a notion of speed of change of a state is used for timing of processes. These three notions are covered by elements in the Network-Oriented Modeling approach based on temporal-causal networks, and form the defining part of a conceptual representation of a specific temporal-causal network model:

- for each connection from a state X to a state Y a *connection weight* $\omega_{X,Y}$ (for the strength of the impact of X on Y)
- for each state Y a *speed factor* η_Y (for the timing of the effect of the impact)
- for each state Y the type of *combination function* $c_Y(..)$ used (to aggregate multiple impacts on a state)

Combination functions can have different forms, as there are many different approaches possible to address the issue of combining multiple impacts. The applicability of a specific combination rule for this may depend much on the type of application addressed, and even on the type of states within an application. Therefore, the Network-Oriented Modeling approach based on temporal-causal networks incorporates for each state, as a kind of label or parameter, a way to specify how multiple causal impacts on this state are aggregated. For this aggregation a number of standard combination functions are made available as options and a number of desirable properties of such combination functions have been identified (see [14], Chapt. 2, Sects. 2.6 and 2.7), some of which are the identity function **id(.)**, the scaled sum function $\text{ssum}_\lambda(...)$:

$$c_Y(V) = \text{id}(V) = V$$
$$c_Y(V_1, ..., V_k) = \text{ssum}_\lambda(V_1, ..., V_k) = (V_1 + ... + V_k)/\lambda$$

where λ is the scaling factor which can be defined as the sum of the incoming weights for the state Y considered, and the variables V and V_i refer to the (single) impacts $\omega_{X,Y}X(t)$ and $\omega_{X_i,Y}X_i(t)$ (also see below). The above three concepts (connection weight, speed factor, combination function) can be considered as parameters or labels representing characteristics in a network model.

A conceptual representation of a temporal-causal network model can be transformed in a systematic or even automated standard manner into an equivalent numerical representation of the model as follows [14], Chap. 2:

- at each time point t each state Y in the model has a real number value in the interval $(0, 1)$, denoted by $Y(t)$
- at each time point t each state X connected to state Y has an impact on Y defined as **impact**$_{X,Y}(t) = \omega_{X,Y} X(t)$ where $\omega_{X,Y}$ is the weight of the connection from X to Y
- The *aggregated impact* of multiple states X_i on Y at t is determined using a *combination function* $c_Y(..)$:

$$\textbf{aggimpact}_Y(t) = c_Y(\textbf{impact}_{X_1,Y}(t), ..., \textbf{impact}_{X_k,Y}(t))$$
$$= c_Y(\omega_{X_1,Y}X_1(t), ..., \omega_{X_k,Y}X_k(t))$$

where X_i are the states with connections to state Y

- The effect of **aggimpact**$_Y(t)$ on Y is exerted over time gradually, depending on speed factor η_Y:

$$Y(t + \Delta t) = Y(t) + \eta_Y(\textbf{aggimpact}_Y(t) - Y(t))\Delta t$$

or $\textbf{d}Y(t)/\textbf{d}t = \eta_Y(\textbf{aggimpactY(t)} - Y(t))$

- Thus, the following *difference* and *differential equation* for Y are obtained:

$$Y(t + \Delta t) = Y(t) + \eta_Y\big(\textbf{c}_Y(\omega_{X_1,Y}X_1(t), \ldots, \omega_{X_k,Y}X_k(t)) - Y(t)\big)\Delta t$$
$$\textbf{d}Y(t)/\textbf{d}t = \eta_Y\big(\textbf{c}_Y(\omega_{X_1,Y}X_1(t), \ldots, \omega_{X_k,Y}X_k(t)) - Y(t)\big)$$

Software templates have been developed (e.g., in Matlab and Excel) that based on a conceptual representation automatically generate these differential equations and an environment to simulate them. These have been used in the work presented here.

For an adaptive network the connection weights are dynamic and treated in a way similar to states, using combination functions and speed factors. The model presented in this paper incorporates social contagion and adaptation of connections by the homophily principle. Since homophily influences the connections a node has with another node, not only is there a connection from one node to the other, but also from that node to the connection weight. The graphical conceptual representation of such a network can be found in Fig. 1. Here X_B represents the state of person B, and $\omega_{A,B}$ (to avoid a notation $\omega_{XA,XB}$) represents the (adaptive) weight of the connection from X_A to X_B.

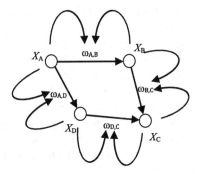

Fig. 1. Contagion and homophily: conceptual representation

Social contagion is based on the scaled sum function where the scaling factor λ is the sum of the incoming weights (which is dynamic over time).

$$X_B(t + \Delta t) = X_B(t) + \eta_B(\textbf{ssum}_\lambda(\omega_{A_1,B}X_{A_1}(t), \ldots, \omega_{A_k,B}X_{A_k}(t)) - X_B(t))\Delta t$$
$$\textbf{d}X_B(t)/\textbf{d}t = \eta_B(\textbf{ssum}_\lambda(\omega_{A_1,B}X_{A_1}(t), \ldots, \omega_{A_k,B}X_{A_k}(t)) - X_B(t))$$

where the X_{A_i} are the incoming connections for X_B. The numerical representation of the principle is as follows:

$$\omega_{A,B}(t+\Delta t) = \omega_{A,B}(t) + \eta_{A,B}(\mathbf{c}_{A,B}(X_A(t), X_B(t), \omega_{A,B}(t)) - \omega_{A,B}(t))\Delta t$$
$$d\omega_{A,B}/dt = \eta_{A,B}(\mathbf{c}_{A,B}(X_A, X_B, \omega_{A,B}) - \omega_{A,B})$$

where the combination function $\mathbf{c}_{A,B}(X_A, X_B, \omega_{A,B})$ is:

$$\mathbf{c}_{A,B}(X_A, X_B, \omega_{A,B}) = \omega_{A,B} + \alpha\,\omega_{A,B}(1 - \omega_{A,B})(\tau^2 - (X_A - X_B)^2)$$

Here $\eta_{A,B}$ is the speed factor of the connection from A to B, τ is the threshold parameter, and α the amplification factor of the homophily combination function.

3 Modeling Segregation of Immigrants

The following situation is assumed: a group of immigrants wants to settle in an English speaking country. The network is considered to only consist of immigrants, which means that the level of English language use is low, but some people do have some level of English. Therefore, the initial state values of the network consists of some values around 0.4–0.5 on a scale from 0 to 1 and some around 0.1–0.2.

Base scenario
The initial connection weights and state values for the base scenario can be found in Table 1. The speed factors for states is set to .4. It is assumed that the immigrants do not know each other, which results in low but nonzero connections weights (between .1 and .3); see Table 1. As we consider all the immigrants to be connected, 0.1 is the lowest initial connection weight. It is assumed that the level of local language use has some persistency. Therefore, each node has a connection to itself, and the initial connection weight thereof has value 5; these connections get speed factor 0, so their weights do not change. The connections between high local language use nodes and others have a speed factor of .3, whereas the connections between low local language use nodes is .6. The speed factor is lower for this group to incorporate the fact that homophily occurs faster for groups with low local language use compared to high local language use. An overview of the parameters can be found in Table 2, second column.

Scenario 2: Adding locals
In the second scenario, two locals are added to the network of immigrants. This results in two nodes with an initial state value of 1 as they are fluent in their own local language, and a speed factor of 0 as this level does not change. The connections between the two local nodes and between the locals and immigrants remain low. The parameters can be found in Table 2, third column.

Scenario 3: Adding a family
A third scenario adds a family of immigrants to the network. Half of the group has a high initial state value, and half has a low initial state value: for example in a family of 4 people, the children (2 people) are more fluent in English as they are being taught it at

Table 1. Initial weights and state values for the Base scenario

Connection	X_1	X_2	X_3	X_4	X_5	X_6	X_7	X_8	X_9	X_{10}
X_1	5	0.26	0.18	0.27	0.12	0.19	0.16	0.23	0.24	0.3
X_2	0.26	5	0.3	0.12	0.18	0.11	0.3	0.11	0.23	0.17
X_3	0.18	0.3	5	0.26	0.14	0.27	0.13	0.14	0.23	0.29
X_4	0.27	0.12	0.26	5	0.29	0.24	0.15	0.1	0.13	0.25
X_5	0.12	0.18	0.14	0.29	5	0.27	0.13	0.15	0.11	0.19
X_6	0.19	0.11	0.27	0.24	0.27	5	0.26	0.19	0.11	0.14
X_7	0.16	0.3	0.13	0.15	0.13	0.26	5	0.23	0.11	0.3
X_8	0.23	0.11	0.14	0.1	0.15	0.19	0.23	5	0.24	0.22
X_9	0.24	0.23	0.23	0.13	0.11	0.11	0.11	0.24	5	0.12
X_{10}	0.3	0.17	0.29	0.25	0.19	0.14	0.3	0.22	0.12	5
State	0.40	0.45	0.50	0.50	0.40	0.10	0.15	0.20	0.15	0.10

Table 2. Parameter values for the scenarios

Scenario	Base	2	3	4
Speed factor states	0.4	0.4	0.4	0.1
Speed factor connections				
same node	0	0	0	0
low-high	0.3	0.3	0.3	0.7
high-high	0.3	0.3	0.3	0.7
low-low	0.6	0.6	0.6	0.7
Threshold homophily	0.04	0.04	0.04	0.04
Amplification homophily	6	6	6	6

school and watch English TV shows/movies, whereas the parents (2 people) have not. The connection weights between the member of the family are high (0.9) and low between the family members and other members of the network; see Table 3 for their values.

Scenario 4: Real world case

Scenario 4 addresses a larger and more realistic network (of 50 people) with more precise values acquired from the literature. The initial state values are created based on the findings of Chiswick *et al.* (2004), who found that 51.5% of newly arrived immigrant are proficient in speaking the new language. Because the state values for the level of new language use is defined as a 0 to 1 scale, proficient is assumed to be a value of 0.5 and above. The initial state values were created with the *normrnd*-function in Matlab, according to a normal distribution with a certain mean and standard deviation(SD). Each node (immigrant) has an initial state value which can either be from a normal distribution with mean 0.7 and SD 0.15 or from a normal distribution with mean 0.3 and SD 0.15. There is a slightly higher chance (51.5%) to have a state value from the first distribution. The initial connection weight for 'not friend' was set to 0.1. The initial connection weights in this scenario were created using the *normrnd*-function with mean 0.1 and SD 0.2, values below 0 excluded. In this scenario there is also an additional learning rate to simulate the fact that by learning immigrants increase their level of language use over time (Chiswick *et al.*, 2004). Some of the parameters were adapted in this scenario; see Table 2, last column. In order to let the contagion effect work a longer duration, all speed factors of the connection weights were raised to 0.7, the connection of the nodes to itself was raised to 8 and the speed factor of the states lowered to 0.1.

Table 3. Initial weights and state values for Scenario 3

Connection	X_1	X_2	X_3	X_4	X_5	X_6	X_7	X_8	X_9	X_{10}
X_1	5	0.9	0.18	0.27	0.12	0.19	0.16	0.23	0.9	0.9
X_2	0.9	5	0.3	0.12	0.18	0.11	0.3	0.11	0.9	0.9
X_3	0.18	0.3	5	0.26	0.14	0.27	0.13	0.14	0.23	0.29
X_4	0.27	0.12	0.26	5	0.29	0.24	0.15	0.1	0.13	0.25
X_5	0.12	0.18	0.14	0.29	5	0.27	0.13	0.15	0.11	0.19
X_6	0.19	0.11	0.27	0.24	0.27	5	0.26	0.19	0.11	0.14
X_7	0.16	0.3	0.13	0.15	0.13	0.26	5	0.23	0.11	0.3
X_8	0.23	0.11	0.14	0.1	0.15	0.19	0.23	5	0.24	0.22
X_9	0.9	0.9	0.23	0.13	0.11	0.11	0.11	0.24	5	0.9
X_{10}	0.9	0.9	0.29	0.25	0.19	0.14	0.3	0.22	0.9	5
State	0.55	0.45	0.50	0.50	0.40	0.10	0.15	0.20	0.15	0.10

4 Simulation Results

Some of the simulation results are outlined here. For the first three scenarios the network consist of 10 nodes, which are all linked to each other. It is an undirected network, so the connection weights are the same no matter the direction of the connection. The nodes also have a loop with themselves, and therefore the total number of edges is 55.

Base scenario
The average degree is 11 (the loop with a node itself counts twice), the modularity is 0, as well as the betweenness centrality. This is due to the fact that all nodes are connected with each other and there are therefore no clusters. In Fig. 2 a visualization made in Gephi[1] is shown. The color of the nodes represents the initial state value; light orange is low and bright purple high. The color of the edges is the connection weight, which is more or less equal for all edges and very low (light color) except for the loop with itself, which is very high (dark blue).

Using the parameters as described in Table 2 the state values and connection weights develop over time as can be seen in Fig. 3. The state values start in two clusters (low value vs high value). From time point 0 to 8, the two clusters move towards each

---

[1] https://gephi.org/

Fig. 2. Visualization of the initial network for the Base scenario (Color figure online)

other: the state values of the high language use decrease, and vice versa for the low language use. After this, however, it is observed that the connections within a cluster move to each other (homophily) and the two clusters stop moving to each other. The connections in the cluster with low language use increase faster to each other than in the cluster with high language use.

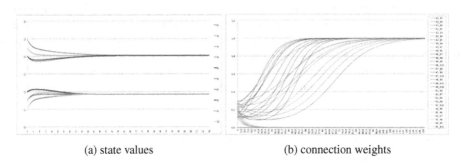

(a) state values   (b) connection weights

Fig. 3. Simulation of the Base scenario (Color figure online)

In Fig. 3(b), the connections between the clusters are in green, the connections within the low language use cluster in red, and the connections within the high language use in blue. The weights of connection between nodes in one cluster all converge to 1 over time, while the weights of connections between the clusters all converge to 0.

The final network visualized using Gephi is shown in Fig. 4. The color of the nodes represents the state value, where orange is low and purple high. The color of the edges is the connection weight, where light orange is a low connection weight, red is a high connection weight and blue extremely high. The visualization illustrates the observations made above: the connection weights within a cluster become high and between the clusters low, and the state values within a cluster are the same. This will be analyzed mathematically in more depth in Sect. 4.

Scenario 2: Adding locals
Simulation results for adding two locals to the network can be found in Fig. 5. Two states start at 1 and stay constant over time (the locals), three states with medium high state values (X_3 and X_4 are identical and therefore only X_4 can be seen) and five with

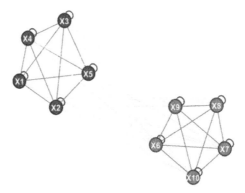

Fig. 4. Visualization of the final network for the Base scenario. (Color figure online)

low initial states. The high initial states first decrease a bit, but then over time increase and grow closer to each other, while the low initial states first have a small increase and then start bending downwards again. The connection weights for the low initial states (red lines) grow faster than of the medium initial states or the locals. The connection weights of the locals is displayed in orange and grows to 1 over time, as well as the connection between X_3 and X_4 (blue line next to the orange line). The connections between X_5 on one side and X_3 and X_4 on the other side, experience a steep decrease at the beginning but then over time increase again. This can also be seen in the state value graph as they first diverge from each other, but then start to converge again. This will be analyzed further in Sect. 4.

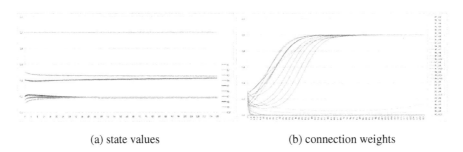

| (a) state values | (b) connection weights |

Fig. 5. Simulation of Scenario 2: effect of locals (Color figure online)

Scenario 3: Adding a family

In the third scenario, a family of four is added to the network (X_1, X_2, X_9 X_{10}). A lot changes; see Fig. 6. First, the state values still end up in two clusters, but converge much more towards each other as well. Second, the family members with low language use start with the lowest initial value, but over time surpass the other nodes in the cluster and have the highest value of the cluster. The same observation but vice versa can be made for the other cluster, in which the family members start with the highest initial value but over time become the ones with the lowest state value of the cluster. When it comes to the connections between and within the clusters, the following can be observed. First, there is no difference anymore between the clusters when it comes to how fast the connections grow. Second, some connections within a cluster (both high and low language use) decrease and eventually become zero. Third, the connections within the family either grow or decrease. The connection between the two family members with the same initial state value grows towards one, while the connection between the family members of different state values decreases and eventually becomes zero.

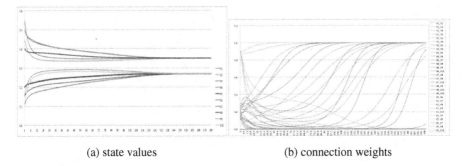

| (a) state values | (b) connection weights |

Fig. 6. Simulation of Scenario 3: effect of family members (Color figure online)

Scenario 4 Real world case

The simulation results of Scenario 4 are displayed in Fig. 7. It shows that here the state values of the nodes become small clusters over time. State values can become higher and lower for this clustering effect, however each cluster still increases its language use level due to the language learning added in this scenario. All connections that go to 1 are connections between nodes that both have initially a proficient language use level (red lines) or initially not a proficient language use level (blue lines). Some of the connection weight go towards 0. None of the connections between nodes with different initial levels of language use (green lines) becomes 1.

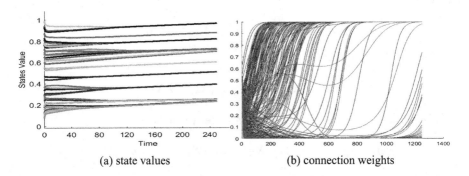

| (a) state values | (b) connection weights |

Fig. 7. Simulation of Scenario 4: real world case (Color figure online)

5 Mathematical Analysis

In this section Mathematical Analysis is used to get some more in depth understanding of the adaptive temporal-causal network model. One way to do this is by considering stationary points and equilibria of the network model. A state Y has a *stationary point* at time t if $dY(t)/dt = 0$; similarly an adaptive connection weight ω has a stationary point at t if $d\omega(t)/dt = 0$. In a temporal-causal model the dynamics are described by a specific format of differential and difference equation; based on that a specific criterion for having a stationary point is (see [14], Chap. 12):

$$\mathbf{c}_Y\big(\boldsymbol{\omega}_{X_1,Y}X_1(t), \ldots, \boldsymbol{\omega}_{X_k,Y}X_k(t)\big) = Y(t)$$

where $X_1, \ldots, X_k$ are the states connected to Y. A similar criterion can be found based on the combination function of an adaptive connection weight. Moreover, an adaptive network model has an *equilibrium state* at t if this criterion is fulfilled for all states and all adaptive connections.

Applying these criteria to the network model considered in this paper yields for an equilibrium state the following equilibrium equations:

$$\mathbf{ssum}_\lambda(\omega_{A_1,A_i}X_{A_1} + \ldots + \omega_{A_n,A_i}X_{A_n}) = X_{A_i}$$
$$\omega_{A,B} + \alpha\omega_{A,B}(1 - \omega_{A,B})(\tau^2 - (X_A - X_B)^2) = \omega_{A,B}$$

First, consider the equations for the connection weights. They are equivalent to

$$\omega_{A,B}(1 - \omega_{A,B})(\tau^2 - (X_A - X_B)^2) = 0$$

and can easily be solved as follows:

$$\omega_{A,B} = 0 \ \text{ or } \ \omega_{A,B} = 1 \ \text{ or } \ |X_A - X_B| = \tau$$

The first two options 0 and 1 have indeed been found all the time in the simulations. It turns out that the third option is a non-attracting equilibrium, so it does not emerge naturally. Next, consider the equilibrium equations for the states. Leaving out the t, their equation

$$(\omega_{A_1,A_i}X_{A_1} + \ldots + \omega_{A_n,A_i}X_{A_n})/\lambda = X_{A_i}$$

can be rewritten as follows

$$\omega_{A_1,A_i}X_{A_1} + \ldots + \omega_{A_n,A_i}X_{A_n} = \lambda X_{A_i}$$

Since $\lambda = \omega_{A_1,A_i} + \ldots + \omega_{A_n,A_i}$ this can be rewritten as

$$\omega_{A_1,A_i}X_{A_1} + \ldots + \omega_{A_n,A_i}X_{A_n} = \omega_{A_1,A_i}X_{A_i} + \ldots + \omega_{A_n,A_i}X_{A_i}$$

This provides a system of n linear equations in n unknowns X_{Ai} with the connection weights as parameters. From this format it is immediately clear that a solution occurs when all state values are equal, say they are all a number a; indeed when a is substituted for all state values, an identity occurs:

$$\omega_{A_1,A_i}a + \ldots + \omega_{A_n,A_i}a = \omega_{A_1,A_i}a + \ldots + \omega_{A_n,A_i}a$$

This does not imply that this type is the only solution. A number of the weights can be 0; recall that the occurring equilibrium values for the connection weights are either 0 or 1. As seen in the simulations, per cluster it can converge to a different state value. This can be analyzed a bit further. A network is called *weakly symmetric* if $\omega_{A,B} > 0 \Leftrightarrow \omega_{B,A} > 0$ for all A, B. It is *fully symmetric* if $\omega_{A,B} = \omega_{B,A}$ for all A, B. Note that since the combination function for homophily is symmetric in its first two arguments V_1 and V_2, when the initial network is fully symmetric, the evolving network will also stay fully symmetric, and hence weakly symmetric.

Lemma. If for some node X_A at time t for all nodes X_B with $X_B(t) > X_A(t)$ it holds $\omega_{A,B}(t) = 0$, then $X_A(t)$ is decreasing at t: $dX_A(t)/dt \leq 0$. If, moreover, a node X_C exists with $X_C(t) < X_A(t)$ and $\omega_{C,A}(t) > 0$ then $X_A(t)$ is strictly decreasing at t: $dX_A(t)(t)/dt < 0$.

In an achieved equilibrium state a *cluster* $\mathcal{C}$ is a maximal subnetwork in which all connections are 1.

Theorem. Suppose the network is weakly symmetric. Consider an equilibrium state. Then the following hold:
(a) For any two nodes X_A and X_B it holds $\omega_{A,B} = 0$ or $\omega_{A,B} = 1$.
(b) For all nodes X_A and X_B with $X_A \neq X_B$ it holds $\omega_{A,B} = 0$ and $\omega_{B,A} = 0$. In any cluster $\mathcal{C}$ the equilibrium values for all the states are equal: $X_A = X_B$ for all X_A and X_B in cluster $\mathcal{C}$.
(c) If $X_A = X_B$, then either $\omega_{A,B} = 0$ or $\omega_{A,B} = 1$.
     If $X_A(t) = X_B(t)$ and $0 < \omega_{A,B} < 1$, then $\omega_{A,B}(t)$ is strictly increasing at time t: $d\omega_{A,B}(t)/dt > 0$, so $\omega_{A,B}(t)$ can never reach the value 0 as long as $X_A(t) = X_B(t)$.

This Theorem proves that it is no coincidence that all simulations (which were fully symmetric) show that in the end the connection weights converge either to 0 or to 1, and that the states within a cluster (that are connected by weights that converge to 1) converge to the same state values. In particular, the third option $|X_A - X_B| = \tau$ does not occur at all in the weakly symmetric case.

6 Discussion

In this paper it was shown how an adaptive temporal-causal network model can simulate the emergence of segregation based on a friendship homophily principle for groups of immigrants based on the level of the new language use. In the most basic scenario with only immigrants, segregation shows both in the level of language use (state value) and the bonding (connection weight) between the immigrants. The addition of locals to the network does not change the behavior of the network much. There are still two clusters (for lower and higher level of language use) that develop over time and the connections within the cluster with the low local language use develop quicker when compared to the other cluster. If there are family connections incorporated in the network, the level of language use of the two clusters are closer together and the speed of the emergence of bonding by the homophily principle is lower. Adding a family of 4 people has the strongest effect on the state values and connections between the nodes. The scenario for a more realistic, larger network shows more emerging clusters.

A previous study reports that the level of language use of newly arrived immigrants shows an increasing trend over time [3]. It was possible to reproduce this effect by adding an autonomous language learning effect, assuming that individual language learning takes place on top of the social contagion, which sounds like a reasonable assumption. Another effect reported in previous studies is that a difference occurs for homophily between low and high local language usage: homophily works faster between immigrants with low language use [13]. The model allows to reproduce this effect by assuming a higher speed of the homophily-based adaptation of the connections for the lower language group. Whether or not this is a reasonable assumption or there are other options that are better can still be an open question for further investigation.

To validate the model more extensively, more detailed numerical empirical data would be needed. Due to the absence of such an empirical data set on this topic, only qualitative empirical information was used for validation. As described above, it was possible to reproduce several qualitatively described patterns reported in the literature, which provides some confidence in the validity. However, if in the future more detailed numerical empirical data can be obtained, still more in-depth validation can be performed.

References

1. Aboud, F., Mendelson, M., Purdy, K.: Cross-race peer relations and friendship quality. Int. J. Behav. Dev. **27**(2), 165–173 (2003)
2. Berry, J.W.: Immigration, acculturation, and adaptation. Appl. Psychol. **46**(1), 5–34 (1997)
3. Chiswick, B.R., Lee, Y.L., Miller, P.W.: Immigrants' language skills: the australian experience in a longitudinal survey. Int. Migr. Rev. **38**(2), 611–654 (2004)
4. Kang, S.-M.: Measurement of acculturation, scale formats, and language competence: their implications for adjustment. J. Cross-Cult. Psychol. **37**(6), 669–693 (2006)
5. Kim, J.: Philosophy of Mind. Westview Press, Boulder (1996)
6. Kuipers, B.J.: Commonsense reasoning about causality: deriving behavior from structure. Artif. Intell. **24**, 169–203 (1984)
7. Kuipers, B.J., Kassirer, J.P.: How to discover a knowledge representation for causal reasoning by studying an expert physician. In: Proceedings of the 8th International Joint Conference on Artificial Intelligence, IJCAI 1983, Karlsruhe, pp. 49–56. William Kaufman, Los Altos, CA (1983)

8. Lorant, V., et al.: A social network analysis of substance use among immigrant adolescents in six european cities. Soc. Sci. Med. **169**, 58–65 (2016)
9. McPherson, M., Smith-Lovin, L., Cook, J.M.: Birds of a feather: homophily in social networks. Ann. Rev. Sociol. **27**(1), 415–444 (2001)
10. Mickelson, R.A., Heath, D.: The effects of segregation on african american high school seniors' academic achievement. J. Negro Educ. **68**(4), 566–586 (1999)
11. Pearl, J.: Causality. Cambridge University Press, New York (2000)
12. Tassier, T., Menczer, F.: Social network structure, segregation, and equality in a labor market with referral hiring. J. Econ. Behav. Organ. **66**(3), 514–528 (2008)
13. Titzmann, P.F., Silbereisen, R.K.: Friendship homophily among ethnic german immigrants: a longitudinal comparison between recent and more experienced immigrant adolescents. J. Family Psychol. **23**(3), 301–310 (2009)
14. Treur, J.: Network-Oriented Modeling: Addressing Complexity of Cognitive, Affective and Social Interactions. Springer, Cham (2016). https://doi.org/10.1007/978-3-319-45213-5
15. Yap, C., Deckert, H., Weiwei, A.: Human flow (movie) (2017). https://www.humanflow.com/synopsis/

Computational Analysis of Bullying Behavior in the Social Media Era

Fakhra Jabeen and Jan Treur[(✉)]

Behavioural Informatics Group, Vrije Universiteit Amsterdam,
Amsterdam, Netherlands
{f.jabeen,j.treur}@vu.nl

Abstract. Internet and cyber-technology have played an important positive role but it also served as a venue for cyber-bullying. This paper presents a computational model for online bullying. Its evaluation was done by simulation experiments and mathematical analysis, in comparison to expected patterns from the empirical literature. This model may provide useful input to build a support system to avoid this negative social behavior within society.

Keywords: Cyberbullying · Temporal causal network · Online perpetration

1 Introduction

Over the past decades, bullying and more recently cyberbullying has increased massively due to the influx of technology and especially social media [1]. The U.S Department of Health and Human Services defines bullying as an act of power imbalance, which results in potentially repetitive negative and aggressive social behavior of a person towards others [2]. On the one hand, cyberspace has given freedom of expression by introducing the concept of anonymity, while, on the other hand, it has indeed opened a new episode for perpetrators to bully, many of them were also reported in form of case-laws [3–5]. Thus cyberbullying is arising as a significant dilemma for our society. It is a cowardly form of bullying in which audience size can grow exponentially due to certain tactics of a bully, but yet it may go unreported [6, 7]. In December 2017, the Cyberbullying Research Centre conducted a survey on cyberbullying, which determined that a person, who has been bullied at school is most likely to be bullied online [8]. Another study revealed that cyberbullying occurs more frequently in adolescents, who were often using internet and social sites more than two hours daily [9]. Cyber-bullying and cyber-misconduct is one of the leading concerns to maintain peace, law, and order within society [3, 4, 10].

Cyberbullies are often visually anonymous, and because of this, they do not fear any personal confrontation or consequences [11]. This not only increases the social influence over a bully-victim, but may also cause depression and anxiety and thus become a reason for poor mental health of a victim [12, 13]. To computationally analyze the behavior of a bully, we propose a temporal-causal network model that addresses the phenomenon and characteristics of a bully. In this work we aim to address what characteristics of a person are related to bullying behavior.

© Springer Nature Switzerland AG 2018
N. T. Nguyen et al. (Eds.): ICCCI 2018, LNAI 11055, pp. 192–205, 2018.
https://doi.org/10.1007/978-3-319-98443-8_18

The remainder of the paper is organized as follows. In Sect. 2 related work is discussed. Section 3 presents the temporal causal model for a bully and its analysis in detail. Sections 4 give its experimental evaluation and simulations results. Section 5 discusses mathematical analysis of the model. Lastly, Sect. 6 concludes the paper.

2 Related Work

According to the principle of guilt, a person feels responsible for his act, if he possesses certain mental attitude towards his action. In cyberspace, this mental attitude is defied. Statistics have shown that 61% of perpetrators are aged between 10–15 years and only 37% had an age more than 17 years [5]. The background literature covers two streams: the neurological and the behavioral aspects of the bullying behavior. Some systematic studies were performed to study a human brain, which aims at intentional harm to others. Strong decrease was reported in the ventromedial PFC, and less coupling was found with amygdala, temporal poles and the reward processing areas [14–16]. During early adolescence of a perpetrator, the prefrontal cortex is consistently active, but the activity is reduced, which reflects his behavioral addictions as compared to controls [17]. Another MRI-based study indicated that biological determinants in the posterior hippocampus of a person determine the depressed or violent behavior [18]. Terms related to a person with such mental impairment causing negative social behavior towards others are, e.g., lunatic, anti-social and psychopath.

Interestingly, the behavior of a bully is closely related to an aggressive conduct disorder or psychopathy. These disorders are associated with pain inflicting behavioral features, like sensation seeking, impulsiveness or aggression towards others. Empathy and sympathy are reactions which hinder such behaviors [19, 20]. A person experiencing such a disorder is not only impaired from the discussed neurological viewpoint but also lack feelings of fear and sympathy for others [21, 22]. Such behavior can be repetitive and is not characterized as acceptable within any society [21]. Therefore, cases with psychopathy were tried to be controlled with education, but not much difference was observed [22]. Moreover, few surveys were conducted for online bullying among adolescents. They showed that cyberbullying has a strong correlation with empathy, delinquent behavior and low-self-esteem [23–25]. As a result, bully-victims face depression and anxiety, which can lead to self-damage [11]. Mostly, cyberbullying related studies were conducted quantitatively [23–26], and didn't address any correlation of mental states such as emotions for this behavior.

Emotions are related to certain mental processes and human behavior in a bidirectional manner [27 (Ch. 3)]. A human is responsible for his behavior or action based on his feelings and internal predictive simulations [28]. A temporal-causal model with positive or negative feelings was presented in which feeling is modeled with its response (emotion) [27 (Ch. 6)]. Likewise, a person can also feel these emotions, and he can predict in how far these emotions are positive or negative; this is known as prediction, and may lead to a decision for an option [27 (Ch. 6)]. Furthermore, a temptation for a particular action is also based on the activation values of associated feelings [29].

In cyberspace, online anonymity may weaken the ability of a person to justify and regulate his behavior [30]. This attribute plays an important role in eliminating empathy and risk or fear from victim's response. Therefore, to study and model the mental states of a bully, we need to consider feelings (like pleasure and influence or risk), emotions, predictions and control states in a unified way. Based on our findings in the literature, we designed our model which is inspired by models related to empathy and emotions [27].

3 The Temporal-Causal Network Model

This section presents a temporal-causal network model designed to study the behavior of bully (see Fig. 1 for a conceptual representation) along with its mathematical representation. A temporal causal model has a conceptual representation, which represents in a declarative manner states along with their connections, which define a causal relationship between states. To illustrate it further, let's assume two states A and B, and there exists a causal relation $A \rightarrow B$. This implies that, if at a time point t, state A occurs then after certain time $t + \Delta t$ state B is affected.

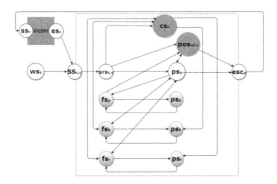

Fig. 1. Temporal-causal network model of a bully: conceptual representation.

Each state in a model is modeled with respect to a particular domain or problem, and has an activation value. This value varies over time, and is computed by three elements: connection weights, speed factors, and combination functions [27]:

Connection weights $\omega_{X,Y}$ specify the strength of the connection from state X to state Y. The value ranges between 0 and 1, but sometimes it may have value below 0 to represent a negative or suppressing effect of X on Y.

Speed factors η_Y indicate how fast a state Y is changing its value upon some causal impact; values range from [0–1].

Combination functions $c_Y (..)$ are used to compute the causal impact of all connected states X_i together on state Y. This impact is called aggregated impact on Y.

Figure 1 displays the conceptual causal representation of the model of the bullying behavior of a person. The bully-victim was modeled by taking a simple situation of bullying to support model for the perpetrator. Table 1 describes the nomenclature of the designed model, depicted in Fig. 1. The model was designed based on the theories addressed in Sect. 2. These theories mainly include the role of pre-frontal cortex and amygdala [14–16, 22]. Our model also incorporate theories related to lack of empathy and power imbalance along with anonymity of a perpetrator [21, 32].

Table 1. Nomenclature of the model

Notation	Description
ws_s	World state stimulus s (s relates to the bully-victim)
$ss_{s,v}$	Sensor state stimulus s
$srs_{s,v}$	Sensory representation state for stimulus
ss_v	Sensor state of victim
es_v	Emotion state of victim
cs_n	Control state with negative feelings
cs_p	Control state with positive feelings
ps_a	Preparation state of action (expression of bullying)
esc_a	Expression of action (bullying)
fs_p	Feeling state for pleasure feeling p
fs_b	Feeling state for his own feeling b
fs_r	Feeling state for risk feeling r
ps_p	Preparation state for pleasure p
ps_b	Preparation state for feeling b
ps_r	Preparation state for risk r

Bullying is not an instantaneous process; the stimulus is received when a victim is spotted. This process depends on the experience based on three feelings studied in the literature. First, his own empathic feeling with respect to the other person (Am I doing right?). Second, the risk feeling (Will someone respond to me?) and third, the pleasure seeking feeling, which he gets by bullying (I am enjoying teasing him). Two kinds of control states are stimulated after receiving signals from these feelings. A negative control state is related to the bully's empathic feeling for the victim and the feeling of risk, while a predicted feeling of pleasure stimulates the prior ownership state.

State fs_b indicates feeling empathy for the other person, which a bully lacks. Similarly state fs_r indicating feeling 'risk', is also absent in a bully, because he feels himself protected by being anonymous. In other words, he lacks such normally occurring feelings; it is assumed that control state cs_n has a suppression effect on such feelings that do not support bullying behavior. Furthermore, getting a pleasure feeling indicated by fs_p (because of power imbalance), encourages the preparation for bullying behavior ps_a via control state cs_p; finally, the bully is ready for perpetration and performs the communication action a: esc_a. The bidirectional arrows between state cs_n and fs_b or fs_r indicate that monitoring and regulation take place in the model.

For computation of the activation levels over time it is assumed that values of speed factors and connection weights are given between 0 and 1, and combination functions are given for all states. Most connection weights are positive. However, there is suppression as well: negative connections $fs_b \rightarrow ps_a$ and $fs_r \rightarrow ps_a$. Furthermore, cs_n also has negative connections towards negative feelings fs_b, fs_r, ps_b, ps_r. To determine the aggregated causal impact of the states, we used different combination functions. For the states ws_s, ss_v, es_v, and $srs_{s,v}$ the identity function $\mathbf{id}(V) = V$ was used. While, for the execution state esc_a the advanced logistic sum function was used:

$$\mathbf{alogistic}_{\sigma,\tau}(V_1, \ldots, V_k) = [(1/(1 + e^{-\sigma(V_1 + \cdots + V_{k-\tau})})) - 1/(1 + e^{\sigma\tau})](1 + e^{-\sigma\tau})$$

Here V and V_i are variables for single impacts $\omega_{X,Y} X(t)$ and $\omega_{X_i,Y} X_i(t)$; see below.

The rest of the states ($ss_{s,v}$, ps_a, fs_b, fs_p, fs_r, cs_n, and cs_p) are mainly the internal states of a bully; for them two options have been explored: the above advanced logistic sum function (Variant-I), and the scaled sum function $\mathbf{ssum}_\lambda(V_1, \ldots, V_k) = (V_1 + \ldots + V_k)/\lambda$, where λ is the scaling factor, which usually gets the value of the sum of the incoming weights (Variant-II).

A conceptual representation of a temporal-causal model can be transformed into a numerical representation as follows [27]:

1. For each time point t every state X has an activation level between 0 and 1, indicated by $X(t)$.
2. The impact of state X on state Y at time point t is computed as
   $\mathbf{impact}_{X,Y}(t) = \omega_{X,Y} X(t)$ where $\omega_{X,Y}$ is the weight of the connection from X to Y.
3. The aggregated impact on Y is determined by the (multiple) states X_1 to X_k with outgoing connections to Y based on the combination function of Y by

$$\mathbf{aggimpact}_Y(t) = \mathbf{c}_Y\left(\mathbf{impact}_{X_1,Y}(t), \ldots, \mathbf{impact}_{X_k,Y}(t)\right)$$
$$= \mathbf{c}_Y(\omega_{X_1,Y} X_1(t), \ldots, \omega_{X_k,Y} X_k(t))$$

4. The effect on Y is exerted gradually using speed factor η_Y to observe the causal effect:

$$Y(t + \Delta t) = Y(t) + \eta_Y[\mathbf{aggimpact}_Y(t) - Y(t)] \text{ and } \mathbf{d}Y(t)/\mathbf{d}t$$
$$= \eta_Y[\mathbf{aggimpact}_Y(t) - Y(t)] \tag{1}$$

Therefore the following difference and differential equations are obtained

$$Y(t + \Delta t) = Y(t) + \eta_Y[\mathbf{c}_Y(\omega_{X_1,Y} X_1(t), , \ldots, \omega_{X_k,Y} X_k(t)) - Y(t)]$$
$$\mathbf{d}Y(t)/\mathbf{d}t = \eta_Y[\mathbf{c}_Y(\omega_{X_1,Y} X_1(t), , \ldots, \omega_{X_k,Y} X_k(t)) - Y(t)]$$

where $\mathbf{c}_Y(\ldots)$ is the combination function of Y. Software environments in Matlab and in Python are available (and have been used) that automate the above steps and enable simulation experiments.

4 Simulation Experiments

This section describes how simulation experiments were set up on the basis of empirical data to evaluate the designed model. Most studies related to cyberbullying were conducted using questionnaires [23–26]. However, these questionnaires were limited to opinions or reactions of a group of people, and didn't provide any numerical empirical data related to mental states of a bully. Due to scarcity of numerical data, we studied predicted qualitative patterns and requirements [33, 35] of a bully from the empirical literature mentioned earlier, and derived expected behavioral patterns for the model enlisted in Table 2. These patterns were considered as requirements for the model. Such patterns are the basis for use cases for the model, a notion from Software Engineering [33]. The requirements were given to a subject matter expert to extract expected patterns for the model (Fig. 1). We took 10 discrete data points for each state for different time. Later on, we populated for more data points using a 'cubic' interpolation technique shown in Fig. 2. It illustrates the requirements, as a use case should do.

Table 2. Requirement patterns based on empirical literature

Empirical literature	Derived pattern
Stimulus triggers emotion [28]	Stimulus is present ($ws_s = 1$), $ss_{s,v}$, $srs_{s,v}$ and ps_a become active
Emotions expressed through actions to express feelings [28]	Feeling related states fs_b, fs_r, fs_p, ps_b, ps_r, ps_p are active
Emotions work when images (feelings) are processed in a brain. [28]	Feeling related states activate before execution of action esc_a
This behavior is sensation seeking [16, 32]	Pleasure states fs_p and ps_p have high values, along with $pos_{self,a}$
Lack fear and empathy [21, 22, 32]	Empathy fs_b, ps_b and fear fs_r, ps_r have low values, along with cs_n
Too much/too little activation give rise to excessive negative affect [38]	High pleasure and low risk and empathy give rise to this behavior, with an increase in esc_a
Emotion regulation [38]	risk and empathy feelings are suppressed but not the pleasure feeling
Predictive and inferential processes contribute to ownership [39]	Prior-ownership state $pos_{self,a}$ is involved in prediction and it contributes to action and vice versa

The negative feelings (bully's own empathic feeling fs_b about the act or the feeling of risk of response: fs_r) weakens, while positive feelings (feeling for pleasure or power: fs_p) strengthen the effect. In Fig. 2, it can be clearly seen that fs_p overplays fs_b and fs_r and as a reaction, the bully is stimulated to perpetrate his victim esc_a.

Given the empirical requirements and the use case discussed above, parameter tuning searches for suitable parameter values, so that the computational model approximates the use case. To find optimum values for speed factors for model variants, we used optimization. Many optimization techniques like gradient descent, simulated annealing, evolution and swarm-based algorithms are well-known for their accuracy and response time [31, 34, 36]. We selected the Simulated Annealing method for optimization of η_Y for all states Y of the model [37]. We performed parameter tuning (Table 3) based on the above data set in fact for both variants, and received values for η after 14909 and 14837 iterations for Variant-I and Variant-II respectively. The corresponding linear error values were RMSE = 0.18, which are reasonable deviations. So, by this method well-tuned parameter values for the speed factors were found. Note that the other parameters were preset by hand: connection weights and parameters steepness $\sigma = 10$ and threshold $\tau = 0.3$ of the logistic combination function.

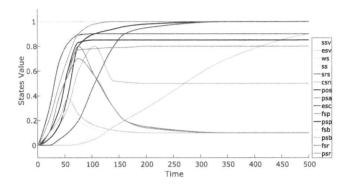

Fig. 2. Specified use case for requirement patterns from the empirical literature

For the different connection weights for all states it was as follows. A state receiving a single input was assigned a connection weight of 1. In Variant-I, all positive connections have connection weights in the 0–1 interval. The suppression connections have connection weights in the −1 to 0 interval. For Variant-II we considered positive connections with weight = 1, and all suppression connection weights as 0.5. Figure 3 presents example simulations of the model based on the parameter values discussed above. The values of speed factor η are taken from Table 2, with step size $\Delta t = 0.5$.

Table 3. Speed factor settings after the parameter tuning

Parameter	ss_v	es_v	ws_s	$ss_{s,v}$	$srs_{s,v}$	cs_n	cs_p	ps_a
Variant-I	0.66	0.65	0.0	0.23	0.2	0.03	0.04	0.14
Variant-II	0.58	0.62	0.0	0.3	0.23	0.01	0.01	0.56
Parameter	esc_a	fs_p	ps_p	fs_b	ps_b	fs_r	ps_r	
Variant-I	0.008	0.32	0.31	0.081	0.08	0.77	0.74	
Variant-II	0.002	0.6	0.6	0.3	0.17	0.7	0.8	

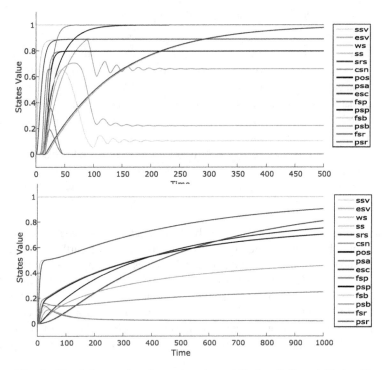

Fig. 3. Simulation results after tuning: (upper) Variant-I, (lower) Variant-II

5   Mathematical Analysis of the Model

Simulation experiments are helpful in studying and verifying the dynamic properties of a model. However, few properties like equilibria, monotonicity and limit cycles, can also be mathematically computed. Equilibrium of the model is one of the useful properties, often used for model verification. A model is in equilibrium at t, if it has stationary points for all states in the model (where $dY(t)/dt = 0$). An equation for stationary points is obtained from (1) as follows [27], Ch. 12:

$$\mathbf{aggimpact}_Y(t) - Y(t) = 0 \Leftrightarrow c_Y\big(\omega_{X_1,Y}X_1(t),\ldots,\omega_{X_k,Y}X_k(t)\big) = Y(t) \qquad (2)$$

For the bullying behavior in Variant-I all states start to increase monotonically. However, after time point $t > 57$, the model starts to converge, and all states start to reach their stationary levels. Equation 2 was used to compute the deviation between left-hand side and right-hand side for simulated data. It was observed that no state had a deviation value higher than 0.02, as indicated in Table 4. In Fig. 3a, it can be seen that as the model receives a stimulus and the values of all states start to increase. The model end up in equilibrium (at $t = 994$), as all states have stationary points, thus model is in equilibrium. For Variant-II, the model ended up in equilibrium after $t = 1598$. The analysis of equilibria of model gives the evidence that the model was correctly implemented.

Table 4. Computation of stationary points for Variant-I

State	ss_v	es_v	$ss_{s,v}$	$srs_{s,v}$	cs_n	cs_p	ps_a
Time t	545	540	43	47	297	161	70
$Y(t)$	0.9854	0.9846	0.8881	0.8886	0.6587	0.9977	0.9991
aggimpact$_Y(t)$	0.9855	0.9847	0.8886	0.8885	0.6558	1	1
Deviation	0.0001	0.0001	0.0005	−0.0001	−0.002	0.002	0.0009

State	esc_a	fs_p	ps_p	fs_b	ps_b	fs_r	ps_r
Time t	565	48	52	275	270	48	48
$Y(t)$	0.9877	0.7972	0.7970	0.1018	0.2209	0.0004	0.0009
aggimpact$_Y(t)$	1	0.7975	0.7975	0.1018	0.2209	0	0
Deviation	0.012	0.0003	0.0005	0.0	0.0	−0.0004	−0.0009

Systems of such equations can be solved numerically and sometimes also algebraically. For a bully, the 11 equilibrium equations have been further analyzed and solved for some cases. For the bully, state $srs_{s,v}$ serves as an input (assumed to have value A in the equilibrium) and state esc_a is generated as output depending on A. State esc_a uses involved connection weights and state values, the steepness σ and threshold parameters τ for the advance logistic combination function, but has no impact on these states. Table 5 shows the relevant connection weights for each state (ω_1 to ω_{24}). From the criterion (2) for equilibria mentioned above the following equilibrium equations were obtained:

Eq1. $srs_{s,v} = A$

Eq2. $(\omega_1 + \omega_2)\, ps_a = \omega_1\, srs_{s,v} + \omega_2\, cs_p + \omega_3\, fs_b + \omega_4\, fs_r$

Eq3. $(\omega_5 + \omega_6)\, fs_p = \omega_5\, ps_a + \omega_6\, ps_p$

Eq4. $(\omega_8 + \omega_9)\, fs_b = \omega_7\, cs_n + \omega_8\, ps_a + \omega_9\, ps_b$

Eq5. $(\omega_{11} + \omega_{12})\, fs_r = \omega_{10}\, cs_n + \omega_{11}\, ps_a + \omega_{12}\, ps_r$

Eq6. $ps_p = \omega_{13}\, fs_p$

Eq7. $ps_b = \omega_{14}\, cs_n + \omega_{15}\, fs_b$

Eq8. $ps_r = \omega_{16}\, cs_n + \omega_{17}\, fs_r$

Eq9. $(\omega_{18} + \omega_{19} + \omega_{20})\, cs_n = \omega_{18}\, srs_{s,v} + \omega_{19}\, fs_b + \omega_{20}\, fs_r$

Eq10. $(\omega_{21} + \omega_{22})\, pos_{self,a} = \omega_{21}\, srs_{s,v} + \omega_{22}\, fs_p$

Eq11. $esc_a = \mathbf{alogistic}_{\sigma,\tau}(\omega_{23}\, pos_{self,a}, \omega_{24}\, ps_a)$

To get more insight, for a number of cases most connection weights were set to 1: the value 1 was assigned to the following connection weights ω_1, ω_2, ω_6, ω_9, ω_{12}, ω_{13}, ω_{15}, ω_{17}, ω_{18}, ω_{21}, ω_{23}, ω_{24}, ω_{25}. Then 12 connections weights are left as parameters: 6 negative weights (ω_3 and ω_4 for suppression of ps_a by negative feelings fs_b and fs_r, ω_7 and ω_{10} for suppression of fs_b and fs_b by cs_n, and ω_{14} and ω_{16} for suppression of ps_b and ps_b by cs_n), and 6 positive weights (ω_5, ω_8, ω_{11} for the feeling associations from the preparation of the action, and ω_{19}, ω_{20}, ω_{22} for the connections from feeling states to

Table 5. Names of connection weights

Connection weight(s)	ω_1 ω_2 ω_3 ω_4	ω_5 ω_6	ω_7 ω_8 ω_9	ω_{10} ω_{11} ω_{12}	ω_{13}	ω_{14} ω_{15}	ω_{16} ω_{17}	ω_{18} ω_{19} ω_{20}	ω_{21} ω_{22}	ω_{23} ω_{24}
From	$srs_{s,v}$ cs_p fs_b fs_r	ps_a ps_p	cs_n ps_a ps_b	cs_n ps_a ps_r	fs_p	cs_n fs_b	cs_n fs_r	$srs_{s,v}$ fs_b fs_r	$srs_{s,v}$ fs_p	$pos_{self,a}$ ps_a
To	ps_a	fs_p	fs_b	fs_r	ps_p	ps_b	ps_r	cs_n	$pos_{self,a}$	esc_a

control state and the ownership state). Then the equilibrium equations become for $\omega_1 = \omega_2 = \omega_6 = \omega_9 = \omega_{12} = \omega_{13} = \omega_{15} = \omega_{17} = \omega_{18} = \omega_{21} = \omega_{23} = \omega_{24} = \omega_{25} = 1$ are:

Eq1. $srs_{s,v} = A$

Eq2. $2\ ps_a = srs_{s,v} + cs_p + \omega_3\ fs_b + \omega_4\ fs_r$

Eq3. $(\omega_5 + 1)\ fs_p = \omega_5\ ps_a + ps_p$

Eq4. $(\omega_8 + 1)\ fs_b = \omega_7\ cs_n + \omega_8\ ps_a + ps_b$

Eq5. $(\omega_{11} + 1)\ fs_r = \omega_{10}\ cs_n + \omega_{11}\ ps_a + ps_r$

Eq6. $ps_p = fs_p$

Eq7. $ps_b = \omega_{14}\ cs_n + fs_b$

Eq8. $ps_r = \omega_{16}\ cs_n + fs_r$

Eq9. $(1 + \omega_{19} + \omega_{20})\ cs_n = srs_{s,v} + \omega_{19}\ fs_b + \omega_{20}\ fs_r$

Eq10. $(1 + \omega_{22})\ pos_{self,a} = srs_{s,v} + \omega_{22}\ fs_p + ps_a$

Eq11. $esc_a = \mathbf{alogistic}_{\sigma,\tau}(pos_{self,a}, ps_a)$

This system of equations has been solved algebraically. For the first 10 linear equations the WIMS Linear Solver[1] was used, and the last equation can be solved by just substituting the values found for $pos_{self,a}$ and ps_a. This provided explicit algebraic solution expressions for the first 10 states; these are shown in Table 6 as Case 1 (cells 1 and 2). It can be seen that the solutions are the value A of $srs_{s,v}$ multiplied by (broken) rational functions (with common denominator) of the connection weights. Some other special cases have been considered and addressed similarly (see Table 6). Case 1 shows the outcomes when all positive weights have the same value ω and all negative weights have the same negative value, indicated by $\underline{\omega}$. As a special case of this, all positive connections were set to 1 and the negative connections to $\underline{\omega}$; see Case 2. Finally, Case 3 shows the case that all positive weights are 1 and all negative weights are -0.5. Here the italic digits indicate the repetitive part. For this case, based on the combination function $\mathbf{alogistic}_{\sigma,\tau}(..)$ with $\sigma = 50$ and $\tau = 1.5$, and assuming $A = 1$, the value of esc_a is 0.999573, which indicates that the bullying action is performed. Values found by this analysis were compared to the values found by simulations and were found the same (up to a deviation $<10^{-2}$), which again provides evidence for the correctness of the implemented model.

---

[1] http://wims.unice.fr/wims/wims.cgi?session=BXC074F7D4.2&+lang=nl&+module=tool%2Flinear%2Flinsolver.en

Table 6. Equilibrium equations: algebraic solutions

Case 1 All positive weights the same ω, all negative weights the same $\underline{\omega}$

$X_1 = \text{srs}_{s,v} = A$

$X_2 = \text{ps}_a = A(9\omega^5 + (6\underline{\omega} - 33)\,\omega^4 + (24 - 4\underline{\omega})\,\omega^3 + (-6\underline{\omega}^2 - 14\underline{\omega} + 12)\,\omega^2 + (6\underline{\omega}^2 - 4\underline{\omega})\,\omega + 12\underline{\omega}^2)$

$/(15\omega^5 + (10\underline{\omega} - 57)\,\omega^4 + (10\underline{\omega} + 54)\,\omega^3 - 54\underline{\omega}\,\omega^2)$

$X_3 = \text{fs}_p = -A(9\omega^4 + (6\underline{\omega} - 15)\,\omega^3 + (8\underline{\omega} - 6)\,\omega^2 + (2\underline{\omega} - 6\underline{\omega}^2)\,\omega - 6\underline{\omega}^2)/(15\omega^5 + (10\underline{\omega} - 57)\,\omega^4 + (10\underline{\omega} + 54)\,\omega^3 - 54\underline{\omega}\,\omega^2)$

$X_4 = \text{fs}_b = -A(9\omega^3 + (5\underline{\omega} - 15)\,\omega^2 + (-4\underline{\omega} - 6)\,\omega - 9\underline{\omega})/(15\omega^4 + (10\underline{\omega} - 57)\,\omega^3 + (10\underline{\omega} + 54)\,\omega^2 - 54\underline{\omega}\,\omega)$

$X_5 = \text{fs}_r = -A(9\omega^3 + (5\underline{\omega} - 15)\,\omega^2 + (-4\underline{\omega} - 6)\,\omega - 9\underline{\omega})/(15\omega^4 + (10\underline{\omega} - 57)\,\omega^3 + (10\underline{\omega} + 54)\,\omega^2 - 54\underline{\omega}\,\omega)$

$X_6 = \text{ps}_p = -A(9\omega^4 + (6\underline{\omega} - 15)\,\omega^3 + (8\underline{\omega} - 6)\,\omega^2 + (2\underline{\omega} - 6\underline{\omega}^2)\,\omega - 6\underline{\omega}^2)/(15\omega^4 + (10\underline{\omega} - 57)\,\omega^3 + (10\underline{\omega} + 54)\,\omega^2 - 54\underline{\omega}\,\omega)$

$X_7 = \text{ps}_b = -A(9\omega^4 + (6\underline{\omega} - 15)\,\omega^3 + (5\underline{\omega} - 6)\,\omega^2 + (-6\underline{\omega}^2 - 31\underline{\omega})\,\omega + 12\underline{\omega}^2)/(15\omega^4 + (10\underline{\omega} - 57)\,\omega^3 + (10\underline{\omega} + 54)\,\omega^2 - 54\underline{\omega}\,\omega)$

$X_8 = \text{ps}_r = -A(9\omega^4 + (6\underline{\omega} - 15)\,\omega^3 + (5\underline{\omega} - 6)\,\omega^2 + (-6\underline{\omega}^2 - 31\underline{\omega})\,\omega + 12\underline{\omega}^2)/(15\omega^4 + (10\underline{\omega} - 57)\,\omega^3 + (10\underline{\omega} + 54)\,\omega^2 - 54\underline{\omega}\,\omega)$

$X_9 = \text{cs}_n = -A(\omega^3 + 9\underline{\omega}^2 + (-6\underline{\omega} - 22)\,\omega + 12\underline{\omega})/(15\omega^4 + (10\underline{\omega} - 57)\,\omega^3 + (10\underline{\omega} + 54)\,\omega^2 - 54\underline{\omega}\,\omega)$

$X_{10} = \text{pos}_{self,a} = A(3\omega^5 + (2\underline{\omega} - 9)\,\omega^4 - 6\omega^3 + (-2\underline{\omega}^2 - 4\underline{\omega} + 24)\,\omega^2 + (4\underline{\omega}^2 - 20\underline{\omega})\,\omega + 6\underline{\omega}^2)/(15\omega^5 + (10\underline{\omega} - 57)\,\omega^4 + (10\underline{\omega} + 54)\,\omega^3 - 54\underline{\omega}\,\omega^2)$

Case 2 All positive weights 1, all negative weights the same $\underline{\omega}$

$X_1 = \text{srs}_{s,v} = A$ $X_2 = \text{ps}_a = -A(6\underline{\omega}^2 - 8\underline{\omega} + 6)/(17\underline{\omega} - 6)$

$X_3 = \text{fs}_p = -A(6\underline{\omega}^2 - 8\underline{\omega} + 6)/(17\underline{\omega} - 6)$ $X_4 = \text{fs}_b = -A(4\underline{\omega} + 6)/(17\underline{\omega} - 6)$

$X_5 = \text{fs}_r = -A(4\underline{\omega} + 6)/(17\underline{\omega} - 6)$ $X_6 = \text{ps}_p = -A(6\underline{\omega}^2 - 8\underline{\omega} + 6)/(17\underline{\omega} - 6)$

$X_7 = \text{ps}_b = A(3\underline{\omega}^2 - 10\underline{\omega} - 6)/(17\underline{\omega} - 6)$ $X_8 = \text{ps}_r = A(3\underline{\omega}^2 - 10\underline{\omega} - 6)/(17\underline{\omega} - 6)$

$X_9 = \text{cs}_n = A(3\underline{\omega} - 6)/(17\underline{\omega} - 6)$ $X_{10} = \text{pos}_{self,a} = -A(4\underline{\omega}^2 - 11\underline{\omega} + 6)/(17\underline{\omega} - 6)$

$X_{11} = \text{esc}_a = \textbf{alogistic} - A(4\underline{\omega}^2 - 11\underline{\omega} + 6)/(17\underline{\omega} - 6), -A(6\underline{\omega}^2 - 8\underline{\omega} + 6)/(17\underline{\omega} - 6)$

Case 3 All positive weights 1, all negative weights -0.5

$X_1 = \text{srs}_{s,v} = A$ $X_2 = \text{ps}_a = 0.79310344827586217A$

$X_3 = \text{fs}_p = 0.79310344827586217A$ $X_4 = \text{fs}_b = 0.2758620689655172A$

$X_5 = \text{fs}_r = 0.2758620689655172A$ $X_6 = \text{ps}_p = 0.79310344827586217A$

$X_7 = \text{ps}_b = 0.017241379310344482A$ $X_8 = \text{ps}_r = 0.017241379310344482A$

$X_9 = \text{cs}_n = 0.517241379310344492A$ $X_{10} = \text{pos}_{self,a} = 0.86206896551724131A$

$X_{11} = \text{esc}_a = \textbf{alogistic}\ 0.86206896551724131A,\ 0.79310344827586217A$

6   Conclusion and Future Work

This paper presents a temporal-causal model, which describes bullying behavior of a person. It seems that in the literature no computational models can be found on cyberbullying. The design of the model and its simulations were based on neurological and behavioral literature, which reported that a bully does not have very correlated PFC and amygdala. Furthermore, he lacks empathy or sympathy for the victim [14, 22]. Two variants were simulated, which showed that the logistic and scaled sum function can approximate each other. Mathematical analysis was also performed on resulting equilibria of the model, by which it was verified that the model was designed and implemented as it was intended. The model is generic so that it can serve as a basis for modeling negative social behavior towards others. It can also be used as a basis to explore different traits of a bully and his victim or ways to support this behavior. Moreover, effects of different social parameters can be observed by different sets of simulations and data.

In future, we aim to use this model within different application contexts, which will act as a solid basis for the knowledgeable intelligent behavior of perpetrator and will be a useful input to provide support to the participants through artificially engineered processes.

References

1. Myburgh, J.-E.: Examining the Relationships between Aggression, Bullying, and Cyber-bullying among University Students in Saskatchewan. Diss. (2017)
2. Chapin, J., Coleman, G.: The cycle of cyberbullying: some experience required. Soc. Sci. J. **54**(3), 314–318 (2017)
3. Ryan v. Mesa unified school district, No. 2: 14-cv-01145 JWS (D. Ariz. July 20, 2016)
4. Diperna v. The chicago school of professional psychology, No. 14-cv-57 (N.D. Ill. Nov. 28, 2016)
5. Bajovic, V.: Criminal proceedings in cyberspace: the challenge of digital era. In: Viano, E.C. (ed.) Cybercrime, Organized Crime, and Societal Responses, pp. 87–101. Springer, Cham (2017). https://doi.org/10.1007/978-3-319-44501-4_5
6. Nuccitelli, M.: Cyber bullying tactics. In: Forensic Examiner, Springfield, vol. 21, no. 3, pp. 24–27 (2012)
7. https://www.ipredator.co/michael-nuccitelli/. Accessed 19 Jan 2017
8. https://cyberbullying.org/new-national-bullying-cyberbullying-data. Accessed 11 Dec 2017
9. Tsitsika, A., et al.: Cyberbullying victimization prevalence and associations with internal-izing and externalizing problems among adolescents in six European countries. Comput. Hum. Behav. **51**, 1–7 (2015)
10. Bosworth, K., et al.: The impact of lead-ership involvement in enhancing high school climate and reducing bullying: an exploratory study. J. Sch. Violence **17**(3), 354–366 (2018)
11. Holmes, R.: Eradicating Cyber Bullying: Through Online Training, Reporting & Tracking System. AuthorHouse (2017)
12. Finkelhor, D., Mitchell, K., Wolak, J.: Online victimisation: a report on the nation's youth. National Center for Missing & Exploited Children. www.unh.edu/ccrc/YouthInternetinfo page.html. Accessed 7 Feb 2007

13. Hinduja, S., Patchin, J.W.: Bullying, cyberbullying, and suicide. Arch. Suicide Res. **14**(3), 206–221 (2010)
14. Decety, J., Porges, E.C.: Imagining being the agent of actions that carry different moral consequences: an fMRI study. Neuropsychologia **49**(11), 2994–3001 (2011)
15. Young, L., et al.: Damage to ventromedial prefrontal cortex impairs judgment of harmful intent. Neuron **65**(6), 845–851 (2010)
16. Schiffer, B., et al.: Disentangling structural brain alterations associated with violent behavior from those associated with substance use disorders. Arch. Gen. Psychiatry **68**, 1039–1049 (2011)
17. Meldrum, R.C., et al.: Brain activity, low self-control, and delinquency: an fMRI study of at-risk adolescents. J. Crim. Justice **56**, 107–117 (2017)
18. Laakso, M.P., et al.: A volumetric MRI study of the hippocampus in type 1 and 2 alcoholism. Behav. Brain Res. **109**, 177–186 (2000)
19. Eisenberg, N.: Age changes in prosocial responding and moral reasoning in adolescence and early adulthood. J. Res. Adolesc. **15**, 235–260 (2005)
20. Hoffman, M.L.: Empathy and Moral Development. Cambridge University Press, New York (2000)
21. Laakso, M.P., Gunning-Dixon, F., Vaurio, O., Repo-Tiihonen, E., Soininen, H., Tiihonen, J.: Prefrontal volumes in habitually violent subjects with antisocial personality disorder and type 2 alcoholism. Psychiatry Res. Neuroimaging **114**, 95–102 (2002)
22. Boccardi, M., et al.: Cortex and amygdala morphology in psychopathy. Psychiatry Res. Neuroimaging **193**, 85–92 (2011)
23. Ildırım, E., Çalıcı, C., Erdoğan, B.: Psychological correlates of cyberbullying and cyber-victimization. Int. J. Hum. Behav. Sci. **3**(2), 7–21 (2017)
24. Barlińska, J., Szuster, A., Winiewski, M.: Cyberbullying among adolescent bystanders: Role of the communication medium, form of violence, and empathy. J. Community Appl. Soc. Psych. **23**(1), 37–51 (2013)
25. Jolliffe, D., Farrington, D.F.: Is low empathy related to bullying after controlling for individual and social background variables? J. Adolesc. **34**(1), 59–71 (2011)
26. Shultz, E., Heilman, R., Hart, K.J.: Cyber-bullying: an exploration of bystander behavior and motivation. Cyberpsychol. J. Psychosoc. Res. Cyberspace **8**(4), Article 3 (2014)
27. Treur, J.: Network-Oriented Modeling. Springer, Cham (2016). https://doi.org/10.1007/978-3-319-45213-5
28. Damasio, A.R.: Self Comes to Mind: Constructing the Conscious Brain. Pantheon Books, New York (2010)
29. Bosse, T., et al.: A computational model for dynamics of desiring and feeling. Cogn. Syst. Res. **19**, 39–61 (2012)
30. McKenna, K., Bargh, J.: Plan 9 from cyberspace: the implications of the Internet for personality and social psychology. Pers. Soc. Psychol. Rev. **4**, 57–75 (2000)
31. Černý, V.: Thermodynamical approach to the traveling salesman problem: an efficient simulation algorithm. J. Optim. Theory Appl. **45**(1), 41–51 (1985)
32. Santana, E.J.: The brain of the psychopath: a systematic review of structural neuroimaging studies. Psychol. Neurosci. **9**(4), 420 (2016)
33. Pohl, K., Rupp, C.: Requirements Engineering Fundamentals (Rocky Nook, 2011)
34. Kirkpatrick, S., Gelatt Jr., C., Vecchi, M.: Optimization by simulated annealing. Science **220** (4598), 671–680 (1983)
35. Ferber, J., Gutknecht, O., Jonker, C.M., Müller, J.P., Treur, J.: Organization models and behavioural requirements specification for multi-agent systems. In: Demazeau, Y., Garijo, F. (eds.) Proceedings of MAAMAW 2001 (2001)
36. Bertsimas, D., Tsitsiklis, J.: Simulated annealing. Stat. Sci. **8**(1), 10–15 (1993)

37. Mollee, J.S., Araújo, E.F.M., Klein, M.C.A.: Exploring parameter tuning for analysis and optimization of a computational model. In: Benferhat, S., Tabia, K., Ali, M. (eds.) IEA/AIE 2017. LNCS (LNAI), vol. 10351, pp. 341–352. Springer, Cham (2017). https://doi.org/10.1007/978-3-319-60045-1_36
38. Davidson, R.J., Putnam, K.M., Larson, C.L.: Dysfunction in the neural circuitry of emotion regulation–a possible prelude to violence. Science **289**(5479), 591–594 (2000)
39. Moore, J., Haggard, P.: Awareness of action: inference and prediction. Conscious. Cogn. **17**(1), 136–144 (2008). p. 142

Recommendation Methods and Recommender Systems

A Hybrid Feature Combination Method that Improves Recommendations

Gharbi Alshammari[1]([✉]), Stelios Kapetanakis[1], Abduallah Alshammari[1],
Nikolaos Polatidis[1], and Miltos Petridis[2]

[1] School of Computing, Engineering and Mathematics, University of Brighton,
Moulsecoomb Campus, Lewes Road, Brighton, BN2 4GJ, UK
{g.alshammari,s.kapetanakis,a.alshammari1,
n.Polatidis}@brighton.ac.uk
[2] Department of Computer Science, Middlesex University London,
The Burroughs, London, NW4 4BT, UK
M.Petridis@mdx.ac.uk

Abstract. Recommender systems help users find relevant items efficiently based on their interests and historical interactions. They can also be beneficial to businesses by promoting the sale of products. Recommender systems can be modelled by applying different approaches, including collaborative filtering (CF), demographic filtering (DF), content-based filtering (CBF) and knowledge-based filtering (KBF). However, large amounts of data can produce recommendations that are limited in accuracy because of diversity and sparsity issues. In this paper, we propose a novel hybrid approach that combines user-user CF with the attributes of DF to indicate the nearest users, and compare the Random Forest classifier against the kNN classifier, developed through an investigation of ways to reduce the errors in rating predictions based on users past interactions. Our combined method leads to improved prediction accuracy in two different classification algorithms. The main goal of this paper is to identify the impact of DF on CF and compare the two classifiers. We apply a feature combination hybrid method that can improve prediction accuracy and achieve lower mean absolute error values compared with the results of CF or DF alone. To test our approach, we ran an offline evaluation using the 1 M MovieLens data set.

Keywords: Recommender systems · Collaborative filtering
Demographic filtering · Hybrid system

1 Introduction

The amount of available information on the Internet has increased exponentially in the last decade and this has led to the problem of information overload. More specifically, the E-commerce industry is presenting a wider range of options, which makes it more difficult for users to shop and find products. Hence, to help customers find new items by means of suggestions, companies need to develop a recommender system. Such systems help their sales to grow by providing relevant options that meet users requirements. For example, in regards to movie recommendation, the Netflix Prize raised the importance of

© Springer Nature Switzerland AG 2018
N. T. Nguyen et al. (Eds.): ICCCI 2018, LNAI 11055, pp. 209–218, 2018.
https://doi.org/10.1007/978-3-319-98443-8_19

the recommender system in attracting more users, and the competition to produce highly developed algorithms led to more accurate results in recommendations [2]. A recommendation system is an information filtering task that is used to predict the items a certain user will like (the prediction problem) or to recommend a set of top items that meet the users preferences [22]. Users have trouble handling large volumes of information, and problems with cognitive and data sparsity when attempting to find appropriate information at the right time [5]. Based on profile data, a recommender system can be categorized into four main stages: similarity computation, neighbourhood selection, prediction and recommendation. The profile can be modelled by content-based, collaborative or demographic filtering. If the user profile contains a set of attributes obtained from the item descriptions that the user has liked, this is content-based filtering (CBF). Demographic filtering (DF) is represented by a set of features in a users profile. Collaborative filtering (CF) can be described as when the profile contains a list of items that have been rated. The CF technique is widely used as a recommender system base method due to its capability and the efficiency of predicting similar neighbour users. However, extensive growth in the number of users and items may cause a sparsity issue in CF techniques when used on their own. Thus, we have made the following contributions:

1. We have developed a method that exploits both ratings and demographic information, by combining demographic attributes with user-item rating CF to solve the problem of data sparsity.
2. This method allows us to efficiently calculate similarities in a large dataset with no pre-calculating or pre-processing.
3. We have evaluated our method using a real dataset and we have shown that it is both practical and effective.

2   Related Work

Recommender systems were first researched in the mid-1990s, relying on the idea that users share similar items or opinions, thereby helping to make recommendations to others [20]. The researchers established a collaborative filtering technique based on ratings structure. Hence, the most common formulation is to calculate ratings for items that have not been seen by a particular user. Recommender systems can be defined as adaptable tools that help users search for, filter and classify information, and then recommend relevant items [21]. Recommendation systems use a number of different techniques. These methods can be implemented based on the domain requirement and are able to identify and predict items that meet the users interests. They also utilize different recommendation algorithms to make suggestions and recommendations.

2.1   Collaborative Filtering

Collaborative filtering is considered to be the most popular technique for recommender systems. It has been widely implemented in different domains to make recommendations. It is a method of information filtering that seeks to predict the rating that a user would give to a particular item based on a similarity matrix. Collaborative filtering provided the foundation for the first recommender systems, which were used to help people make choices based on the opinions of other people [10]. It helps users to find relevant items and makes suggestions based on similar users tastes. It has been applied in a variety of areas, such as in regards to movies, books and research articles. In this approach the similarity calculation is based on the users peers. User-based CF: this method looks for similarity between users based on the same rating pattern [23]. It makes a recommendation based on the similarity between the target user and other users. The idea is that, for a given user, the preferences of similar users (neighbours) can serve as recommendations. A user-user approach was proposed as an appropriate method for recommending items based on expert opinions [3]. In addition, another example is provided by [29], where mobile activities were recommended to users based on their locations. Item-based CF: this method recommends items based on similarities between items shared with similar users [23]. In [14] an item to item collaborative filtering approach was designed that matched items rated or purchased by the user with other similar items.

2.2 Demographic Filtering

It is possible to identify the type of person that likes a particular item by referencing their demographic details [18]. User attributes are incorporated into demographic recommender systems and this demographic data is used as a basis for arriving at suitable recommendations, sometimes relying on pre-generated demographic clusters [25]. This information is gathered either explicitly through user registrations or implicitly via navigation of the system they use [17]. Subsequently, demographically similar users are identified by means of the recommendation algorithm. Recommendations are based on how similar people (in terms of their demographics) rated a particular item [24]. In [25] a hybrid algorithm was presented that keeps the core ideas of two existing recommender systems and enhances them with relevant information extracted from demographic data. The authors in [18] presented an approach that considers user profiles as vectors constructed from demographic attributes such as age, gender or postcode to find relationships with other users and calculate similarities between users, in order to generate the final prediction. Demographic-based filters are similar to collaborative filters in the sense that both are able to identify similarities between users. In this case, demographic features are used to determine similarity rather than the users previous ratings of items [24]. Demographic attributes are added as meta-data to help the neighbourhood algorithm find similar users. The author in [16], presented the importance of these meta-data in producing significant results and providing better recommendations.

In [19], the authors stated that demographic information helps to address the cold-start problem. This is because this approach does not require a detailed history of user ratings before making recommendations, unlike the content-based and collaborative approaches [9]. studied the importance of demographic information (age, gender) in a research paper recommender system [4]. The authors showed that demographic information had a significant impact on recommendations. The combination of collaborative filtering and the demographic base can enrich user preferences and more accurately identify their interests.

2.3 Hybrid Recommendation Approaches

More recently, the hybrid recommendation approach has become a widely debated issue. A possible way to combine the recommendation methods was introduced in [8]. Authors in [27] also introduced a hybrid approach for solving the problem of finding the rating of unrated items in a user-item matrix through a weighted combination of user-based and item-based collaborative filtering. These methods addressed the two major challenges of recommender systems, the accuracy of recommendations and sparsity of data, by simultaneously incorporating the correlation of users and items.

Because of data sparsity, finding nearest neighbours is becoming more of a challenge, with the fast growth in users and items. In [1] a switching hybrid approach was proposed to solve the long tail problem in recommendations. A hybrid approach was applied that utilised clustering and genetic algorithms to reduce data sparsity in movie recommendations. The results showed that this approach improves recommendation accuracy [26]. In [7] a hybrid framework was proposed that utilized collaborative filtering relying on user/item metadata and demographic data. The framework benefits from the similarity between users via correlation in terms of demographic attributes. This improves prediction and is able to solve the cold start problem compared with the baseline. The author points out the importance of item metadata in overcoming the challenge in which user and item have little information. In [25] the discussion explored the usefulness of demographic data as an enhancing factor, by employing a hybrid algorithm to improve collaborative filtering in terms of both algorithms, user-based and item-based. In [11] a combination algorithm was proposed using demographic attributes based on a clustering approach in a weighted scheme. It solved the cold-start problem by assigning a new user to the nearest cluster based on demographic similarity. Our proposed method is beneficial for exploring the effect of demographic data and makes a comparison between the kNN and the Random Forest classifiers.

3 Proposed Method

In this section, the proposed method is defined that combines CF and DF. The main idea of the method, which is not found in other works in the literature is to have a hybrid recommendation approach that can be easily used for the evaluation of different classifiers in order to identify which classifier performs better when demographic data are integrated into the recommendation process.

The sparsity issue is a major challenge for recommender systems in terms of producing the right recommendations for the right users. This issue has been further expanded due to the growth of items available and of users with few ratings and little user information. This leads to difficulty in finding similarity between two users. In this section, we propose how a feature combination hybrid approach solves the sparsity problem and reduces the error rate through using two classifiers. It combines matching user demographic attributes with the user rating CF method as shown in Fig. 1.

In order to evaluate the proposed method we conducted an experiment on a real dataset that is publicly available from MovieLens [12]. In this paper we used the 1 M dataset which contains 1,000,209 ratings that were assigned by 6,040 users on around

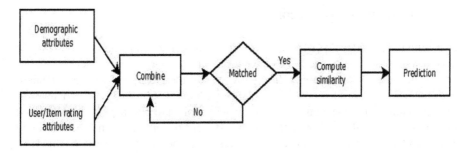

Fig. 1. The Architecture of the proposed method.

3,900 movies. We utilized demographic information that includes age, gender and occupation. We combined the demographic information for each user with user-item ratings. Hence, each user was defined as a vector with those features.

The attributes used in this filtering system are age, gender and occupation. Those attributes are defined as categorical, and represent each user in a group. They can help in finding similar users, in order to improve rating prediction accuracy. The profile vector is represented at the attribute-level to compute the similarity.

Then, we calculated the similarity between the active user and the nearest one. Next is the final step of calculating the predicted rating. We ran this experiment using Orange 3.7.0, which is a data mining and machine learning tool. We conducted a cross-validation with number of folds = 10. In summary, the steps of the proposed method are:

1. CF is combined with demographic attributes such as age, gender and occupation to find more similar users. The combination was made through matching the user ID from user-item ratings data with user demographics data as detailed in algorithm 1 below. Where row[0] and line[0] represent user-id. And, row[0, 1, .. , N] represent the attributes in CF and line[1, 2, ... , N] represent the attributes in DF.
2. After matching each user with the demographic attributes, the similarity is computed using two classifier kNN and Random Forest.
3. The final step is, predicted rating is calculated and compared with the actual rating to calculate the differences.

Algorithm 1. Combined algorithm

```
1: Input: user-item rating attributes file (f1). demographic attributes (f2).
2: Output:  user demographic attributes with item rating (f3).
3: for <row in f2> do
4:     for <line in f1> do
5:         if row [0] == line [0] then
6:             f3 = row [0,1,..,N] + line [1,2,...,N]
7:         end if
8:     end for
9: end for
```
---

4   Experimental Evaluation

Evaluation metrics play an important role in measuring the quality and performance of a recommender system. Since 1994 [20], the accuracy of the recommender system has been evaluated in the literature in different ways. Furthermore, as there is no standard for evaluation, it is hard to compare the results with other published articles. However, there are main evaluation metrics that are widely applied to benchmark the results and compare them with the proposed algorithms. Most of the empirical studies examining recommender systems have focused on appraising the accuracy of these systems [13]. This insight is useful for evaluating the quality of the system and its ability to forecast the rating for a particular item.

4.1   Predictive Accuracy Metrics

This measures similarity between true user ratings and the predicted ratings. This research applies accuracy metrics to measure the performance of the proposed methods. Both the mean absolute error (MAE) in Eq. 1 and the root mean squared error (RMSE) in Eq. 2 is used to evaluate the prediction accuracy of the different recommendation techniques, where p_i is the predicted rating and r_i is the actual rating. Prediction accuracy is enhanced when MAE and RMSE are lower. Here we detail those similarity measures

Mean Absolute Error (MAE) takes the mean of the absolute difference between the predicted rating and actual rating for all the ratings as follows:

$$\text{MAE} = \frac{1}{n} \sum_{i=1}^{n} |p_i - r_i| \tag{1}$$

Root Mean Squared Error (RMSE) represents the sample standard deviation of the differences between predicted values and the actual values:

$$\text{RMSE} = \sqrt{\frac{1}{n} \sum_{i=1}^{n} (p_i - r_i)^2} \tag{2}$$

4.2   Classification Algorithms

In order to find out which classifier is the most appropriate one to use for this dataset and to make a good prediction for the movie domain, we ran an experiment into those two that are widely applied in movie recommendation for evaluating the results. Next, we describe in detail each classifier.

k-Nearest Neighbour (kNN) classifier finds its k nearest neighbours. The given user is assigned to a similar users that shares the most common features of its k nearest neighbour users. Certain factors need to be considered, such as the similarity measurement, which calculates the distance between two vectors p_i and r_i for $i = 1, 2, \ldots, k$ representing the neighbourhood, which needs to be positive number. For that purpose, we use Euclidean distance as follows:

$$d(p, q) = \sqrt{\sum_{i=1}^{k} (p_i - r_i)^2} \tag{3}$$

Random Forest is an ensemble learning classifier that builds a set of decision trees. Each tree is developed from a bootstrap sample from training data. It is more robust with respect to noise [6]. This method has been successfully approved as an accurate machine learning classifier [15]. We set the number of trees to be 10, 20, 50 and 100, which is the most likely change between this range [28].

4.3 Results

As we can see in Fig. 2, the MAE accuracy metrics were made through applying different kNN with $k = 3, 10, 30, 50$ and 100. We then performed the experiment with the Random Forest classifier using a different set of trees. It is clear that performance is improved when we combined the demographic attributes (CF+DF). However, it is noticeable that the improvement in Random Forest is much higher than in kNN. For example, in Fig. 2b, when $T = 10$ the improvement is 5%. Whereas in Fig. 2a, when $k = 3$ the improvement is only 1%.

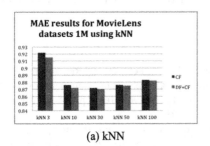
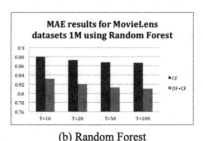

(a) kNN   (b) Random Forest

Fig. 2. MAE results for the MovieLens dataset using kNN and Random Forest

Figure 3 shows the results using RMSE. There is significant improvement in the Random Forest compared to kNN in all sets. For instance, in Fig. 3b when $T = 10$ the collaborative filtering performed 1.11 whereas the combined method it was 1.05. By contrast, in Fig. 3a when $k = 3$ the enhancement is only 0.01.

In general, the Random Forest significantly outperforms the kNN. Therefore, this experiment proves the quality of the Random Forest compared with the kNN classifier using this large dataset. We also ran all the data to make a reliable evaluation and produce an accurate result, and compared it to the benchmark. Our hybrid method outperformed the baseline CF method.

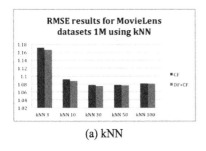
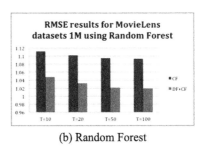

(a) kNN   (b) Random Forest

Fig. 3. RMSE results for the MovieLens dataset using kNN and Random Forest

5   Conclusion

In this paper, we propose a novel hybrid method for recommender systems based on simultaneous combination of user-based collaborative filtering and demographic attributes. The results suggest that demographic filtering can effectively improve the overall recommendation. Moreover, the proposed method addresses two common challenges of recommendation systems, namely sparsity of data and improved accuracy of recommender systems, by combining the hidden relations between users and comparing two different classifiers with a large dataset. The proposed method is a comparison between the Random Forest and kNN classifiers. In future work, we may consider specific users who rate only a few items, or possibly other attributes relating to item representation. In addition, other different classifiers could be used to add a comprehensive comparison.

References

1. Alshammari, G., Jorro-Aragoneses, J.L., Kapetanakis, S., Petridis, M., Recio-García, J.A., Díaz-Agudo, B.: A hybrid CBR approach for the long tail problem in recommender systems. In: Aha, D.W., Lieber, J. (eds.) ICCBR 2017. LNCS (LNAI), vol. 10339, pp. 35–45. Springer, Cham (2017). https://doi.org/10.1007/978-3-319-61030-6_3
2. Amatriain, X.: Beyond data: from user information to business value through personalized recommendations and consumer science. In: Proceedings of the 22nd ACM International Conference on Information & Knowledge Management, pp. 2201–2208. ACM (2013)
3. Amatriain, X., Lathia, N., Pujol, J.M., Kwak, H., Oliver, N.: The wisdom of the few: a collaborative filtering approach based on expert opinions from the web. In: Proceedings of the 32nd International ACM SIGIR Conference on Research and Development in Information Retrieval, pp. 532–539. ACM (2009)
4. Beel, J., Langer, S., Nürnberger, A., Genzmehr, M.: The impact of demographics (age and gender) and other user-characteristics on evaluating recommender systems. In: Aalberg, T., Papatheodorou, C., Dobreva, M., Tsakonas, G., Farrugia, C.J. (eds.) TPDL 2013. LNCS, vol. 8092, pp. 396–400. Springer, Heidelberg (2013). https://doi.org/10.1007/978-3-642-40501-3_45

5. Bobadilla, J., Ortega, F., Hernando, A., Gutiérrez, A.: Recommender systems survey. Knowl. Based Syst. **46**, 109–132 (2013)
6. Breiman, L.: Random Forests, pp. 1–33 (2001)
7. Bremer, S., Schelten, A., Lohmann, E., Kleinsteuber, M.: A framework for training hybrid recommender systems, pp. 30–37 (2017)
8. Burke, R.: Hybrid recommender systems: survey and experiments. User Model. User Adap. Inter. **12**(4), 331–370 (2002)
9. Burke, R.: Hybrid web recommender systems. In: Brusilovsky, P., Kobsa, A., Nejdl, W. (eds.) The Adaptive Web. LNCS, vol. 4321, pp. 377–408. Springer, Heidelberg (2007). https://doi.org/10.1007/978-3-540-72079-9_12
10. Goldberg, D., Nichols, D., Oki, B.M., Terry, D.: Using collaborative filtering to weave an information tapestry. Commun. ACM **35**(12), 61–70 (1992)
11. Gupta, J.: Performance analysis of recommendation system based on collaborative filtering and demographics (2015)
12. Harper, F.M., Konstan, J.A.: The movielens datasets: History and context. ACM Trans. Interact. Intell. Syst. (TiiS) **5**(4), 19 (2016)
13. Herlocker, J.L., Konstan, J.A., Borchers, A., Riedl, J.: An algorithmic framework for performing collaborative filtering. In: Proceedings of the 22nd Annual International ACM SIGIR Conference on Research and Development in Information Retrieval, pp. 230–237. ACM (1999)
14. Linden, G., Smith, B., York, J.: Amazon. com recommendations: item-to-item collaborative filtering. IEEE Internet Comput. **7**(1), 76–80 (2003)
15. Louppe, G.: Understanding random forests: from theory to practice. arXiv preprint arXiv: 1407.7502 (2014)
16. Mittal, P.: Metadata based recommender systems, pp. 2659–2664 (2014)
17. Moldovan, A.N., Muntean, C.H.: Personalisation of the multimedia content delivered to mobile device users. In: IEEE International Symposium on Broadband Multimedia Systems and Broadcasting, BMSB 2009, pp. 1–6. IEEE (2009)
18. Pazzani, M.J., Billsus, D.: Content-based recommendation systems. In: Brusilovsky, P., Kobsa, A., Nejdl, W. (eds.) The Adaptive Web. LNCS, vol. 4321, pp. 325–341. Springer, Heidelberg (2007). https://doi.org/10.1007/978-3-540-72079-9_10
19. Redpath, J.L.: Improving the performance of recommender algorithms. Ph.D. thesis, University of Ulster (2010)
20. Resnick, P., Iacovou, N., Suchak, M., Bergstrom, P., Riedl, J.: GroupLens: an open architecture for collaborative filtering of netnews. In: Proceedings of the 1994 ACM Conference on Computer Supported Cooperative Work, pp. 175–186 (1994)
21. Resnick, P., Varian, H.R.: Recommender systems. Commun. ACM **40**(3), 56–58 (1997)
22. Shah, L., Gaudani, H., Balani, P.: Surv. Recomm. Syst. **137**(7), 43–49 (2016)
23. Spiegel, S.: A hybrid approach to recommender systems based on matrix factorization. Department for Agent Technologies and Telecommunications, Technical University Berlin (2009)
24. Tintarev, N.: Explaining recommendations. In: User Modeling 2007, pp. 470–474 (2009)
25. Vozalis, M., Margaritis, K.G.: Collaborative filtering enhanced by demographic correlation. In: AIAI Symposium on Professional Practice in AI, of the 18th World Computer Congress (2004)
26. Wang, Z., Yu, X., Feng, N., Wang, Z.: An improved collaborative movie recommendation system using computational intelligence. J. Vis. Lang. Comput. **25**, 667–675 (2014)

27. Wei, S., Zheng, X., Chen, D., Chen, C.: A hybrid approach for movie recommendation via tags and ratings. Electron. Commer. Res. Appl. **18**, 83–94 (2016)
28. Zhang, H., Min, F., Wang, S.: A random forest approach to model-based recommendation. J. Inform. Comput. Sci. **11**(15), 5341–5348 (2014)
29. Zheng, V.W., Cao, B., Zheng, Y., Xie, X., Yang, Q.: Collaborative filtering meets mobile recommendation: a user-centered approach. In: AAAI, vol. 10, pp. 236–241 (2010)

A Neural Learning-Based Clustering Model for Collaborative Filtering

Grzegorz P. Mika[✉] and Grzegorz Dziczkowski

University of Economics, Katowice, Poland
grzegorz.mika@edu.uekat.pl, grzegorz.dziczkowski@ue.katowice.pl

Abstract. In this paper we present a neural learning-based clustering method for collaborative filtering. Collaborative filtering is an important task in recommender systems and has been investigated extensively in the past. Traditional approaches often require preprocessing steps, standard conditions or manually set gain. Our method is automatic, fast and robust towards cold start often seen in recommender systems. Furthermore, it can easily be trained to be used with any kind of data. The recommendation task is formulated as hybrid learning problem over two levels: artificial neural networks and clustering. Following the learning paradigm the detection on each level is performed by a trained classifier. First results of collaborative filtering using neural networks and clustering are presented and future additions are discussed.

Keywords: Recommender systems · Collaborative filtering
Clustering · User-profiling · Neural networks · Cold start problem

1 Introduction

Recommendation systems are difficult, repetitive problems and challenges in the scientific community. It is a computing process aimed at searching for the best, significant character of items. Depending on an application, you can take different approaches in implementation. Task of such a system is to predict user ratings on invaluable items or recommending based on existing ratings. Techniques in recommendation systems can be divided into three categories: content-based - CB, collaborative filtration - CF [1] and hybrid recommender system - HRS (see Fig. 1).

Vector space model is a standard algebraic model commonly used in searching for information. It is classified as one of the most popular techniques in the implementation of content-based recommendation models [4]. In this approach text document is a bag of words, ignoring grammar, and even verbal order. It usually uses TF-IDF (Frequency Inverse Document Frequency). Then each document is represented as a weight vector. At this point, it is worth noting that inquiries are also considered as documents. Most commonly, cosine measures are used to calculate the similarity between the document vector and query vector. This technique is most effective when recommending items with a high

© Springer Nature Switzerland AG 2018
N. T. Nguyen et al. (Eds.): ICCCI 2018, LNAI 11055, pp. 219–227, 2018.
https://doi.org/10.1007/978-3-319-98443-8_20

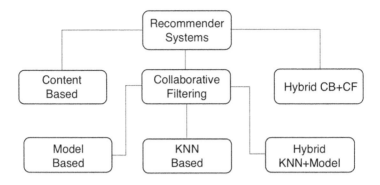

Fig. 1. Overview of recommender systems.

content of words, such as documents, websites or publications [7,8]. It is recommended to use items that are mainly associated with positively evaluated items. In addition to the vector space model, CB uses different types of models to find similarity between documents. These include probabilistic models such as the naive Bayesian classifier [11], decision trees [10] or artificial neural networks [6] to model relationships between different documents.

Collaborative filtering consists in building a database (matrix of items and users) of item preferences by users. Then it matches users to the relevant interests and preferences, calculating the similarities between their profiles to present recommendations. For identifying similarities, it uses the behavior of users which can be direct, allowing for the sorting of products. It can also be indirect, based on the behavior of inspection and purchase [16]. Recommendations developed in this technique can be both predictions and recommendations [7]. Scientifically speaking, CF algorithms can be divided into two main methods, the first based on memory (memory-based), the other on the model (model-based) [7].

The two main techniques of collaborative filtering, methods based on memory and on model, can be combined into one hybrid solution. Currently, it is a very often used approach. Model is used to only data preparation. Calculations leading to recommendations are already built based on memory.

This paper explores a new approach with the use of artificial neural networks and clustering to try to solve the cold start and scalability problems. In our approach introduces a new machine-based approach to implementation of hybrid recommendation techniques. The suggested recommendation method can introduce automatic calculations and recommendations for new users.

Thus, the primary goal will be to verify how the model based on traditional neural networks will behave. At the same time keeping in mind the problem of cold start and scalability in recommendation systems, we also tried to solve these problems in the implemented model.

This article is organized as follows. Section 2 describes our approach using neural networks. Section 3 describes the results of experiments of our approach to determine if our method is developmental. Finally, Sect. 4 presents our conclusions.

2 Our Approach

The Fig. 2 is an overview of our approach, and the detail procedure is described as follows:

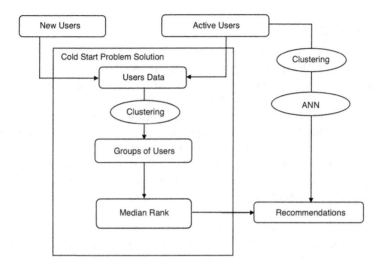

Fig. 2. Overview of our approach.

1. Apply clustering algorithm to group the users, then use the result, which is represented by median value.
2. Make a prediction for a user by performing a median from all grades.
3. Train new dataset (include groups) using neural networks.
4. Make a prediction for a user.

2.1 Clustering

User profile is a set of information representing it by rules, needs, interests, behaviors and preferences [2,12,18]. The content and amount of information contained in the profile may vary depending on the application area. Nevertheless, the accuracy of profile modeling is based on the method of collecting and organizing user information and how exactly this information reflects the user [2].

In our approach, we used well-known clustering - the k-means algorithm to solve the problem of cold start and scalability.

The purpose of clustering is to divide original dataset (clusters) into disjoint groups. The learning process in this algorithm is based solely on information contained in the data describing objects and their relations [6,23]. Most existing clustering algorithms, such as the k-means algorithm that is popular and liked for its simplicity, are sensitive to initial parameters such as the number of clusters and initial centroid positions. To limit these deficiencies, the third step takes into account the global coefficient k [17], the deterministic approach of the traditional k-means grouping algorithm. Finding the right number of groups for a given dataset is essentially a trial and error procedure, obstructed by the subjective nature of the decision what is the correct grouping [15,24].

2.2   Scalable Problem

K-means Clustering Algorithm is a simple and relatively efficient method. However, the number of k clusters should be determined in advance, where the final results are sensitive to initialization and often the local optimum occurs, and operating on larger dataset may have scalability problems. The difference is that we apply more scalability algorithm to operate on larger dataset.

Arthur and Vassilvitskii have implemented a pre-release version of k-means++ to evaluate the performance of this algorithm. It uses a careful sowing method to optimize time and precision [3,5]. The results of the experiment carried out by the authors themselves have shown that this algorithm is compatible with optimal clustering. In addition, they also showed a better time - by almost 70% and precision, where the potential value obtained is better with coefficients from 10 to 1000, in comparison with the traditional k-means algorithm. However, it should be noted that the central disadvantage of initializing k-means++ from the point of view of scalability is its inherent sequential nature. The choice of the next center depends on the current set of centers. To solve this problem, k means|| have been implemented. Intuitively similar to k means++, except for using the oversampling factor and some linear functions k [18].

2.3   Cold Start Problem

In traditional collaborative filtering approach when a new user appears in the system, there is a problem with showing recommendations. Main reason of new user problem is the lack of an earlier user activity. New items cannot be recommended until some users rate it, and new users are unlikely given good recommendations because of the lack of their rating or purchase history [22]. However, there is a simpler solution not based on complex algorithms. Many websites immediately after registration ask new users to rate a group of items in order to facilitate the preparation of recommendations in the future [22]. Thanks to such a procedure, the data can be obtained with a proper density, and the common filtration algorithms can be decidedly simpler.

In our approach, based on data of active users and user group demographic information and item group information, we can make predictions for the new users. Using for this purpose the previously discussed k-means with the optimal number k.

2.4   Traditional Neural Networks

The most popular choice in the applications of neural networks are multilayer perceptron networks, trained using the backward propagation algorithm [20]. In our approach for active users, in order to obtain better results, an attempt was made to train the artificial neural networks classifier with backward propagation.

We do not intend to go into the details, because the main goal was to solve the problems of cold start and scalability. However, in order to effectively apply the artificial neural networks model to clustering, the data should be transformed into the appropriate form for the algorithm. Based on the received clusters to the model, the input data was a user-item matrix with grades (within one group) and content of element content, such as species and description in binary values. As an alternative to the model, you can use the full data set with an additional feature - the cluster number.

3   Experiments

3.1   Datasets

Currently, we perform experiments on two datasets. First, real data collected from the MovieLens web-based recommender system. The dataset contained 100,000 ratings, with each user rating at least 20 items. Second, 20,000 random data.

3.2   Evaluation Metrics

Several indicators were proposed to assess the quality of methods in recommendation systems in order to verify the error. The indicators are divided into two main categories, i.e. statistical accuracy and accuracy metrics for decision support [21]. The suitability of each metric depends on the characteristics of the dataset and type of tasks the system performs [11]. Most often, two main methods of measurement stand out. The first of these is often called Mean Absolute Error - MAE or average deviation [9,13,14,19].

$$\frac{1}{n}\sum_{t=1}^{n}|e_t| \tag{1}$$

where n is the number of errors, $\sum$ is summation symbol (add them all up) and $|e_t|$ is absolute value of error at each observation point.

MAE has been widely used in evaluating the accuracy of a recommender system by comparing the numerical recommendation scores against the actual user ratings in the test data. The MAE is calculated by summing these absolute errors of the corresponding rating-prediction pairs and then computing the average. The second technique is Root Mean Squared Error - RMSE [9,13]:

$$\sqrt{\frac{1}{n}\sum_{t=1}^{n}e_t^2} \tag{2}$$

where n is the number of errors, $\sum$ is summation symbol (add them all up) and e is error value at each observation point.

In both cases, the error is returned based on the differences between the predicted value for the user and the observed value. In the case of RMSE, the emphasis is placed on large errors due to the increase of this difference to the square. In each of these measures, the lower the return score (the smaller the error), the higher the precision of the algorithm.

In the case of the ideal method, this value should be equal to zero or be as close as possible to zero. When creating the model and its matching, the ideal model will have historical predictions as little as possible from the predictions. Ultimately, their average will be as small as possible. Of course, no model or method will return the value for the MAE and RMSE equal to zero. Each model is burdened with an error, even very small. Thus, it is usually sought to achieve the lowest possible coefficient.

3.3 Models

The main assumption of using the grouping algorithm in the collaborative filtering method is to try to solve the cold start problem and scalability. The first version of our approach attempts to test different versions of the k-means algorithm, which improves efficiency on large dataset. We empirically tested solutions to the problem of scalability by iterating the grouping of a random dataset.

In the first step, the experiments were carried out for a different k value (Table 1). three clustering algorithms for modeling user profiles. For each parameter, 10 attempts were made, which made it possible to average all results. Evaluating the quality of the unsupervised learning algorithm is difficult to identify. The number of clusters for the random set of data we generate is estimated to be approximately 4 (see Fig. 3).

In our approach, we used the time of performing the algorithm on the increasing sample and the number of k. We took into account execution time with the increasing sample, and at the same time testing the resistance to scalability (Table 2). As you can see the results for larger datasets, it is economical to implement a scalable k-means‖ algorithm.

The time to train all training data using each algorithm takes about one day. Testing takes about 4 s. The above testing was performed on a smaller data set, which translated into the final evaluation results in each cluster. It can be noticed that the results are promising already at the first stage of the project. The proposed solution has achieved better results compared to the most popular KNN algorithm (Table 3).

Table 1. Our evaluations of k-means based on k value.

	K = 2	K = 3	K = 4	K = 5
k-means	0.02	0.05	0.21	0.40
k-means++	0.01	0.04	0.16	0.46
k-means‖	0.02	0.05	0.18	0.38

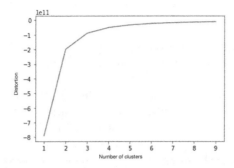

Fig. 3. Estimate the number of clusters.

Table 2. Our evaluations of k-means based on n value for k = 4.

K = 4	N = 500	N = 1500	N = 5000	N = 20000
k-means	0.50	0.40	0.42	0.63
k-means++	0.26	0.31	0.38	0.60
k-means‖	0.26	0.30	0.37	0.57

Table 3. Our evaluations of ANN (Random dataset).

Algorithms	MAE	RMSE	Precision	F-Measure
Content-based	2.18	2.78	0.44	0.42
KNN	2.07	2.53	0.45	0.44
Naive Bayes	1.99	1.75	0.65	0.46
Our approach	1.99	2.40	0.50	0.51

In traditional collaborative filtering, it is not possible to predict probable ratings for users, due to the insufficient number of ratings. In our approach, the inclusion of clustering algorithm allows us to overcome the cold start problem, achieving similar evaluation metrics.

Table 4. Our evaluations of ANN (MovieLens dataset).

Algorithms	MAE	RMSE	Precision	F-Measure
Content-based	1.75	1.52	0.69	0.65
KNN	1.77	2.15	0.68	0.67
Naive Bayes	1.35	1.75	0.65	0.65
Our approach	1.17	1.44	0.67	0.66

4   Conclusion

In this paper, we apply the neural learning-based clustering technique, which improves the correctness of collaborative predication, and as well as solves the scalable and the cold start problems. Our work indicates that system gives a new look at the possible use of machine learning and it can be developed using advanced algorithms (Table 4) within recommendation systems. Of course, the numerical results are specific for our approach (Table 3); however, we expect that it is worth experiments with deep neural networks in further phases. We have not yet considered deep learning architectures, but we expect that the results achieved could be significant. It cannot be concluded clearly how effective the results of clustering. This is a well-known problem of unsupervised learning. We expect that it is worth experiments with data from a well-known field to realistically verify the effectiveness of the model.

References

1. Adomavicius, G., Tuzhilin, A.: Toward the next generation of recommender systems: a survey of the state-of-the-art and possible extensions. IEEE Trans. Knowl. Data Eng. **17**(6), 734–749 (2005)
2. Araniti, G., De Meo, P., Iera, A., Ursino, D.: Adaptively controlling the QoS of multimedia wireless applications through user profiling techniques. IEEE J. Sel. Areas Commun. **21**(10), 1546–1556 (2003)
3. Arthur, D., Vassilvitskii, S.: k-means++: the advantages of careful seeding. In: Proceedings of the 18th Annual ACM-SIAM Symposium on Discrete Algorithms, SIAM, New Orleans, LA, pp. 1027–1035 (2007)
4. Baeza-Yates, R., Ribeiro-Neto, B.: Modern Information Retrieval. Addison Wesley Longman, New York City (1999)
5. Bahmani, B., Moseley, B., Vattani, A., Kumar, R., Vassilvitskii, S.: Scalable k-means++. Proc. VLDB Endow. **5**(7), 622–633 (2012)
6. Bishop, C.: Pattern Recognition and Machine Learning, 1st edn. Springer, New York (2006)
7. Bobadilla, J., Ortega, F., Hernando, A., Gutiérrez, A.: Recommender systems survey. Knowl. Based Syst. **46**, 109–132 (2013)
8. Burke, R.: Hybrid recommender systems: survey and experiments. User Model. User-Adap. Inter. **12**(4), 331–370 (2012)
9. Claypool, M., Gokhale, A., Miranda, T., Murnikov, P., Netes, D., Sartin, M.: Combining content-based and collaborative filters in an online newspaper. In: Proceedings of ACM SIGIR Workshop on Recommender Systems: Algorithms and Evaluation, Berkeley, CA (1999)
10. Duda, R., Hart, P., Stork, D.: Pattern classification, 2nd edn. Wiley-Interscience, New York City (2000)
11. Friedman, N., Gieger, D., Goldszmidt, M.: Bayesian network classifiers. Mach. Learn. **29**(2–3), 131–163 (1997)
12. Henczel, S.: Creating user profiles to improve information quality. Factiva **28**(3), 30 (2004)
13. Goldberg, T., Roeder, D., Gupta, C., Perkins, E.: A constant time collaborative filtering algorithm. Inf. Retrieval **4**(2), 133–151 (2001)

14. Gong, S., Chongben, H.: Employing fuzzy clustering to alleviate the sparsity issue in collaborative filtering recommendation algorithms. In: Proceeding of 2008 International Pre-Olympic Congress on Computer Science, World Academic Press, Nanjing, China, pp. 449–454 (2008)
15. Han, J., Kamber, M., Pei, J.: Data Mining: Concepts and Techniques, 3rd edn. Morgan Kaufmann, Waltham (2011)
16. Isinkaye, F., Folajimi, Y., Ojokoh, B.: Recommendation systems: principles, methods and evaluation. Egypt. Inform. J. **16**(3), 261–273 (2015)
17. Likas, A., Vlassis, N., Varbeek, J.: The global k-means clustering algorithm. Pattern Recogn. **36**(2), 451–461 (2003)
18. Kuflik, T., Shoval, P.: Generation of user profiles for information filtering-research agenda. In: 23rd Annual International ACM SIGIR conference on Research and Development in Information Retrieval, Athens, Greece, pp. 313–315. ACM (2000)
19. Huang, Q.H., Ouyang, W.M.: Fuzzy collaborative filtering with multiple agents. J. Shanghai Univ. (English Edition) **11**(3), 290–295 (2007)
20. Rumelhart, D., Hinton, G., Williams, R.: Learning Representations by Back? Propagating Errors. Nature **323**, 533–536 (1986)
21. Sarwar, B., Karypis, G., Konstan, J., Reidl, J.: Item-based collaborative filtering recommendation algorithms. In: Proceedings of the 10th International Conference on World Wide Web, Hong Kong, China, pp. 285–295. ACM (2001)
22. Su, X., Khoshgoftar, T.: A survey of collaborative filtering techniques. Adv. Artif. Intell. **4**, 1–19 (2009)
23. Theodoridis, S., Koutroumbas, K.: Pattern Recognition, 3rd edn. Academic Press Inc, Orlando, FL (2006)
24. Witten, I., Frank, E., Hall, M., Pal, C.: Data Mining, 4th edn. Morgan Kaufmann, San Francisco, CA (2016)

Influence Power Factor for User Interface Recommendation System

Marek Krótkiewicz[1]([✉]) [iD], Krystian Wojtkiewicz[1] [iD], and Denis Martins[2] [iD]

[1] Faculty of Computer Science and Management,
Wroclaw University of Science and Technology, Wroclaw, Poland
`marek.krotkiewicz@pwr.edu.pl`
[2] ERCIS, University of Muenster, Leonardo-Campus 3, 48149 Muenster, Germany

Abstract. User interface is an important element of software system since it provides the means for utilizing applications' functionalities. There is number of publications that propose guidance for proper interface design, including adaptive approach. Following paper introduces general idea for definition of interface design in a way that allows for easy computing of user interface effectiveness. The introduced factor can be used for recommendation of interface changes and adjustment.

Keywords: User interface design · Interface recommendation
Interface components

1 Introduction

The quality and effectiveness of user interfaces relies mostly on the technical knowledge, experience, and talent of the designer. However, with the increasing need for personalization in the Web, particularly after the advent of the Web 2.0, a growing population of non-technical, unexperienced users is required to perform user interface design tasks (e.g., Blogs and personal Website personalization). Unfortunately, these users are commonly not aware of design principles that may assist them during the conceptualization and implementation of user interfaces. Moreover, due to the intrinsic creative nature, this process is frequently characterized as semi-structured (i.e., there is no methodology that can successfully serve as a standard to every design case) and highly interactive, which potentially hinders productivity.

In this work, the authors propose the idea of building and evaluating a recommender system that supports unexperienced designers in an efficient way, allowing to design high-quality interfaces for user-centric applications. Such system considers the appropriate content to be delivered to the user as well as the user interface structure in which content is presented. The process of recommending appropriate design insight is based on ranking and reordering items in a meaningful way in order to both differentiate items by means of an importance degree, and increase the time end-users spend on the platform.

Recommender Systems have been widely used to improve user experience and content personalization in many domains [2,4]. In general, user-centric

© Springer Nature Switzerland AG 2018
N. T. Nguyen et al. (Eds.): ICCCI 2018, LNAI 11055, pp. 228–237, 2018.
https://doi.org/10.1007/978-3-319-98443-8_21

recommendations aim to promote the discovery of unexpected items that are likely to be relevant to users. For providing such personalization, collaborative filtering has become one of the most representative methods in Recommender Systems. This approach consists of gathering information about several users [13], such as interests, habits, goals, behaviors, preferences and usage statistics in order to select a subset of relevant items to be present to the user [3,15]. The assumption behind collaborative filtering is that users with similar interests and preferences potentially like similar items [9].

Providing such personalization to user interface designers empower them with tools that enhance their capabilities of adapting system's functionality, appearance, or information representation to better suit user needs [11,12,14].

However, at the beginning the designer has little knowledge about users and their preferences as such. Therefore, there is a need to build a system that allows to change the design in time and evaluate it in terms of systems functional assumptions. To build such a system one has to define a proper and arbitrary method to determine whether the system works properly or does it need to be tuned in some way [1]. In order to do that the method presented in this paper was designed. The aim of it is to provide means of design evaluation based on user experience. The method assumes that there are multiple interface (system) states and there are many connections among those states. The system is evaluated on the basis of users states changes. Such an approach is already known [10], however we extend it showing that not only a change from one state to another matters, but rather which component has been used.

In the following section the problem of building the factor is presented in the formal way as well as implementation of structures describing interface using AssoBase – Association-Oriented Metamodel (AOM) is presented. Section 3 presents the algorithm for computing influence power of individual elements that will be used for evaluation of interface design. The summary concludes the paper.

2 Problem Formalization

User interface definition is a demanding task and it relies on many factors e.g. technology, purpose of the system, target user group etc. The intent of authors was to build an abstraction that would allow to introduce an universal factor for assessment of interface definition, i.e. its usability. In order to do that we propose an unambiguous introduction of elements that are further used to describe the proposed factor.

2.1 Structures

Following formalisms are used to identify heterogeneous structures:

$\{e\} = \{e_1, \ldots, e_n\}$ – set of e elements (not changeable, sequence is irrelevant),
$(e) = (e_1, \ldots, e_n)$ – tuple of e elements (not changeable, sequence is relevant),
$\langle e \rangle = \langle e_1, \ldots, e_n \rangle$ – list of e elements (changeable, sequence is relevant).

2.2   Formal Interface Definition

For the purpose of this paper the user interface I is defined as a set of interface states:

$$I = \{s\} = \{s_1, \ldots, s_n\} \tag{1}$$

where each state s is defined as tuple built from set of components C and a template t.

$$s = (C, t) \tag{2}$$

where

$$C = \{c\} = \langle c_1, \ldots, c_n \rangle \tag{3}$$

and

$$t = \langle r \rangle. \tag{4}$$

The template t, in scope of user interface, is understood as set of regions, where each region r is defined by its position on the screen x and y, width w, height h and influence power i.

$$r = (x, y, w, h, i). \tag{5}$$

The influence power i is assigned to each region based on its location in the list according to the function defined by the designer. However, the concept of the influence power will be evaluated in another section of the paper, thus it will not be explained here.

The component c is understood as tuple built from reference to region r, applied design characteristics d and a link l:

$$c = (r, d, l) \tag{6}$$

Where the design characteristics d of interface component is defined as following tuple:

$$d = (P, a_a, a_m) \tag{7}$$

where P is a list of properties, while a_a is additive influence power adjustment factor and a_m multiplicative influence power adjustment factor. The set of properties is defined as:

$$P = \{p\} = \langle p_1, \ldots, p_n \rangle \tag{8}$$

where each property is a tuple of feature list F and value list V:

$$p = (F, V) \tag{9}$$

where:

$$F = \langle f_1, \ldots, f_n \rangle \tag{10}$$

$$V = \langle v \rangle = \langle v_1, \ldots, v_n \rangle. \tag{11}$$

The link l points to either interface state $l \in I$ or another component $l \in C$, thus $l \in C \cup I$.

2.3   Interface Definition in AssoBase Environment

The proposed method for assessment of interface definition usability aims at possibility not only to define the interface in a formal way but also to provide computational platform for measuring the actual factors. In order to do that authors have used AssoBase, formerly known as AODB [6] as a database layer for a system. The implementation of the structure in AssoBase proves its cohesion. Following the definition of the system in AssoBase formalism [5,7] is provided. Figure 1 presents the definition of the system by the use of AML [8,16] which is a modeling language for AssoBase.

$$FEATURE \left\{ \left| \left\langle \begin{array}{l} +name : unicode(32), \\ +type \ \ : ascii(2) \end{array} \right\rangle \right| ; \right\} \tag{12}$$

$$VALUE \left\{ \left\langle +value : byte \right\rangle ; \right\} \tag{13}$$

$$CHARACTERISTICS \left\{ \left| \left\langle \begin{array}{l} +add\_power\_adj\_factor \ \ : double, \\ +mult\_power\_adj\_factor : double \end{array} \right\rangle \right| ; \right\} \tag{14}$$

$$REGION \left\{ \left\langle \begin{array}{ll} +x & : double, \\ +y & : double, \\ +width & : double, \\ +height & : double, \\ +influence : double \end{array} \right\rangle ; \right\} \tag{15}$$

$$Property \left\{ \left| \begin{array}{l} [*] \xrightarrow{+Feature} [1] \ \Box FEATURE, \\ [1] \xrightarrow{+Value} [*] \ \Box VALUE \end{array} \right| ; \right\} \tag{16}$$

$$Characteristics \left\{ \left| \begin{array}{l} [1] \xleftrightarrow{+Characteristics} [1] \ \Box CHARACTERISTICS, \\ [1] \xrightarrow{+Property} [*] \ \Diamond Property \end{array} \right| ; \right\} \tag{17}$$

$$Component \left\{ \left| \begin{array}{l} [*] \xrightarrow{+Characteristics} [1] \ \Diamond Characteristics, \\ [*] \xrightarrow{+Placement} [1] \ \Box REGION, \\ [*] \xrightarrow{+Link(\Diamond\{State,Component\})} [1] \ \Diamond State \\ \xrightarrow{f^v} State; \end{array} \right| \right\} \tag{18}$$

$$State \left\{ \left| \begin{array}{l} [*] \xrightarrow{+Template} [1] \ \Diamond Template, \\ [1] \xrightarrow{+Component} [*] \ \Diamond Component \end{array} \right| ; \right\} \tag{19}$$

$$Template \left\{ \left\langle [1] \xrightarrow{+Region} [1..*] \ \Box REGION \right\rangle ; \right\} \tag{20}$$

$$UserInterface \left\{ \left\langle [1] \xrightarrow{-State} [1..*] \ \Diamond State \right\rangle ; \right\} \tag{21}$$

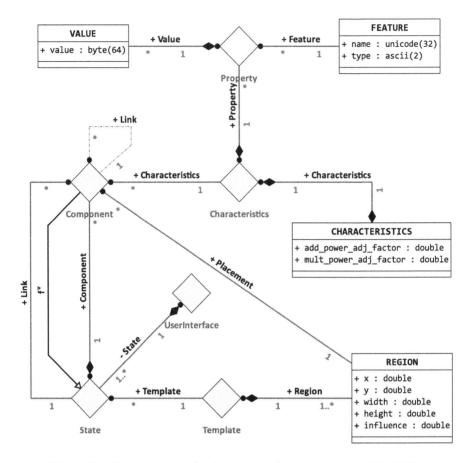

Fig. 1. Database structure for data retrieval system expressed in AML

3   Component Influence Power

The algorithm used for assessing *component influence power* is mainly based on computing it from both design characteristic of the component and the position of the component on interface canvas. These two interlate with each other but they are also quite different. The interface is defined as a set of interface states that are organized as graph. Individual interface states are understood as graph nodes, while links connecting these states are edges. In most common approach, i.e. traffic statistics, advertisement statistics, this level of abstraction is sufficient. However, for the purpose of interface design evaluation it is important to assess performance of each individual element that directs to another state of the interface. It is due to the fact that it is the only way to understand how users interact with system and how to enhance desired users activities.

The proposed factor is built according to following formula:

$$cip = pip + dip \tag{22}$$

where:

cip – component influence power,
pip – position influence power,
dip – design characteristic influence power.

As it was pointed out earlier, the influence power derived from the position of the element and its physical characteristic have different origin and properties. The idea how to understand them is presented in following subsections.

3.1   Position Influence Power

Each system that have human interaction component uses some kind of interface. For the sake of further deliberation the graphical interface will only be evaluated, since other types of interfaces are not in the thematic scope of presented method. It is assumed that the individual interfaces are build with the use of templates. Each template provide means for component distribution over interface canvas by the use of regions. The regions are used as containers for components and it is assumed that each and every component assigned to a given container have the same influence power, i.e. pip. The power distribution that defines influence power values assigned to region lays in discretion of the designer. However, Fig. 2 provides an example, where vertical and horizontal power distribution were shown.

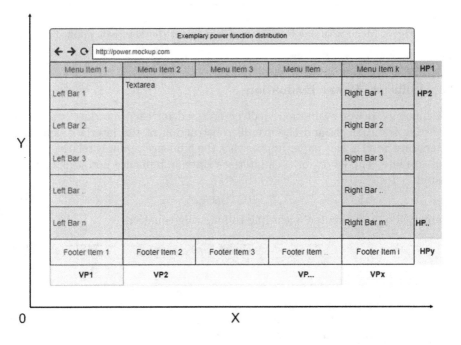

Fig. 2. Exemplary template with power distribution assignments

The function that defines power of each element may be defined according to either absolute position of this element on canvas or its relative position in reference to other elements. Those two approaches might be mixed for optimal power assignment.

3.2 Design Characteristic Influence Power

The second element dip in Eq. 22 refers to component design characteristics. It is assumed that physical appearance of components affect the influence that this component have on user decision, whether or not to use it. The definition of components (Eq. 6) assume that its design characteristics is derived from properties assigned to this component. For each property there might be additive or multiplicative power adjustment factor defined. Values of those factors are in the discretion of the designer, however the basic idea of why those should be applied is presented in the Fig. 3. The picture consists of five items aligned horizontally. One of the item differs from the other in spite of its color and an image positioned next to it. Those properties may imply presence of adjustment factors. The actual values of them depend on designer.

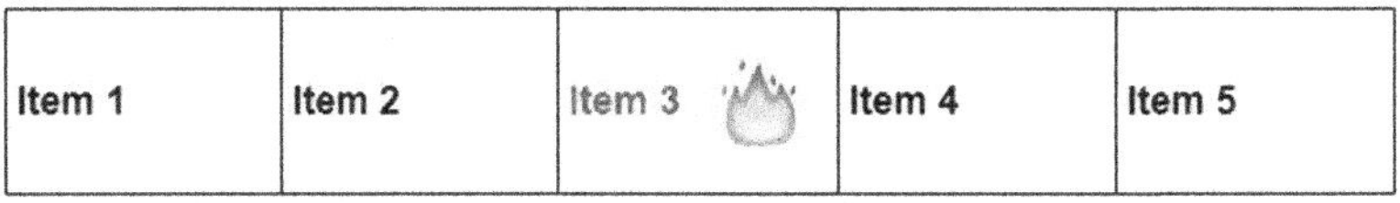

| Item 1 | Item 2 | Item 3 | Item 4 | Item 5 |

Fig. 3. Menu with one item having adjustment characteristic applied

3.3 Influence Power Evaluation

The influence power evaluation shall be assessed for each interface separately since the definition 22 and the overall construction of the interface definition system presented in this paper implies that the influence power is relative, rather than absolute. Therefore, for each interface state an influence power matrix L is created.

$$L = [l_{ij}] \tag{23}$$

where each l_{ij} is computed according to following equation

$$l_{ij} = (c_s, c_e, cip) \tag{24}$$

where:

c_s – start component in the link,
c_e – end component in the link,
cip – computed influence power.

3.4   Algorithm Definition

The complete algorithm for assessing and evaluating of interface design is presented in the Fig. 4. The first phase consist of several steps, each of which focuses on definition of elements building the interface, namely templates, regions, characteristics, components and states. Once the system is defined in terms of its states the second phase of algorithm is set in motion. The power of edges is computed and formed as influence power matrices, according to formula 23. Following the user experience data is acquired and stored. It is used for computing *Interface Design Effectiveness – IDE* matrix for each of states as:

$$IDE = [ide_{ij}] \qquad (25)$$

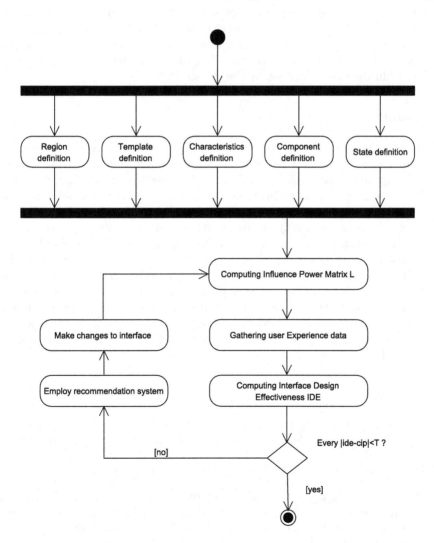

Fig. 4. Algorithm for user interface effectiveness assessment

where each ide_{ij} is computed according to following equation

$$ide_{ij} = (c_s, c_e, uu) \tag{26}$$

where:

c_s – start component in the link,
c_e – end component in the link,
uu – user usage.

User usage is calculated as number of times the link was used divided by the sum of all outgoing links ware used from a given state. Having both L and IDE matrices it is possible to conclude whether the difference between computed influence power cip and interface design effectiveness ide go beyond the scope of threshold set by the designer. In case of that the optional phase of employing the recommendation system is executed. The algorithm is concluded if all components in all states are defined in the way that their *interface design effectiveness* states within the boundaries set by the threshold.

4   Summary

The paper is dedicated to the user interface definition in the aspect of building a recommendation system that would allow easy adaptation of interface design to obtain optimal effectiveness. This is important due to the fact that user interface gives means to utilize applications' functionalities. There is number of publications that propose guidance for proper interface design, including adaptive approach. However, authors in the paper did not put the main emphasis on recommendation system, but rather on building the framework that may be used for evaluating of interfaces design and algorithms that aim at its recommendation. Thus, the paper introduces general idea for definition of interfaces based on elements such as templates, regions, characteristics components and interface states. Furthermore, the component influence power – cip and interface design effectiveness ide are defined in order to assess the effectiveness of interface design. Last but not least, the general algorithm for interface recommendation system evaluation has been proposed.

References

1. Beel, J., Breitinger, C., Langer, S., Lommatzsch, A., Gipp, B.: Towards reproducibility in recommender-systems research. User Model. User Adap. Inter. **26**(1), 69–101 (2016)
2. Bobadilla, J., Ortega, F., Hernando, A., Gutiérrez, A.: Recommender systems survey. Knowl. Based Syst. **46**, 109–132 (2013)
3. Cooley, R., Mobasher, B., Srivastava, J.: Web mining: information and pattern discovery on the world wide web. In: Proceedings Ninth IEEE International Conference on Tools with Artificial Intelligence, pp. 558–567, November 1997. https://doi.org/10.1109/TAI.1997.632303

4. Isinkaye, F., Folajimi, Y., Ojokoh, B.: Recommendation systems: principles, methods and evaluation. Egypt. Inform. J. **16**(3), 261–273 (2015)
5. Krótkiewicz, M.: A novel inheritance mechanism for modeling knowledge representation systems. Comput. Sci. Inform. Syst. (2017). https://doi.org/10.2298/CSIS170630046K
6. Krótkiewicz, M.: Association-oriented database model – n-ary associations. Int. J. Softw. Eng. Knowl. Eng. **27**(02), 281–320 (2017). https://doi.org/10.1142/S0218194017500103
7. Krótkiewicz, M.: Cyclic value ranges model for specifying flowing resources in unified process metamodel. Enterp. Inform. Syst., 1–23 (2018). https://doi.org/10.1080/17517575.2018.1472810
8. Krótkiewicz, M., Jodłowiec, M.: Modeling autoreferential relationships in association-oriented database metamodel. In: Świątek, J., Borzemski, L., Wilimowska, Z. (eds.) ISAT 2017. AISC, vol. 656, pp. 49–62. Springer, Cham (2018). https://doi.org/10.1007/978-3-319-67229-8_5
9. Kukla, E., Nguyen, N.T., Sobecki, J., Danilowicz, C., Lenar, M.: Determination of learning scenarios in intelligent web-based learning environment. In: Orchard, B., Yang, C., Ali, M. (eds.) IEA/AIE 2004. LNCS (LNAI), vol. 3029, pp. 759–768. Springer, Heidelberg (2004). https://doi.org/10.1007/978-3-540-24677-0_78
10. Malski, M.: A Method for web-based user interface recommendation using collective knowledge and multi-attribute structures. In: Jędrzejowicz, P., Nguyen, N.T., Hoang, K. (eds.) ICCCI 2011. LNCS (LNAI), vol. 6922, pp. 346–355. Springer, Heidelberg (2011). https://doi.org/10.1007/978-3-642-23935-9_34
11. Mican, D., Tomai, N.: Association-rules-based recommender system for personalization in adaptive web-based applications. In: Daniel, F., Facca, F.M. (eds.) ICWE 2010. LNCS, vol. 6385, pp. 85–90. Springer, Heidelberg (2010). https://doi.org/10.1007/978-3-642-16985-4_8
12. Montaner, M., López, B., de la Rosa, J.L.: A taxonomy of recommender agents on the internet. Artif. Intell. Rev. **19**(4), 285–330 (2003). https://doi.org/10.1023/A:1022850703159
13. Shahabi, C., Banaei-Kashani, F.: A framework for efficient and anonymous web usage mining based on client-side tracking. In: Kohavi, R., Masand, B.M., Spiliopoulou, M., Srivastava, J. (eds.) WebKDD 2001. LNCS (LNAI), vol. 2356, pp. 113–144. Springer, Heidelberg (2002). https://doi.org/10.1007/3-540-45640-6_6
14. Sobecki, J.: Ant colony metaphor applied in user interface recommendation. New Gener. Comput. **26**(3), 277 (2008). https://doi.org/10.1007/s00354-008-0045-9
15. Srivastava, J., Cooley, R., Deshpande, M., Tan, P.N.: Web usage mining: discovery and applications of usage patterns from web data. SIGKDD Explor. Newsl. **1**(2), 12–23 (2000). https://doi.org/10.1145/846183.846188
16. Wojtkiewicz, K., Jodłowiec, M., Krótkiewicz, M.: Association-oriented database metamodel: modelling language (2017). https://doi.org/10.13140/RG.2.2.18483.73769

A Generic Framework for Collaborative Filtering Based on Social Collective Recommendation

Leschek Homann[1], Bernadetta Maleszka[2($\boxtimes$)], Denis Mayr Lima Martins[1], and Gottfried Vossen[1]

[1] ERCIS, University of Münster, Leonardo-Campus 3, 48149 Münster, Germany
{leschek.homann,denis.martins,vossen}@wi.uni-muenster.de
[2] Faculty of Computer Science and Management, Wroclaw University of Science and Technology, Wybrzeze Wyspianskiego 27, 50-370 Wroclaw, Poland
Bernadetta.Maleszka@pwr.edu.pl

Abstract. Collaborative filtering has been considered the most used approach for recommender systems in both practice and research. Unfortunately, traditional collaborative filtering suffers from the so-called cold-start problem, which is the challenge to recommend items for an unknown user. In this paper, we introduce a generic framework for social collective recommendations targeting to support and complement traditional recommender systems to achieve better results. Our framework is composed of three modules, namely, a User Clustering module, a Representative module, and an Adaption module. The User Clustering module aims to find groups of users, the Representative module is responsible for determining a representative of each group, and the Adaption module handles new users and assigns them appropriately. By the composition of the framework, the cold-start problem is alleviated.

Keywords: Recommender system · Cold-start problem · User profile
Social networks · Collaborative filtering

1 Introduction

Recommender systems are basically divided into content-based, knowledge-based, collaborative-based, and hybrid approaches [7]. Content-based recommender systems predict recommendations for a customer's interest in an item by considering its characteristics. Collaborative filtering approaches are either user-based or item-based. The former first identifies similar users by calculating their ratings for items rated by both of them and then considering their ratings to the item to predict. The latter first identifies similar items by their ratings and then by considering the users rating for those similar items. As the name suggests, hybrid systems aim to combine different approaches to improve the achieved recommendations. Although substantial research has been conducted on collaborative filtering conducted in recent years, the problem of generating suitable

© Springer Nature Switzerland AG 2018
N. T. Nguyen et al. (Eds.): ICCCI 2018, LNAI 11055, pp. 238–247, 2018.
https://doi.org/10.1007/978-3-319-98443-8_22

recommendations for an unknown user is still open. This paper presents a fresh approach based on a generic framework for social collective recommendations.

Current approaches try to tackle the cold start problem by considering social characteristics of a user as part of the recommendations. Nevertheless, these approaches are limited, as they basically only consider one aspect of the problem. Therefore, we separate the problem into three different modules: a Clustering module, a Representative module, and an Adaption module, which compose a generic framework to solve the cold-start problem. At the same time, we introduce our idea of how a social collective can be represented. The most obvious source for social collectives nowadays are social networks [4,6]. As an example, we propose to use a clique [11], known from graph theory, to describe a social collective, as it provides a very intuitive and easily understandable representation of groups. Furthermore, we introduce an evaluation concept for our generic framework.

The remainder of this paper is structured as follows: In Sect. 2 we describe related work to recommender systems. In Sect. 3 the problem definition is stated. Afterwards, we specify our framework and its modules in Sect. 4. Section 5 describes the evaluation idea of the framework. We conclude our paper in Sect. 6 by describing limitations and giving an outlook of future work.

2   Related Work

Recommender systems become more and more important in everyday life. When a user wants to find or buy an item, he or she usually first checks information about the product, e.g., in the Internet (content-based approach), asks some specialists about the quality of the product, or simply asks his or her friends for an opinion on the product (collaborative filtering approach). Information about similar people or tags is often used in tagged recommendation systems [16].

The increasing interest in recommender systems research is confirmed by the substantial literature produced in this area. Beel et al. [2] noticed that 55% of over 200 papers from the last few years are focusing on content-based approaches, while almost 35% are connected to collaborative filtering or graph-based recommendation. This demonstrates that content (description) of products, documents or items are really important, but also the opinion of different users about items should be taken into account. A similar result is presented by Anjumol et al. [1]. They claim that psychological and sociological studies have shown dependencies between individual preferences and interpersonal influences: Users' decisions and absorptions of information are affected by other users.

With growing interest in social networks, information about the community of a user can be obtained easily. Social recommender systems were introduced by Ma et al. [12]. A social network is considered as a set of users (e.g., nodes of the network) and links (e.g., the ties) by one or more relations [3]. One can differentiate weak and strong ties which influence user opinions [17]. Among different types of social networks we can distinguish the following: dedicated networks (e.g., networks of business, friends, graduates, etc.); indirect (e.g., online communities) or networks connected with common activities (e.g., co-authors, co-organizers); local networks (e.g., neighborhood, family, employees). Kywe et al. [9] differentiate two types of social recommendation approaches: The first one is connected with other users' recommendation based on their social network and the collaborative or content-based approach, while the second one uses information about social connections and the collaborative approach to recommend new items [18]. In our paper, we focus on the second approach.

The most important challenges are determined by several disadvantages and weaknesses of existing approaches [15]. The first aspect is correlated with data sparsity – determining complete sets of similar users is a problem in NP. Quality of recommendations is mostly biased by the quality of members of the group. The second key issue is connected with a new user (cold-start problem) – the system cannot classify him or her into a proper group until he or she rates some items. The next problem is to overfit the profile and recommending items from a very narrow scope of user interests. The last problem is a more sociological problem: People trust recommendations which were generated in a more transparent way, e.g., by his or her friends (not by anonymous people). Another approach for recommender systems is deep collaborative filtering recommendation [5,19]. Deep learning techniques become more popular in recommendation systems due to the problem of boosting scalability of the system and improving activity among users. The method could be effectively used to enrich collaborative filtering based recommender systems.

In this paper, we develop a framework for a social recommendation system which aims to avoid or fix the above-mentioned problems by combining the collaborative filtering approach with the social collective of users.

3 Problem Description

Traditional recommender systems produce recommendations by analyzing historical data about items and/or users. More specifically, a recommender system tries to estimate the relevance of an item i from an itemset I for a user u from a set U of users using a predictive method. In collaborative filtering, this prediction method exploits patterns in previous ratings $r_{u,i}$, $u \in U = \{u_l : l = 1, 2, \ldots, n_u\}$ and $i \in I = \{i_k : k = 1, 2, \ldots, n_i\}$, where n_u is the number of users and n_i is the numbers of items in the system, respectively. Nevertheless, these traditional approaches usually produce poor recommendations for users that are new to the system. As no historical data about these users is available, it is not possible to find users with alike interests in the past. Hence, the predictive method outputs inappropriate estimations. In the literature, this problem is known as the *cold-start problem*, which we try to tackle with the design of a generic, adaptable framework that considers social-collective knowledge.

4 Generic Framework: Social Recommendation System

In this section, we introduce a generic framework for alleviating the cold-start problem by considering social-based recommendations.

A general idea of our approach is to consider a set of registered users. By processing the available knowledge about them it is possible to discover groups of users with similar interests and to determine a representative of each group. Based on only little information about a new user, we can classify him or her into a proper group and recommend relevant items directly at first use of the system. In this paper, we consider social aspects as a source of knowledge about the users.

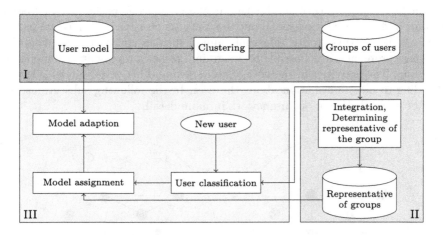

Fig. 1. Conceptual schema of the proposed generic framework.

Our framework which is depicted in Fig. 1 consists of three modules, namely, User clustering, Representative, and Adaptive module [13]. An overall idea of the collaborative social recommendation method is presented in Algorithm 1. The first module is connected with a base of users models. A single user model should contain information about users preferences, demographic data, usage

---

Algorithm 1. Idea of method for collective social recommendation.

Input: A set of N users profiles in social community, a new user u.
Output: A profile for the new user.
1. Cluster users according to social and usage data stored in profiles.
2. Determine a representative profile for each group.
3. Determine significant demographic data for each group of users.
4. **if** *A new user u join the system* **then**
   - (a) Ask him or her about significant demographic data.
   - (b) Collect information about social data.
   - (c) Classify him or her to the most appropriate group.
   - (d) Assign him or her the representative profile of the proper group.
   - (e) Adapt his or her profile according to his or her current activities.

---

data, or social connections to other users in the system. Based on this user model database it is possible to cluster similar user models into groups taking different aspects into account, where each group is stored separately. These groups are updated regularly to account with recent changes into user preferences.

The second module aims to determine a representative model (e.g., characteristics) of a group. When users in the groups change (i.e., when the set of users in the group or the models of users changes) it is necessary to recalculate the representative model.

The last module is correlated with a new user that enters the system. Based on some additional information about the user it is possible to assign him or her to one of the existing groups. The main aim of such an approach is to avoid the cold-start problem – it is possible to assign him or her a representative model of the group. The activities of the user are observed and his or her model is tuned according to new information about the user. In the following subsections, we describe each module of our framework in more detail.

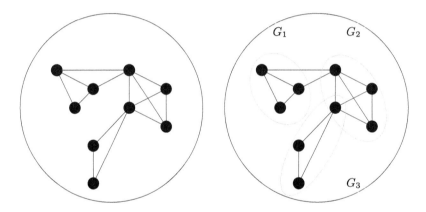

Fig. 2. A community in which user relations are represented as edges (left). A community with three groups G_1, G_2 and G_3 given as cliques (right).

4.1   User Clustering Module

The user clustering module aims to divide the entire set of users (i.e., user model) into groups so that user with similar interests are part of the same group. In the context of social-based recommendations, the user model is represented as a graph where each node portrays a user and each edge depicts any kind of connection or relation between two users. For instance, relations can describe friendship, collaboration (e.g., coauthors of papers), or interactions in social media channels (e.g., commenting on Facebook or Twitter posts). Given this social representation, user groups can be considered as communities (i.e., a subgraph) within the graph, as Fig. 2 illustrates.

A possible strategy to find communities is to determine cliques in the social graph. A clique is a special graph in which each node is connected to every other node in the graph. In other words, each user in a clique is directly related to all other users of the clique. Although the problem to find cliques within a given graph is NP-complete, this strategy is suitable for finding close related users, and therefore, helps to clarify the idea of a social collective approach. Nevertheless, heuristic algorithms exist to calculate cliques based on a given graph in a more acceptable runtime. Additionally, other graph algorithm approaches to find communities can also be included such as betweenness-oriented [11] or density-oriented algorithms. Figure 2 shows possible cliques and indicates the possibility of assigning a given user to multiple cliques or, more generally, to multiple groups.

4.2   Representative Module

The task of the representative module is to determine the representatives of the clustered user groups from the user clustering module in the first place and to store the representatives in a database for further classifications. Here, multiple aspects could be considered to determine a representative. To determine a model of the entire group, it is necessary to check similarity between each pair of users; in particular, some classical methods to determine the centroid or median object can be used.

As we are considering the social collective among users for our recommendation system, our main focus is to identify possible representatives by their importance within the community. Note that the importance could be measured by the number of possible relations each user in a group has to other users in the community, i.e., the number of incoming and outgoing edges of the user. Another possible approach is to weight the edges between users within a clique by the number of communications among them. Thereby, the most communicative user is considered as representative of the group. Finally, a representative could be chosen by its (node) betweenness, which indicates that he or she is important for the interaction between separate groups in the community.

4.3   Adaptation Module

The adaption module tends to be the core module to solve the well-known cold-start problem of recommender systems. The cold-start problem describes the situation in which a so-far unknown new user interacts with the recommender system. In this case, a collaborative recommender system fails to recommend items, as no similarities to other users can be determined. Some solutions for this problem involve interactions of the user, e.g., selecting or rating specific items after setting up an account.

More sophisticated methods for the user classification problem were presented in [14]. As we can gather demographic data about users, it is possible to analyze which attributes are significant for the classification process. In the first step, users are clustered into groups of similar interests (based on usage data). Next, in each group, we can determine a subset of demographic data that represents the group in the most suitable way. When a new user is entering our system, it is sufficient to ask him or her only about a few demographic features to classify him or her into a proper group.

In a social collective recommendation system, this situation can be simplified by the fact that in social networks most users directly define relations to their friends and interests without the influence of a system. For this reason, our approach is as follows: A new user is classified by the already calculated groups of users in the user clustering module. After the user has been classified, the model assignment takes place based on the representatives of the different groups. In the last step, the model is adapted or updated with the new user to recalculate the user model and clustering, as they might have changed.

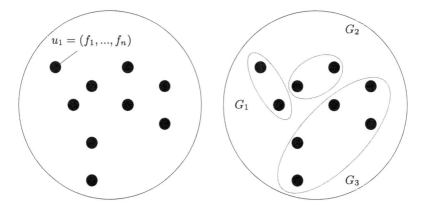

Fig. 3. A set of users in which each user is described by a user profile with features (left). Three user groups G_1, G_2 and G_3 discovered by some clustering algorithm based on user profile information (right).

5   Proposal of an Experimental Evaluation

In this section, we describe how to evaluate our approach based on comparing user profiles and relations among users for the clustering process. Since the introduced framework comprises only the theoretical aspects of a social-based recommender system, we highlight a concrete experimental scenario is outside the scope of this paper. Nevertheless, a description of future empirical analysis and experimentation is provided in the following.

An empirical evaluation of the benefits brought by our social-based recommendation approach can be assessed by measuring the quality of the returned recommendations via several evaluation metrics such as root mean squared error (RMSE), mean absolute error (MAE), precision, recall, and F-measure. For instance, a baseline approach composed by a k-Means algorithm may be applied to cluster the entire set of user considering only user-related attributes such as age, gender, and location, all stored in a user profile. Thereby, a user profile u is characterized by a number of n attributes so that $u = (f_1, \ldots, f_n)$ with

$u \in U$, as illustrated in Fig. 3. Additionally, in this approach, representatives of the clusters may be determined by, for instance, the centroid of a cluster.

On the other hand, our social-based approach exploits a different data source (i.e., a social database) to find user groups (i.e., communities) and afterward select representatives to alleviate the cold-start problem, as previously shown in Fig. 2. A social database is represented as a graph in which users are nodes and edges are relations among users. For instance, relations may depict users that rate or interact with a particular item (e.g., movie, picture, article). Since a social-based grouping may not depend on any other information besides user-user relations, an empirical evaluation may examine the amount of data needed to generate appropriate results.

We highlight that different clustering or community-based algorithms can be applied to calculate groups of users and find appropriate representatives for the new user. These groups or communities are afterward used to calculate either item-item matrices in the case of item-based collaborative filtering or user-user matrices in the case of user-based collaborative filtering. Especially in the latter case, the storage complexity can be reduced since the size of a user-user matrix is smaller for each group of users – either representing a cluster or community – found by the algorithms. In this sense, an empirical analysis may focus on the reduction of computation runtime due to the fact that groups of users can be employed to speed up similarity calculations during the rating prediction process, for instance, by considering only users within the current representative's group.

Another important analysis consists of evaluating the different representative selection strategies. Since finding appropriate representatives has a substantial impact system's performance, distance or similarity metrics between user profiles or member of a community may be applied as proxies to evaluate the quality of produced recommendations. Finally, we argue that, in order to conduct a complete analysis of a social-based recommender system designed from the principles described in this work, the following data is required:

- User data: Age, gender, date of birth, address, height, weight etc.
- User relations: Relations between users (e.g., friend, family, colleague).
- Usage data: User ratings, comments, reviews, and interactions with items.
- Geographic data: Location of his or her interactions, places that he or she would like to visit etc.

6 Summary and Future Work

The importance of recommender systems is still ubiquitous. Social-based recommender systems – especially those exploiting social relationships among users (e.g., communities) to improve personal recommendations – are a new approach in this field. In this paper, we have introduced a generic framework to support the design of collaborative filtering based recommender systems. The proposed framework describes how social collective information can be incorporated in traditional collaborative filtering approaches to alleviate the cold-start problem.

In this sense, our framework considers simultaneously the clustering, the representative determination, and adaption to handle new users. Additionally, we presented our idea how to evaluate the framework's performance compared to a collaborative filtering approach using user profiles.

As our framework is a theoretical one, the main future task is to evaluate its efficiency using real-world datasets. Therefore, it is important to define a specific application for which user profiles, items, ratings and graph representations of users are either already given or can be created from the data at hand. As the quality of user groups have a huge impact on the prediction results, different graph algorithm approaches to determine distinct groups have to be investigated, e.g., based on betweenness, density and so on. Furthermore, the complexity of these algorithms has to be taken into account, as many algorithms are computation-intensive and approximation have to be used. Additionally, the framework should be implemented in a big-data infrastructure, considering the storage of the different models in an appropriate graph database.

We also plan to investigate how the introduced framework can alleviate other traditional collaborative filtering problems such as grey-sheep users (i.e., users who present rating patterns mismatching any other users, and hence, receiving poor recommendations) [8] or shilling attacks (i.e., when fake user profiles are created by malicious users to manipulate item ratings and distrust recommendations) [10].

Acknowledgment. This research was partially supported by the Wrocław University of Science and Technology under Polish-German cooperation program between the Ministry of Science and Higher Education and the German Academic Exchange Service, Project No. 0401/0115/18; by statute research grant of Ministry of Science and Higher Education, Project No. 0402/0071/17; by DAAD under grant PPP 57391625; and by the Brazilian National Council for Scientific and Technological Development (CNPq) - Science without Borders Program.

References

1. Anjumol, M., Ancy, K.S.: A survey on semantic based social recommendation. Int. Res. J. Eng. Technol. (IRJET) **3**(6), 415–420 (2016)
2. Beel, J., Gipp, B., Langer, S., Breitinger, C.: Research-paper recommender systems: a literature survey. Int. J. Digit. Libr. **17**(4), 305–338 (2016)
3. Ben-Shimon, D., Tsikinovsky, A., Rokach, L., Meisles, A., Shani, G., Naamani, L.: Recommender system from personal social networks. In: Wegrzyn-Wolska, K.M., Szczepaniak, P.S. (eds.) Advances in Intelligent Web Mastering. AINSC, vol. 43, pp. 47–55. Springer, Heidelberg (2007). https://doi.org/10.1007/978-3-540-72575-6_8
4. Bernardes, D., Diaby, M., Fournier, R.: A social formalism and survey for recommender systems. SIGKDD Explor. **16**(2), 20–36 (2014)
5. Betru, B.T., Onana, Ch.A, Batchakui, B.: Deep learning methods on recommender system: a survey of state-of-the-art. Int. J. Comput. Appl. **162**(10), 17–22 (2017)
6. Gorripati, S.K., Vatsavayi, V.K.: A community based content recommender systems. Int. J. Appl. Eng. Res. **12**(22), 12989–12996 (2017). ISSN 0973–4562

7. Jannach, D., Zanker, M., Felfernig, A., Friedrich, G.: Recommender Systems: An Introduction. Cambridge University Press, Cambridge (2011)
8. Khusro, S., Ali, Z., Ullah, I.: Recommender systems: issues, challenges, and research opportunities. Information Science and Applications (ICISA) 2016. LNEE, vol. 376, pp. 1179–1189. Springer, Singapore (2016). https://doi.org/10.1007/978-981-10-0557-2_112
9. Kywe, S.M., Lim, E.-P., Zhu, F.: A survey of recommender systems in Twitter. In: Aberer, K., Flache, A., Jager, W., Liu, L., Tang, J., Guéret, C. (eds.) SocInfo 2012. LNCS, vol. 7710, pp. 420–433. Springer, Heidelberg (2012). https://doi.org/10.1007/978-3-642-35386-4_31
10. Lam, S.K., Riedl, J.: Shilling recommender systems for fun and profit. In: Proceedings of the 13th ACM International Conference on World Wide Web, pp. 393–402 (2004)
11. Leskovec, J., Rajaraman, A., Ullman, J.D.: Mining of Massive Datasets (2014)
12. Ma, H., Yang, H., Lyu, M.R., King, I.: SoRec: social recommendation using probabilistic matrix factorization. In: International Conference on Information and Knowledge Management (CIKM) (2008)
13. Maleszka, M., Mianowska, B., Nguyen, N.T.: A method for collaborative recommendation using knowledge integration tools and hierarchical structure of user profiles. Knowl. Based Syst. **47**, 1–13 (2013)
14. Mianowska, B., Nguyen, N.T.: A method for collaborative recommendation in document retrieval systems. In: Selamat, A., Nguyen, N.T., Haron, H. (eds.) ACIIDS 2013. LNCS (LNAI), vol. 7803, pp. 168–177. Springer, Heidelberg (2013). https://doi.org/10.1007/978-3-642-36543-0_18
15. Nagwekar, K.: A survey on recommendation systems based on online social communities. Int. J. Innov. Res. Comput. Commun. Eng. **4**(12), 21857–21863 (2016)
16. Ricci, F., Rokach, L., Shapira, B., Kantor, P.B.: Recommender Systems Handbook. Springer, Boston (2011). https://doi.org/10.1007/978-0-387-85820-3
17. Wang, X., Hoi, S.C.H., Ester, M., Bu, J., Chen, Ch.: Learning personalized preference of strong and weak ties for social recommendation. In: Proceedings of International World Wide Web Conference, pp. 1601–1610 (2017)
18. Yang, X., Guo, Y., Liu, Y., Steck, H.: A survey of collaborative filtering based social recommender systems. Comput. Commun. **41**, 1–10 (2014)
19. Zhang, S., Yao, L., Sun, A.: Deep learning based recommender system: a survey and new perspectives. ACM J. Comput. Cult. Herit. **1**(1), Article 35 (2017). https://dblp.uni-trier.de/rec/bibtex/journals/corr/ZhangYS17aa

Recommender System Based on Fuzzy Reasoning and Information Systems

Martin Tabakov[(✉)]

Faculty of Computer Science and Management, Department of Computational Intelligence, Wroclaw University of Science and Technology, Wroclaw, Poland
martin.tabakow@pwr.edu.pl

Abstract. In this research a recommender system with possible applications in e-commerce, based on rule induction mechanism and fuzzy reasoning, is presented. The theoretical concept proposed assume the application of fuzzy sets in a procedure of rule induction, as an information generalization, in purpose to predict the degree of subjective customer satisfaction with respect to his previous reviews. The innovative idea lays in the transformation of decision rules into fuzzy rules, regarding to the basic Mamdani reasoning model. The research was verified on real data, i.e. customer reviews of different products.

Keywords: E-commerce · Recommender systems · Fuzzy systems
Rule induction

1 Introduction

In recent years, e-commerce systems have developed rapidly as customers purchase regularly products from online stores. What more, there are many ways and tools to monitor the activity of potential customers, for example through social media and many others applications. As part of the business information flow, recommender systems have proven to be a successful tool to assist customers, by advising and finding the most suitable products to facilitate online decision-making [10, 15–18]. There are many successful technologies for recommender systems, such as collaborative filtering (CF) systems, which have already been applied by many commercial web sites such as Amazon.com, Netflix, and so on [1, 2, 7, 8]. The basic concept of these systems lays in the use of historical data related to user preferences or behavior to predict how new users will act [6, 9]. There is a lot of development of this concept using fuzzy sets as well [4, 5, 12, 20].

In the research proposed, a new concept combining rule induction and fuzzy reasoning is proposed. The concept introduced uses learning set, which presents customer preferences and generalize information with fuzzy sets to predict degree of customer interest with new products. What distinguishes the research proposed, is the interpretation of the information function, applied in the rule induction process, by defining it's values as fuzzy sets, which allows to generate fuzzy rule base from the data itself. The rest of the paper is organized as follows: in Sect. 2 the necessary theoretical background is briefly explained, in Sect. 3 the recommender system process flow diagram is

© Springer Nature Switzerland AG 2018
N. T. Nguyen et al. (Eds.): ICCCI 2018, LNAI 11055, pp. 248–259, 2018.
https://doi.org/10.1007/978-3-319-98443-8_23

presented, in Sect. 4 the data set used is explained, in Sect. 5 experiments and results are given and finally conclusions are introduced.

2 Theoretical Background

In this section, the pre preliminaries of the Mamdani fuzzy model [11] and Pawlak's information systems rule induction procedure [19], are briefly explained.

2.1 Fuzzy Sets

A fuzzy set A consists of a domain X of real numbers together with a function $\mu_A: X \rightarrow [0, 1]$, [21] i.e.:

$$A =_{df} \int_X \mu_A(x)/x, x \in X \tag{1}$$

here the integral denotes the collection of all points $x \in X$ with associated membership grade $\mu_A(x) \in [0, 1]$. The function μ_A is also known as the membership function of the fuzzy set A, as its values represents the grade of membership of the elements of X to the fuzzy set A. The idea is to use membership functions as characteristic functions (any crisp set is defined by its characteristic function) to describe imprecise or vague information. This possibility along with the corresponding defined mathematical apparatus, initiated a number of applications.

2.2 Fuzzy Reasoning – Mamdani Model

Figure 1 shows the schematic diagram of a type-1 fuzzy controller. The main idea is that all input information are fuzzified and then processed with respect to the assumed knowledge base, inference method and the corresponding defuzzification method.

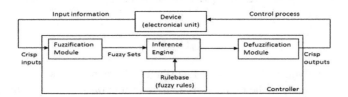

Fig. 1. Information flow within a typical type-1 fuzzy controller

Let consider the rule base of a fuzzy logic controller consisting of N rules which take the following form:

$$R^n :_{df} IF\left(x_1 \text{ is } X_1^n\right)o. \ldots o\left(x_i \text{ is } X_i^n\right) THEN \; y \text{ is } Y_n \tag{2}$$

Where X_i^n ($i = 1, \ldots, I$; $n = 1, \ldots, N$) are fuzzy sets defined over corresponding domains and Y_n is an output information, which in the Mamdani model [11] is assumed as fuzzy set as well, defined over some domain Y. The binary operator 'o' is the t- or s-*norm* (o $\in \{\otimes, \oplus\}$; $\otimes, \oplus$: $[0, 1]^2 \rightarrow [0, 1]$) which have the commutative, associative and the monotonic properties, and have the constants 1 and 0 as unit elements, respectively. In fuzzy logic, the t-norm operator provides the characteristic of the AND operator, while the s-norm provides the characteristic of the OR operator [3].

Assuming an input vector $\bar{x} = \{x_1', x_2', x_3', \ldots, x_i'\}$, typical computations of a fuzzy system consist of the following steps:

(1) Compute the membership grades of x_i' on each X_i^n, $\mu_{X_i^n}(x_i')$, $i = 1, \ldots, I$; $n = 1, \ldots, N$

(2) Compute the firing value of the nth rule, f^n:

$$f^n(\bar{x}) =_{df} \mu_{X_1^n}(x_1') o \ldots o \, \mu_{X_i^n}(x_i') \in [0, 1] \tag{3}$$

(3) Compute defuzzification output. The most common method is the centre of gravity (COG) method with assumed relation between the premise and the conclusion of the fuzzy rules as the *min* operator:

$$Y_{COG}(\bar{x}) =_{df} \frac{\sum_{y \in Y} \mu_{\cup_n Y_n'}(y) \cdot y}{\sum_{y \in Y} \mu_{\cup_n Y_n'}(y)} \tag{4}$$

$$\text{where}: \forall_{y \in Y} \mu_{Y_n'}(y) =_{df} \min\{f^n, \mu_{y_n}(y)\}, n = 1, \ldots, N \tag{5}$$

The output value is directly related to the control process.

2.3   Rule Induction

In the eighties and nineties Zdzislaw Pawlak introduced the fundamentals of information systems [13] and rough sets [14]. An information system (IS) in defined by the following elements:

$$IS =_{df} (U, A, V, f) \tag{6}$$

where U is a universe, A is a set of attributes, V represents attributes domains: $V =_{df} \cup_a V_a$, where V_a is the domain of the ath attribute ($a \in A$) and f is the so called information function: $f: U \times A \rightarrow V$, $\forall_{x \in U, a \in A} f(u, a) \in V_a$. Important role in the theory plays the introduced *indiscernibility binary relation*, defined over U: $\text{IND}(B) =_{df} \{(x, y) \in U \times U : \forall_{a \in B} f(x, a) = f(y, a)\}$, under which the *lower* and *upper* approximations of any subset of U can be defined, respectively:

$$B \downarrow X =_{df} \{x \in U : [x]_{\text{IND}} \subseteq X\}, B \uparrow X =_{df} \{x \uparrow U : [x]_{\text{IND}} \cap X \neq \emptyset\}, \tag{7}$$

where $B \subseteq A$ and $X \subseteq U$.

Information systems can be interpreted as a decision tables if a decision attribute is introduced. With this assumption, a decision making approach was introduced by Skowron and Suraj [19], which generates a set of rules for any decision attribute value. A detailed explanation is omitted here, but briefly the procedure consists of the following steps:

1. Define an information system with decision attribute,
2. Eliminate object conflicts (i.e. objects with same information function values, but different decision values), applying *lower* or *upper* approximation precision analysis,
3. Provide *attribute reduct* using *discernibility matrix*,
4. Apply rule induction algorithm on the so defined new information system (completing step 2 and 3) and thus, define set of rules for each decision attribute value, which correctly cover the decision problem.

Below, an example is introduced which illustrates the input and the output of the rule induction algorithm, assuming that step 2 and 3 are completed.

Input: information system $(x_1, x_2, \ldots, x_5 \in U; \{a, b, c\} \subseteq A; a*$ - decision attribute; $V_a = V_b = V_c = V_{a*} = \{0, 1, 2\})$

	a	b	c	a^*
x_1	1	0	1	0
x_2	0	0	0	1
x_3	2	0	1	0
x_4	0	0	1	2
x_5	1	1	1	0

Output:
$Rule_1$ (concern decision value 0): $f(x_1,a) \lor f(x_3,a) \lor (f(x_5,a) \lor f(x_5, b)) \Rightarrow$ (decision: 0),
$Rule_2$: $f(x_2,c) \Rightarrow$ (decision: 1),
$Rule_3$: $f(x_4,a) \land f(x_4,c) \Rightarrow$ (decision: 2).

The only disadvantage with this decision making approach is that the induced rules are very crisp, i.e. $rule_2$ is interpreted as follows: *if object of x_2 type has information function value for the attribute 'c' exactly equal to '0' then make decision '1'.*

2.4 Fuzzy Information System

The above mentioned disadvantage of the rule induction process, lays in the basic of the research proposed. It is enough to involve fuzzy sets in the process of rule induction and thus, to achieve information generalisation. Therefore, the new proposal is to modify the information function introduced, by defining all attribute values as fuzzy sets. What more, for any numerical attribute, the basic fuzzy sets *low*, *medium* and *high* can be defined by using directly the data, assuming normal distributions.

So, for any attribute $A_i =_{df} \{a_1^i, a_2^i, \ldots, a_n^i\}$ with respect to all considered objects, a normal distribution of attribute values may be considered in purpose to define the fuzzy set *medium*, as follows:

$$\mu_{medium}(a) =_{df} e^{\frac{-(a-a_0)}{2\sigma^2}} \qquad (8)$$

determining the expected value (a_0) and the standard deviation (σ) directly from the set A_i. Next, the fuzzy sets *low* and *high* attribute values can be easily defined as well:

$$\mu_{low}(a) =_{df} \begin{cases} 1 - e^{\frac{-(a-a_0)}{2\sigma^2}} & : a < a_0 \\ 0 : a \geq a_0 \end{cases}; \mu_{high}(a) =_{df} \begin{cases} 0 : a < a_0 \\ 1 - e^{\frac{-(a-a_0)}{2\sigma^2}} & : a \geq a_0 \end{cases} \quad (9)$$

Therefore, an exemplary fuzzy information system could take the following form:

	attribute$_1$	attribute$_2$	decision attribute
object$_1$	low	medium	D_1
object$_2$	medium	low	D_1
object$_3$	high	low	D_2

with the interpretation of the information function for example pair object/attribute (object$_1$ and attribute$_1$), as follows: f_{fuzzy}(object$_1$, attribute$_1$) = *low*, meaning f(object$_1$, attribute$_1$) has the highest degree of membership to the fuzzy set *low*, defined over the domain of the attribute$_1$.

The decision attribute values can be defined as fuzzy sets as well, representing the degree of decision. So, under this interpretation of a fuzzy information system, the rule induced are very simple to be transformed into fuzzy rules. For example, if a rule is induced as follows:

$$f(\text{object}_1, \text{attribute}_1) \wedge f(\text{object}_2, \text{attribute}_1) \Rightarrow \text{decision} : D_1,$$

it can be naturally transformed into the fuzzy rule:

IF (f(object$_1$, attribute$_1$) is '*small*') $\otimes$ (f(object$_2$, attribute$_1$) is '*medium*') THEN (decision is D_1) .

In the provided experiments in this research, this concept of fuzzy information system and rule induction mechanism, was applied.

3   Recommender System Proposal

The aim of the system proposed is to apply fuzzy sets during the assumed rule induction procedure with fuzzification defined directly from the data used and therefore, to induce in automatic manner fuzzy rules suitable for further reasoning process with respect to the Mamdani model. The chart below shows the process flow diagram of the recommender mechanism proposed considering the learning phase and the system use as well.

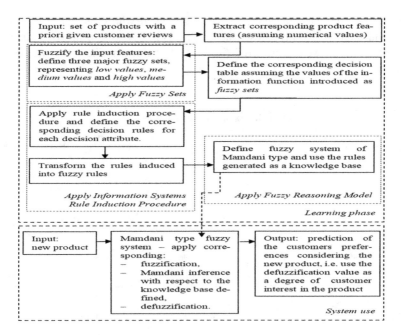

Fig. 2. Process flow diagram

4 The Data Set Used

The experiments provided, were carried out by polling 45 potential customers (IT students) of electronic products, such as: *laptops, cell phones* and *tablets*. Each of the customer was given the opportunity to review a list of product items during a survey. A customer had the possibility to review a product and to mark it with one of four numerical values related to a certain level of interest with the item: 0 – '*not interested*', 1 – '*very little interested*', 2 – '*interested*' and 3 – '*very interested*'. Degree of interest 0 and 1 could be interpreted as negative and degree of interest 2 and 3 as positive. These four possible customer evaluation values are defined by corresponding fuzzy sets, as they determine the conclusions in the Mamdani reasoning model proposed. The fuzzy sets were defined as Gaussian distributions (but without the factor before the exponential function to ensure the achievement of 1), representing the degree of customer interest, with expected values = 10, 20, 30 and 40 and equal standard deviation $\sigma = 25$ (see Fig. 3).

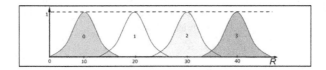

Fig. 3. Representation of the degree of customer interest with fuzzy sets

Below, an exemplary evaluation table is presented with customer review of cell phones with eight randomly chosen items[1]. It should be noted, however, that only numerical attributes were considered (in the table below, attribute OS was ignored) (Table 1).

Table 1. Exemplary evaluation table.

	Screen diagonal	Screen resolution	Built-in memory	Battery	Weight	RAM	Camera (Mpix)	Thickness (mm)	Clock speed	OS	Number of cores	Price	Customer evaluation
Apple iPhone 5S	4"	1080 × 1920	32 GB	1560 mAh	112 g	1 GB	8 Mpix	7,6	1.3 GHz	iOS 7	2	3 149 zl	0
Samsung Galaxy S5	5.1"	1080 × 1920	16 GB	2800 mAh	146 g	2 GB	16 Mpix	8.3 mm	2.45 GHz	Android	4	2 999 zl	3
HTC One Max	5.9"	1080 × 1920	16 GB	3300 mAh	217 g	2 GB	2.1 Mpix	10,28	1.7 GHz	Android	4	2 976 zl	2
Sony Xperia Z2	5.2"	1080 × 1920	16 GB	3200 mAh	163 g	3 GB	20.7 Mpix	8,2	2.36 GHz	Android	4	2 799 zl	2
Nokia Lumia 1020	4.5"	768 × 1280	32 GB	2000 mAh	160 g	2 GB	41.0 Mpix	10,5	1.5 GHz	Windows Phone 8	2	2 099 zl	0
Samsung Galaxy S4	5"	1080 × 1920	16 GB	2600 mAh	130 g	2 GB	13 Mpix	7,9	1,9	Android	4	1 899 zl	1

Summarizing the data used, 45 potential customers were asked to provide reviews of three groups of electronic products: *laptops*, *cell phones* and *tablets*, with 25 item in each product group. Therefore, each customer has reviewed 25 × 3: 75 product items, with total number of reviews: 45 × 75: 3375 items. The set of attributes describing the product items considered is listed in Table 2.

Table 2. The attributes used.

Laptops	Cell phones	Tablets
− screen diagonal (inches)	− screen diagonal (inches)	− screen diagonal (inches)
− screen resolution (pixels)	− screen resolution (pixels)	− screen resolution (pixels)
− hard drive (GB)	− built-in memory (GB)	− built-in memory (GB)
− number of USB 3.0 connectors	− battery (mAh)	− battery (mAh)
− weight (kg)	− weight (g)	− weight (g)
− RAM	− RAM (GB)	− RAM (GB)
− video graphic card memory (MB)	− camera (MPix)	− camera (MPix)
− clock speed (GHz)	− thickness (mm)	− clock speed (GHz)
− number of cores	− clock speed (GHz)	− number of cores
− price (PLN)	− number of cores	− price (PLN)
	− price (PLN)	

---

[1] The price attribute may not be up-to-date, but it doesn't change the concept proposed.

5 Experiments and Results

The main assumption of the conducted experiments was to investigate the accuracy of the recommender concept proposed. As the considered case is not a classification problem, but rather requires a certain interpretation of the results achieved as they are very subjective, the following two accuracy tests were proposed:

1. *Test one* – all product from each customer list were processed by the fuzzy system proposed and sorted with respect to the system outputs.
2. *Test two* – products from each customer list were divided into two parts: learning set (19 items) and test set (6 randomly selected items) for validation purpose.

In both cases, for each customer, a knowledge base was generated with respect to his specific product review under the concept proposed (see Fig. 2). Therefore, for each customer a fuzzy reasoning system under the recommender concept proposed was developed which provides personalization.

Test one – results

The results are shown in the infographics below, with corresponding explanation (Fig. 4).

The results are shown in the infographics below, with corresponding explanation.

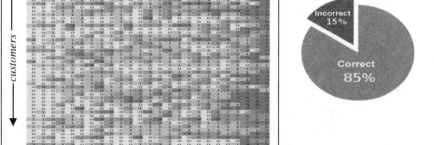

Fig. 4. Infographic presenting the results achieved for product *laptops* with corresponding classification

To explain the infographic above, let consider the first row. It contains the first customer product reviews highlighted with increasing colour saturation with respect to the review values. The first customer marked the chosen products with values: 0, 1 and 2. The values in the columns were sorted in ascending order with respect to the fuzzy recommender system output. Therefore, high system accuracy is related to more intense red colours on the right part of the infographic, see Fig. 5 below.

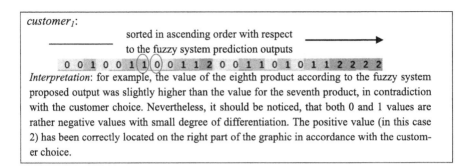

Fig. 5. Infographic interpretation. (Color figure online)

So, it is expected that more saturated colours will be located on the right part of the graphic. Furthermore, a classification of the results in terms of correct system prediction was introduced, assuming that if the average of customer review values of the half of the products from the right site of the row is higher than the average for all products, which indicates that the most of the products with higher degree of customer interest are on the right side of the graphic, it is classified as correct, else it is classified as incorrect. For example, the above system result regarding to *customer₁* reviews, is classified as correct system answer, as the average of the half products on the right side (12 product) of the row is equal to ≈1.08, which is higher than the average value for all products: 0.8. For example, the second and the third rows (*customer₂* and *customer₃* results) are recognized as incorrect. Below in Fig. 6, results for products *tablets* and *cell phones* are presented in the same manner.

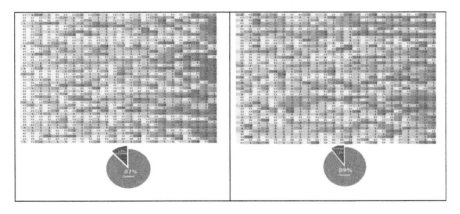

Fig. 6. *Cell phones* and *tablets* infographics and classification results, respectively.

As it can be noticed, the most saturated colours are mostly located on the right side of the infographics, which indicates correct system predictions.

Test two – results

The idea of the second test was more classic, i.e. for each customer a learning set and a test set was provided, and next a fuzzy system with respect to the concept proposed (see Fig. 2) was designed using the learning set. The system was applied on the test data set with the idea that products with higher customer interest should have higher system output values. A test for a certain customer was recognized as correct, if the sum of the customer reviews of the half of the products with higher system scores was more or equal to the sum of customer reviews with lowest system scores (Fig. 7).

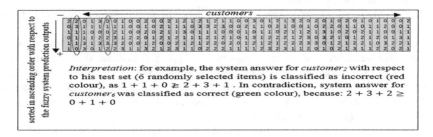

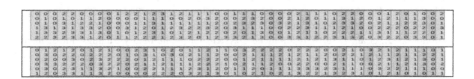

Interpretation: for example, the system answer for *customer₂* with respect to his test set (6 randomly selected items) is classified as incorrect (red colour), as $1 + 1 + 0 \not\geq 2 + 3 + 1$. In contradiction, system answer for *customer₆* was classified as correct (green colour), because: $2 + 3 + 2 \geq 0 + 1 + 0$

Fig. 7. Results for product: *laptops*

Below in Fig. 8, results for products *tablets* and *cell phones* are presented in the same manner.

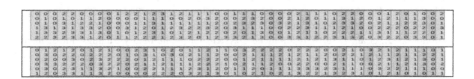

Fig. 8. Results for products: *cell phones* and *tablets* respectively

As it can be noticed, for all considered products, satisfactory results were achieved, which proves good system accuracy.

6 Conclusions

In this research a new recommender system was introduced. The theoretical concept presented combines two major approaches: rule induction procedure based on information systems and fuzzy reasoning. The involvement of fuzzy sets into the rule induction process allowed information generalization which represents subjective observations and opinions very accurate. The experiments provided, illustrated with corresponding infographics, proved good system accuracy and possibility of further development.

Acknowledgement. I would like to thank to my student Jakub Salamon for the experiments provided.

References

1. du Boucher-Ryan, P., Bridge, D.: Collaborative recommending using formal concept analysis. Knowl.-Based Syst. **19**(5), 309–315 (2006)
2. Breese, J., Heckerman, D., Kadie, C.: Empirical analysis of predictive algorithms for collaborative filtering. In: UAI98, pp. 43–52 (1998)
3. Bronstein, I.N., Semendjajew, K.A., Musiol, G., Mühlig, H.: Taschenbuch der Mathematik, p. 1258. Harri Deutsch (2001)
4. Cao, Y., Li, Y.: An intelligent fuzzy-based recommendation system for consumer electronic products. Expert Syst. Appl. **33**(1), 230–240 (2007)
5. Cheng, L.-C., Wang, H.-A.: A fuzzy recommender system based on the integration of subjective preferences and objective information. Appl. Soft Comput. **18**, 290–301 (2014)
6. Gediminas, A.: Toward the next generation of recommender systems: a survey of the state-of-the-art and possible extensions. IEEE Trans. Knowl. Data Eng. **17**(6), 734–749 (2005)
7. Greg, L., Brent, S., Jeremy, Y.: Amazon.com recommendations: item-to-item collaborative filtering. IEEE Int. Comput. **7**(1), 76–80 (2003)
8. Herlocker, J.L., Konstan, J.A., Borchers, A., Riedll, J.: An algorithmic framework for performing collaborative filtering. In: Proceedings of the 1999 Conference on Research and Development in Information Retrieval (1999)
9. Hofmann, T.: Latent semantic models for collaborative filtering. ACM Trans. Inf. Syst. **22**(1), 89–115 (2004)
10. Liu, D.R., Lai, C.H., Lee, W.J.: A hybrid of sequential rules and collaborative filtering for product recommendation. Inf. Sci. **179**(2), 3505–3519 (2009)
11. Mamdani, E.H., Assilian, S.: An experiment in linguistic synthesis with a fuzzy logic controller. Int. J. Man-Mach. Stud. **7**(1), 1–13 (1975)
12. Morawski, J., et al.: A fuzzy recommender system for public library catalogs. Int. J. Intell. **32**, 1062–1084 (2017)
13. Pawlak, Z.: Information systems theoretical foundations. Inf. Syst. **6**, 205–218 (1981)
14. Pawlak, Z.: Rough Sets: Theoretical Aspects of Reasoning About Data. Kluwer Academic Publishers Group, Dordrecht (1991)
15. Porcel, C., Herrera-Viedma, E.: Dealing with incomplete information in a fuzzy linguistic recommender system to disseminate information in university digital libraries. Knowl. Based Syst. **23**(1), 32–39 (2010)
16. Resnick, P., Iacovou, N., Suchak, M., Bergstrom, P., Riedl, J.: GroupLens: an open architecture for collaborative filtering of netnews. In: Proceedings of ACM Conference on Computer-Supported Cooperative Work, pp. 175–186 (1994)
17. Sarwar, B., Karypis, G., Konstan, J., Reidl, J.: Item-based collaborative filtering recommendation algorithms. In: Proceedings of the 10th International Conference on World Wide Web, Hong Kong, pp. 285–295 (2001)
18. Schafer, J.B., Konstan, J.A., Riedl, J.: E-commerce recommendation applications. Data Min. Knowl. Disc. **5**(1), 115–153 (2001)

19. Skowron, A., Suraj, Z.: A rough set approach to real-time state identification for decision making. Institute of computer science research, Report 18/93, Warsaw University of Technology (1993)
20. Year, R., Martínez, L.: Fuzzy tools in recommender systems: a survey. Int. J. Comput. Intell. Syst. **10**, 776–803 (2017)
21. Zadeh, L.: Fuzzy sets. Inf. Control **8**(3), 338–353 (1965)

Proposal of a Recommendation System for Complex Topic Learning Based on a Sustainable Design Approach

Xanat Vargas Meza$^{(\boxtimes)}$ (ID) and Toshimasa Yamanaka (ID)

The University of Tsukuba, Tsukuba, Japan
kt_designbox@yahoo.com,
tyam@geijutsu.tsukuba.ac.jp

Abstract. There are several issues compromising the educational role of social networks, particularly in the case of video based online content. Among them, individual (cognitive and emotional), social (privacy and ethics) and structural (algorithmic bias) challenges can be found. To cope with such issues, we propose a recommendation system for online video content, applying principles of sustainable design. Precision and recall in English were slightly lower for the system in comparison to YouTube, but the variety of recommended items increased; while in Spanish, precision and recall were higher. Expected results include fostering learning and adoption of complex thinking by taking on account a user's objective and subjective context.

Keywords: Sustainable design · Social networks · Computational intelligence
Information retrieval · Multicriteria decision making

1 Introduction

Social Networking Sites have affected educational methods. Some educators create Facebook groups, upload presentations in SlideShare or send assignments through Coursera. Those who choose a self-learning approach can try websites like Khan Academy or TED. At the university level, the adoption rate of Social Networking Sites by faculty is over 90% [18], being Facebook and YouTube the most cited. YouTube has been widely explored in relationship with education [6, 10, 12] and most studies note its benefits. However, YouTube is not exempted of issues.

The first question to ask is, is the video content suitable for the user? Are balanced remarks used to explain a concept? Did the user have a bad day and is not in the mood to watch a maths class? These are examples of cognitive and psychological aspects that not even the best data tracking and recommendation systems can predict and react to with accuracy in the case of open video data. Nevertheless, probably the most worrying trend is how YouTube creates addiction in adults and youngsters [5].

Having a hypothetical full access to track a user's behaviors, educational progress and emotions in a closed network environment would also imply issues like leakage of personal information or data-based discrimination. Moreover, the way companies collect and manage citizens data is often unclear, although the user frequently skips

N. T. Nguyen et al. (Eds.): ICCCI 2018, LNAI 11055, pp. 260–269, 2018.
https://doi.org/10.1007/978-3-319-98443-8_24

reading the terms of service when they are too long. In the case of YouTube, unclear copyright terms and a system that treats all matching videos as infringing content [23] have lead to the muting and removal of videos and user accounts. Therefore, the balance between an individual's rights and their usage of social networks is often at odds with other stakeholders' interests.

The structure of Social Networks has also been shown to lead to echo chambers, particularly in the case of divisive topics [3, 4]. Although these studies argue that individual choices have a stronger effect than algorithms, the impact of advertising and propaganda posing as information in videos should be noted. Recommendation algorithms have often been fooled to fuel harmful content, as the 2017 Presidential Election in U.S. has recently shown [16]. For all the aforementioned reasons, the objective of the present study is to explore alternative ways to manage YouTube's content through a recommendation system, enhancing it for educational purposes.

2   State of the Art

Adomavicius and Tuzhilin [2] described limitations of recommendation systems and discussed areas of improvement, including users and items understanding, incorporation of contextual information, multicriteria ratings, and more flexible and less intrusive recommendations. Among possible methods for improvement, soft computing approaches have been widely researched. Such approaches can be classified in five types [24]: markov models, bayesian approaches, neural networks, genetic algorithms, and fuzzy logic. Another relevant aspect of recommender systems is the type of data input. Collaborative filtering systems employ user's ratings, content-based systems are more focused in the items attributes, and social or knowledge-based systems manage data from more than one user at a time. A review of recommender systems from the last type can be found in [9].

The YouTube recommendation algorithm is assumed to be based on neural networks, which model implicit feedback to provide fresh, time consuming content at a large scale [7]. However, human behaviour and knowledge processes can be considered as fuzzy, as fuzzy logic represents the real world based on a degree of membership for elements into a given set. Fuzzy logic in recommender systems has been mostly focused in item profiling through tags and rates. Yera and Martinez [24] surveyed over 100 algorithms, finding that many have not been tested with real world data, and when they are, it has been on movie recommendations. Moreover, diversity of recommendations is not explored. Leung et al. [15] were the only ones who considered level of interest in their algorithm. Also, Trung et al. [21] modelled fuzzy propagation for opinion mining, providing evidence of a relationship between specific sentiments and topics.

3   Problem Definition: Sustainable Design as a Case of Complex Content

Complexity is measured in terms of depth of knowledge, understanding, awareness and strategy; consequently, complex problems can have multiple solutions or no solution at all. Complex thinking implies that students have to deal with uncertainty and count with the ability to visualize relationships.

As a correct approach to Sustainable Design involves taking in account the whole product/service lifecycle and other stakeholders during the design process, it can be considered as an exemplary case of complexity. A previous study [22] found an acceptance of complexity by English and Spanish speaking viewers of videos about sustainable design in YouTube. Such acceptance might be aided by several factors, including topics related to design (its context) and emotions. We extracted English and Spanish comments from top videos in [22], calculated frequent keywords and word co-occurrence with ConText [8], and performed modularity test to find word clusters in Gephi. This clustering method was more precise than LDA (which was also applied with ConText). Sentiment was analyzed with Sentistrength [20] and the Zyushet package for R [17].

Also, a multiple rank regression analysis was performed in SPSS to find out how much a words cluster (hence a topic) could predict sentiment. Because relations between words and sentiments tend to be small [14], we considered an R^2 equal or higher than 0.15 to be significant. The results in Table 1 indicate that people and money related topics were better predictors for emotions than engineering and design topics (the last ones in the table), confirming the importance of the human dimension in Sustainable Design. An exception would be the Spanish Bullfighting cluster, which was a campaign pushed by grassroots associations.

Table 1. Regression analysis for top word clusters and sentiments.

English				Spanish			
Cluster name	F	R^2	Adj. R^2	Cluster name	F	R^2	Adj. R^2
People	459.389	.228	.228	Jorge B.	58.618	.282	.278
Money	397.567	.205	.205	System	53.481	.264	.259
American	397.567	.204	.203	Bullfighting	51.402	.256	.252
System	372.83	.194	.193	Economy	31.952	.177	.171
Sci-Tech	350.388	.184	.184	Latinamericans	30.211	.169	.163
Holistic	343.372	.181	.181	Earth	25.173	.145	.139
Ignorance	332.252	.176	.176	Chile-US	20.02	.118	.113
Information	327.32	.174	.174	Housing	13.716	.084	.078
She	323.556	.173	.172	Lucia G.	12.245	.076	.070
James B.	310.23	.167	.166	Energy	11.454	.071	.065
Jacque F.	240.613	.134	.134	House	10.131	.064	.057
Energy	225.672	.127	.126	Mexico	9.467	.06	.053

Independent variables: Positive Polarity, Negative Polarity, Anger, Anticipation, Disgust, Fear, Joy, Sadness, Trust. Regression = 9, P < 0.005. English N = 13976, Residual = 13966. Spanish N = 1351, Residual = 1341.

A qualitative analysis on comments classified according to Zyushet scores also revealed barriers for Sustainable Design behaviors associated with negative feelings (Table 2). Coincidences between the two languages were marked with asterisks.

Table 2. Summary of barriers for Sustainable Design behavior.

English comments	Spanish comments
Environmental	**Environmental**
– Natural disasters	– Dirty nature
Systemic	**Systemic**
– Money*	– Money*
– Politics/Government*	– Politics/Government*
– Religion	
– War	
Psycho-Social	**Psycho-Social**
– Ignorance*	– Ignorance*
– People*	– Incompetence
– Racism	– People*
– Resistance to change	
Structural	**Structural**
– Ineffectiveness*	– Conventional methods
– Infeasible	– Ineffectiveness*
– Information (quality and quantity)*	– Information (quality and quantity)*
– Ugliness*	– Ugliness*
Hierarchical	
– Loss of autonomy	
– Loss of freedom	

4 Proposed Approach

Taking on account the individual, social and structural issues reviewed, strategies based on Ito [13] and Acaroglu [1] were considered. They propose a view of sustainable technology as extended, adaptive and regenerative intelligence that also aids psychological restorage. The emotions treatment should vary according to the classification of possible users in interested, indecisive and uninterested in the topic (based on [11]). Moreover, negative emotions towards people and nature could be managed through environmental psychology and sociology of emotions [19].

Regarding testing, precision and recall for type of speaker, design area and continent were calculated for 20 items in 3 search cases (for interested, indecisive and uninterested types of user). The results were compared with a search in YouTube in the two languages.

4.1 Design Concept

We propose a YouTube based application to watch educative videos with a reduction of distractions (e.g. ads, viral videos). The application does not store user information permanently to enhance anonymity and flexibility of recommendations. It aims to be a calculator that takes on account the user's objective characteristics and their sentiment towards a topic to optimize recommendations. Because the system should also offer varied content to avoid eco chambers, some degree of randomization in the items was preferred. As the videos are called through the YouTube API, they still receive views and are not manipulated in a way that infringes the social network policies.

The content of curated video databases was classified according to language, uploading date, type of speaker, type of design shown, and continents of countries of the YouTube channel and/or disclosed countries of speakers. Polarity (average of the sum of Sentistrength negative and positive scores) and average Zyushet scores (anger, anticipation, disgust, fear, joy, sadness and trust) for each video were also stored in the databases. Moreover, an emotionality score of comments was computed using the following formula:

$$e = (np^* - 1) + pp - 2 \tag{1}$$

Where e stands for emotionality, np stands for negative polarity and pp stands for positive polarity. The average of emotionality was considered as a proxy for interest in the video, given that arousing content is more likely to provoke a reaction in the user. Maximum fuzzy similarity of topics found in videos comments was also included in the databases. It was calculated with Excel Fuzzy Lookup ad-on, which compares how similar a text is to groups of words (in this case, the 14 top groups found with the modularity test). The advantage of this method is that the video is classified into many overlapping topic groups, besides being quick and easy to compute.

4.2   Recommendation Method

The input for the system is occupation, expertise field, country, sentiment and interest. The output is a list of recommended videos for the average user with the same characteristics. System processing is as follows:

(a) **User:** a student, professional or professor selects the desired language.
    System: activates the corresponding database. Emotionality scores are scaled from [0, 4] to [0, 10] and Polarity scores are scaled from [–4, 4] to [–5, 5].
(b) **User:** inputs variables for occupation, expertise field, country, interest (emotionality) related to the topic and sentiment about the topic (polarity).
    System: defines maximum list size as 20. Classifies user in interested (Emotionality [6, 10], Polarity [–5, 5]), indecisive (Emotionality [0, 5], Polarity [1, 5]) or uninterested (Emotionality [0, 5], Polarity [–5, 0]). Assigns a sequence start (shown underlined in Table 3) for Zyushet scores.

Sadness is related to love and according to previous literature, interested users might be able to stand small doses of negative emotions. Anticipation related comments tended to evaluate pros and cons of sustainable design in an emotionally neutral way, so this emotion was assigned as starting point for indecisive users. Previous literature also argues that urgent and concrete arguments are better for uninterested users, so fear was chosen as their starting point.

Table 3. Sequence of video sentiment scores according to user classification.

User	Joy	Trust	Anticipation	Sadness	Disgust	Anger	Fear
Interested	1	2	3	4	5	6	7
Indecisive	1	2	3	4	5	6	7
Uninterested	1	2	3	4	5	6	7

(c) **System:** Listing of videos is done according to calculation of similarity based on Euclidean distance. It is also divided in two processes. The first video is selected according to polarity and most recent uploading year. Other videos are selected based on most recent uploading year, occupation/type of speaker, expertise field/type of design shown, country/continent and emotionality.

(d) **User:** they click a video from the list in the left side of the first video.
System: there are four ways to handle the user request, shown in Table 4.

Table 4. Handling of new video request.

For English Interested_User and Indecisive_User if watch_time 10 $\leq$ seconds go forward with Zyushet scores sequence else go backward with Zyushet scores sequence	For Spanish Interested_User and Indecisive_User if watch_time 7 $\leq$ seconds go forward with Zyushet scores sequence else go backward with Zyushet scores sequence
For English Uninterested_User if watch_time 10 $\leq$ seconds jump to Sadness and then to Trust, else go backward with Zyushet scores sequence	For Spanish Uninterested_User if watch_time 7 $\leq$ seconds jump to Sadness and then to Trust, else go backward with Zyushet scores sequence

If the user(s) watch time is short, it means they are bored. On that case, Interested and Undecisive users are exposed to more negative emotions, while Uninterested users are exposed to more positive emotions. We also considered that Spanish speakers tend to click faster than English speakers. When handling a new video request, listing is based on most recent uploading year, occupation/type of speaker, expertise field/type of design shown, country/continent and emotionality.

(e) **User:** they type keywords in the search bar.
System: The recommendation method is the same as described above, adding the topic variable.

(f) **User:** to quit the program, the user exits the browser.

4.3   Structure

Figure 1 shows the basic structure of the recommendation system. After the user chooses a language, a data input interface appears. The data is stored in a temporary database that communicates with other two databases (Country/Continent and Video Information). Then, a third interface appears with the video recommendations. If the user wants to search by keywords, the system will list videos again, using the additional Topics database.

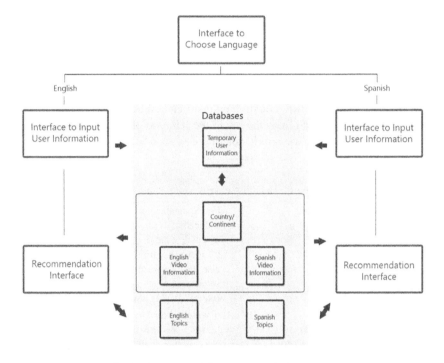

Fig. 1. Structure of the recommendation system.

5 Testing Results

Table 5 shows search cases for YouTube compared to the recommendation system. We calculated average precision and recall scores based on whether the items matched a student of architecture located in top countries related to sustainable design according to [22]. We only included the values for student and architecture in US and Spain because they were the same for other countries with the same language. Average scores in English were lower for the system, but content from more continents was included among the recommended items, which was an intended effect to increase variety. In the case of Spanish, average scores were higher, with content from unclear locations considerably less present among recommended items.

Table 5. Precision and recall for YouTube; and system interested, indecisive and uninterested users.

	YT Eng Prc	YT Eng Rec	Sys Int Prc	Sy Int Rec	Sys Ind Prc	Sys Ind Rec	Sys Uni Prc	Sys Uni Rec
US student	0.075	0.075	0	0	0	0	0	0
US architecture	0.375	0.375	0.450	0.450	0.375	0.375	0.375	0.375
US n. america	0.375	0.375	0.500	0.500	0.450	0.450	0.450	0.450
US avg.	0.275	0.275	**0.316**	**0.316**	0.275	0.275	0.275	0.275
UK europe	0.350	0.350	0.150	0.150	0.125	0.125	0.125	0.125
UK avg.	**0.266**	**0.266**	0.200	0.200	0.166	0.166	0.166	0.166
Australia oceania	0	0	0	0	0.100	0.100	0.100	0.100
Australia avg.	0.150	0.150	0.150	0.150	**0.158**	**0.158**	**0.158**	**0.158**
India s. asia	0	0	0.100	0.100	0.050	0.050	0.050	0.050
India avg.	0.150	0.150	**0.183**	**0.183**	0.141	0.141	0.141	0.141
	YT Spa Prc	YT Spa Rec	Sys Int Prc	Sy Int Rec	Sys Ind Prc	Sys Ind Rec	Sys Uni Prc	Sys Uni Red
Spain student	0.300	0.300	0.125	0.125	0.050	0.050	0.050	0.050
Spain architecture	0.225	0.225	0.350	0.350	0.600	0.600	0.600	0.600
Spain europe	0.025	0.025	0.375	0.375	0.150	0.150	0.150	0.150
Spain avg.	0.183	0.183	**0.283**	**0.283**	0.266	0.266	0.266	0.266
Colombia l. america	0.650	0.650	0.550	0.550	0.650	0.650	0.650	0.650
Colombia avg.	0.391	0.391	0.341	0.341	**0.433**	**0.433**	**0.433**	**0.433**

6 Discussion and Conclusions

We proposed a video recommendation system for complex topic learning with a sustainable design approach in this paper. This is an application that can be downloaded and used anywhere with an internet connection, does not store personal information and takes on account the users feelings about a topic, advancing technology for the democratization and personalization of educative materials.

There are some limitations to our approach. Because of the wide range of analysis that were required to obtain the video databases, it is not possible to add more videos. Later application versions could include a special module to suggest videos, so that human experts can evaluate them before adding them to the databases. Automatizing the semantic, sentiment analyses and keywords classification would accelerate the process of adding videos to the databases. Another aspect that was out of this version's scope was videos in Asian languages, in order to cover the greater number of potential users in the world. Also, testing of the recommendation system with human subjects is highly recommended for future work.

Acknowledgements. The authors wish to thank Akio Sakai, Alejandra Vilaplana, Anna Bogdanova, Eiji Onchi, Emika Okumura, Felix Dollack, Gustavo Ruiz, Imme Arce, Marusia Flores, Nikos Fragkiadakis, Shane Williamson, Suomiya Bao, and our reviewers for their contributions. This paper was written with the support of the Rotary Yoneyama Scholarship Foundation.

References

1. Acaroglu, L.: A Manifesto for Design-Led Systems Change. Medium (2018). https://medium.com/disruptive-design/a-manifesto-for-design-led-systems-change-28ac240db6dd
2. Adomavicius, G., Tuzhilin, A.: Toward the next generation of recommender systems: a survey of the state-of-the-art and possible extensions. IEEE Trans. Knowl. Data Eng. **17**(6), 734–749 (2005)
3. Bakshy, E., Messing, S., Adamic, L.A.: Exposure to ideologically diverse news and opinion on Facebook. Science **348**(6239), 1130–1132 (2015). https://doi.org/10.1126/science.aaa1160
4. Barberá, P., Jost, J.T., Nagler, J., Tucker, J.A., Bonneau, R.: Tweeting from left to right: Is online political communication more than an echo chamber? Psychol. Sci. **26**, 1531–1542 (2015). https://doi.org/10.1177/0956797615594620
5. Burroughs, B.: YouTube kids: the app economy and mobile parenting. Social Media+ Society **3**(2) (2017). https://doi.org/10.1177/2056305117707189
6. Burke, S.C., Snyder, S.L., Shonna, L.: Students' perceptions of YouTube usage in the college classroom. Int. J. Instr. Technol. Distance Learn. **5**(11), 13–23 (2008)
7. Covington, P., Adams, J., Sargin, E.: Deep neural networks for YouTube recommendations. In: Proceedings of the 10th ACM Conference on Recommender Systems, Boston, pp. 191–198 (2016)
8. Diesner, J.: ConText: software for the integrated analysis of text data and network data. In: Conference of International Communication Association (ICA), Seattle (2014)
9. Grad-Gyenge, L., Kiss, A., Filzmoser, P.: Graph embedding based recommendation techniques on the knowledge graph. In: Adjunct Publication of the 25th Conference on User Modeling, Adaptation and Personalization, pp. 354–359 (2017). https://doi.org/10.1145/3099023.3099096
10. Ham, J.J., Schnabel, M.A.: Web 2.0 virtual design studio: social networking as a facilitator of design education. Archit. Sci. Rev. **54**(2), 108–116 (2009). https://doi.org/10.1080/00038628.2011.582369
11. Hine, D.W., et al.: Preaching to different choirs: how to motivate dismissive, uncommitted, and alarmed audiences to adapt to climate change? Glob. Environ. Change **36**, 1–11 (2016). https://doi.org/10.1016/j.gloenvcha.2015.11.002
12. Juhasz, A.: Learning from YouTube. MIT Press, Cambridge (2011)
13. Ito, J.: Resisting Reduction: A Manifesto. Designing our Complex Future with Machines (2017). https://pubpub.ito.com/pub/resisting-reduction

14. Kuster, D., Kappas, A.: Measuring emotions in individuals and internet communities. In: Benski, T., Fisher, E. (eds.) Internet and Emotions, vol. 22, pp. 48–64. Routledge, New York (2013)
15. Leung, C.W.K., Chan, S.C.F., Chung, F.L.: A collaborative filtering framework based on fuzzy association rules and multiple-level similarity. Knowl. Inf. Syst. **10**(3), 357–381 (2006). https://doi.org/10.1007/s10115-006-0002-1
16. Lewis, P.: Fiction is Outperforming Reality: How YouTube's Algorithm Distorts Truth. The Guardian   (2018).   https://www.theguardian.com/technology/2018/feb/02/how-youtubes-algorithm-distorts-truth
17. Mohammad, S.M., Turney, P.D.: Crowdsourcing a word–emotion association lexicon. Comput. Intell. **29**(3), 436–465 (2013). https://doi.org/10.1111/j.1467-8640.2012.00460.x
18. Moran, M., Seaman, J., Tinti-Kane, H.: Teaching, Learning, and Sharing: How Today's Higher Education Faculty Use Social Media. Babson Survey Research Group (2011)
19. Norgaard, K.M.: Living in Denial: Climate Change, Emotions, and Everyday Life. MIT Press, Cambridge (2011)
20. Thelwall, M., Buckley, K., Paltoglou, G., Cai, D., Kappas, A.: Sentiment strength detection in short informal text. J. Assoc. Inf. Sci. Technol. **61**(12), 2544–2558 (2010). https://doi.org/10.1002/asi.21416
21. Trung, D.N., Jung, J.J., Vu, L.A., Kiss, A.: Towards modeling fuzzy propagation for sentiment analysis in online social networks: a case study on TweetScope. In: IEEE 4th International Conference on Cognitive Infocommunications (CogInfoCom), pp. 331–338. https://doi.org/10.1109/CogInfoCom.2013.6719266
22. Vargas-Meza, X., Yamanaka, T.: Sustainable design in YouTube: a comparison of contexts. Int. J. Affect. Eng. (IJAE-D) (2017). https://doi.org/10.5057/ijae.IJAE-D-17-00010
23. Von Lohmann, F.: YouTube's january fair use massacre. Deep Links (2009)
24. Yera, R., Martinez, L.: Fuzzy tools in recommender systems: a survey. Int. J. Comput. Intell. Syst. **10**(1), 776–803 (2017). https://doi.org/10.2991/ijcis.2017.10.1.52

A Group Recommender System
for Selecting Experts to Review a Specific
Problem

Dinh Tuyen Hoang[1], Ngoc Thanh Nguyen[2], and Dosam Hwang[1(✉)]

[1] Department of Computer Engineering, Yeungnam University,
Gyeongsan, South Korea
`hoangdinhtuyen@gmail.com`, `dosamhwang@gmail.com`
[2] Faculty of Computer Science and Management, Wroclaw University of Science
and Technology, Wroclaw, Poland
`Ngoc-Thanh.Nguyen@pwr.edu.pl`

Abstract. With the increase in the number of publications and scientific projects, its quality requirements are increasingly needed. Reviewing is the most important step in accrediting the quality of scientific work. Criteria such as independence, competence, and lack of conflicts of interest in an expert are essential in the reviewer selection process. However, we also know that experts have limited knowledge, experience, and opinions about the work of others, so they might misunderstand the viewpoints of the authors, which may lead to rejection of an excellent scientific work or an implicitly successful project proposal. Manually selecting reviewers can be a biased and time-consuming process. In order to solve these problems, we developed a recommender system to choose a group of experts to evaluate a specific problem, such as a research proposal or paper. Our recommender system consists of three main modules: data collection, expert detection, and expert prediction. The data collection module is to collect data from various sources to create a database of scientist profiles. The expert detection module is used to determine the experts on each particular topic. The expert prediction module is to provide a list of experts to answer the query. We conducted experiments with the DBLP Computer Science Bibliography dataset, and the results show that our system is an up-and-coming selection process.

Keywords: Experts finding · Experts' group recommendation
Group experts

1 Introduction

When written work needs to be published or a project needs to be evaluated or a question simply needs to be answered, we usually think of finding experts. The scientific community is expanding, and research issues are increasingly complex and challenging. The number of publications and scientific projects is also

ⓒ Springer Nature Switzerland AG 2018
N. T. Nguyen et al. (Eds.): ICCCI 2018, LNAI 11055, pp. 270–280, 2018.
https://doi.org/10.1007/978-3-319-98443-8_25

increasing. Finding reviewers and experts to evaluate scientific work is a matter of great concern. Reviewers play a significant role in determining the quality of scientific products. However, most publishers and journals choose reviewers from people they already know.

Criteria such as independence, lack of conflicts of interest, and competence are essential to evaluating the quality of scientific products. However, we should also realize that people have limited knowledge, experience, and opinions about other people's work. These restrictions may affect a person's point of view and may result in the rejection of a scientific work or project proposal that potentially could be fruitful. Besides, the person selecting an expert often has some biases that may affect selecting an evaluator or expert. Therefore, the selection process should be supported by unbiased and automated methods that will make the processing time of applications as short as possible. Fortunately, with the increasing amount of scientific data available for collection, it is possible to build a group recommender system to support the selection of reviewers. Recommendation systems create a list of suggestions to the target user inquiry by means of collaborative or content-based filtering methods. A collaborative-based filtering method aims to develop a system to predict items that the user may be interested in, taking into account the user's behavior in the past (i.e., purchased items or rated lists for things) and by considering similar choices made by other users. The main advantage of a collaborative filtering method is that it is not based on content that can be analyzed on the computer. Therefore, it is capable of precisely proposing complex categories (such as movies) that do not require an understanding of the item. A content-based filtering method is based on a description of the item and a profile of the user's favorites. In a content-based recommender system, keywords are used to represent the items, and a user profile is developed to show the kind of item the user likes. In other words, these algorithms try to suggest items that are similar to those that a person liked in the past. This method has its origins in information filtering and information retrieval.

Recent research has presented a variety of work that addresses the problem of selecting suitable experts' or reviewers' papers, conferences, assigned projects, etc. [2,4]. The matching of experts to problems can be considered as a kind of optimization task to create a ranked list of expert candidates for a specific problem. However, data related to the experts are available but cannot be analyzed manually without assistance due to the big size. Various complex algorithms have solved these problems.

Mainly, previous approaches relate to human support [10,12,15]. This leads to have some biases that may affect selecting an evaluator or expert. Besides, most of previous works used traditional machine learning algorithms that usually have a problem when dealing with a big dataset. Also, Criteria such as diversity, competence and independence should be considered in the selection process [7].

In order to solve the above problems, we built a framework to automatically select experts to review a specific problem. The contribution of this paper can be summarized as follows:

(1) We build a module to collect data from open-source databases, such as the DBLP Computer Science Bibliography, ResearchGate, etc., and creates the user's profiles.
(2) An expert detection module is generated to classify experts and non-experts.
(3) An experts group prediction module is developed for matching a specific problem with a suitable group of experts.
(4) Criteria such as diversity, competence and independence are considered in order to find the most appropriate group of experts.

The remainder of this paper is constructed as follows. In Sect. 2, related work on finding expert groups is reviewed briefly. The proposed method is presented in Sect. 3. In Sect. 4, experiments are explained. Lastly, the conclusions and some directions for future work are described in Sect. 5.

2   Related Works

The reviewing of scientific work was discussed a long time ago. Various efforts to resolve the problem of finding an expert who is entirely able to satisfy a specific assignment have been presented in the research. In summary, some approaches are especially worth noting, such as machine learning, decision support systems, and recommender systems.

The machine learning approach is effective when dealing with large datasets that humans cannot handle manually without help. We also know that many useful data related to scientists are available from a variety of sources. Mori et al. [6] built a system to find new business partners, such as suppliers and customers, and developed correlative relationships among them with the help of a machine learning technique. Several features that characterize customer–supplier relationships were represented by a large dimensional vector. The support vector machine method was used to learn a model of customer–supplier relationships. Pavlov and Ichise [8] proposed a method to find experts by link prediction in co-authorship networks. They extracted structural properties from a graph of past collaborations and used them to train a set of predictors by means of supervised learning algorithms.

A decision support system was also used for tasks related to scientific works. Xu et al. [14] proposed a group decision support model to evaluate reviewers' performances in the peer review process. They considered the competitiveness and relevance of reviewers in order to assess them. Tian et al. [11] proposed an organizational decision support system (ODSS) for R&D project selection. Fan et al. [2] introduced a method for proposal grouping, where knowledge rules are determined to deal with proposal identification and classification. A genetic algorithm was developed to find the expected groupings.

Recommender systems are also used to suggest experts to solve a specific problem. Protasiewicz [9] built a support system aimed at the selection of reviewers (experts) to assess research proposals or articles. The author extracted the keywords from the publication of researchers to compute cosine similarity between the query and the researchers' profiles. Hoang et al. [3] developed a recommender system to recommend academic events to the target scientist based on research similarity and by exploring the interactions between scientists.

The related works mentioned above point out the need for more research into the selection of experts and reviewers. In this work, we develop a system to automatically select experts without human intervention. The word-embedding method is used to solve the problem of matching a specific problem to experts' records. In addition, the diversity and competence criteria are taken into account to improve the accuracy of the system.

3   Experts' Group Recommender System

This section presents how we developed the recommender system for selection of reviewers or experts to evaluate a specific problem. The issue can be determined by finding a ranking of reviewers who can give the best review or answer for a particular problem. Our system consists of three modules: data collection, expert detection, and expert group prediction (Fig. 1).

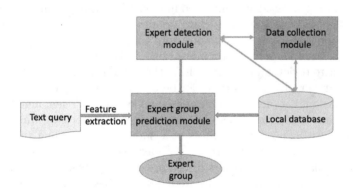

Fig. 1. The framework of the experts' group recommender system

3.1   Data Collection Module

The amount of scholarly data has rapidly increased in recent years. There are many scholarly data resources, such as such as the DBLP Computer Science Bibliography, ResearchGate, CiteSeer, and Google Scholar, which give users easy access. Our data collection module generates a database of scientists' profiles by collecting and importing from different resources data related to scientists. Scholarly data can include information about authors, co-authors, papers, citations,

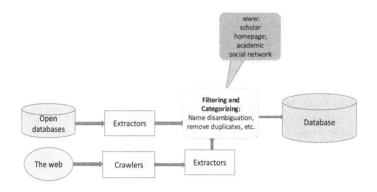

Fig. 2. The data collection module of the proposed recommender system

academic networks, etc. Exploring scholarly data has many significant advantages. Figure 2 shows the structure of the data collection module. There are two main types of data source for academic data collection: open databases and the web. Open databases, such as the DBLP Computer Science Bibliography, are well defined and structured; thus, an extractor process can be used to obtain useful information.

Besides, a very rich source of unstructured data to collect is the web. We used Scrapy[1] to crawl data from the web and put them into an extractor process. All the data were preprocessed and imported into our database system. Author name disambiguation and duplication checking are necessary phases for the creation of a complete profile of scientists to import into the database. There may be more than one person with a particular name, or one author with multiple names, and it is important to review ambiguity in the authors' names. This can be considered by using a number of features that can be extracted from papers, such as co-authors' names, affiliations, research topics, and keywords. Then, matching rules are applied sequentially for each author name obtained from the database. If any rule sufficiently recognizes the authors of a paper, the remaining rules are not enforced. After applying all the rules, the next paper is considered on a non-discriminatory basis for all authors in the previously published paper, whether disambiguated or not. Matching rules were designed based on the available data and experiments. Order consideration is (1) first and last names, (2) co-authors, (3) affiliations, (4) scientific fields and keywords.

3.2 Expert Detection Module

In this subsection, expert detection is introduced by taking into account the researchers' behaviors and the researchers' scientific achievements to determine whether a researcher is an expert or a non-expert. We consider the expert detection module an important step in filtering out new researchers who have not

---

[1] https://scrapy.org/.

Table 1. Features used in detecting experts

Feature name	Description
author_hindex	H-index
author_citation_count	Cumulative citation count
author_mean_citations_per_paper	Mean number of citations per paper
author_mean_citations_per_year	Mean number of citations per year
author_papers	Number of papers published
author_coauthor	Number of coauthors collaborated with
author_age	Career length (years since the first paper was published)
author_max_single_paper_citations	Max number of citations for any of the author's papers

been published before and who do not have much influence on the scientific community. We extract features to detect experts, as shown in Table 1. There are a lot of features that can be used to determine if a researcher is an expert or a non-expert. In our case, we implement the features that we can collect from an open database. A transformation process is implemented to convert data so it is ready for machine learning by engineering the data features. We implemented this process with several machine learning algorithms and chose the best algorithm by using an ensemble method. The details about the expert detection module are shown in Fig. 3.

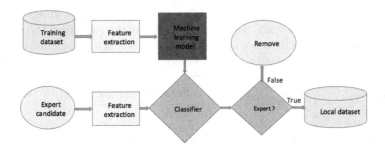

Fig. 3. The expert detection module of the proposed recommender system

3.3 Expert Group Prediction Module

The expert group prediction module is the essential part of the recommender system. When a request is submitted to the system, the expert prediction module will aggregate the information related to potential experts to find a list of experts who best match the request. The request can be given to the system as a text query. An indexing method is used to convert it into a vector that can be sent directly to the recommender system. The expert's profile is also indexed as a vector by using the deep learning method proposed by Le and Mikolov [5]. Let Q be a vector that represents the text query. Let E_i be a vector that represents the profile of expert i. The content similarity between Q and E_i can be computed as follows:

$$Similarity(Q, E_i) = \frac{|Q.E_i|}{|Q|.|E_i|} \tag{1}$$

The expert candidates list can be found by sorting results in descending order according to the similarity between query Q and the experts' profiles.

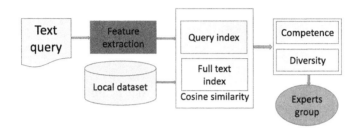

Fig. 4. The expert group prediction module of the recommender system

A list of N expert candidates may be selected to review or answer problem Q. The system has to choose a group of K experts $(1 \leq K \leq N)$ to solve the problem. In our method, we consider three important factors for selecting a group of experts: diversity, competence, and suitableness.

The diversity criterion is preferred when selecting an expert group. In order to meet the diversity criterion, we set a rule whereby the experts in the group do not have the same affiliations or co-authors. After removing the groups that do not meet the rule, the diversity value of each group (D_g) is computed as follows:

$$D_g = \frac{\sum_{i=1}^{K} \sum_{j=1}^{K} (1 - Similarity(E_i, E_j))}{\frac{1}{2} \times K \times (K - 1)} \tag{2}$$

where $Similarity(E_i, E_j)$ is the content similarity of experts E_i and E_j.

Competence in an expert is an important factor affecting the quality of the reviewing process. In this case, we consider three features: the number of publications (P_{E_i}), the number of citations (C_{E_i}) and the h-index (H_{E_i}). Usually, the number of citations of each author is higher than the number of papers. According to Thomson Reuters[2], the number of citations for the average paper is four. Thus, we add $\frac{1}{4}$ as a constant to normalize the competence function. The competence of the group (C_g) can be computed as follows:

$$C_g = \frac{1}{1 + e^{-\sum_{i=1}^{K} \frac{1}{3}(P_{E_i} + \frac{1}{4}C_{E_i} + H_{E_i})}} \tag{3}$$

---

[2] https://www.timeshighereducation.com/news/citation-averages-2000-2010-by-fields-and-years/415643.article.

The sigmoid function is used to convert the value to fit in the range $(0, 1)$ which is the same domain as the diversity value.

The suitableness of an expert's group is calculated by summing the similarity value between query Q and each expert member in the group.

$$Sui\_g = \frac{1}{1 + e^{-\sum_{i=1}^{K} Similarity(Q, E_i)}} \tag{4}$$

The recommendation score for the group of experts is calculated by combining the three components' values as follows:

$$R\_score = \alpha \times D_g + \beta \times C_g + \gamma \times Sui\_g \tag{5}$$

where $\alpha + \beta + \gamma = 1$ to control the weight of D_g, C_g and Sui_g in the recommendation score. The value of R_score is within the range $(0, 1)$. The experts' group with the highest R_score is selected for reviewing or answering problem Q.

4 Experiments

4.1 Datasets

We collected data from two kinds of source: open databases and the web. From the open databases, we chose DBLP to start collecting the dataset, mainly for two reasons: (i) with more than 3.4 million publications by over 1.7 million authors and from 4896 conferences, DBLP is a good site to find authors and publications; and (ii) all bibliographic records are stored in the file $dblp.xml^3$ which is well defined and structured; thus, the extractor process can be used to extract useful information. The data were downloaded in January 2018 and contained the following attributes: titles, authors, years, types, names of conferences, and digital object identifier (DOI) links. The web is a rich source of unstructured data to collect from. We used Scrapy to crawl data from the web and put them into the extractor process. We also used an open source on Github[4] that made it easy to collect scholarly data. There is no labeled dataset available, so we manually labeled the dataset to get ground truth. We collected around 400 papers and more than 1000 experts to evaluate our proposed method. Each paper was considered a query to the system.

4.2 Evaluation

There are several methods for evaluating the performance of a recommender system. In this case, we needed to evaluate the list of experts returned. That requires evaluating all the members of the group with ranked relevance to the problem. Therefore, normalized discounted cumulative gain ($nDCG$) [13] is a

---

[3] http://dblp.uni-trier.de/xml/.
[4] https://github.com/Lucaweihs/impact-prediction.

good choice. The formulations of $nDCG$ and discounted cumulative gain (DCG) [1] are defined as follows:

$$nDCG_{K,Q} = \frac{DCG_{K,Q}}{Ideal\_DCG_{K,Q}} \tag{6}$$

$$DCG_{K,Q} = \sum_{i=1}^{K} \frac{2^{rel_i} - 1}{log_2(1 + i)} \tag{7}$$

where rel_i is the binary relevant of the recommended results at the Kth ranking, which is set to a value of 1 if the expert is suitable to review or answer problem Q, and 0 otherwise. $Ideal\_DCG_{k,u}$ is the ideal obtained value as ground truth.

In order to compute the $nDCG$ value of a given group, we need to compute the $nDCG$ value of each group's member, and get the average $nDCG$ values of all group members. The value of $nDCG$ is greater for correspondingly higher relevance of the recommended group to problem Q. In this case, the online system may return many experts that can review specific problem Q. However, some of them are not usually willing to become a reviewer due to their situation. Therefore, in the offline scenario, we only evaluated the data that experts have reviewed. It helps us evaluate the performance of the recommender system in the most accurate way.

4.3   Results and Discussions

We considered each paper a query to the recommender system. The result from the system is the average of all 400 queries. We set the number of experts in group K at 2, 3, 4, or 5. The values of α, β, γ are flexible, where $\alpha + \beta + \gamma = 1$. We chose a good paper proposed by Protasiewicz [9] as a baseline for our system. In the baseline, a reviewer is described by feature vector C_r. Query Q was also represented by a vector. A recommendation list of reviewers was generated by sorting results in descending order according to the content similarity between vector Q and feature vectors C_r, computed as follows:

$$R\_score\_b = \frac{|Q \times C_r|}{|Q| \times |C_r|} \tag{8}$$

As shown in Figs. 5 and 6, our system achieves better results than the baseline in both evaluation methods. Simply, the baseline method depends on the quality of feature vectors. Extraction keywords from researchers' publications to create the vectors without context will be lost in the semantics of the vectors. Using a word-embedding method to create the vectors has been proven to produce better results. Moreover, the diversity and competence criteria are taken into account in our proposed method. They help our system avoid overfitting and make the system more accurate.

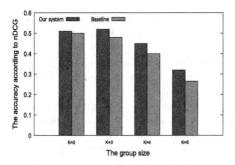

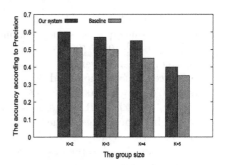

Fig. 5. The results use nDCG   **Fig. 6.** The results use Precision

5 Conclusion and Future Work

In this work, we develop a recommender system for selecting a group of experts to evaluate a specific problem, such as a research proposal or paper. The system consists of three main modules: data collection, expert detection, and expert prediction. The data collection module collects data from various sources to create a database of scientists' profiles. The expert detection module determines who the experts are on each particular topic. The expert prediction module provides a list of experts to answer the query. The experimental results show that our system achieves better results, compared to a baseline.

So far, we do not consider the reputation of experts. That factor should have weight in distinguishing the differences among professors, doctors, and research students. Professors should be selected as reviewers, instead of research students. It is an important factor to improve this work.

Acknowledgment. This research was supported by the Basic Science Research Program through the National Research Foundation of Korea (NRF) funded by the Ministry of Science, ICT & Future Planning (2017R1A2B4009410).

References

1. Burges, C., et al.: Learning to rank using gradient descent. In: Proceedings of the 22nd International Conference on Machine Learning, pp. 89–96. ACM (2005)
2. Fan, Z.P., Chen, Y., Ma, J., Zhu, Y.: Decision support for proposal grouping: a hybrid approach using knowledge rule and genetic algorithm. Expert Syst. Appl. **36**(2), 1004–1013 (2009)
3. Hoang, D.T., Hwang, D., Tran, V.C., Nguyen, N.T., et al.: Academic event recommendation based on research similarity and exploring interaction between authors. In: 2016 IEEE International Conference on Systems, Man, and Cybernetics (SMC), pp. 004411–004416. IEEE (2016)
4. Kolasa, T., Krol, D.: A survey of algorithms for paper-reviewer assignment problem. IETE Tech. Rev. **28**(2), 123–134 (2011)
5. Le, Q., Mikolov, T.: Distributed representations of sentences and documents. In: International Conference on Machine Learning, pp. 1188–1196 (2014)

6. Mori, J., Kajikawa, Y., Kashima, H., Sakata, I.: Machine learning approach for finding business partners and building reciprocal relationships. Expert Syst. Appl. **39**(12), 10402–10407 (2012)

7. Nguyen, N.T.: Advanced Methods for Inconsistent Knowledge Management. Advanced Information and Knowledge Processing. Springer, London (2008). https://doi.org/10.1007/978-1-84628-889-0

8. Pavlov, M., Ichise, R.: Finding experts by link prediction in co-authorship networks. In: Proceedings of the 2nd International Conference on Finding Experts on the Web with Semantics, vol. 290, pp. 42–55. CEUR-WS.org (2007)

9. Protasiewicz, J.: A support system for selection of reviewers. In: 2014 IEEE International Conference on Systems, Man and Cybernetics (SMC), pp. 3062–3065. IEEE (2014)

10. Rodriguez, M.A., Bollen, J.: An algorithm to determine peer-reviewers. In: Proceedings of the 17th ACM Conference on Information and Knowledge Management, pp. 319–328. ACM (2008)

11. Tian, Q., Ma, J., Liang, J., Kwok, R.C., Liu, O.: An organizational decision support system for effective R&D project selection. Decis. Support Syst. **39**(3), 403–413 (2005)

12. Wang, F., Zhou, S., Shi, N.: Group-to-group reviewer assignment problem. Comput. Oper. Res. **40**(5), 1351–1362 (2013)

13. Wang, Y., Wang, L., Li, Y., He, D., Chen, W., Liu, T.Y.: A theoretical analysis of ndcg ranking measures. In: Proceedings of the 26th Annual Conference on Learning Theory (COLT 2013) (2013)

14. Xu, W., Du, W., Ma, J., Wang, W., Liu, Q.: An integrated decision support model to assess reviewers for research project selection. In: 2012 45th Hawaii International Conference on System Science (HICSS), pp. 1414–1423. IEEE (2012)

15. Xu, Y., Ma, J., Sun, Y., Hao, G., Xu, W., Zhao, D.: A decision support approach for assigning reviewers to proposals. Expert Syst. Appl. **37**(10), 6948–6956 (2010)

Agents and Multi-Agent Systems

An Agent-Based Collective Model to Simulate Peer Pressure Effect on Energy Consumption

Fatima Abdallah, Shadi Basurra$^{(\boxtimes)}$, and Mohamed Medhat Gaber

School of Computing and Digital Technology,
Birmingham City University, Birmingham, UK
{fatima.abdallah,shadi.basurra,mohamed.gaber}@bcu.ac.uk

Abstract. This paper presents a novel model for simulating peer pressure effect on energy awareness and consumption of families. The model is built on two well-established theories of human behaviour to obtain realistic peer effect: the collective behaviour theory and the theory of cognitive dissonance. These theories are implemented in a collective agent-based model that produces fine-grained behaviour and consumption data based on social parameters. The model enables the application of different energy efficiency interventions which aim to obtain more aware occupants and achieve more energy saving. The presented experiments show that the implemented model reflects the human behaviour theories. They also provide examples of how the model can be used as an analytical tool to interpret the effect of energy interventions in the given social parameters and decide the optimal intervention needed in different cases.

1 Introduction

Increased energy consumption generated from fossil fuels is causing high carbon emissions and increased global temperature which is mainly attributed to human actions rather than nature [1]. A significant part of the human effect is accounted for the residential sector which consumes high percentages of the world's electricity consumption (23–31%) [2]. Although many technological and structural improvements are suggested to decrease energy consumption, occupants' behaviour plays an important role in this matter [3]. A human solution is based on peer pressure, knowing that human actions are mostly affected by the behaviour of others [4]. Hence, it is suggested that policy makers work on stimulating peer pressure to encourage energy efficient behaviour.

This paper presents an Agent-Based Model (ABM) that studies the collective peer pressure effect on energy consumption in a family environment (hereafter family pressure). The occupant agent's peer effect behaviour is inspired by two theories of human social behaviour: *collective behaviour* by Granovetter [5] and *cognitive dissonance* by Festinger [6]. The model then adds two types of interventions that aim to enhance the occupants' energy awareness and thus reduce their

© Springer Nature Switzerland AG 2018
N. T. Nguyen et al. (Eds.): ICCCI 2018, LNAI 11055, pp. 283–296, 2018.
https://doi.org/10.1007/978-3-319-98443-8_26

consumption. The presented model offers a tool that enables analysing the outcomes of energy efficiency interventions in different social conditions. The paper is organised as follows. The next section presents related work including similar ABMs. The used human behaviour theories and available energy interventions are presented in Sect. 3.1. Section 4 presents the ABM that simulates family pressure and energy efficiency interventions, and explains how the behaviour theories were adapted to the application at hand. Section 5 presents the results of simulating a number of scenarios showing how the model can be used to determine the efficiency of interventions in these scenarios. Finally, Sect. 6 concludes the paper with a summary and pointers for future directions.

2   Related Work

Agent-based modelling is considered the most suitable technique to simulate social interaction [7]. An agent-based model is composed of a group of autonomous software components, called agents, which take decisions based on their state and rules of behaviour. The collective agents' decisions cause changes in the environment which is observed and analysed [8]. The technique has been widely used to study occupants energy consumption behaviour.

Among existing ABMs, there are few that simulate occupants' behaviour change due to peer effect. Azar and Menassa [9] propose a model that adds occupants' energy consumption characteristics and interaction to traditional energy simulation tools. The peer effect model is based on the level of influence of individuals and the number of occupants in each level of consumption. However, the used behaviour change model is not theoretically grounded. Models that involve human behaviour simulations need to be validated using huge amounts of real data, and if not available, need to be based on well established and accepted human behaviour theories. Another ABM that simulates social interactions is Chen et al. [10] who explore the effect of peer network structures on the energy consumption in a residential community. The occupant agents decrease their consumption when the consumption of connected occupants is less than that of the agent. On the other hand increasing the agent's consumption is based on a constant probability that represents the percentage of occupants who increase their consumption with no effect from peers. However, it is more logical that peer effect happens in both directions so that high energy consumers may affect others and cause them to increase their consumption in the same way low energy consumers may affect others. Network structures were also studied in Azar and Menassa [11] which is applied in an office environment. The model uses the relative agreement theory which is applied in a community of heterogeneous culture and values. Thus, behaviour change starts between close individuals. However, in a family environment, which is the case in the current paper, it is common that family members have similar culture and values. Therefore, other behaviour change theories need to be applied which will be detailed in Sect. 3.1.

Studies in [10,11] vary the structure of peer networks based on the fact that not all individuals in a community are connected. While in a family environment, family members are always connected at least at night. Therefore, in the current model, the agents are structured in a fully connected network. Another difference between the currently proposed model and existing models [9–11] is related to the occupant awareness modelling. Existing models characterise occupants by one attribute which is the average yearly/monthly consumption. This attribute does not only reflect the awareness of occupants, but also the time they spend in the building. Hence, it is hard to distinguish if high energy consumption is due to low awareness or daily occupancy. However, the proposed model separates daily human behaviour of occupants (which is based on social parameters) from their energy awareness. More details will follow in Sect. 4.

3 Background: Behaviour Change Theories and Energy Interventions

3.1 Behaviour Change Theories

Humans beings can be highly affected by the behaviour of others. Based on this observation, the theory of collective behaviour was formalised in Granovetter's threshold model [5] to explain the diffusion of a behaviour due to social contagion. The model follows a simple decision rule, where individuals choose to adopt a behaviour when the percentage of others doing the behaviour exceeds a threshold. This threshold represents a complex combination of norms, values, motives, beliefs, etc. Once the threshold is exceeded, it is considered that the net benefit of the behaviour exceeds the perceived costs. The threshold model has been widely used in several applications such as effective targets to influence collective behaviour [12]. The other human behaviour theory used in this model is cognitive dissonance by Festinger [6]. Dissonance is defined as the inconsistency that happens between the individual's knowledge, opinion, beliefs, or attitudes, which are the cognitive factors that drive behaviour. Based on the fact that dissonance is uncomfortable, Festinger [6] proves that humans try to reduce it by adapting their behaviour or changing one or more of the cognitive factors. One of the major sources of dissonance are social groups. Therefore, observing others doing a behaviour that is very different from the individual's behaviour or spreading a general belief that a specific behaviour is not accepted, drives members of a social group to adapt their behaviour, thus reducing the uncomfortable dissonance. Besides, as the magnitude of dissonance increases, it is expected that the tendency to reduce it will increase. The magnitude of dissonance is affected by (1) the number of others who hold a different behaviour, and (2) the level of difference between the individuals' behaviours.

3.2 Energy Efficiency Interventions and Peer Pressure

Given the high percentage of energy consumption in residential buildings, research and policy makers efforts have been focused on promoting energy efficient behaviour, technologies, and structural improvements. This paper is focused on the behavioural aspect by modelling energy efficiency interventions. The target of interventions is to motivate occupants to adopt energy efficiency behaviour by working on their values, attitudes, beliefs, and knowledge [13]. Interventions can be of many forms such as goal setting, information (workshops, mass media campaigns, and home audits), rewards, and feedback [13]. In many occasions, these interventions take advantage of the peer pressure effect by comparing ones behaviour with the behaviour of others. Peer pressure is the influence that members of the same community have on each other which leads to change in behaviour. This effect is shown to be the most influential reason of environmental behaviour change [4]. This is because information received from personal relationships are better recognised and remembered than other sources of information [14].

4   Methodology

4.1   The Agent-Based Model

The proposed family pressure model is based on the ABM developed in Abdallah et al. [15,16]. The model simulates energy consumption behaviour of families. Every occupant is represented by an agent that acts in a house environment and interacts with appliances. The inputs of the model are the social parameters including family size, ages, and employment types (full/part-time job, unemployed, retired and school). Besides, the energy awareness type of occupants determines the probability of performing energy saving actions (e.g. turning off devices when not in use). This can be one of four types: 'Follower Green', 'Concerned Green', 'Regular Waster', and 'Disengaged Waster'. Each of these types is reflected in the model as a continuous attribute called 'energy awareness' between 0 and 100 based on a normal distribution as shown in the 2nd and 3rd column of Table 1.

Table 1. Mean and standard deviation of awareness types

Awareness type	Mean μ	Standard deviation σ	Value (a)	Abbreviation	Category
Follower Green	0.74	0.041	1	F	Green
Concerned Green	0.72	0.043	2	C	Green
Regular Waster	0.41	0.033	3	R	Waster
Disengaged Waster	0.25	0.057	4	D	Waster

The ABM is supported by probability distributions from an integrated probabilistic model based on large sets of real data. The distributions are used to generate realistic occupancy and activities based on the given social parameters. The simulation time is determined by the day of the week (d) and 144 time steps per day (t) each representing 10 min. During the simulation, the occupant agent selects an occupancy state (os_{td}) which can be *away*, *active* at home, or *sleeping*, for a duration (dr). The occupancy state is selected based on the occupant's previous state $os_{(t-1)d}$, *age*, employment type (emp), day (d), and time (t) as shown in functions (1) and (2). When the occupant agent is *active* at home, it performs activities from the following set {*Using the computer, Watching television, Listening to music, Taking shower, Preparing food, Vacuum cleaning, Ironing, Doing dishes, Doing laundry*}. The decision of doing an activity for a specific duration (dr) depends on the occupant's *age*, employment type (emp), day (d), and time (t) as shown in function (3).

$$OS : age, emp, os_{(t-1)d}, t, d \rightarrow os_{td} \tag{1}$$

$$age, emp, os_{td}, t, d \rightarrow dr \tag{2}$$

$$AC : age, emp, t, d \rightarrow ac_{td}, dr \tag{3}$$

Every activity that the occupant performs is associated to an appliance a. Appliances are modelled as dummy agents that only react to occupant agents actions (turn ON and OFF). When the occupant agent starts an activity, it turns the associated appliance ON. When the activity ends, it chooses to turn the appliance ON or OFF based on its energy awareness attribute (ea) and any other occupant (O_a) who is sharing the same appliance according to functions 4 and 5. For more details about the previous model, readers are referred to [15, 16].

$$TO_a : ac_{td} \rightarrow turnOn_a \tag{4}$$

$$ac_{td}, O_a, ea \rightarrow \{keepOn, turnOff\}_a \tag{5}$$

4.2   The Family Pressure Model

The family pressure model is composed of two sub-models: behaviour change sub-model, and energy efficiency interventions sub-model.

Behaviour Change Sub-Model. The occupants behaviour change is motivated by Granovetter's threshold model [5] such that the occupant agents change their behaviour when a threshold is exceeded. Although Granovetter's model explains the effect of social pressure on behaviour, it does not fit to the family pressure effect on energy efficient behaviour for two reasons. First, the model is applied in a public community which has different values and motives, therefore different thresholds. However in a family setting, we consider that family members have similar values and motives based on the fact that they have chosen to live together or were raised together. Therefore, when adapting Granovetter's model to the application at hand, we consider one global threshold for the whole family. This does not revoke the fact that people react differently because we have set the global threshold as a probabilistic one [17] – so once the threshold is exceeded the individuals adopt the behaviour with a probability. Second, the threshold model considers binary decisions. However, energy consumption behaviour is a continuous behaviour that is performed at different levels.

This difference led us to explore the well-established theory of cognitive dissonance by Festinger [6] which is used to adapt the threshold model to the energy consumption application. Based on the two factors that affect the magnitude of dissonance outlined in Sect. 3.1, we adapt the definition of the threshold to fit the energy consumption behaviour. The first factor goes along with Granovetter's threshold definition such that more adopters of a given behaviour leads to changing others' behaviour. The second factor is used to overcome the inapplicability of the threshold model with the energy efficiency behaviour being continuous. Therefore, we define the threshold as the difference between the individual's awareness type and the average of other's awareness types.

The time step in this model is set to 4 weeks of simulation time since individuals usually take time to observe the behaviour of others. In order to express awareness types in numerical values, every awareness type is given an integer value as shown in the 4th column of Table 1. For a family composed of N occupants, every time step T, each occupant agent i calculates the difference $diff_{Ti}$ between its awareness type a_i and the average awareness types of others a_j, where $j \in [1, N] : j \neq i$ using Eq. (6).

$$diff_{Ti} = a_i - (\sum_{j=1, j \neq i}^{N} a_j)/(N-1) \tag{6}$$

Behaviour change happens if $|diff_{Ti}|$ exceeds the global threshold d where $d \in [0, 4]$. A high threshold implies low sensitivity to cognitive dissonance and a low threshold implies high sensitivity to cognitive dissonance. The global threshold d is a probabilistic threshold such that the occupant changes behaviour with probability p where $p \in [0, 1]$. This attribute is referred to as threshold lag [18] which explains the stochastic nature of human behaviour due to uncertainty and differences in the speed of reaction, where a higher value of p means a higher rate of change. p is set to 0.5 as a middle point between high and low rate of change throughout the simulations in this paper. Once behaviour change is decided, the awareness type of the occupant changes towards the average of other's awareness types assuming that the occupant is adapting her/his behaviour to be similar to others. Behaviour change is done by stepping between the awareness types one step at a time either to the green side (green effect) or the waster side (waster effect). The behaviour change process step is outlined in Algorithm 1 which is repeated for every agent i at every time step T.

Energy Efficiency Interventions Sub-Model. This paper distinguishes between family-level interventions, and occupant-level interventions. Each of these interventions can be of any form as outlined in Sect. 3.2, but they differ in the number of occupants to target. The family-level intervention targets the family in general by changing its overall norms, values and beliefs. It can be applied by promoting the energy efficient behaviour such as giving financial incentives or repressing the wasting behaviour such as incurring charges [12]. The occupant-level intervention targets the least aware occupant/s in the family

Algorithm 1: Behaviour Change Step

calculate $diff_{Ti}$ using equation (6)
if $\lvert diff_{Ti} \rvert \geq d$ **then**
   select random number $rand$
   if $rand \leq p$ **then**
      if $diff_{Ti} > 0$ **then**
         $\lvert$ $a_i = a_i - 1$
      else
         $a_i = a_i + 1$

Algorithm 2: Intervention Behaviour Change Step

calculate $diff_{Ti}$ using equation (6)
if $diff_{Ti} > 0$ **then**
   if $\lvert diff_{Ti} \rvert \geq d_g$ **then**
      select random number $rand$
      if $rand \leq p$ **then**
         $a_i = a_i - 1$
 	
if $diff_{Ti} < 0$ **then**
   if $\lvert diff_{Ti} \rvert \geq d_w$ **then**
      select random number $rand$
      if $rand \leq p$ **then**
         $a_i = a_i + 1$

and leads to increasing their awareness levels. These two types of interventions are considered to observe how the collective family pressure can help in achieving more aware occupants, thus less energy consumption. It also allows policy makers to decide the needed combination and intensity of interventions based on each family composition (in terms of awareness levels and social parameters).

When the family-level intervention happens, the overall norms, values and beliefs of the family change. The family-level intervention has two intensities which represent the efficiency or effort made to achieve better results. Therefore, $I_p \in [1,4]$ is defined as the promotion intensity and $I_r \in [1,4]$ as the repression intensity. These two types of family-level interventions are reflected by two thresholds: one that affects the promotion of green effect $d_g \in [0,4]$ and another that affects the repression of waster effect $d_w \in [0,4]$. Therefore, the intervention increases d_w by I_r thus increasing the cost to adopt waster behaviour and/or decreases d_g by I_p thus increasing the benefit of adopting the green behaviour as outlined in Granovetter [5]. d_g and d_w change in effect of the intervention based on Eqs. (7) and (8) given the initial threshold d. For deciding behaviour change, d_g is checked when there is a possibility to change towards the green side ($diff_{Ti} > 0$), and d_w is checked when there is a possibility to change towards the waster side ($diff_{Ti} < 0$) as shown in Algorithm 2. The occupant-level intervention does not change the threshold of the family because it targets specific occupants. It aims to change the awareness of occupants while the regular behaviour change step in Algorithm 1 is applied. The intervention can have an intensity $I_o \in [1,3]$ and can be applied to a member of the family i at a specific time step T according to Eq. (9).

$$d_g = d - I_p \quad : \quad d_g \in [0,4] \tag{7}$$

$$d_w = d + I_r \quad : \quad d_w \in [0,4] \tag{8}$$

$$a_{i(T+1)} = a_{iT} - I_o \quad : \quad a_{iT} \in [0,4] \tag{9}$$

5   Experiments and Discussion

This section presents a number of experiments with different input parameters to show how varying these inputs can result in different intervention outcomes. It is worth to mention that this paper only presents a number of significant scenarios as a proof-of-concept while achieving the purpose of the paper. Abbreviations of awareness types (5th column of Table 1) are used to identify the initial awareness of the family, such that a four occupant family with one 'Follower Green' and three 'Disengaged Wasters' is denoted by *FDDD*. In every simulation run, 100 households were simulated to capture the stochastic effect of the threshold lag[1]. The scenarios are run for a year and the resulting average yearly consumption and converged awareness types are recorded. These types were categorised based on the number of *Green* occupants in the family (represented in the figures by different colours in the bars). The categories of the awareness types are determined by the last column of Table 1.

5.1   Family Pressure Convergence

The aim of this experiment is to observe the resulting awareness types as an effect of family pressure based on different thresholds. Figure 1 shows the results of three scenarios: (a) *FFFD*, (b) *FCRD*, and (c) *FDDD*. The last scenario of every bar graph (d = 4) shows the initial category of the family because $diff_{Ti}$ can be maximum 3, thus no change in awareness types.

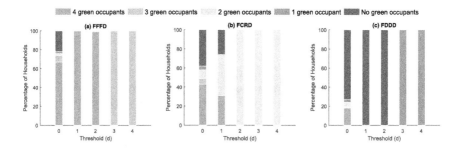

Fig. 1. Family awareness types convergence (Color figure online)

In scenario (b), the family remained with two green occupants at thresholds 2 and 3, besides, in (a) and (c) the family remained the same at threshold 3 and changed only one occupant at threshold 2. This indicates that the family does not change significantly when the threshold is high (d = 2 and 3). However, at low thresholds (d = 0 and 1), the family converged mainly towards the dominant awareness type. For example, in (a) the convergence was mostly towards '4

---

[1] The model was validated by running a number of scenarios with different random numbers seed where the results came out to be similar.

green occupants', because initially there were three green occupants. A similar observation was noticed in (c). In scenario (b) where there is no dominant awareness type, the convergence was with equal probabilities either to all green occupants or all waster occupants ('no green occupants' category) with higher convergence to the extremes at threshold 0. These results indicate that the proposed model reflects the theory of cognitive dissonance and collective behaviour which agree that people tend to change their behaviour to conform with the behaviour of others. It is worth noting that in (a) and at threshold 0, around 20% of the households converged to 'no green occupants'. This means that the only waster occupant succeeded to change the behaviour of the other three green occupants. This phenomenon is explained in the cognitive dissonance theory which states that dissonance can be reduced by either adapting with others, or convincing the others to adapt with the individual. This explains how the three green occupants converged to wasters in effect of one waster occupant as in (a) and vice versa in (c). Festinger [6] mentions that in this case, the overall cognitive elements of the surrounding environment change, but this is easy when the individual can find others who hold the same behaviour, which explains the low percentage of this convergence (20% in our experiment).

5.2   Family-Level Intervention

In this experiment, family-level interventions are applied to scenario (c) of the experiment 1 (FDDD) as it has the most waster occupants after convergence. For each threshold, the possible intensities of family-level interventions are applied keeping the thresholds d_g and d_w in their limits $[0, 4]$. The aim of this experiment is to show the effect of promotion and repression interventions when varying their intensities. Figure 2 shows the results with initial thresholds 0, 1 and 2.

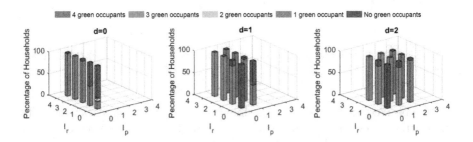

Fig. 2. Family-level intervention convergence (Scenario FDDD)

It is noticed at thresholds 1 and 2 that the number of green occupants increases as the promotion intensity (I_p) increases, which is not the case with repression intensity (I_r) where most of the occupants stayed wasters. This indicates that repression intervention is less efficient than the promotion intervention. This is attributed to the high number of waster occupants, such that encouraging them to adopt the green behaviour is more effective than repressing the only green occupant from getting affected by waster occupants. Another indication from varying intervention intensities is inferring the minimum intensity needed to increase the possibility of getting 4 green occupants. For example at threshold 0, repression intensity 2 is enough to get '4 green occupants' with probability more than 0.95. This allows to identify the minimum effort needed while achieving the maximum number of green occupants.

5.3   Occupant-Level Intervention

This experiment studies the effect of occupant-level interventions which directly change the awareness of least aware occupants. Scenario FFFD with threshold 0 is selected to get the minimum intensity required to prevent the 'no green occupants' convergence (as shown in scenario (a) in Sect. 5.1). As the family initially has one waster occupant, the intervention is applied for one occupant with different intensities. Besides, the intervention can be applied at specific times of the year, therefore it can be an 'early intervention' at $T = 2$, 'mid-year intervention' at $T = 6$ or 'late intervention' at $T = 9$. This determines the best intervention time just before the waster occupant affects other green occupants. Figure 3 shows the results while varying the intervention time and intensity.

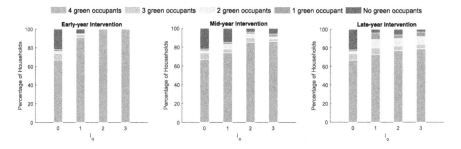

Fig. 3. Occupant-level intervention convergence (Scenario FFFD d = 0) (Color figure online)

It is observed that as earlier the intervention and as higher its intensity, as more green occupants are obtained. The early interventions with intensities 2 and 3 are the most effective with no waster occupants after a year. This is expected because the waster occupant is affected by the external intervention at an early stage, thus leading to 4 green occupants. However, in all other scenarios, waster occupants are observed even at higher intensities. This shows that one intervention per year is not enough to make an impact on families with only one waster occupant. This suggests to perform continuous interventions to maintain the green effect and combine them with family-level interventions. Note the this experiment was performed with very low threshold of the family $(d = 0)$ so occupants can easily influence each other.

5.4 Effect of Interventions on Families with Varied Social Parameters

In our previous paper [16], it was concluded that social parameters affect the energy waste of the family. Although the previous model does not simulate family pressure, we showed that energy waste in large families is less than small families. On the basis of this conclusion, the current experiment tests if a family-level intervention is more efficient in big families than small families. For this purpose, the family-level intervention is applied on (a) a two-occupant family and (b) a four-occupant family. Figures 4a and b show the awareness types convergence of scenarios (a) and (b) respectively with an equivalent initial numbers of green and waster occupants (FD and FFDD) and threshold $d = 0$. Figure 4c shows the resulting energy saving percentage when compared to the no-intervention scenario ($I_r = 0$) and the convergence time which is the time it takes the family to reach a stable state where the occupants are no more affected by each other.

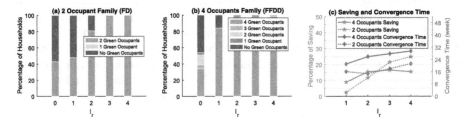

Fig. 4. Effect of Family-level intervention on two and four occupant families ($d = 0$) (Color figure online)

In Fig. 4c at intervention intensities 1 and 2, the percentages of saving for big families are 9% and 16% respectively, which are more than that of small families (i.e. 1% and 11%). This is also observed in the awareness types convergence (Figs. 4a and b) where the '4 green occupants' category is more dominant in (a) than the '2 green occupants' category in (b). However, at intensities 3 and 4, the savings of small families are 21% and 25% respectively, which dominates that of big families (i.e. 16% and 15%) (Fig. 4c). Besides, all of the occupants in scenarios (a) and (b) converged to green occupants as shown in Figs. 4a and b. This is explained by the lower convergence time of small families (Fig. 4c). This means that a higher intensity intervention converges small families quicker than big families which consequently leads to higher saving. Thus, the family-level intervention can result in maximum saving at low intensity in big families as opposed to small families. While a high intensity intervention is more efficient in small families as it leads to a larger and quicker saving than big families. This experiment can be repeated with varied social parameters, thresholds, and intervention types to obtain the most efficient intervention in every case.

5.5   Discussion

The model proposed in this paper simulates peer pressure effect on energy awareness levels and consumption of families. The peer effect behaviour of occupants is based on two human behaviour theories opposed to other models that do not use existing theories [9]. The behaviour theories were adapted to comply with the energy consumption behaviour and family environment, while other models use different theories that simulate office environments [11]. Beside, the current model offers different options of input including social parameters (family size, employment types, ages), awareness levels, values and beliefs that affect the energy consumption behaviour, and intervention options. We proved in the experiments that these inputs affect the outcome of interventions. The experiments focused on demonstrating the application of the model in pre-specified scenarios. The model can ideally be used to study the impact of any intervention planned by governing bodies on the outcome (i.e. energy saving). This can be done by estimating unknown parameters, running the model with initial parametrisation of known and unknown parameters. Then a search mechanism (e.g. grid search) is applied to best estimate the unknown parameters, minimising the difference between the model's synthesised data and the observed real data. If the search space is large, in case of having too many unknown parameters, computational intelligence methods like *Genetic Algorithm* can be applied. Revealing these unknown parameters can help in determining the reason why interventions are effective in some cases, but not in others.

6   Conclusion and Future Work

This paper presented an ABM that simulates energy awareness peer pressure in a family setting. The model uses the *collective behaviour theory* and the *theory of cognitive dissonance* to reflect realistic peer effect. Different energy efficiency interventions can be applied and the resulting awareness types and savings are observed. The presented experiments show that the human behaviour theories are well-reflected in the model. Besides, they show how the model can offer an analytical tool for governing bodies to analyse the effect of interventions and make decisions of how to target different families to get the best results.

A variation of this model is to make the effect of members depend on how often they are in contact in the house, which makes the interaction more realistic. This can be easily achieved because the ABM simulates individuals' daily availability at home in a 10-minute time step. The current model have not considered a weighting attribute which determines the level of relation between the occupants which affects the level of influence. This attribute can be added in the future where the intervention may be targeted at a specific relationship if it proves efficient. Also the modelling of behaviour change can be done at the energy awareness level, not at the awareness type level. This can enhance model's capability to simulate more fine-grained behaviour change. These enhancements are expected to produce an even more realistic model that reflects the quality and rate of daily interactions among the family members.

References

1. US Global Cilmate Change Research Program (USGCRP): Global Climate Change. In: Global Climate Change Impacts in the United States, pp. 13–26. Cambridge University Press, New York (2009)
2. Internation Energy Agency (IEA): Electricity Information Overview 2017. Technical report, IEA (2017)
3. Zipperer, A., Aloise-Young, P.A., Suryanarayanan, S., Roche, R., Earle, L., Christensen, D., Bauleo, P., Zimmerle, D.: Electric energy management in the smart home: perspectives on enabling technologies and consumer behavior. Proc. IEEE **101**(11), 2397–2408 (2013)
4. Nolan, J.M., Schultz, P.W., Cialdini, R.B., Goldstein, N.J., Griskevicius, V.: Normative social influence is underestimated. Pers. Soc. Psychol. Bull. **34**(7), 913–923 (2008)
5. Granovetter, M.: Threshold models of collective behaviour. Am. J. Sociol. **83**(6), 1420–1443 (1978)
6. Festinger, L.: A Theory of Cognitive Dissonance, vol. 2, 2nd edn. Stanford University Press, Stanford (1962)
7. Epstein, J.M.: Agent-based computational models and generative social science. Complexity **4**(5), 41–60 (1999)
8. Axtell, R.: Why agents? on the varied motivations for agent computing in the social sciences. The Brookings Institution, Technical report, Center on Social and Economics Dynamics (2000)
9. Azar, E., Menassa, C.C.: Agent-based modeling of occupants and their impact on energy use in commercial buildings. J. Comput. Civil Eng. **26**(4), 506–518 (2012)
10. Chen, J., Taylor, J.E., Wei, H.H.: Modeling building occupant network energy consumption decision-making: the interplay between network structure and conservation. Energy Build. **47**, 515–524 (2012)
11. Azar, E., Menassa, C.C.: A comprehensive framework to quantify energy savings potential from improved operations of commercial building stocks. Energy Policy **67**, 459–472 (2014)
12. Hu, H.H., Lin, J., Cui, W.T.: Intervention strategies and the diffusion of collective behavior. J. Artif. Soc. Soc. Simul. **18**(3) (2015). Paper 16. http://www.jasss.soc.surrey.ac.uk/18/3/16.html
13. Abrahamse, W., Steg, L., Vlek, C., Rothengatter, T.: A review of intervention studies aimed at household energy conservation. J. Env. Psychol. **25**(3), 273–291 (2005)
14. Costanzo, M., Archer, D., Aronson, E., Pettigrew, T.: Energy Conservation Behavior. The Difficult Path From Information to Action. Am. Psychol. **41**(5), 521–528 (1986)
15. Abdallah, F., Basurra, S., Gaber, M.M.: A hybrid agent-based and probabilistic model for fine-grained behavioural energy waste simulation. In: IEEE 29th International Conference on Tools with Artificial Intelligence (ICTAI), pp. 991–995. IEEE (2017)
16. Abdallah, F., Basurra, S., Gaber, M.M.: Cascading probability distributions in agent-based models: an application to behavioural energy wastage. In: Rutkowski, L., Scherer, R., Korytkowski, M., Pedrycz, W., Tadeusiewicz, R., Zurada, J.M. (eds.) ICAISC 2018. LNCS (LNAI), vol. 10842, pp. 489–503. Springer, Cham (2018). https://doi.org/10.1007/978-3-319-91262-2_44

17. Kiesling, E., Günther, M., Stummer, C., Wakolbinger, L.M.: Agent-based simulation of innovation diffusion: a review. Central Eur. J. Oper. **20**(2), 183–230 (2012)
18. Bohlmann, J.D., Calantone, R.J., Zhao, M.: The effects of market network heterogeneity on innovation diffusion: an agent-based modeling approach. J. Prod. Innov. Manag. **27**(5), 741–760 (2010)

Agents' Knowledge Conflicts' Resolving in Cognitive Integrated Management Information System – Case of Budgeting Module

Marcin Hernes[1(✉)], Anna Chojnacka-Komorowska[1],
Adrianna Kozierkiewicz[2], and Marcin Pietranik[2]

[1] Wrocław University of Economics, ul. Komandorska 118/120,
53-345 Wrocław, Poland
{marcin.hernes,anna.chojnacka-komorowska}@ue.wroc.pl
[2] Wroclaw University of Science and Technology,
Wybrzeże Wyspiańskiego 27, 50-370 Wrocław, Poland
{adrianna.kozierkiewicz,marcin.pietranik}@pwr.edu.pl

Abstract. Nowadays management is supporting by using integrated management information systems, including multi-agent systems, where most often the relational or object databases are used. However, it becomes necessary not only to register, by IT systems, the values of economic phenomena' attributes but also to automatically analyze their meaning. These functions can be realized by using, among others, the cognitive agents running in the frame of a multi-agent system. More often the knowledge of such agents is represented by using semantic methods. However, it often occurs that a multi-agent integrated management information system generates conflicts of knowledge among the agents. These conflicts result from the fact that agents may generate different decisions or solutions to the user, which, in turn, may result from different methods of decision making employed by the agents, different, heterogeneity data sources or different agents' goals. The aim of this paper is to analyze the knowledge conflicts of cognitive agents and to develop a heuristic algorithm for these conflicts resolving in a Cognitive Integrated Management Information Systems.

Keywords: Budgeting · Cognitive agents
Integrated management information systems · Knowledge conflicts

1   Introduction

Nowadays companies use integrated management information systems, including multi-agent systems, for supporting management, when the registration of an economic phenomena is most often performed with the use of relational or object databases. Analysis of meaning of these phenomena by decision-makers is a very time-consuming process. However, in a turbulent economic environment, business decisions should be taken near real time to achieve a high efficiency of business organizations' functioning.

© Springer Nature Switzerland AG 2018
N. T. Nguyen et al. (Eds.): ICCCI 2018, LNAI 11055, pp. 297–308, 2018.
https://doi.org/10.1007/978-3-319-98443-8_27

Therefore, it becomes necessary not only to register by IT systems the values of phenomena's attributes but also to automatically analyze their meaning. It is important to interpret phenomena in the context of supporting decisions, and realizing unexpected information management needs [1]. These functions can be realized by using, among others, the cognitive agents [2] running in the frame of multi-agent system. More often the knowledge of such agents is represented by using semantic methods (e.g. ontologies, semantic nets).

However, it often occurs that a multi-agent integrated management information system generates conflicts of knowledge among the agents [3]. These conflicts result from the fact that agents may generate different decisions or solutions to the user, which in turn may result from different methods of decision making employed by the agents, different, heterogeneity data sources or different goals of agents. If a conflict of knowledge occurs in the system, the system is unable to generate a correct decision and, consequently, the decision maker will then have to decide without help from the system, which is time-consuming, requires much work and can lead to a decision that is inaccurate and made with incomplete information.

The aim of this paper is to analyze the knowledge conflicts of cognitive agents and to develop heuristic algorithm for these conflicts resolving in a Cognitive Integrated Management Information System (CIMIS). Agents' knowledge in this system is represented by a semantic net with nodes' and links' activation levels. These activation levels allows to take into a consideration the risk and uncertainty related to economic-decisions. Such type of knowledge representation has not been yet formally defined.

Developed in this work algorithm has been verified on the budgeting module of CIMIS. Budgeting is one of the best known and most commonly used financial management tools. Budgets can fulfill one or more of the following functions: mapping, controlling, coordinating, communicating, instructing, authorizing, motivating, performance measurement, decision making. Therefore, their efficient functioning in enterprises is extremely important. The main contribution of paper is, therefore, related to formal definition of semantic net with nodes' and links' activation levels and developing algorithm for resolving the knowledge conflicts between cognitive agents.

The paper has been divided as follows: the first part characterizes a state-of-the-art in considered field. Next, agents' knowledge representation method is described and formally defined. The third part of paper presents analysis of knowledge conflicts and algorithm for their resolving, developed by authors. The research experiment, performed in order to verify developed algorithm, is presented in the last part of the paper.

2 Related Works

In order to interpret economic phenomena by cognitive agents, their knowledge must be represented by semantic methods, such as ontologies or semantic nets. The existing formal definitions of such structures (e.g. [4, 5]) in very small degree take into consideration the uncertainty of economic decisions. The better solution is to use the semantic net with nodes' and links' activation level (these activation levels allow for representing the uncertainty of economic events). In addition it enables processing knowledge represented in a symbolic manner as well as knowledge represented in a numerical manner. The semantic net with nodes and links activation level also enables processing both structured and unstructured knowledge. Thus, it is possible to determine the certainty level of nodes and semantic relations between these nodes. The first solution used such a structure has been called "slipnet" in the "Copycat" project [6]. However, the existing papers lack the formal definition of this structure and do not consider issues related to instances and relations between nodes and instances. This definition is however necessary mainly in order developed methods for the agents' knowledge conflict resolving.

The knowledge conflict resolving in multi-agent systems is performed by using different methods. For example negotiations [7] (they enable effective knowledge conflicts resolving by reaching a compromise, however they require exchanging a large number of messages between agents, which results in decreased efficiency of system), or deduction-calculation methods [8] (these methods, e.g. ones based on the theory of games, classical mechanics, or methods of choice, enable one to obtain a great computational capacity of a system, however they do not guarantee a proper result of knowledge conflicts resolving [3]) can be used.

Existing papers on resolving conflicts of unstructured knowledge mostly involved presentation of documents in the form of Term Frequency Matrix (TFM) [e.g. 9], and ontology [e.g. 10]. The work [11] puts attention on conflicts in the budgeting process in ERP systems and considers these conflicts at the activity level and the contradiction at the structural level. However most of related works are concerned to structured knowledge, stored in relational or object databases. The knowledge conflicts on the semantic level was analyzed, for example in [12], however they don't take into the consideration a risk and uncertainty related to economic decisions.

The budgeting has to be oriented to such purposes as economic stabilization and planning. Blumentritt [11] explained that budgeting is the process of allocating an organization's available financial resources to its units, activities and investments and to monitor the performance of managers and employees. The budgeting as a process implemented by a larger number of centers means that there is a certain dynamic of change taking place from forming, storming, norm forming and finishing on pre-forming [13]. The forming phase is completing the team's composition, storming is a struggle for power, friction, and attempts to adapt to others. In the next phase, a consensus is established, necessary to approve the overall budget of the company, while the last phase is the practical implementation of the goal set for the team [14].

3   Agents' Knowledge Representation

The CIMIS has been detailed described in [2]. This is a multi-agent system (based on LIDA cognitive agent architecture [15]) consists of following sub-systems: fixed assets, logistics, manufacturing management, human resources management, financial and accounting, controlling, CRM, business intelligence. The agents running in CIMIS for a knowledge representation use a semantic net with nodes and links activation level taking into consideration instances and relations between nodes and instances. This structure is called a "slipnetplus". It should be noted, that this type of representation, allows processing both knowledge represented in a symbolic way, and knowledge represented in a numerical way. Simultaneously a structured and unstructured knowledge can be processed. Thus, it is possible to determine a certainty level of semantic relations between nodes. In case of the economic knowledge, it is a very important issue because the decisions-making process based on the type of knowledge usually takes place in conditions of risk and uncertainty. Figure 1 presents the example of a graphical representation of the "slipnetplus".

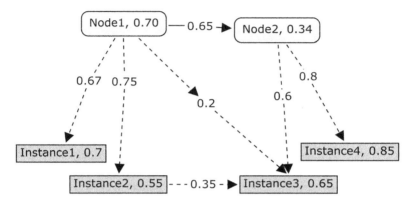

Fig. 1. The example of a graphical representation of the "slipnetplus".

The arrows drawn with a continuous line represent links, while the arrows drawn dotted lines denote the axioms. The presented "slipnetplus" consist of two nodes with their activation levels. The interpretation of the *Node1* is as follows: The *Node1* exists in the real world with a probability level 0.70. Interpretation of node 2 is similar. The *Node1* is connected with *Node2* by links with the level of probability 0.65.

The interpretation of the *Instance1* is as follows: The *Instance1* exists in the real world with a probability level 0.7. Interpretation of other instances is similar. The *Node1* is connected with *Instance1* by axiom with the level of probability 0.67, with *Instance2* by axiom with the level of probability 0.75 and with *Instance3* with the level of probability 0.2. Interpretation of other axioms is similar.

For formal definition of agents' knowledge structure, we assume the real world is represented by a pair $\langle A, V \rangle$ where A is a finite set of attributes, V is the domain of A and

$$V = \bigcup_{a \in A} V_a \tag{1}$$

where V_a is a domain of attribute a.

The knowledge structure is called a slipnetplus and is defined as follows:

Definition 1. We consider the "slipnetplus" referring to the real world $\langle A, V \rangle$ - such "slipnetplus" is called $\langle A, V \rangle$-based. The "slipnetplus" is called a quadruple:

$$SN = \langle N, I, L, Z \rangle \tag{2}$$

where:

N – a set of nodes,
I – a set of instances of nodes,
L – a set of links defined as a set of fuzzy relations over the set N,
Z – a set of axioms defining a connection between instances of nodes and nodes ♦

Let us define elements of the "slipnetplus".

Definition 2. A node of an <A, V>-based the "slipnetplus" from its set N is defined as a tuple:

$$Node = \langle n, A^n, V^n, al^n \rangle \tag{3}$$

where n is the unique name of the node, $A^n \subseteq A$ is a set of attributes describing the node, $V^n \subseteq V$ is the attributes domain $V^n = \bigcup_{a \in A^n} V_a$ and al^n is an activation level of the node, defined as a value from the range $[0, 1]$.

For simplicity, in further parts of this article we will treat it simply as a value from the range $[0, 1]$.

The nested triple $\langle A^n, V^n, al^n \rangle$ is called the structure of the node n. It is obvious that all nodes belonging to the same "slipnetplus" are different from one another. However, notice that within the "slipnetplus" there may be two or more nodes with the same structure. Such a situation may take place, for example, for nodes "person" and "body". For expressing the relationship between them the links from set L will be very useful.

Definition 3. An instance of a node n within the slipnetplus is described as a pair:

$$instance = \langle i, v \rangle \tag{4}$$

where: i is a unique identifier of the instance and v is a tuple of type A^n. Value v is also called a description of the instance within the node. We can then write $i \in n$ for presenting the fact that i is an instance of the node n. We also accept an auxiliary notion $I(i)$ to get a description of the instance with identifier i from the set I of the particular slipnetplus.

In the "slipnetplus" within a pair of nodes there may be defined one or more links. Links between nodes describe the relationships between them. For example, between two nodes there may be defined such relations as a Synonym relation or an Antonym relation. Links between the nodes are included in set L of the "slipnetplus" definition.

Definition 4. A set N of nodes is given. The link is a fuzzy set with an activation function $l : N \times N \rightarrow [0, 1]$:

$$L : \{(\langle n1, n2 \rangle, l(n1, n2)) | \langle n1, n2 \rangle \in N \times N\} \tag{5}$$

The set Z of axioms can be interpreted as a connection between instances and nodes.

Definition 5. The set N of nodes and set I of instances are given. The axiom is a following fuzzy set with an activation function $z : N \times I \rightarrow [0,1]$:

$$Z : \{(\langle n,i \rangle, z(n,i)) | \langle n,i \rangle \in N \times I\} \tag{6}$$

Generally speaking, a conflict between two or more agents appears, when budgets generated by these agents differ i.e. their knowledge is represented by different slip-netpluses. Analysis of these conflicts and the algorithm for their resolving is presented in next section of this paper.

4 Knowledge Conflicts and Their Resolving

Knowledge conflicts are reflected in the agents' knowledge structure representation. Generally speaking, slipnetpluses of two or more agents differ. Further part of this section describes conflicts between slipnetpluses on particular levels.

Nodes Level
On this level it is assumed that two nodes in different slipnetpluses differ from each other in their structures (attributes or values of attributes, which their nodes represents, or in the activation level). That means that these slipnetpluses may contain the same node, but its structure is different in each slipnetpluses. The reason of this phenomenon is that these slipnetpluses come from different autonomous cognitive agents.

Definition 6. Let nodes $node1 = \langle n^{node1}, A^{node1}, V^{node1}, al^{node1} \rangle \in SN_1$ and $node2 = \langle n^{node2}, A^{node2}, V^{node2}, al^{node2} \rangle \in SN_2$ are in <A, V>-based slipnetpluses SN_1 and SN_2. The conflict on the node level takes place if $n^{node1} = n^{node2}$, but $A^{node1} \neq A^{node2}$ or $V^{node1} \neq V^{node2}$ or $al^{node1} \neq al^{node2}$ ◆

Links Level
Two kinds of slipnetplus conflicts on the level of links may occur. The first kind of conflict refers to situations where between the nodes n and n' in different slipnetpluses, the different links are assigned. As an example, let's consider links Commodity and Customer, in one slipnetplus they are in link Buy, in the another slipnetplus they are in link Complaint. Notice also that within the same slipnetplus two nodes may be in more than one link. Besides, the same pair of nodes can belong to the same link, but with different activation levels in different slipnetpluses. Let $l_1(node1, node2) \in SN_1$ and $l_2(node1, node2) \in SN_2$ denote the links between nodes $node1$ and $node2$ in <A, V>-based slipnetpluses SN_1 and $\in SN_2$.

We assume also that the set of nodes is the same for all slipnetpluses to be integrated.

Definition 7. The conflict on links level take place if $l_1(node1, node2) \neq l_1(node1, node2)$ ◆

Axioms Level
The conflicts on the axioms level is defined similar to the links level.

Instances Level

On this level, it is assumed that two instances in different slipnetpluses differ from each other in their values. That means these slipnetpluses can contain the same instance but its value is different in each slipnetpluses.

Definition 8. Let instances $instance1 = \langle i^{instance1}, v^{instance1} \rangle$ and $instance2 = \langle i^{instance2}, v^{instance2} \rangle$ are in <A, V>-based slipnetpluses SN_1 and $\in SN_2$. The conflict on the instance level takes place if $i^{instance1} = i^{instance2}$, but $v^{instance1} \neq v^{instance2}$ ◆

Taking into consideration the example of the enterprise's budget preparation, it carries out a highly developed business activity employing hundreds of employees, is often associated with many problems. The budgeting module should examine the interrelationships and dependencies among different parts of the organization, and in the process, to identify and resolve conflicts [16]. Many organizational units are involved in the process of creating the overall corporate budget, representing conflicting individual interests, and at the same time the overall budget of the enterprise must ensure the implementation of the overall objectives "imposed" by the shareholders while ensuring liquidity and the value of the enterprise. Thus, one can speak in this case about a decision-making conflict that concerns individual organizational units of the company. Examples of conflicts in this area can be as follows:

- differences in waiting for the number of employees employed in the production department and the amount of their remuneration on the line of the production department-sales department;
- a distribution of the overall sales planned for the enterprise between its various branches;
- limiting the costs of material consumption (e.g. fuel) with simultaneous increase of sales plan.

The budgeting also involves the problem of various expectations on the line of the company's management and its shareholders [17]. When talking about a decisions conflict on the line of shareholders and management of an enterprise, it is worth quoting an example illustrating the problem described. Thus, the company's shareholders expect attractive profits and dividends. The way to achieve such goals may be, for example, unjustified increase in sales prices, the reduction of manufacturing costs using cheaper raw materials or employment of less-qualified employees with lower salary expectations, but also through deteriorating working conditions. The agents that accomplish stakeholders' goals must take into consideration above issues. As we can see, the conflict between the management and shareholders' goals can lead to a conflict between the management goals and the clients' goals or the management goals and staff goals. For this reason, quick resolution of conflicts between agents seems extremely important.

In addition to the problems related to decision-making conflicts described above, the time-barring problem arises during the budgeting process, which involves the following factors [18]:

- the need to quickly and efficiently prepare the overall budget of the enterprise,
- the managers' need to set up individual organizational units for partial budgets and the efficiency of these sub-processes determines the efficiency of the entire process,

- errors made by preparing partial budgets and the need to correct them by creating a global budget,
- the need to conduct many organizational meetings that will allow reaching a consensus,
- unidentified threats, such as holidays or sick leave related to the implementation of the budgeting process.

In order to knowledge conflicts between cognitive agents resolving we propose the slipnetplus integration heuristic algorithm (Fig. 2). We define an auxiliary function ids which for a set of instances of a selected slipnetplus returns a set of instance identifiers used within it. Formally $ids(I) = \{i | \langle i, v \rangle \in I\}$.

Algorithm 1 Slipnetplus integration algorithm

Require: $SN_1 = < N_1, I_1, L_1, Z_1 >$, $SN_2 = < N_2, I_2, L_2, Z_2 >$

1: Set $SN^* = < N^* = \phi, I^* = \phi, L^* = \phi, Z^* = \phi >$
2: **for all** $(< n_1, A^{n_1}, V^{n_1}, al^{n_1} >, < n_2, A^{n_2}, V^{n_2}, al^{n_2} >) \in N_1 \times N_2$ **do**
3:     **if** $n_1 == n_2$ **then**
4:         $A^{n^*} = A^{n_1} \cup A^{n_2}$
5:         $V^{n^*} = \bigcup_{a \in A^{n^*}} V_a$
6:         $N^* = N^* \cup \{< n_1, A^{n^*}, V^{n^*}, min(al^{n_1}, al^{n_2}) >\}$
7:     **else**
8:         $N^* = N^* \cup \{< n_1, A^{n_1}, V^{n_1}, al^{n_1} >, < n_2, A^{n_2}, V^{n_2}, al^{n_2} >\}$
9:     **end if**
10: **end for**
11: $I^* = \{< i, v > | < i, v > \in I_1 \wedge i \in ids(I_1) \setminus ids(I_2)\}$
12: $I^* = I^* \cup \{< i, v > | < i, v > \in I_2 \wedge i \in ids(I_2) \setminus ids(I_1)\}$
13: **for all** $i \in ids(I_1) \cap ids(I_2)$ **do**
14:     $i^* = < avg(I_1(i)[k], I_2(i)[k])\ for\ k \in card(I_1(i)) >$
15:     $I^* = I^* \cup \{i^*\}$
16: **end for**
17: $L^* = L_1 \cup L_2$
18: $Z^* = Z_1 \cup Z_2$
19: **return** SN^*

Fig. 2. Slipnetplus integration algorithm.

The proposed heuristic algorithm takes two slipnetpluses as its input and returns identically defined slipnetplus as an output. At first it integrates sets of nodes. Initially (Line 3), nodes, that share the same unique name, are merged - sets of their attributes are summed, the attributes' domain is created using Eq. 3, and the lower (more strict) activation level is chosen. Secondly - nodes with unique names are simply added to the resulting set (Line 8).

The next step is the integration of instances which is done by adding those instances which identifiers don't overlap within merged slipnetpluses. Then, the algorithm creates new instances from instances which identifiers where used in both inputs. Practical applications show that instances share the same structure, therefore, this step (Lines 14–15) is done by crawling through instances structures and (if the domains of partial elements of instances are numerical) calculating an average value which is eventually added to the output sets. For simplicity, Line 14 is using a list comprehensions notation inspired by Python programming language.

As explained, nodes are merged based on shared names. Therefore, the pool of such names does not expand beyond names used within input slipnetpluses. Due to this fact, we can eventually (Lines 17–18) integrate sets of links and axioms using a fuzzy sum of sets operator.

The algorithm is easy to implement. Using a modern programming language, it is also simple to parallelize integrations of sets of nodes, instances, links and axioms. Thus, the algorithm can be effective in a production environment and real world applications.

5   Research Experiment on Budgeting Module

In order to verify the developed algorithm, a research experiment on the case of budgeting module, has been performed. The budgeting module is placed in controlling sub-system. Cognitive agents perform following functions:

- generating partial budgets for various responsibility centers,
- an aggregation of partial budgets,
- providing management with information on the level of revision of budget assumptions for the entire enterprise and the deviations created,
- providing information to individual managers of organizational units about the degree of implementation of budgetary plans by their unit,
- calculation of ratios showing the financial situation of the company: liquidity, profitability or debt indicators.

The following assumptions were made for this experiment:

1. Budgeting processes in 17 enterprises has been analyzed.
2. There are from 3 to 5 responsibility centers in considered enterprises.
3. The partial budgets are generated by agents and next for each enterprise a knowledge conflict resolving agent (KCRA - running on the basis of algorithm developed in this paper) generates aggregated budgets.
4. The budgets of enterprises have been standardized (for the needs of experiment) and consist of following items: Amortization and depreciation, consumption of materials and energy, external services, taxes and charges, payroll, social security and other benefits, value of bought goods and materials, other costs, net revenues from sales of goods and materials and profit.
5. Budgets generated by KCRA has been compared by budgets aggregated manually by humans by using the following measures: average deviation in [%] (between budgets generated by KCRA and budgets aggregated by humans), standard deviation and variance (related to percentage deviation for 17 enterprises). The results are presented in Table 1.
6. During consultations with the enterprises owners and management board it has been stated, that percentage deviations lower than 5% are not important.

Table 1. Deviations between budgets generated by KCRA and budgets aggregated manually by humans.

Item	Average deviation for 17 enterprises [%]	Standard deviation for 17 enterprises	Variance for 17 enterprises
Amortization and depreciation	1.32	1.77	0.31
Consumption of materials and energy	2.49	0.61	0.04
External services	0.91	0.59	0.04
Taxes and charges	4.96	1.63	0.26
Payroll	3.25	2.67	0.71
Social security and other benefits	2.74	2.91	0.84
Other costs by type	6.39	2.72	0.74
Value of purchased goods and materials	7.02	7.08	5.01
Net revenues from sales of goods and materials	0.45	0.26	0.01
Average	**3.28**	**2.25**	**0.88**

Taking into consideration particular items, they are characterized by high fluctuations of percentage deviation. Such items as "Net revenues from sales of goods and materials" and "External services" are characterized by low percentage deviation, standard deviation and variance. Instead, such items as "Value of purchased goods and materials" and "Other costs by type" are characterized by high deviation (over 5%) and high standard deviation and variance.

The values of standard deviations and variances allow to draw a conclusion, that deviations between budgets generated by KCRA and budgets aggregated manually by humans are below average deviation.

Taking into consideration average values of measures it can be state, that the percentage deviation (3.28%) is lower than 5%. Therefore, budgets generated based on algorithm developed in this paper meet the requirements of enterprises owners and management board. At the same time, budgets are generated by KCRA near real time, while humans aggregate the budgets manually on average for several days.

6   Conclusions

The process of making economic decisions is very complex, especially when conducted under uncertainty and risk. The decision maker may then be certain of the consequences that the decision will bring. Supporting the decision-making process by using multi-agent systems is effective only when the decision maker receives a credible solution from the system. However, if a conflict of knowledge occurs among the agents, it considerably lowers the credibility of the budget decision generated by the system. The conflict needs to be resolved, then, so that the decision maker receives the best suggestion from the system and, consequently, makes the right decision which will improve the operation of an organization. Employing in this case a developed algorithm for resolving conflicts of knowledge in budgeting module may lead to a result that brings satisfactory benefit to the investor. Another advantage is shortening the time necessary to making a decision, and decreasing the risk associated with the budgeting process.

The developed algorithm does not guarantee the optimal decision, but it guarantees a certain level of satisfaction. Future works may be related, for example, to formal proving if the presented algorithm belongs to consensus methods and to developing a target function for the integration (e.g. the profit to achieve after integration).

References

1. Owoc, M.L., Weichbroth, P., Zuralski, K.: Towards better understanding of context-aware knowledge transformation. In: Proceedings of FedCSIS 2017, pp. 1123–1126 (2017)
2. Hernes, M.: A cognitive integrated management support system for enterprises. In: Hwang, D., Jung, J.J., Nguyen, N.-T. (eds.) ICCCI 2014. LNCS (LNAI), vol. 8733, pp. 252–261. Springer, Cham (2014). https://doi.org/10.1007/978-3-319-11289-3_26
3. Hernes, M., Sobieska-Karpińska, J.: Application of the consensus method in a multi-agent financial decision support system. Inf. Syst. e-Bus. Manag. **14**(1), 167–185 (2016)
4. Zeng, Z.: Construction of knowledge service system based on semantic web. J. China Soc. Sci. Tech. Inf. **24**(3), 336–340 (2005)
5. Atanasova, T.: Towards semantic-based process-oriented control in digital home. In: Federated Conference on Computer Science and Information Systems (FedCSIS), pp. 1133–1137 (2014). https://doi.org/10.15439/2014f317
6. Hofstadter, D., Mitchell, M.: The copycat project: a model of mental fluidity and analogy-making. In: Hofstadter, D. (ed.) The Fluid Analogies Research Group, Fluid Concepts and Creative Analogies. Basic Books (1995). Chap. 5
7. Dyk, P., Lenar, M.: Applying negotiation methods to resolve conflicts in multi-agent environments. In: Zgrzywa, A. (ed.) Multimedia and Network Information Systems, MISSI 2006, Oficyna Wydawnicza PWr, Wrocław (2006)
8. Barthlemy, J.P.: Dictatorial consensus function on n-trees. Math. Soc. Sci. **25**, 59–64 (1992)
9. Hernes, M.: Deriving consensus for term frequency matrix in a cognitive integrated management information system. In: Núñez, M., Nguyen, N.T., Camacho, D., Trawiński, B. (eds.) ICCCI 2015. LNCS (LNAI), vol. 9329, pp. 503–512. Springer, Cham (2015). https://doi.org/10.1007/978-3-319-24069-5_48
10. Kozierkiewicz-Hetmańska, A., Pietranik, M.: The knowledge increase estimation framework for ontology integration on the relation level. In: Nguyen, N.T., Papadopoulos, G.A., Jędrzejowicz, P., Trawiński, B., Vossen, G. (eds.) ICCCI 2017. LNCS (LNAI), vol. 10448, pp. 44–53. Springer, Cham (2017). https://doi.org/10.1007/978-3-319-67074-4_5
11. Blumentritt, T.: Integrating strategic management and budgeting. J. Bus. Strateg. **27**(6), 73–79 (2006)
12. Nguyen, N.T.: Inconsistency of knowledge and collective intelligence. Cybern. Syst. **39**(6), 542–562 (2008)
13. Schick, A.: Off-budget expenditure: an economic and political framework. OECD J. Budg. **7**(3), 7 (2007)

14. Lohan, G.: A brief history of budgeting: reflections on beyond budgeting, its link to performance management and its appropriateness for software development. In: Fitzgerald, B., Conboy, K., Power, K., Valerdi, R., Morgan, L., Stol, K.-J. (eds.) LESS 2013. LNBIP, vol. 167, pp. 81–105. Springer, Heidelberg (2013). https://doi.org/10.1007/978-3-642-44930-7_6

15. Franklin, S., Patterson, F.G.: The LIDA architecture: adding new modes of learning to an intelligent, autonomous, software agent. In: Proceedings of the International Conference on Integrated Design and Process Technology. Society for Design and Process Science, San Diego (2006)

16. Chojnacka-Komorowska, A.: Principles of modelling the controlling system in enterprise. In: Hittmar, S. (ed.) Theory of Management 3. The Selected Problems for the Development Support of Management Knowledge Base, pp. 327–331. Faculty of Management Science and Informatics, University of Zilina (2011)

17. Rashidi-Bajgan, H., Rezaeian, J., Nehzati, T., Ismail, N.: A genetic algorithm for capital budgeting problem with fuzzy parameters. In: International Conference on Computer Applications and Industrial Electronics, Kuala Lumpur, pp. 233–238 (2010)

18. Amthor, M., Rodner, E., Denzler, J.: Impatient DNNs - Deep Neural Networks with Dynamic Time Budgets. CoRR abs/1610.02850 (2016)

Agent-Based Decision-Information System Supporting Effective Resource Management of Companies

Jarosław Koźlak[1], Bartłomiej Śnieżyński[1], Dorota Wilk-Kołodziejczyk[1,2(✉)],
Stanisława Kluska-Nawarecka[2], Krzysztof Jaśkowiec[2],
and Małgorzata Żabińska[1]

[1] AGH University of Science and Technology, Al. Mickiewicza 30,
30-059 Kraków, Poland
dwilk@agh.edu.pl
[2] Foundry Research Institute in Krakow, Al. Zakopiańska 73, Kraków, Poland

Abstract. The aim of the work is to propose a universal multi-agent environment for resource management in the enterprise. The system being developed is to be useful for employees of various divisions of the company: device operators, engineering staff optimizing the production process and senior management. The paper describes the architecture of the solution, which has a layered structure. The environment uses advanced techniques of artificial intelligence, including machine learning and negotiation algorithms. In the evaluation part, an implementation of a pilot version of the foundry management system is presented and a study of selected test scenarios is carried out.

Keywords: Agent-based system · Resource management
Machine learning

1 Introduction

The information and decision-support system, which would collect on-line data from the units in the foundry and use this for carrying out subsequent processing stages, which in turn would offer advice on how to optimize the use of resources, is a system with a high degree of complexity and heterogeneous structure. To obtain the reliable analysis and results, it is necessary to consider information close to reality, which comes from the different levels of detail perception (for example, it is necessary to consider both the execution of processes on given production units and general rules regarding costs and parameters of materials, labour costs, thermal conditions at the level of the whole plant). Dealing with such complexity, heterogeneity and variability, and taking into account the necessary dependences is very difficult. Similarly, data analysis and formulation of conclusions requires advanced programming methods. It seems therefore that a useful approach is the application of multi-agent system methods, that have

© Springer Nature Switzerland AG 2018
N. T. Nguyen et al. (Eds.): ICCCI 2018, LNAI 11055, pp. 309–318, 2018.
https://doi.org/10.1007/978-3-319-98443-8_28

been continuously developed already for about 30 years, and which are designed precisely to solve problems of this kind. During the development of the multi-agent approach a number of methods and algorithms that can be employed to solve the given problem have been developed. In particular, the multi-agent approach aims at the development of systems with a high degree of complexity, consisting of cooperating intelligent entities. In order to ensure the intelligent behavior, we can use algorithms from different areas of artificial intelligence, such as knowledge representation, planning, reasoning and machine learning. Additionally, to ensure the cooperation between such entities the solutions for the following issues are proposed: conversational protocols, organizational models, negotiations between different elements, setting mutual commitments and norms for which they should follow.

Multi-agent systems are efficient architectures for decentralized problem solving [18]. Decision support systems may be very complex. Therefore such architectures are applied in such systems development [21,23]. An agent-based technology allows to mix various processing techniques, such as simulation, reasoning and machine learning and allows to distribute processing.

Our main contribution is an agent-based architecture for decision support systems. The originality of the proposed solution is focused on the following areas. A direct cooperation with available process units involved in the manufacturing process and participating staff dedicated to fulfill the needs of the foundry industry. A designed system dynamically collects data on-line and performs a number of significant information and decision-support functions for different categories of users (workers/operators, engineers-technologists, managers and directors). The application of a flexible multi-agent architecture that alongside artificial intelligence algorithms, also use and profit from solutions specific to the domain (organizational models, commitments, norms, negotiation protocols), and at the development level design patterns for multi-agent systems engineering. The data analysis and advice to support decision-making use various artificial intelligence techniques. The main innovative feature in the proposed system is a use of a rich set of different artificial intelligence methods (machine learning, data mining, reasoning, planning), separately or in a cooperation with other methods, for a variety of selected scopes of data and time periods.

In the following sections related research is discussed, the agent-based model of the system is presented. Next, implementation and evaluation results in a domain of casting plant are described. Finally, conclusions and further are presented.

2   Related Research

Different techniques of AI and MAS are applied in the developed system. The methods we consider of particular importance in this project are described briefly in Subsect. 2.1. In Sect. 2.2 selected solutions concerning resource management in the production process using AI and multi-agent approaches are presented.

2.1   AI and MAS Approaches

Organizations in multi-agent systems consist of a set of roles, relationships and authority structures) [8]. Different types or organizations were distinguished, such as hierarchies, holarchies, coalitions, teams, congregations, federations, matrix organizations, compound organizations and characteristic properties of each of them were identified and analyzed.

In the 1970s, supervised learning algorithms with symbolic knowledge representation emerged. Later, other learning strategies were developed such as unsupervised learning, reinforcement learning and inductive logic programming. Machine learning algorithms are often used in agent-based systems [14].

Reasoning methods enables the processing of a knowledge base and facts to infer some conclusions. Processing is usually based on logics and applies deduction-like operators. These techniques are commonly used in practical AI applications, like Expert Systems. In case of distributed systems, like the one considered here, multi-agent reasoning systems are created [1,7].

Planning process determines sequences of actions so as to achieve the desired objective, which may be a system state or an ability to perform the desired activity. In a dynamically changing environment, the important parameters describing the planning process are: a robustness and flexibility. The planning algorithm should be based on specific information expressed in such a way as to build a feasible plan leading to the goal to be achieved [6].

Intelligent entities – agents can have goals in mutual contradiction. In such a situation, it is necessary to carry out the negotiation process. Negotiations are usually a complex process, described by the type of conflict, participants, description of how to collect necessary information, how the participants analyze the situation, protocols of conversation, possible learning algorithms of participants in order to better understand the situation, and the renegotiation protocol [5]. To describe the decision process of negotiation participants different approaches are used, primarily based on game theory and heuristics [9].

Commitments are considered an important way of determining the interactions between agents. They help to represent the status of the relationship between two or more agents. Several agent systems were prepared using architectures based on the commitments, for example [2].

Norms in agent systems are aimed at determining the expected behaviour of agents. They can be used to define undesirable behaviour or to identify violations of the behaviour and determine ways to respond to these violations [20]. Norms may be set by the designers of the system or may emerge as a result of an automatic process of social learning of agents.

2.2   Resource Management in Production Using MAS

The multi-agent system we have developed here refers to several classes of systems and solutions. These are cyber-physical software systems, distributed resource management systems and agent-based decision support systems.

Cyber-physical software systems are characterized by combining software elements, that being, – computing systems with elements directly related to physical devices [12]. In [13], the applications of intelligent agents in various cyber-physical systems have been discussed, falling into three categories: Intelligent production systems, application systems in Smart Electric Grids and applications in intelligent logistics systems. They put forward in the paper the many benefits related to the use of agent systems in such systems, including support for autonomy, use of artificial intelligence techniques, the ability to research systems operation strategies with the help of specific agent simulations, as well as support for the integration of complex systems and support of human-computer integration. Paper [22] is focused on distributed decision making and negotiation mechanisms in multi–agent systems for self-organized manufacturing systems.

An important field of application of such types of systems, which are also convergent with significant problems appearing in the system we create, are energy management systems [17]. The elements included in such systems, in particular agents with their possible evolution and social networks created by them, are considered. In [4], the multi-agent systems used to integrate energy management systems (Smart Microgrids) were reviewed, and agent and multi-agent architectures are discussed. The multi–agent approach was studied to obtain feasible and reasonable performance. The authors of [10] emphasize the trend of implementing MAS for the management of real-time systems, with the use of low-cost based technologies, which posses enough power to be embedded with high performance algorithms using intelligent techniques. In [16] the authors discuss significant aspects related to Industrial Systems, Multi–Agent Systems and Cloud Systems. The role of agents is to coordinate activities of energy management. One of the important reasons of applying multi–agent approaches in decentralized management systems, and particularly in systems concerned energy management, is the higher robustness of such systems and lower risk of total system break-down [11].

Other types of agent systems we refer to in our work are agent-based decision support systems. In [21] the application of agent-based decision support systems is presented, discussing various types of agent models and agent organizations in the field review. For example, in [23] an agent-based expert system to make strategic decisions using fuzzy calculations is described. In [19], an example of an agent-based decision support system for the needs of energy market management is presented, describing the types of agents used and various market management algorithms.

3   Agent-Based Decision-Information System

In this section our agent-based architecture for decision-information systems is introduced. At the beginning general architecture and agents are introduced. Next, the first layer is described. At the end, the most important issues related to system implementation are discussed.

General Architecture and Agents. Architecture of the system is presented in Fig. 1. Agents are monitoring devices used in production using dedicated hardware connected to the system. Some data has to be entered manually by operators on a keyboard interface. Registered data is stored in a database and may be used later by other agents. Interaction agents are responsible for these activities. They cooperate with more advanced agents responsible for data analysis and decision support. They are supported by Artificial Intelligence Techniques. We are planning to apply agent related techniques, like commitments, negotiations, norms, organization techniques and also more general, like reasoning, machine learning and planning. Analysis results in the form of reports or recommendations will be presented to the users.

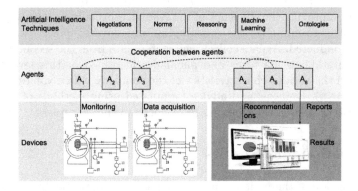

Fig. 1. The architecture of the solution

One can distinguish three logical layers in the developed system. They are described below.

First layer: Interactions with Users and Devices Layer – this layer consists of agents representing the various production processes cooperating with the users (agent-industrial processes) and batches of materials produced together and having significant part of characteristics common (agent-product batches). They acquire the necessary information by reading data from the devices/units involved in the industrial process or as a result of data entered by authorized users. The units communicate with one another according to the flow of resources (products) between the industrial processes and the necessary information exchanged between them. Agent-industrial processes represent, among others, processes of preparing the forms, smelting, casting, grinding, etc. They may be described by types and quantities of resources provided at the input, types and quantities of resources available at the output, process configuration parameters, process control parameters obtained as a result by available controlling devices. Employees/Operators with relevant permissions have access and may introduce or modify this data. The second type of agents operating at this layer are agents-batches. They represent a set of products that are manufactured together and are subject to the same processes (e.g. they come from the same material and are melted in the same furnace). Agents of the first layer provide data on the production process for the upper layer.

Second Layer: Data Processing and Optimization Layer – on this level, aggregated information concerning currently executed or historical production processes are processed, taking into consideration the interactions between them. As a result of the performed analysis with the use of various algorithms, different kinds of regularities, such as patterns (often repetitive) and anomalies (abnormal situations), are identified. Agents at this level represent the different data sources and different techniques for the analysis. The processing may be influenced and managed by different norms, either manually set or learned and automatically generated.

Third Layer: Inference and Recommendation Layer – on this level the analysis aimed at improving the efficiency of the system, especially formulating advice, which can help to minimize the consumption of resources, power, water, heat, air, etc., is conducted. Agents collect aggregated data from the second layer and the representations of patterns and anomalies identified at this level. Agents represent (a) the various characteristics of the company whose values should be optimized or (b) diagnosis and advice formulated using methods of reasoning, data analysis or heuristic techniques in response to the sent query.

4   Evaluation and Results

The pilot version of the multi-agent environment was implemented using a universal JADE platform [3]. The system is based on three logical layers, as shown in the chapter devoted to the model: the interaction layer with users and devices, the data processing and optimization layer as well as the application and recommendation layer (Fig. 2).

Interaction with Users and Devices Layer:

– *Sensor Interface Agent (SIA)* – agent responsible for gathering characteristics of the production process directly from the selected device. Its communication activities contain initialization of the new production process, configuration of the sensor and its connection with a production process, as well as starting, ending and configuration of the production process.
– *User Interface Agent (UIA)* – agent responsible for entering the characteristics of the production process by the user – operator. Its communication activities are similar to *Sensor Interface Agent's*, but contain also information about used and obtained products.
– *Query Interface Agent (QIA)* – is responsible for querying for data about production process, production of various sets of charts and requesting recommendations. Its communication activities contain configuration of agents for optimization of the company characteristics, recommendations concerning optimization of the production process, analyzed data, and data about process production and products.

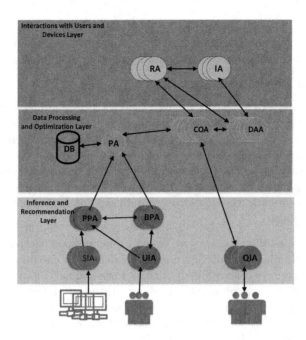

Fig. 2. Organization of the system and interactions between agents

- *Production Process Agent (PPA)* – the agent which represents the course of the production process and its features. Its communication activities contain initialization, configuration, starting and ending of the production process, associations with a group of products, creation of the products, receiving information of produced products.
- *Batch of Products Agent (BPA)* – Agent represents a batch of processed products, their relationship with other products and sets of attributes describing them. Communication activities contain configuration of the link with *Persisted Agent*, querying about information about lately produced products, creation of a group of products with a given type.

Data Processing and Optimization Layer:

- *Persistence Agent (PA)* – responsible for the communication of agents with the database and storing information in it. Its communication activities contain providing information about products and production processes.
- *Enterprise Characteristics Optimization Agent (COA)* – an agent trying to improve the characteristics of the production process, reducing the consumption of resources and energy, and improving the quality of the obtained products. The communication activities contain answering to queries about available inference and data analysis algorithms, recommendations concerning the optimization of the given production parameters, providing information about process productions for a group of products, queries for results of data analysis and advices based on analyzed data.

– *Data Analysis Agent (DAA)* – an agent looking for dependencies in current data on products and the production process, identifying patterns and anomalies. Its communication activities contain providing results of data analysis.

Inference and Recommendation Layer:

– *Inference Agent (IA)* – agent for inference for query implementation, data optimization and recommendation preparation. Its communication activities contain sending processed and data and identified patterns in the data.
– *Recommendation Agent (RA)* – an agent conducting the recommendation process in response to the request of the user. We distinguish several special kinds of *Recommendation Agents*: *Energy Agents*, *Resource Agents* or *Quality Agents*. Its communication activities contain sending recommendations based on analyzed data.

Two *Recommendation Agents* were created: *EnergyAgent* checking energy efficiency and *QualityAgent* predicting number of defects in castings. Weka implementations of supervised learning algorithms were applied.

EnergyAgent collects observations about status of chosen machines (e.g. induction furnace, casting line or dryers) and energy consumption. These observations are analyzed by engineers and energy efficiency class is assigned (low, medium, high). After collecting enough data, supervised learning algorithm is used to build a knowledge base K_e. In experiments we have applied J48 algorithm (Java implementation of C4.5 [15]), but other algorithms can be used too. After learning, the agent uses K_e to classify current state of the plant. In case of low efficiency, appropriate message is displayed on a console. Example of a decision tree is presented in Fig. 3. In this experiment there are only 18 examples in the training data. Therefore, accuracy is measured on the training data and is equal to 83,3%. Symbolic representation of the knowledge allows engineers to interpret it.

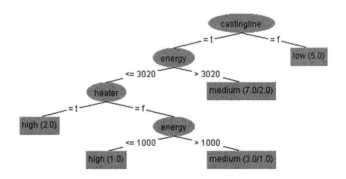

Fig. 3. Example of a decision tree learned by EnergyAgent

QualityAgent collects observation about the casting process (e.g. overheating temperature, form temperature or percentage of carbon in the metal) and a percentage of defective products. Here supervised regression algorithms are applied to build a model K_q that predicts the percentage of defective products representing quality. If K_q applied to the current status gives a number of defective products above a user-defined threshold, appropriate message is displayed on a console. There are 61 examples collected in the training data therefore 10-fold cross-validation may be applied. LinearRegression has Mean absolute error equal to 27.1 and Root mean squared error equal to 31.4, while for MultilayerPerceptron Mean absolute error is 13.8 and Root mean squared error is 20.5. This result suggests that MultilayerPerceptron works well.

5 Conclusions

Current experiments have a form of a proof of concept. Therefore in the future more experiments have to be done to compare the results with other systems.

In the paper the concept of multi-agent environment for optimization of resource consumption during the production process was presented The environment uses machine learning and negotiations techniques to obtain recommendations regarding efficient strategies. A pilot version of the system was prepared and research related to the optimization of selected action scenarios was conducted.

In the further work, it is expected to expand the elements of the system focused on the identification of patterns and anomalies in the data, and the use of norms and commitments to organize the decision-making process.

References

1. Arsene, O., Dumitrache, I., Mihu, I.: Expert system for medicine diagnosis using software agents. Expert Syst. Appl. **42**(4), 1825–1834 (2015)
2. Baldoni, M., Baroglio, C., Capuzzimati, F.: 2COMM: a commitment-based MAS architecture. In: Cossentino, M., El Fallah Seghrouchni, A., Winikoff, M. (eds.) EMAS 2013. LNCS (LNAI), vol. 8245, pp. 38–57. Springer, Heidelberg (2013). https://doi.org/10.1007/978-3-642-45343-4_3
3. Bellifemine, F., Poggi, A., Rimassa, G.: Developing multi-agent systems with JADE. In: Castelfranchi, C., Lespérance, Y. (eds.) ATAL 2000. LNCS (LNAI), vol. 1986, pp. 89–103. Springer, Heidelberg (2001). https://doi.org/10.1007/3-540-44631-1_7
4. Coelho, V., Cohen, M., Coelho, I., Liu, N., Guimarães, F.: Multi-agent systems applied for energy systems integration: State-of-the-art applications and trends in microgrids. Appl. Energy **187**, 820–832 (2017)
5. Fatima, Sh., Kraus, S., Wooldridge, M.: Principles of Automated Negotiation, 1st edn. Cambridge University Press, New York (2014)
6. Ghallab, M., Nau, D., Traverso, P.: Automated Planning: Theory & Practice. Morgan Kaufmann Publishers Inc., San Francisco (2004)
7. Haghighi, P., Burstein, F., Zaslavsky, A., Arbon, P.: Development and evaluation of ontology for intelligent decision support in medical emergency management for mass gatherings. Decis. Support Syst. **54**(2), 1192–1204 (2013)

8. Horling, B., Lesser, V.: A survey of multi-agent organizational paradigms. Knowl. Eng. Rev. **19**, 4 (2004)
9. Jennings, N., Faratin, P., Lomuscio, A., Parsons, S., Sierra, C., Wooldridge, M.: Automated negotiation: prospects, methods and challenges. Int. J. Group Decis. Negot. **10**(2), 199–215 (2001)
10. Kantamneni, A., Brown, L., Parker, G., Weaver, W.: Survey of multi-agent systems for microgrid control. Eng. Appl. Artif. Intell. **45**, 192–203 (2015)
11. Karavas, C., Kyriakarakos, G., Arvanitis, K., Papadakis, G.: A multi-agent decentralized energy management system based on distributed intelligence for the design and control of autonomous polygeneration microgrids. Energy Convers. Manag. **103**, 166–179 (2015)
12. Leitao, P., Colombo, A., Karnouskos, S.: Industrial automation based on cyber-physical systems technologies: prototype implementations and challenges. Comput. Ind. **81**, 11–25 (2016)
13. Leitao, P., Karnouskos, S., Ribeiro, L., Lee, J., Strasser, T., Colombo, A.W.: Smart agents in industrial cyber-physical systems. Proc. IEEE **104**, 1086–1101 (2016)
14. Panait, L., Luke, S.: Cooperative multi-agent learning: the state of the art. Auton. Agents Multi-Agent Syst. **11**, 387–434 (2005)
15. Quinlan, J.: C4.5: Programs for Machine Learning. Morgan Kaufmann, San Francisco (1993)
16. Rahman, M., Oo, A.: Distributed multi-agent based coordinated power management and control strategy for microgrids with distributed energy resources. Energy Convers. Manag. **139**, 20–32 (2017)
17. Rai, V., Robinson, S.: Agent-based modeling of energy technology adoption: empirical integration of social, behavioral, economic and environmental factors. Environ. Model Softw. **70**, 163–177 (2015)
18. Ricci, A., Santi, A.: Agent-oriented computing: agents as a paradigm for computer programming and software development. Int. J. Adv. Softw. **5**, 36–52 (2012)
19. Sueyoshi, T., Tadiparthi, G.: An agent-based decision support system for wholesale electricity market. Decis. Support Syst. **25**, 225–237 (2009)
20. Vázquez-Salceda, J., Aldewereld, H., Dignum, F.P.M.: Int. J. Comput. Syst. Sci. Eng. **20**(4), 225–236 (2005)
21. Wagner, N., Agrawal, V.: An agent-based simulation system for concert venue crowd evacuation modeling in the presence of a fire disaster. Expert Syst. Appl. **41**, 2807–2815 (2014)
22. Wang, S., Wan, J., Zhang, D., Li, D., Zhang, C.: Towards smart factory for industry 4.0: a self-organized multi-agent system with big data based feedback and coordination. Comput. Netw. **101**, 158–168 (2016)
23. Fazel Zarandi, M., Tarimoradi, M., Shirazi, M., Turksan, I.: Fuzzy intelligent agent-based expert system to keep Information Systems aligned with the strategy plans: A novel approach toward SISP. In: Proceedings of the 5th World Conference on Soft Computing (WConSC), 2015 Annual Conference of the North American (2015)

Evolutionary Multi-Agent System
in Planning of Marine Trajectories

Maciej Gawel[1], Tomasz Jakubek[1], Aleksander Byrski[1(✉)],
Marek Kisiel-Dorohinicki[1], Kamil Pietak[1], and Daniel Hernandez[2]

[1] Department of Computer Science, AGH University of Science and Technology,
Al. Mickiewicza 30, 30-059 Krakow, Poland
{mgawel,olekb,doroh,kpietak}@agh.edu.pl, jakubek.t@gmail.com
[2] SIANI Research Institute, Computer Science and Systems Department,
University of Las Palmas de Gran Canaria, 35017 Las Palmas, Spain
daniel.hernandez@ulpgc.es

Abstract. The paper considers application of agent-based computing system, namely Evolutionary Multi-Agent System, to solving a difficult yet interesting problem of a marine glider path planning. Different version of mutations are compared both for EMAS and evolutionary algorithm parametrized in the most possibly similar manner to EMAS and the observed results show that the EMAS is better in most of the experiments.

1   Introduction

Tackling difficult search problems calls for applying unconventional methods. This necessity is imposed by having little or no knowledge of the intrinsic features of the problem, topology of search space etc. In such cases, approximate techniques, like metaheuristics become the methods of last resort.

Having a plethora of metaheuristics to choose from, those population-based (as opposed to single solution oriented) seem to be the best choice, both at algorithmic and implementation level, and as they process more than one solution at a time, they can evade local extrema easier than single-solution approaches. Moreover, it is easy to implement them efficiently using ubiquitous parallel systems, such as multi-core processors, graphical processing units, clusters and grids.

An interesting and already empirically proven as efficient (cf. e.g., [1]) Evolutionary Multi Agent-System [2], combines evolutionary and agent-based computing paradigms in order to create a decentralized computing system. EMAS has also a sound formal background proving its correctness (ergodicity of a dedicated Markov chain modeling the EMAS search) [3]. Therefore the system has been chosen as a tool for solving the problem described in this paper.

Agents in EMAS represent solutions to a given optimization problem. They are located on islands representing distributed structure of computation. The islands constitute local environments, where direct interactions among agents

© Springer Nature Switzerland AG 2018
N. T. Nguyen et al. (Eds.): ICCCI 2018, LNAI 11055, pp. 319–328, 2018.
https://doi.org/10.1007/978-3-319-98443-8_29

may take place. In addition, agents are able to change their location, which makes it possible to exchange information and resources all over the system [4].

In EMAS, phenomena of inheritance and selection—the main components of evolutionary processes—are modeled via agent actions of *death* and *reproduction* (see Fig. 1). As in the case of classical evolutionary algorithms, inheritance is accomplished by an appropriate definition of reproduction. Core properties of the agent are encoded in its genotype and inherited from its parent(s) with the use of variation operators (mutation and recombination). Moreover, an agent may possess some knowledge acquired during its life, which is not inherited. Both inherited and acquired information determines the behavior of an agent. It is noteworthy that it is easy to add mechanisms of diversity enhancement, such as allotropic speciation (cf. [5]) to EMAS. It consists in introducing population decomposition and a new action of the agent based on moving from one evolutionary island to another (migration) (see Fig. 1).

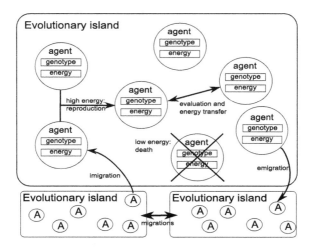

Fig. 1. Evolutionary multi-agent system (EMAS)

Assuming that no global knowledge is available and the agents being autonomous, selection mechanism based on acquiring and exchanging non-renewable resources [2] is introduced. It means that a decisive factor of the agent's fitness is still the quality of solution it represents, but expressed by the amount of non-renewable resource it possesses. In general, the agent gains resources as a reward for "good" behavior, and looses resources as a consequence of "bad" behavior (behavior here may be understood as, e.g. acquiring sufficiently good solution). Selection is then realized in such a way that agents with a lot of resources are more likely to reproduce, while a low level of resources increases the possibility of death. So according to classical Franklin's and Graesser's taxonomy—agents of EMAS can be classified as Artificial Life Agents (a kind of Computational Agents) [6].

Many optimization tasks, which have already been solved with EMAS and its modifications, have yielded better results than certain classical approaches. They include, among others, optimization of neural network architecture [7], multi-objective optimization [8], multi-modal optimization [9] and financial optimization [10]. EMAS has thus been proved to be a versatile optimization mechanism in practical situations. A summary of EMAS-related review has is given in [1].

This paper focuses on applying EMAS and comparing it to Evolutionary Algorithm tuned to use as similar search parameters as possible (similar selection, the same crossover and mutation and their parameters) to an interesting real-life problem of optimization of marine glider path planing along with detailed description of the mutation mechanisms used. In the next section the problem is explained, later the considered problems are described followed by the presentation of the experimental results. In the end the paper is concluded.

2 Glider Path Planning

A glider is an Autonomous Underwater Vehicle (AUV) that operates by modifying its buoyancy in a cyclic pattern [11]. It submerges, gathers research data and returns to the surface. A target direction (bearing) is commanded for each single submergence transect or stint, and followed by the glider in closed loop by means of its control surfaces.

The underwater glider path planning (UGPP) problem consists of finding the best route from one point to another. The glider's path is represented by points of resurfacing. The path is affected by constantly changing ocean currents and obstacles such as mainland, islands, reef. Since the nominal glider speed is quite low (approx. 1 Km per hour), the existence of strong currents forces these vehicles to have an achievable path computed in advance. The problem is non-trivial because of varying spatio-temporal domain and other restrictions.

In the past, different approaches and solutions to the glider path planning problem were proposed. Most researchers have approached the UGPP problem by applying different heuristics and metaheuristics (see, e.g. [12–16]). In this paper we would like to test the EMAS metaheuristic and compare it with classic EA, in order to evaluate the agent-based computing approach in this complex and interesting problem.

There are many variations of UGPP problem. In this paper it is assumed that we are given a map of ocean currents with obstacles, a glider model, a starting and target point, a starting date, and a number of times the glider submerges.

Genotype for both algorithms, EMAS and classic EA, is defined as a series of angles in degrees that correspond to subsequent glider bearings. The above-mentioned algorithms can be altered by using various operators. We tried several combinations of variation operators. Besides classical mutations like *Normal* or *Uniform* mutations, we designed and implemented other, more complex ones, based on the idea of local search.

Single Point Glider mutation procedure can be described in a few steps (Fig. 2). Initially, the genotype is randomly divided into two parts. The first segment does not change and the second is mutated in n different ways (again, each gene with probability P using normal distribution with radius R). This creates n new genotypes from which one with the best fitness is selected. It consists of one phase in which each gene is mutated with a probability P. The new value is generated from normal distribution with standard deviation (radius) R and mean equal to the gene value. In our experiments $n = 5$.

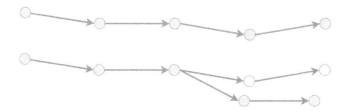

Fig. 2. Single Point Glider mutation. Upper graph represents path before mutation, lower—after.

SDLS mutation stands for *Steepest Descent Local Search*. It moves to the best solution in the neighborhood until it reaches a local minimum or a termination condition. The SDLS mutation is performed in the following way: for each gene a new genotype is created as a copy of the original. Then i-th genotype mutates only its i-th gene (normal distribution, radius R). At the end, genotype with the best fitness is chosen. In our experiments, the described process is repeated $n = 3$ times.

Greedy Glider mutation (Fig. 3) is based on the fact that glider can evaluate the distance to the target after each surfacing, which suggests the possibility of greedy algorithm. The idea of this mutation is as follows: glider simulates a process of submerging for a few different bearings and chooses the one with the shortest distance to the target. This continues until the specified number of surfacings is completed. From the genotype and single step perspective it mutates (with probability P) the i-th gene n times using normal distribution with standard deviation (radius) R and mean equal to the gene value. The generated values are the suggested bearings for i-th submerging. In our experiments $n = 5$.

3   Considered Problems

Experiments were conducted for 12 test cases where each of them is composed of a starting and a target point. Each point is described by latitude and longitude (Table 1).

Figure 4 presents a part of the sea around the Gran Canaria Islands with islands and currents for problems presented in Table 1. Starting and target points are marked as dots, islands are presented as the green areas. Black arrows indicate strength of the current and its direction.

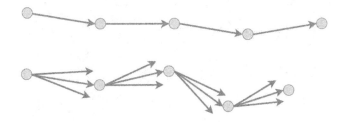

Fig. 3. Greedy Glider mutation. Upper graph represents path before mutation, lower—after.

Table 1. Problems.

Number	Starting point		Target point	
	latitude	longitude	latitude	longitude
S1	29.584800	−18.346770	28.83660	−17.153230
S2	28.836260	−17.153230	29.584800	−18.346770
S3	26.508770	−18.233870	27.421050	−17.104840
S4	27.421050	−17.104840	26.508770	−18.233870
S5	26.871350	−14.056450	27.760230	−14.782260
S6	27.760230	−14.782260	26.871350	−14.056450
S7	29.736840	−12.540320	29.175440	−13.975810
S8	29.175440	−13.975810	29.736840	−12.540320
S9	28.964910	−16.653230	27.549710	−16.717740
S10	27.549710	−16.717740	28.964910	−16.653230
S11	28.684210	−15.330650	27.269010	−15.282260
S12	27.269010	−15.282260	28.684210	−15.330650

In Fig. 4 are presented all problems considered in our experiments. Each scenario is represented by a pair of points of the same color. These points were chosen to test different cases of paths which glider should go through. There are scenario on open sea, along or around island and with or against the current.

4   Experimental Results

We have used a supercomputer called Prometheus installed in Academic Computer Centre Cyfronet AGH Cracow (HP Apollo 8000 platform working under Linux CentOS 7, 53 568 cores Intel Haswell and 279 TB DDR4 RAM). Each node has ability to run 24 tasks. The application was built using AgE component-oriented flexible computing platform implemented in Java[1]

---

[1] http://age.agh.edu.pl.

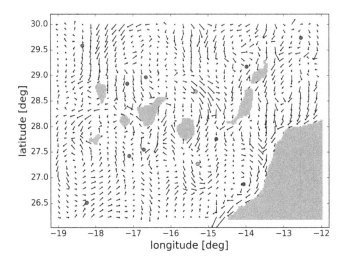

Fig. 4. Map of currents for all problems (Color figure online)

The tested algorithms, namely classic EA and EMAS, have a common set of parameters. Both algorithms are run as single population—no parallel model was used. In order to find the best configuration for each mutation, we tested several combinations of different values. Because of the complexity and time needed to perform a single simulation, we decided to optimize each parameter separately. Finally, we have chosen the values that would reflect the behavior of the mutations the best.

We have done an extensive testing of different parameters of the metaheuristics, in order to prepare for proper approaching the available problems. The best results obtained are presented in the Table 2. Each simulation presented in the experimental results was repeated 24 times. In this table Rad. stands for radius of mutation, Prob. for probability of mutation occurring, M.Prob. for probability of the mutation for a single gene.

Table 2. The best parameters for different variants of the tested algorithms

Alg.	Mut.	Rad.	Prob.	M.Prob	Alg.	Mut.	Rad.	Prob.	M.Prob
	Normal	1	0.2	0.5		Normal	10	0.5	0.5
	Uniform	30	N/A	0.9		Uniform	30	N/A	0.9
EA	SPG	10	0.8	0.9	EMAS	SPG	10	0.5	0.9
	Greedy	10	0.2	0.1		Greedy	1	0.2	0.1
	SDLS	30	N/A	0.5		SDLS	30	N/A	0.1

Before we delve into actual experimental results, we would like to present a visualization of a final values found by different variants of the algorithm, namely the visualization of the final populations attained by a single run of the algorithms (see Figs. 5 and 6). One can see that currents affected by the presence of obstacles make these problems especially difficult, and the visualized versions of the algorithms did not attain the final destination, although the first two were quite close (the destination is marked with a red dot).

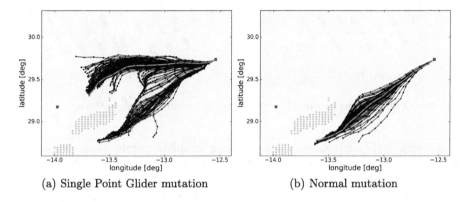

(a) Single Point Glider mutation   (b) Normal mutation

Fig. 5. Final population of EMAS algorithm for scenario 7 and 16 bearings. The best individual is colored orange. (Color figure online)

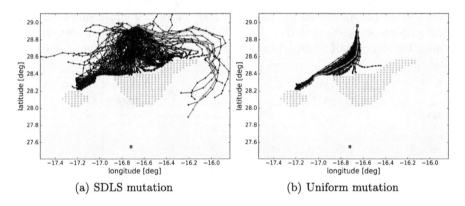

(a) SDLS mutation   (b) Uniform mutation

Fig. 6. Final population of EMAS algorithm for scenario 9 and 20 bearings. The best individual is colored orange. (Color figure online)

Now we compare the results achieved by algorithms for selected scenarios. Only the most interesting scenarios were selected, e.g. scenarios 1–6 were similar and relatively simple.

In scenario 1 (Fig. 7a) results of most algorithms are comparable and in a range 4000–8000. Simplicity of this problem shows in SDLS mutation which gives varying results. Very similar results can be seen in scenario 6 (Fig. 7b).

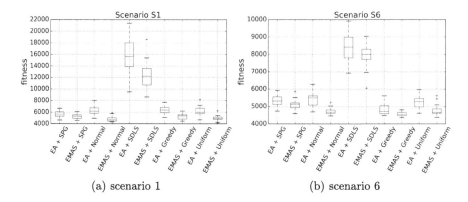

Fig. 7. Visualization of the best results for different algorithms.

Scenario 7 (Fig. 8a) shows varying results for different mutations. For example, SPG mutation converges very well to some minimum, while SDLS mutation still shows differing rates of fitness decrease. Additionally, a boxplot for a combination of EA and Uniform mutation shows that even though half of the results can be described as comparable to those of other "good" mutations, the other half is spread up to the point of 40000. It suggest that some solutions fall into the "wrong" local minimum (other side of the island).

Results for scenario 9 (Fig. 8b) are particularly interesting. Most results of half of the algorithms cannot get past the 85000 mark. Visualization of the map of the currents suggests that they fall in the same local minimum between the islands. The other half of algorithms gets past it and finds on the east of the island.

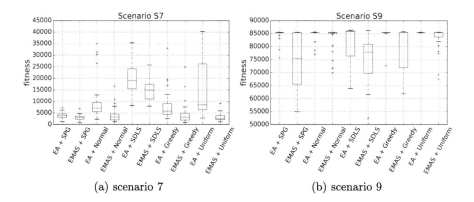

Fig. 8. Visualization of the best results for different algorithms.

Box plots for scenario 10 (Fig. 9)—the opposite of S9—show overall relatively small deviations in results. This suggests that currents on the south side of the

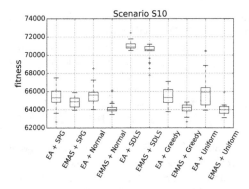

Fig. 9. Visualization of the best results for scenario 10 and different algorithms

island are not chaotic but rather regular and consistent which makes algorithms choose the path along them.

5   Conclusions

In the paper we have tackled an interesting problem of finding bearings for the marine glider using agent-based computing approach compared to classic one (evolutionary).

The obtained results can be treated as additional proof of quality of the EMAS metaheuristic. This algorithm turned out to be better in majority of the tests—of course not in all the cases (cf. No Free Lunch Theorem [17]). Therefore, further exploration of the considered problem, looking for even more efficient EMAS instances, and possibly new hybrid algorithms is planned.

In this paper we have tackled the comparison of EMAS and appropriately parameterized EA, to make them both similar from the point of view of e.g. the variation operators used. The obtained results prove the attractiveness of the agent-based paradigm in the computing domain.

Acknowledgment. The research presented in this paper was partially supported by the AGH University of Science and Technology Statutory Project.

References

1. Byrski, A., Dreżewski, R., Siwik, L., Kisiel-Dorohinicki, M.: Evolutionary multi-agent systems. Knowl. Eng. Rev. **30**(2), 171–186 (2015)
2. Cetnarowicz, K., Kisiel-Dorohinicki, M., Nawarecki, E.: The application of evolution process in multi-agent world (MAW) to the prediction system. In: Tokoro, M. (ed.) Proceedings of the 2nd International Conference on Multi-Agent Systems (ICMAS 1996). AAAI Press (1996)
3. Byrski, A., Schaefer, R., Smołka, M.: Asymptotic guarantee of success for multi-agent memetic systems. Bull. Pol. Acad. Sci. Tech. Sci. **61**(1), 257–278 (2013)

4. Kisiel-Dorohinicki, M.: Agent-oriented model of simulated evolution. In: Grosky, W.I., Plášil, F. (eds.) SOFSEM 2002. LNCS, vol. 2540, pp. 253–261. Springer, Heidelberg (2002). https://doi.org/10.1007/3-540-36137-5_19
5. Cantú-Paz, E.: A summary of research on parallel genetic algorithms. IlliGAL Report No. 95007. University of Illinois (1995)
6. Franklin, S., Graesser, A.: Is It an agent, or just a program?: a taxonomy for autonomous agents. In: Müller, J.P., Wooldridge, M.J., Jennings, N.R. (eds.) ATAL 1996. LNCS, vol. 1193, pp. 21–35. Springer, Heidelberg (1997). https://doi.org/10.1007/BFb0013570
7. Byrski, A., Kisiel-Dorohinicki, M., Nawarecki, E.: Agent-based evolution of neural network architecture. In: Hamza, M. (ed.) Proceedings of the IASTED International Symposium Applied Informatics. IASTED/ACTA Press (2002)
8. Siwik, L., Dreżewski, R.: Agent-based multi-objective evolutionary algorithms with cultural and immunological mechanisms. In: dos Santos, W.P. (ed.) Evolutionary Computation, pp. 541–556. In-Teh (2009)
9. Dreżewski, R.: Co-evolutionary multi-agent system with speciation and resource sharing mechanisms. Comput. Inf. **25**(4), 305–331 (2006)
10. Dreżewski, R., Sepielak, J., Siwik, L.: Classical and agent-based evolutionary algorithms for investment strategies generation. In: Brabazon, A., O'Neill, M. (eds.) Natural Computing in Computational Finance. Studies in Computational Intelligence, vol. 185, pp. 181–205. Springer, Heidelberg (2009)
11. Rudnick, D.L., Davis, R.E., Eriksen, C.C., Fratantoni, D.M., Perry, M.J.: Underwater gliders for ocean research. Mar. Technol. Soc. J. **38**(2), 73–84 (2004)
12. Davis, R.E., Leonard, N.E., Fratantoni, D.M.: Routing strategies for underwater gliders. Deep Sea Res. Part II: Top. Stud. Oceanogr. **56**(3), 173–187 (2009)
13. Isern-González, J., Hernández-Sosa, D., Fernández-Perdomo, E., Cabrera-Gámez, J., Domínguez-Brito, A.C., Prieto-Marañón, V.: Path planning for underwater gliders using iterative optimization. In: 2011 IEEE International Conference on Robotics and Automation, pp. 1538–1543, May 2011
14. Inanc, T., Shadden, S.C., Marsden, J.E.: Optimal trajectory generation in ocean flows. In: Proceedings of the 2005, American Control Conference, June 2005, pp. 674–679 (2005)
15. Alvarez, A., Caiti, A., Onken, R.: Evolutionary path planning for autonomous underwater vehicles in a variable ocean. IEEE J. Ocean. Eng. **29**(2), 418–429 (2004)
16. Zamuda, A., Hernández Sosa, J.D., Adler, L.: Constrained differential evolution optimization for underwater glider path planning in sub-mesoscale eddy sampling. Appl. Soft Comput. **42**(1), 93–118 (2016)
17. Wolpert, D., Macready, W.: No free lunch theorems for optimization. IEEE Trans. Evol. Comput. **67**(1), 67–82 (1997)

Airplane Boarding Strategies Using Agent-Based Modeling and Grey Analysis

Camelia Delcea[1](✉) ⓘ, Liviu-Adrian Cotfas[1] ⓘ, and Ramona Paun[2]

[1] Bucharest University of Economic Studies, Bucharest, Romania
camelia.delcea@csie.ase.ro
[2] Webster University, Bangkok, Thailand

Abstract. The cost pressure is still one of the main concerns of the airline companies, and one of the possible means to reduce these costs can be done my minimizing the turn time of their fleet. Three processes are included in the turn time: the deplanation process, aircraft cleaning and passenger boarding. Among these, the passenger boarding is the part that takes the longest time and therefore is the most important one when reducing the turn time and its associated cost. Trying to minimize the time needed by the boarding procedure, a series of boarding techniques have been developed. As no complete agreement has been made in the literature over the best boarding technique, the present paper considers some of the most used techniques and simulates them on an A320 aircraft. To this extent, a NetLogo program is created and several situations are considered. Some of them, such as, whether the passengers are traveling with no luggage or with hand luggage are often considered in the literature. Besides them, a third case in which the passengers are delaying other passengers due to the fact that they are loading their luggage is implemented as we believe is closer to the reality. Different passengers loading are also considered ranging between 60%–100% aircraft occupancy in order to determine the boarding time. Starting from the determined durations, the grey incidence is used in order to determine the main factors influencing the airplane passengers boarding time, which could allow each company to decide the most appropriate boarding method.

Keywords: Agent-based modeling · Boarding strategies · Airplane simulation
Grey systems theory · Grey analysis

1 Introduction

According to [1] the airplane turn time is dependent of the three processes: the deplanation process, aircraft cleaning and passenger boarding, while [2] identifies more processes to be included in the turn time, such as: fueling, airplane maintenance, cargo loading, baggage loading, passengers boarding, passengers deplaning, baggage unloading, cargo unloading. Considering all of them, the authors state that the air transportation companies should pay more attention to the boarding process as here they have at least a limited control over the passengers and therefore, they can make changes in order to increase the operational efficiency of this process. Also, considering all the processes involved in the airplane tur time, the authors believe that the passenger

N. T. Nguyen et al. (Eds.): ICCCI 2018, LNAI 11055, pp. 329–339, 2018.
https://doi.org/10.1007/978-3-319-98443-8_30

boarding is the part that takes the longest time and is the most important one when reducing the turn time and its associated cost.

According to [3], among the traditional boarding methods, the following average times (in minutes) have been recorded: by half-row (27.4 min), by half-block (27.9 min), by block (31.5 min) and by row (34.5 min).

Five boarding strategies (Steffen, random, outside-in, back-to-front on groups, reverse pyramid) are considered by [4] and the fact that the passengers are carrying with them, in the cabin, different pieces of luggage. After simulation, the authors conclude that the normal time for boarding is 25.30 min for their proposed method (with a high of 25.62 and a low of 25.15), then Steffen method with 25.73 min (high: 26.61, low: 25.51), reverse pyramid with 28.14 (high: 30.58, low: 28.08), outside-in with 28.74 min (high: 30.58, low: 28.11), random with 31.40 min (high: 34.05, low: 30.54) and last: back-to-front on groups with 32.35 min (high: 36.82, low: 31.61).

Boeing Enplane/Deplane Simulation (PEDS) made by the Boeing company found that on a 757-200 jetliner with 100% capacity usage and having only one door and one central aisle, the enplane process takes 26 min when using the traditional boarding method. This result is consistent with the study of [3] on an A320 and a 737. More, the Boeing study showed that boarding simultaneously by the two doors of the airplane is reducing the time by 5 min, while boarding on the two doors and considering an outside-in strategy had the potential to save 17 min, for an average boarding time of 13 min.

A half-filled aircraft (60 passengers) is considered in [5] with different seats preferences and the determined mean boarding time is 15 units for each passenger with different overall seats allocation depending on considering or non-considering the seat preference. [6] used an airplane with 30 rows with 6 seats per row and a boarding percentage of 51.1% and determined a boarding time of 7.33 min. In terms of costs, [6] estimated a cost of 30 $ per minute for each minute an airplane sits in the airport.

A recent study made by [7] ranked the methods according to the average time of boarding and found the following order: the best performing method is Steffen, followed by: outside-in, reverse pyramid, by half-block, back-to-front, by half-row, by row, random and last: the modified optimal method.

Thus, in the present paper the seven best performing methods found by [7] have been used for the airplane passenger boarding time, as presented in Table 1, in order to establish the best boarding technique and which are its main triggers.

In addition, we have also considered the fact that the passengers can carry with them different hand luggage items which slows them down and they may lose time while loading their luggage (which can also slow down the other passengers). These cases are implemented in NetLogo 6.0.2, which allows us the use of an agent-based modeling. More, as [1] considers that the efficiency of the boarding strategies may depend on the aircraft occupancy and different strategies may produce better results than other on the same aircraft model depending on the occupancy percentage, the simulations have been performed using different airplane boarding loadings ranging between 60% and 100%.

Table 1. Short description of the considered boarding methods regarding the seat assignment

Classification	Boarding method	Boarding strategy
By group	Outside-in/WilMA	Or the Window-Middle-Aisle method. The first class passengers are boarded firstly, then the passengers in window seats, followed by the middle seats and last the aisle seats. Basically, here we have the sequence of seats: (F, A), (E, B), (D, C)
	Reverse pyramid	Window seats first, followed by middle, then aisle and loading diagonally
	Half-block back-to-front	Passengers of some consecutive rows on one side of the aisle board together
	Back-to-front	It consists in boarding the first class firstly. Then, the rest of the passengers are boarded by groups, starting from the rear of the aircraft and moving forward, about 1/5 of the rows at a time
	By row back-to-front	By rows, starting at the rear of the aircraft and moving forward, a row at a time
	By half-row back-to-front	Starting at the rear of the aircraft and moving forward in just one part of the aircraft, a half-row at a time
	Modified optimal	As Steffen admits that the Steffen method is hard to realize in practice, he develops the modified optimal method. This method consists in four boarding groups with each group encompassing every second half-row of one side of the aisle. Therefore, the passengers who sit in a row with an even number on one side of the aisle board together. In the same way, those who sit in a row with an odd number on the same side board as a group. The same works for the passengers sitting on the other side of the aisle
By seat	Steffen	The first seated passenger is near the window in the rear. The second passenger is placed two rows apart from the first one near the window. The placement continues on the same manner from the back to the front and from the window to the aisle by seating the adjacent passengers two rows apart from each other in the corresponding seats

2   Modeling with NetLogo

According to ISI Web-of-Science [8] in 2003–2017 period, 368 papers have been written on different topics using NetLogo. Most of them are in areas such as: computer science, engineering, education and educational research, social sciences, business economics, mathematics, environmental sciences and ecology, operations research and management, etc.

Analyzing their content, the following categories of papers succeed in best applying NetLogo and agent-based modeling in their research: production control in manufacturing systems [9], distributed recommendation in social networks [10], social simulation [11, 12], information diffusion in social networking services [13], attitude change [14], organizational simulation [15], resources allocation [16], trade and development theory [17], etc. From the social sciences area, the following have been found: measurement of the social morality levels [18], the role of evangelism in consensus formation [19], the evolution of knowledge sharing behaviour in social commerce [20], etc.

As for the transportation area, two papers can be mentioned: a macroscopic model and simulation analysis of air traffic flow in airport terminal area [21] and intelligent traffic information system based on integration of internet of things and agent technology [22]. Other papers from different areas, include MarSim - a simulation of the MarsuBots Fleet [23], modelling academics as agents [24] and SimDrink - An Agent-Based NetLogo Model of Young, Heavy Drinkers for Conducting Alcohol Policy Experiments [25].

3   Grey Systems Theory Incidence Analysis

Over the years, the methods proposed by the grey systems theory have been successfully used in different social and economic area, physics, neuroscience, telecommunications, agriculture, geology, etc. [26–30]. From these methods, two of them are often used in applications: the grey model when making predictions and the grey analysis when conducting diagnosis or calculating the incidence of a phenomenon to another phenomenon/phenomena. For the incidence degree among two or more considered data series, the following degrees can be determined ranging between 0 and 1. The closer the value of the grey degree is to the 1 to greater is the incidence among the considered series.

3.1   The Absolute Degree of Grey Incidence

There are considered two sequences of data with non-zero initial values and with the same length, data X_0 and X_j, j = 1…n, with t = time period and n = variables [27, 31, 32]:

$$X_0 = \left(x_{1,0}, x_{2,0}, \dots, x_{t,0}\right) \tag{1}$$

$$X_j = \left(x_{1,j}, x_{2,j}, \dots, x_{t,j}\right) \tag{2}$$

The zero-start points' images are:

$$X_j^0 = \left(x_{1,j} - x_{1,j}, x_{2,j} - x_{1,j}, \ldots, x_{t,j} - x_{1,j}\right) = \left(x_{1,j}^0, x_{2,j}^0, \ldots, x_{t,j}^0\right) \tag{3}$$

The absolute degree of grey incidence is:

$$\varepsilon_{0j} = \frac{1 + |s_0| + |s_j|}{1 + |s_0| + |s_j| + |s_0 - s_j|} \tag{4}$$

with $|s_0|$ and $|s_j|$ computed as follows:

$$|s_0| = \left| \sum_{k=2}^{t-1} x_{k,0}^0 + \frac{1}{2} x_{t,0}^0 \right| \tag{5}$$

$$|s_j| = \left| \sum_{k=2}^{t-1} x_{k,j}^0 + \frac{1}{2} x_{t,j}^0 \right| \tag{6}$$

3.2 The Relative Degree of Grey Incidence

There are two sequences of data with non-zero initial values and with the same length, X_0 and X_j, $j = 1 \ldots n$, with $t =$ time period and $n =$ variables [31, 32]:

$$X_0 = \left(x_{1,0}, x_{2,0}, \ldots, x_{t,0}\right) \tag{7}$$

$$X_j = \left(x_{1,j}, x_{2,j}, \ldots, x_{t,j}\right) \tag{8}$$

The initial values images of X_0 and X_j are:

$$X_0' = \left(x_{1,0}', x_{2,0}', \ldots, x_{t,0}'\right) = \left(\frac{x_{1,0}}{x_{1,0}}, \frac{x_{2,0}}{x_{1,0}}, \ldots, \frac{x_{t,0}}{x_{1,0}}\right) \tag{9}$$

$$X_j' = \left(x_{1,j}', x_{2,j}', \ldots, x_{t,j}'\right) = \left(\frac{x_{1,j}}{x_{1,j}}, \frac{x_{2,j}}{x_{1,j}}, \ldots, \frac{x_{t,j}}{x_{1,j}}\right) \tag{10}$$

The zero-start points' images calculated based on (9) and (10) for X_0 and X_j are:

$$X_0^{0'} = \left(x_{1,0}' - x_{1,0}', x_{2,0}' - x_{1,0}', \ldots, x_{t,0}' - x_{1,0}'\right) = \left(x_{1,0}^{0'}, x_{2,0}^{0'}, \ldots, x_{t,0}^{0'}\right) \tag{11}$$

$$X_j^{0'} = \left(x_{1,j}' - x_{1,j}', x_{2,j}' - x_{1,j}', \ldots, x_{t,j}' - x_{1,j}'\right) = \left(x_{1,j}^{0'}, x_{2,j}^{0'}, \ldots, x_{t,j}^{0'}\right) \tag{12}$$

The relative degree of grey incidence is computed as:

$$r_{0j} = \frac{1 + |s'_0| + |s'_j|}{1 + |s'_0| + |s'_j| + |s'_0 - s'_j|} \tag{13}$$

with $|s'_0|$ and $|s'_j|$:

$$|s'_0| = \left| \sum_{k=2}^{t-1} x'^0_{k,0} + \frac{1}{2} x'^0_{t,0} \right| \tag{14}$$

$$|s'_j| = \left| \sum_{k=2}^{t-1} x'^0_{k,j} + \frac{1}{2} x'^0_{t,j} \right| \tag{15}$$

3.3   The Synthetic Degree of Grey Incidence

The synthetic degree of grey incidence is based on both the absolute and the relative degrees of grey incidence [31, 33]:

$$\rho_{0j} = \theta \varepsilon_{0j} + (1 - \theta) r_{0j} \tag{16}$$

with $j = 2, \ldots, n, \theta \in [0, 1]$ and $0 < \rho_{0j} \le 1$.

Grey incidence analysis is increasingly more used in economic and social sciences due to the fact that focuses on the closeness of relations between factors, being based on the similarity level of the geometrical patterns of sequence curves [27]. The size of the degree of incidence is directly proportional to the degree of similarity between the considered variables' curves.

4   Airplane Passengers Boarding Modeling and Simulation

In order to simulated the considered boarding strategies, a series of agents have been created in NetLogo 6.0.2 [34, 35], each of them having their own characteristics, such as the speed of passing through the aircraft cabin, the assigned seat, carrying a small or a large luggage, etc. After determining the best boarding method based on the aircraft passenger occupancy and the passengers' characteristics, the overall needed boarding time has been determined and its main triggers have been identified using the grey incidence analysis as presented below.

4.1   Agent Based Modeling and Simulation

Using NetLogo 6.0.2. and the A320 configuration, the eight considered boarding methods (see Table 1) have been implemented. Additionally, we have also modeled in the program the possibility of passengers to carry with them hand luggage in the cabin. Having the luggage and carrying it through the cabin will determine the slowing down of the passengers. Thus, their speed is amended by a random number having the lower bound of 0.1 and the upper bound of 0.4 (the values have been determined in accordance with [36]), conducting to a diminish of the passengers' speed.

Even though most of the studies in the airplane boarding strategies are just considering the simple case in which the passengers are boarding without a luggage, or the case in which they are carrying a luggage, we have considered in this study another situation, closer to the reality, in which the passengers are also losing time due to luggage storage. We found this aspect important as the passengers are not only spending time for their luggage storage, but they are also delaying the other passenger behind them, which can affect the overall boarding time. Even this aspect is modeled using a random interval number, based on [36].

Thus, the three considered situations are:

Case 1: the passengers alone are occupying their assigned seats (their speed in this case is one patch per tick);
Case 2: the passengers are carrying with them a luggage which is slowing them down (in this case the agents' speed is varying between 0.6 patches to 0.9 patches per tick);
Case 3: the passengers are carrying with them a luggage and they are losing time storing it (the agents' speed is varying between 0.6 patches to 0.9 patches per tick and they are also losing time for storing the luggage between 3 and 7 ticks). The simulation results and the grey analysis is presented in the following.

As the time in NetLogo is expressed in ticks where a tick is equal to the time needed by and agent to move between two patches, based on literature review it has been determined that in the real boarding situations, it can be assimilated to 5.4 s (the same result has also been used in [36]).

4.2   Simulation Results

In order to determine the best boarding method for the A320 airplane, different passengers loading have been used, ranging between 60% of the aircraft to 100%. Tables 2, 3 and 4 are presenting the average simulation results (each case has been simulated ten times in order to obtain the average value).

Table 2. Simulation results without luggage (in ticks)

Without luggage	Passengers loading percent				
	60%	70%	80%	90%	100%
Outside-in/WilMA	160.2	179	196.4	212.7	229.7
Reverse pyramid	135.5	181.2	194.8	195.7	213.5
Half-block back-to-front	165.6	170.8	177.6	182.1	188.7
Back-to-front	141.3	146.8	164.8	170.5	188.3
By row back-to-front	**133**	**145**	**157**	**169**	**181**
By half-row back-to-front	157	163	169	175	**181**
Modified optimal	160.6	175.5	192.8	213.2	231.4
Steffen method	163	180	195	208	223

Table 3. Simulation results with luggage (in ticks)

With luggage	Passengers loading percent				
	60%	70%	80%	90%	100%
Outside-in/WilMA	235.2	261.1	289.3	312.2	342.7
Reverse pyramid	203.9	268.1	292.6	292.5	316.9
Half-block back-to-front	248.2	251.7	268.8	278.8	286.6
Back-to-front	212.4	226.7	251.3	259.1	283
By row back-to-front	**199.6**	**224.3**	**242.2**	**257.1**	277.4
By half-row back-to-front	238.3	249.6	250.6	266.4	**277.3**
Modified optimal	239.2	260	289.6	318.8	347.5
Steffen method	243.2	260.1	289.5	309.3	332.4

Table 4. Simulation results with luggage and loading time (in ticks)

With luggage and loading time	Passengers loading percent				
	60%	70%	80%	90%	100%
Outside-in/WilMA	256.1	284.3	320.1	343.8	377.7
Reverse pyramid	217.6	289	316.1	322	351.1
Half-block back-to-front	270.5	272.8	287.8	298.3	318.3
Back-to-front	218.3	232.2	257.2	278.4	320.4
By row back-to-front	**205.2**	**229.9**	**247.8**	**262.2**	**307.8**
By half-row back-to-front	259.2	269.9	271.3	286.3	**307.8**
Modified optimal	248.6	276.3	315.5	347.4	382.5
Steffen method	263.3	286.4	315.1	334.3	362.6

The first conclusion which can be drawn after analyzing the data is that for an A320 aircraft, the best boarding method is by row back-to-front no matter the passenger loading degree or the fact that they are carrying luggage, nor that loading the luggage takes additional time which can delay other passengers.

Considering all the methods it can be seen that the order is almost the same in all the three considered situations with different loadings (from 70% to 100%): the second-best method is back-to-front, followed by: by half row back to front, half-block back to front, reverse pyramid, Steffen, WilMA and the last on the modified optimal method.

A slight difference can be seen in the 60% loading cases: here the best method is still by row back-to-front, while the second one changes and is the reverse pyramid.

4.3 Grey Incidence Analysis

Considering the simulation results and using the grey incidence analysis on all the cases and situations presented above, it has been determine that the dependent variable (the needed boarding time in this case) is more related to the number of passenger carrying luggage and which needs to load them (an incidence of 0.811), then with the number of passengers boarding in the airplane (an 0.686 synthetic incidence degree) and last with the number of passengers carrying luggage without needed time for loading it (an incidence degree of 0.509). As all the three grey incidence degree are above 0.5 we can conclude that all the three considered variables are influencing the overall boarding time.

Companies can analyses in this case their policy of cabin luggage and decide on their storage policy.

5   Concluding Remarks

The present study starts from the results gathered by [7] and analysis the boarding methods using an agent-based modeling approach. For this, a series of individual agents are created in NetLogo 6.0.2 having their own characteristics such as: speed, assigned seat, carrying or not a luggage, losing time for storing the luggage. In order to model the agents' speed, random numbers have been used, which offers the possibility to each agent to differentiate from another one.

Considering the simulation on an A320 aircraft and a grey incidence analysis it has been shown that depending on the type of travelers a company has, different policies related to the hand baggage cabin allowance of boarding method can be adopted by a company in order to reduce the airplane turn time.

As further research we aim to include a random passengers loading with and without assigned seats and compare the results with the one found above. For the random situation without assigned seats we are planning to build a questionnaire in order to determine each passenger's preference for a specific part of the aircraft (rear, middle, front) and for a specific place (located near the window, aisle or middle). Nevertheless, we think that other qualitative aspects should also be included in the agent-based simulation using random loading such as: the preference for free rows; ensuring that there are no better places in the next few rows; passengers in a queue can lose their patience and accept more unsuitable seat then in normal conditions; accepting a non-optimal seat given the other seats occupancy, etc.

The NetLogo 6.0.2 and the simulations made in this paper can be accessed at the following address: https://github.com/liviucotfas/ase-2018-iccci-bristol.

References

1. Ferrari, P., Nagel, K.: Robustness of efficient passenger boarding strategies for airplanes. Transp. Res. Rec. J. Transp. Res. Board **1915**, 44–54 (2005)
2. Soolaki, M., Mahdavi, I., Mahdavi-Amiri, N., Hassanzadeh, R., Aghajani, A.: A new linear programming approach and genetic algorithm for solving airline boarding problem. Appl. Math. Model. **36**, 4060–4072 (2012)
3. Van Landeghem, H., Beuselinck, A.: Reducing passenger boarding time in airplanes: a simulation based approach. Eur. J. Oper. Res. **142**, 294–308 (2002)
4. Qiang, S.-J., Jia, B., Xie, D.-F., Gao, Z.-Y.: Reducing airplane boarding time by accounting for passengers' individual properties: a simulation based on cellular automaton. J. Air Transp. Manag. **40**, 42–47 (2014)
5. Steffen, J.H.: A statistical mechanics model for free-for-all airplane passenger boarding. Am. J. Phys. **76**, 1114–1119 (2008)

6. Notomista, G., Selvaggio, M., Sbrizzi, F., Di Maio, G., Grazioso, S., Botsch, M.: A fast airplane boarding strategy using online seat assignment based on passenger classification. J. Air Transp. Manag. **53**, 140–149 (2016)
7. Kierzkowski, A., Kisiel, T.: The human factor in the passenger boarding process at the airport. Procedia Eng. **187**, 348–355 (2017)
8. WoS: Web of Science. webofknowledge.com
9. Zbib, N., Pach, C., Sallez, Y., Trentesaux, D.: Heterarchical production control in manufacturing systems using the potential fields concept. J. Intell. Manuf. **23**, 1649–1670 (2012)
10. Koohborfardhaghighi, S., Kim, J.: Using structural information for distributed recommendation in a social network. Appl. Intell. **38**, 255–266 (2012)
11. Bollinger, L.A., van Blijswijk, M.J., Dijkema, G.P.J., Nikolic, I.: An energy systems modelling tool for the social simulation community. J. Artif. Soc. Soc. Simul. **19**, 1 (2016)
12. Izquierdo, L.R., Olaru, D., Izquierdo, S.S., Purchase, S., Soutar, G.N.: Fuzzy logic for social simulation using NetLogo. J. Artif. Soc. Soc. Simul. **18**, 1 (2015)
13. Jung, J.J.: Measuring trustworthiness of information diffusion by risk discovery process in social networking services. Qual. Quant. **48**, 1325–1336 (2014)
14. Chattoe-Brown, E.: Using agent based modelling to integrate data on attitude change. Sociol. Res. Online **19**, 16 (2014)
15. O'Neil, D.A., Petty, M.D.: Organizational simulation for model based systems engineering. Procedia Comput. Sci. **16**, 323–332 (2013)
16. Kosmann, W.J., Sarkani, S., Mazzuchi, T.: Optimization of space system development resources. Acta Astronaut. **87**, 48–63 (2013)
17. Gulden, T.R.: Agent-based modeling as a tool for trade and development theory. J. Artif. Soc. Soc. Simul. **16**, 1 (2013)
18. Liu, H., Chen, X., Zhang, B.: An approach for the accurate measurement of social morality levels. PLoS ONE **8**, e79852 (2013)
19. Sharma, I., Chourasia, B., Bhatia, A., Goyal, R.: On the role of evangelism in consensus formation: a simulation approach. Complex Adapt. Syst. Model. **4**, 16 (2016)
20. Jiang, G., Ma, F., Shang, J., Chau, P.Y.K.: Evolution of knowledge sharing behavior in social commerce: an agent-based computational approach. Inf. Sci. **278**, 250–266 (2014)
21. Zhang, H., Xu, Y., Yang, L., Liu, H.: Macroscopic model and simulation analysis of air traffic flow in airport terminal area. Discrete Dyn. Nat. Soc. **2014**, 1–15 (2014)
22. Darbari, M., Yagyasen, D., Tiwari, A.: Intelligent traffic monitoring using internet of things (IoT) with semantic web. In: Satapathy, S., Govardhan, A., Raju, K., Mandal, J. (eds.) Emerging ICT for Bridging the Future - Proceedings of the 49th Annual Convention of the Computer Society of India (CSI) Volume 1. AISC, vol. 337, pp. 455–462. Springer, Cham (2015). https://doi.org/10.1007/978-3-319-13728-5_51
23. Leal Martínez, D., Halme, A.: MarSim, a simulation of the MarsuBots fleet using NetLogo. In: Chong, N.-Y., Cho, Y.-J. (eds.) Distributed Autonomous Robotic Systems. STAR, vol. 112, pp. 79–87. Springer, Tokyo (2016). https://doi.org/10.1007/978-4-431-55879-8_6
24. Gu, X., Blackmore, K., Cornforth, D., Nesbitt, K.: Modelling academics as agents: an implementation of an agent-based strategic publication model. J. Artif. Soc. Soc. Simul. **18**, 10 (2015)
25. Scott, N., Livingston, M., Hart, A., Wilson, J., Moore, D., Dietze, P.: SimDrink: an agent-based NetLogo model of young, heavy drinkers for conducting alcohol policy experiments. J. Artif. Soc. Soc. Simul. **19**, 10 (2016)
26. Delcea, C.: Grey systems theory in economics – a historical applications review. Grey Syst. Theor. Appl. **5**, 263–276 (2015)

27. Liu, S., Yang, Y., Xie, N., Forrest, J.: New progress of Grey system theory in the new millennium. Grey Syst. Theor. Appl. **6**, 2–31 (2016)
28. Delcea, C.: Grey systems theory in economics – bibliometric analysis and applications' overview. Grey Syst. Theor. Appl. **5**, 244–262 (2015)
29. Xie, N., Liu, S.: Novel methods on comparing grey numbers. Appl. Math. Model. **34**, 415–423 (2010)
30. Delcea, C.: Not black not even white definitively grey economic systems. J. Grey Syst. **26**, 11–25 (2014)
31. Liu, S., Lin, Y.: Grey Systems. Springer, Heidelberg (2011). https://doi.org/10.1007/978-3-642-16158-2
32. Cotfas, L.-A., Delcea, C., Segault, A., Roxin, I.: Semantic web-based social media analysis. In: Nguyen, N.T., Kowalczyk, R. (eds.) Transactions on Computational Collective Intelligence XXII. LNCS, vol. 9655, pp. 147–166. Springer, Heidelberg (2016). https://doi.org/10.1007/978-3-662-49619-0_8
33. Lian, Z.W., Dang, Y.G., Wang, Z.X., Song, R.X.: Grey distance incidence degree and its properties. In: 2009 IEEE International Conference on Grey Systems and Intelligent Services (GSIS 2009), pp. 37–41 (2009)
34. Wilensky, U., Rand, W.: An Introduction to Agent-Based Modeling: Modeling Natural, Social, and Engineered Complex Systems with NetLogo. The MIT Press, Cambridge (2015)
35. Delcea, C., Bradea, I.A.: Economic Cybernetics. An Equation-Based Modeling and Agent-Based Modeling Approach. Editura Universitara (2017)
36. Milne, R.J., Kelly, A.R.: A new method for boarding passengers onto an airplane. J. Air Transp. Manag. **34**, 93–100 (2014)

Agent-Based Optimization of the Emergency Exits and Desks Placement in Classrooms

Camelia Delcea[1]([✉]) ⓘ, Liviu-Adrian Cotfas[1] ⓘ, and Ramona Paun[2]

[1] Bucharest University of Economic Studies, Bucharest, Romania
camelia.delcea@csie.ase.ro
[2] Webster University, Bangkok, Thailand

Abstract. Even though the average number of structure fires in educational properties have fallen by 67% since 1980, the National Fire Protection Association has still recorded an average of 4980 structure fires (2011–2015), causing annual damages of 1 death, 70 injuries and $70 million in direct property damage. A series of studies have been conducted over the time in order to minimize the loses, with a particular focus on saving humans life. Thus, in order to reduce the evacuation time and the causalities, factors such as: the distance to exit, the density around the exit, room information, the presence of individuals with disabilities, heterogeneous population and obstacles have been considered. In this context, the present paper aims to determine if there is any connection between the structure of the classroom in terms of placing the exits and the desk placement. For this, a simulation is made using heterogeneous agents and a classroom with two exits. As the position of the exits is rather less changeable in real life as it depends directly on the building's characteristics, the desk placement can be easily modified inside the classroom with effects on the evacuation time.

Keywords: Agent-based modeling · Agent-based optimization
Agent-based simulation · Classroom evacuation · Emergency exits

1 Introduction

The study of crowd dynamics and the complex pedestrian behavior in emergency situations has gained significance lately in the literature associated to public buildings evacuation [1] and the models developed over the time have addressed this issue from different angles, such as: evacuation behavior [2–4], the route choosing decision making [5, 6], considering the distance and local density [7, 8], predicting the physical congestion [9], considering obstacle in the evacuation processes [7, 10], evacuation behavior involving individuals with disabilities [11]. As the pedestrian behavior is complex when dealing with the evacuation issue, the individual property can intervene in such situations [12, 13]. Thus, [14] considers three dimensions of the individual property: speed, density and flow, while [15] uses a questionnaire and determined that each pedestrian held different preferences and probabilities of choosing a particular exit for evacuation due to diversity of social background.

© Springer Nature Switzerland AG 2018
N. T. Nguyen et al. (Eds.): ICCCI 2018, LNAI 11055, pp. 340–348, 2018.
https://doi.org/10.1007/978-3-319-98443-8_31

Among the public buildings evacuation, the classroom evacuation problem has attracted a series of researchers [11, 16–18] and some of the particulars factor influencing the evacuation are presented in the following section.

2   Theoretical Background

2.1   Factors Influencing the Evacuations in Classrooms

According to the report of the National Fire Protection Association [19] published in September 2017: between 2011–2015 an average of 4980 structure fires have been recorded in educational properties, 3430 of them occurred in nursery, elementary, middle or high schools. Among them, 49% were is middle school and high school (causing 43 injuries and $20 million direct property damage), 33% in elementary school (causing 11 injuries and $20 million direct property damage), while the rest of them have been recorded in unclassified non-adults school and preschool. From the first category, 13% have been reported in college classrooms and adult education centers (counting for 670 fires).

The leading cause was the intentional fire set (45% of the cases), followed by cooking equipment (29%), playing with heat source (20%), heating equipment (10%), electrical distribution and lighting equipment (9%) and smoking materials (3%). More, it has been established that four of five fires in schools (80% of the total) occurred between 6 a.m. and 6 p.m., while two of five fires occurred between noon and 6 a.m. [19], when people were around. Thus, the structure fires in educational properties have fallen by 67% since 1980, but there is still place for improvement.

As for the classroom evacuation simulations, a series of papers have tried to better simulate what is happening with the people dealing with an ignite classroom and how they are making their decision in such moments. The most considered factors when dealing with the evacuation problem have been the distance to exit and the density around the exit [2, 3, 20], room information [3, 9, 20, 21], individuals with disabilities [11, 22], heterogeneous population [23] and obstacles [1]. Considering all the above, the present study tries to use a heterogeneous population of agents and to establish the connection between the classroom's characteristics in term of exit door and the desk placement. As in most of the cases, each class needs to have two exits and their place is given by the building's characteristics, the desk placement can be adjusted based on the simulation's results.

2.2   Modeling with NetLogo

NetLogo is a Multi-Agent-based model which uses the logo language in order to create the desired models and now it got to the 6.0.2 version [24, 25] and has been extensively used in agent-based modeling over the past years: according to ISI Web-of-Science [26] in 2003–2017 period, 368 papers have been written on different topics. In the following, a short literature review is provided based on the evacuation problem solving in NetLogo is provided.

Trying to enhance the importance of conducting fire drills with students, [16] considers two scenarios for a classroom evacuation: students evacuating without instructions and students evacuating in good order, both of them implemented in NetLogo. In addition the authors consider two classroom layouts: with one exit and with two exits. After conducting the simulations, the results have shown that the classroom with two exits shortens students' evacuation time, while another one suggests that the students who follow presented instructions escape faster and insure better their safety during the evacuation process.

On the contrary, [27] believe that that the fire drills are not a realistic situation if one wants to understand people behaviour, thus they use Fira de Barcelona pavilions and a 62820 runs in NetLogo to observe a large number of agents' behaviour.

Emotional contagion for emergency evacuation problem is discussed in [28]. Different types of agents are used representing both children (having two types of emotions: either calm or hysteria), adults (having emotions such as: calm, alarm, fear, terror, panic and hysteria) and security persons (with only one level: calm). For situations have been considered and based on simulations on each of them, the authors have concluded that the needed proportion in order to save as much people as possible in 1/35 (between the number of security members and the number of adults and children).

Households evacuation warning is studied by [29] in their paper, concluding that in order to have an efficient evacuation in the residential areas, one should try to motivating the public to inform the opposite houses about the imminent danger.

Tsunami evacuation has been considered in [30]. As a result, the authors conclude that the mortality rate is sensitive to the decision-making time, the variations in walking speed has significant impacts on the number of causalities and that the mortality rate increases as the number of evacuees who used automobile to evacuate increases as a result of congestion and bottleneck effects.

3   Simulation Design, Optimization and Results

According to ASSIST (Australian School Science Information Support for Teachers and Technicians) [31]: laboratories and classrooms should have at least two separate means of egress; at least one opening directly to the outside or a corridor that has external egress; the other may open into another room with an exit to the outside or corridor that has an external egress.

As for the small laboratories sub-compartments they may have only one access door provided that the distance of travel to the door from any point in the room does not exceed 7 m. In the case there are two or more doors, the distance between them should be at least 12.5 m, or 20% of the perimeter of the room (whichever is the lesser).

Considering a class with only two exits, the following situations can be possible for placing the doors as depicted in Fig. 1.

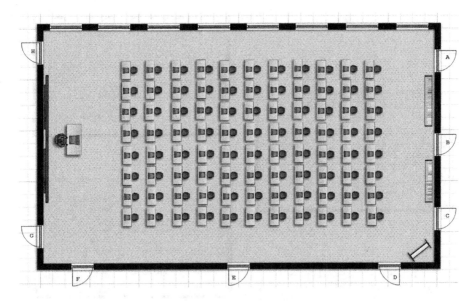

Fig. 1. The general plan of one-block desk classroom (with 8 possible placements for the doors)

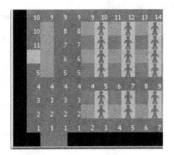

Fig. 2. The "cone exit"

We have marked the exits from A to H (A: right-top, B: right-middle, C: right-bottom, D: bottom-right, E: bottom-middle, F: bottom-left, G: left-bottom and H: left-top), while not any combination of them is doable according to the above rules (for example, one cannot pick both C and D exits for a classroom due to their too close neighborhood, the same for G and F). As for the desks, 6 situations have been considered: one-block, two-column-blocks, two-row-blocks, two-column-and-two-row-blocks, three-column-blocks and three-column-and-two-row-blocks (see Figs. 1 and 3).

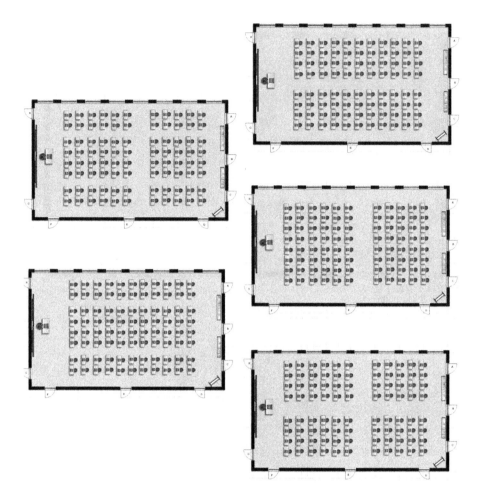

Fig. 3. The three-column-blocks, three-column-and-two-row-blocks, two-column-blocks, two-row-blocks and two-column-and-two-row-blocks desks' setting

3.1 Agent Based Modeling and Simulation

Using NetLogo 6.0.2. [24, 25] and the 88 desks' configurations an agent-based model has been created Fig. 4. The interface is perfectly configurable in terms of exits placement, exits width and the number of students attending the course.

For the minimum distance travelled by each agent for reaching one of the two exits and for its decision making process, we have implemented a "cone exit" as proposed by [32], with the only difference that we have started with smaller values for the exits, increasing the value of the adjacent patches at each iteration - Fig. 2.

Also, the patches representing the desks are seen as obstacles in the modeling, thus these patches haven't received a number and agents have no reason to pick them up, but rather to take a detour on their way to the exits.

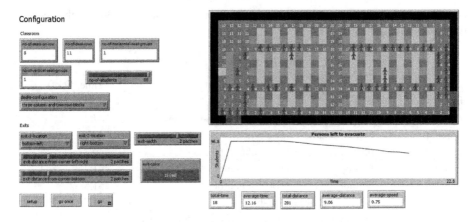

Fig. 4. The proposed model in NetLogo 6.0.2

As for real-life distance associated to each patch, we have used the results provided by [16] who concluded that in NetLogo, a patch has the length of 1, equivalent to 0.3 m × 0.3 m in the real world.

3.2   Optimization and Results

In order to determine the best match between the exits and desks placement, Behavior search tool offered by NetLogo has been used and several indicators such as the total evacuation time, the average evacuation time, the total distance, the average distance and the average speed. The simulations have been made on 17 from all the 28 possible cases as some of them as mentioned above have produced similar results (e.g. bottom-middle with bottom-left was similar to bottom-middle with left-bottom). The simulation has been conducted for all the seats in the classroom being fully occupied and the width of all the entrances was kept firm, measuring 2 patches (0.6 m).

The results have shown that the quickest evacuation is made for the classrooms having the exits placed one of them in right-middle, while the other one in bottom-left or left-top. For both situations, the best desks placement is two-column-and-two-row-blocks - Table 1.

Considering all the situations in which the best desk placement is one-block, one can observe that in six of the seven possible cases, one of the exits has been placed in the bottom-middle. Thus, if the building require such an exit, the desk placement within the classroom should be one-block.

More, the most popular desk placement two-column-blocks, while three of the considered designs, namely: two-row-blocks, three-column-blocks and three-column-and-two-row-blocks desks' setting haven't produce any good results no matter the exits' placement.

Last, by observing the best cases in the table above, one can observe that greater evacuation times are obtained when the two exists are place next to each other. On the contrary, as the two evacuation exits are as far as possible one from another, the needed time for evacuation decrease.

Table 1. Simulation results

Exits location		Desks placement	Time (ticks)
Exit 1	Exit 2		
right-top	right-bottom	two-column-and-two-row-blocks	47.2
right-top	bottom-middle	one-block	40.1
right-top	bottom-left	two-column-blocks	40.3
right-top	left-top	two-column-blocks	41.2
right-middle	bottom-right	two-column-blocks	51.5
right-middle	bottom-middle	one-block	41.7
right-middle	bottom-left	two-column-and-two-row-blocks	39.1
right-middle	left-top	two-column-and-two-row-blocks	39.3
right-bottom	bottom-middle	two-column-blocks	45.4
right-bottom	bottom-left	two-column-blocks	40.6
right-bottom	left-top	two-column-blocks	40.5
bottom-right	bottom-middle	one-block	45
bottom-right	bottom-left	two-column-blocks	39.8
bottom-right	left-top	two-column-and-two-row-blocks	40.3
bottom-middle	bottom-left	one-block	47.1
bottom-middle	left-top	one-block	42.1
bottom-left	left-top	two-column-blocks	51.7

4   Concluding Remarks

The present study aims in determining the best fit between the evacuation doors and desks' placement within a classroom by considering six different desks placement and eight possible doors' location. After conducting the simulation using a multi-agent model created in NetLogo 6.0.2 it can be concluded that there can be determined a connection among the doors' location and desks' placement. Based on the conditions given by the building, one can use the present model in order to decide which desks' placement he/she should consider in order to minimize the evacuation time in case of an emergency. The model is configurable in order to match different classroom sizes and spaces.

As further research we aim to include (besides the agents' emotions, mood and emotions' contagion) classrooms with elevated floors which can delay the evacuation time, different obstacles due to evacuation (including here fallen people) and the presence of persons with disabilities (both visual and wheelchair). For the wheelchair persons we aim to determine the best place for placing their desks within the classroom for an easy and safe evacuation.

The application in NetLogo 6.0.2 and the simulations made in this paper can be accessed at the following address: https://github.com/liviucotfas/ase-2018-iccci-bristol-2.

References

1. Han, Y., Liu, H., Moore, P.: Extended route choice model based on available evacuation route set and its application in crowd evacuation simulation. Simul. Model Pract. Theor. **75**, 1–16 (2017)
2. Duives, D., Mahmassani, H.: Exit choice decisions during pedestrian evacuations of buildings. Transp. Res. Rec. **2316**, 84–94 (2012)
3. Haghani, M., Ejtemai, O., Sarvi, M., Sobhani, A., Burd, M., Aghabayk, K.: Random utility models of pedestrian crowd exit selection based on SP-off-RP experiments. Transp. Res. Procedia **2**, 524–532 (2014)
4. Papakonstantinou, D., Benardos, A., Kallianiotis, A., Menegaki, M.: Analysis of the crowd evacuation modeling approaches for the case of urban underground spaces. Procedia Eng. **165**, 602–609 (2016)
5. Hoogendoorn, S.P., Bovy, P.H.L.: Pedestrian route-choice and activity scheduling theory and models. Transp. Res. Part B Methodol. **38**, 169–190 (2004)
6. Kneidl, A., Borrmann, A.: How Do Pedestrians find their Way? Results of an experimental study with students compared to simulation results. 7
7. Eng Aik, L., Wee Choon, T.: Simulating evacuations with obstacles using a modified dynamic cellular automata model. J. Appl. Math. **2012**, 12 (2012). https://www.hindawi.com/journals/jam/2012/765270/
8. Stubenschrott, M., Kogler, C., Matyus, T., Seer, S.: A dynamic pedestrian route choice model validated in a high density subway station. Transp. Res. Procedia **2**, 376–384 (2014)
9. Guo, R.-Y., Huang, H.-J., Wong, S.C.: Collection, spillback, and dissipation in pedestrian evacuation: a network-based method. Transp. Res. Part B Methodol. **45**, 490–506 (2011)
10. Varas, A., et al.: Cellular automaton model for evacuation process with obstacles. Phys. A: Stat. Mech. Appl. **382**, 631–642 (2007)
11. Gaire, N.: A study on human evacuation behavior involving individuals with disabilities in a building. Grad. Thesis Dissertation (2017)
12. Jiang, G., Ma, F., Shang, J., Chau, P.Y.K.: Evolution of knowledge sharing behavior in social commerce: an agent-based computational approach. Inf. Sci. **278**, 250–266 (2014)
13. Tang, T.-Q., Shao, Y.-X., Chen, L.: Modeling pedestrian movement at the hall of high-speed railway station during the check-in process. Phys. Stat. Mech. Appl. Complet. **467**, 157–166 (2017)
14. Wong, S.C., Leung, W.L., Chan, S.H., Lam, W.H.K., Yung, N.H.C., Liu, C.Y., Zhang, P.: Bidirectional pedestrian stream model with oblique intersecting angle. J. Transp. Eng. **136**, 234–242 (2010)
15. Li, Z.: An emergency exits choice preference model based on characteristics of individual diversity. Adv. Mech. Eng. **9**, 168781401769354 (2017)
16. Liu, R., Jiang, D., Shi, L.: Agent-based simulation of alternative classroom evacuation scenarios. Front. Archit. Res. **5**, 111–125 (2016)
17. Buehlmann, M., Beitler, A.: Trails of Irchelpark How students walk. 30
18. Kimura, M., Sime, J.D.: Exit choice behaviour during the evacuation of two lecture theatres. Fire Saf. Sci. **2**, 541–550 (1989)
19. Campbell, R.: Structure Fires in Educational Properties. Natl. Fire Prot. Assoc. (2017)

20. Liu, S., Yang, L., Fang, T., Li, J.: Evacuation from a classroom considering the occupant density around exits. Phys. Stat. Mech. Appl. **388**, 1921–1928 (2009)
21. Fu, S.: Three-parameter interval grey number multi-attribute decision making method based on information entropy. Math. Comput. Appl. **21**, 17 (2016)
22. Sørensen, J.G., Dederichs, A.S.: Evacuation characteristics of visually impaired people - a qualitative and quantitative study. Fire Mater. **39**, 385–395 (2015)
23. Sørensen, J.G., Dederichs, A.S.: Evacuation from a complex structure – the effect of neglecting heterogeneous populations. Transp. Res. Procedia **2**, 792–800 (2014)
24. Wilensky, U., Rand, W.: An Introduction to Agent-Based Modeling: Modeling Natural, Social, and Engineered Complex Systems with NetLogo. The MIT Press, Cambridge, Massachusetts (2015)
25. Delcea, C., Bradea, I.A.: Economic Cybernetics. An Equation-Based Modeling and Agent-Based Modeling Approach. Editura Universitara, Bucuresti (2017)
26. WoS: Web of Science. webofknowledge.com
27. Gutierrez-Milla, A., Borges, F., Suppi, R., Luque, E.: Individual-oriented model crowd evacuations distributed simulation. Procedia Comput. Sci. **29**, 1600–1609 (2014)
28. Faroqi, H., Mesgari, M.-S.: Agent-based crowd simulation considering emotion contagion for emergency evacuation problem. Int. Arch. Photogramm. Remote Sens. Spat. Inf. Sci. **XL-1-W5**, 193–196 (2015)
29. Nagarajan, M., Shaw, D., Albores, P.: Informal dissemination scenarios and the effectiveness of evacuation warning dissemination of households — a simulation study. Procedia Eng. **3**, 139–152 (2010)
30. Wang, H., Mostafizi, A., Cramer, L.A., Cox, D., Park, H.: An agent-based model of a multimodal near-field tsunami evacuation: decision-making and life safety. Transp. Res. Part C Emerg. Technol. **64**, 86–100 (2016)
31. Karin, K.: Number of exit doors in labs and prep rooms. https://assist.asta.edu.au/print/2699
32. Dario, B., Brun, N.: Evacuation bottleneck: simulation and analysis of an evacuation of a lecture room with MATLAB. https://www.ethz.ch/content/dam/ethz/special-interest/gess/computational-social-science-dam/documents/education/Fall2011/matlab/projects/Evacuation_Simulation-Biner_Brun-2.pdf

Agents Interaction and Queueing System Model of Real Time Control of Students Service Center Load Balancing

Malika Abdrakhmanova[✉], Galimkair Mutanov, Zhanl Mamykova,
and Ualsher Tukeyev

Al-Farabi Kazakh National University, Al-Farabi Avenue 71,
Almaty, Kazakhstan
malika.berikovna@mail.ru, zhmamykova@gmail.com,
ualsher.tukeyev@gmail.com

Abstract. The problem of effective organization of Students service center (SSC) activities is considered. In this paper is proposed combine agents interaction and queuing system model for creation real time control of SSC load balancing. The developed combined model allows to minimize the number of required personnel resources and their idle time and to create adaptive, modular, well scalable system.

Keywords: Electronic university · Students Service Center · Queueing system
Load balancing · Multi-agent modeling

1 Introduction

Informatization of society involves radical social changes therefore high-quality expansion of the informatization demands comprehensive accounting of a human factor, relevant there is a problem of adaptation of the person to life in the conditions of the new information environment. Transformation of university activities with creating favorable conditions for consumers of its services through elimination of administrative barriers and preservation of valuable free time is possible when following to experience of creation of Public Service Centers and provide services by the principle of "single window" in the uniform Students Service Center (SSC).

The practical importance of SSC is directed to creation of socially important conditions for high-quality stay of students in the territory of a campus, having provided them conditions for receiving consultation on the organization of educational process, receiving high-quality socially significant services in one place.

Main activities of SSC: zones of providing administrative and consulting services in educational process; access points for "narrow experts" consultations: financier, lawyer; uniform center of the youth organizations; center of providing medical services; access point to electronic services of the university; zone of providing administrative services of the state character.

For effective management of SSC activities it is necessary to predict future load in various zones of service. Significant for consideration for the purpose of load

N. T. Nguyen et al. (Eds.): ICCCI 2018, LNAI 11055, pp. 349–359, 2018.
https://doi.org/10.1007/978-3-319-98443-8_32

optimization is zone of administrative and consulting services in educational process. The list of the functions realized in this zone by employees of registrar office and student's department is very extensive, also level of demand on these services raised. Due to the reorganization of their activities the new view on their services is created. In these departments in addition to consulting support of students, functions on administration of educational process are provided. This fact allows to divide these zones into sectors with the reduced number of functions. Functions of these sectors can be crossed on some questions, for this purpose registrar office and student's department employees are given with the corresponding functions of role policy of access, that allowing to serve as necessary any application with the crossed requests in these categories of service.

Now we are faced by a problem of modeling of registrar office and student's department zones activities. As SSC is the socially oriented system having signs of queueing systems, parameters of efficiency of its functions can be determined in terms of the queueing theory. Considering the fact that the documents generated in one departments are widely used in work of others, and functions are crossed thus between the specified departments and between sectors in departments, it is possible to consider in the system the possibility of transfer of requests to serve in other sectors at emergence of load balancing needs.

The field of social service including service of students, assumes constant need for service expansion, addition of new types of service. In view of need of fast and effective scaling of the system, the best approach to modeling of interaction between sectors for the purpose of load balancing will be multiagent approach.

2   Related Works

The research and development of queueing systems models have gained distribution in connection with tasks from the most various spheres of services. The detailed analysis of various models of queueing systems, their mathematical characteristics and opportunities of application are presented by Adan and Resing [1].

The social queueing systems represent the greatest difficulties for simulation owing to difficulty of prediction not only characteristics of an arrival steam of requests for service, but also prediction characteristics of service process. The reason for that is the impossibility of creation of process exact protocols in which it is difficult to provide requests in strictly predetermined template. The research of behavior of social queueing systems has found reflection in works of Yuan and Hwarng [2], where they have conducted a research of dynamic behavior of queueing system under impact of social interactions.

The complexity of prediction of social queueing system load requires to analyse of possibility of load balancing in the presence of several parallel operating queues for the purpose of maximizing use of resources and minimization of wait time. Jiang and Li in their work [3] consider 2 approaches to distribution of tasks, the value of this work is the analysis of talent-based allocation approach, following which tasks are distributed according to skills of agents. The algorithm of load balancing in parallel systems of customer service is offered by Down and Lewis [4].

In social simulation and modeling have been increasingly applied recent years agent-based approach. This approach allows to model the systems which are difficult to formalization by standard mathematical tools, it allows to build adaptive, modular, well scalable systems [5]. Researches in the field of agent-based approach represent an assessment of various theories of agent-based modeling of social systems, main applied fields of agent-based approach and offer various rules of agents interaction [5–7]. The important direction of agents interaction is cooperation for achievement of a common goal, various models of cooperation of agents were considered in works [8, 9].

Within the article considered the model of load balancing control of students service center in which the queueing theory and agent-based approach are combined. The control system is presented as multiagent system, where agents interact for providing effective service by redistribution of load, each agent achieves the local objectives proceeding from rules of the queueing theory.

3 Description of the Offered Approach

3.1 Modeling of the Students Service Center as a Queueing System

Students service center is an example of queueing system with several queues (zones) with non-uniform arrivals of model (M/M/c): (GD/∞/∞), on Kendall's classification [1]. This is the model with exponential interarrival times with mean $1/\lambda$, exponential service times with mean $1/\mu$ and c parallel identical servers. Customers are served in order of arrival. We suppose that the occupation rate per server,

$$\rho = \frac{\lambda}{c\mu}, \tag{1}$$

is smaller than one.

An important quantity is the probability that a job has to wait. Denote this probability by ΠW. It is usually referred to as the delay probability [1].

$$\Pi_w = \frac{(c\rho)^c}{c!} \left((1-\rho) \sum_{n=0}^{c-1} \frac{(c\rho)^n}{n!} + \frac{(c\rho)^c}{c!} \right)^{-1} \tag{2}$$

From the equilibrium probabilities we directly obtain for the mean queue length

$$L = \Pi_W \cdot \frac{\rho}{1-\rho} \tag{3}$$

The mean waiting time,

$$W = \Pi_W \cdot \frac{1}{1-\rho} \cdot \frac{1}{c\mu} \tag{4}$$

In general, arrival stream of queueing system can be non-uniform when the requests of several categories differing from each other in laws of distribution either intervals of receipt, or service time, or serving priorities come to system. Very often in the analysis of such queues initial non-uniform load comes down to equivalent uniform. This process includes the following transformations of initial parameters:

1. intensity of the integrated stream

$$\lambda_{gen} = \sum_{k=1}^{H} \lambda_k, \tag{5}$$

where H is the number of requests categories

2. average holding time of requests of the integrated stream

$$t_{av} = \frac{1}{\lambda_{gen}} \sum_{k=1}^{H} \lambda_k \cdot t_k \tag{6}$$

3.2 Description of the Method of SSC Input Streams Balancing Dynamic Control with Use of Multiagent Approach

The primary activity of SSC is represented by functions on the organization of educational process. Considering the fact that within the academic year the intensity of addresses on many functions considerably changes, it is important to organize work of SSC with minimization of employees' idle time and customers waiting time in queue. In traditional approach to the organization of educational process, the main functions are distributed between registrar office and student's department. Functions of these departments can be subdivided into 4 categories:

1. functions of registrar office on consultation of students about the organization of educational process – f_1;
2. functions of registrar office on administration of educational process – f_2;
3. functions of student's department on consultation of students about the organization of educational process – f_3;
4. functions of student's department on administration of educational process – f_4.

Social systems are characterized by unpredictable behavior, therefore it is impossible to calculate ideal parameters of effective work of sectors as a queueing systems with a big share of reliability and to base long-term work of system on these parameters. For balancing of changing load in sectors, the possibility of the decentralized redistribution of an arrival requests flow with use of multiagent approach is considered. Use of multiagent approach will allow to develop module system, which can be scaled easily: to add new sectors with new sets of functions and new nature of interaction with other sectors.

Agents in system will be parallel program modules which accept the sector requests flow as an entrance, count parameters of work of the sector according to the queueing theory and interact with other agents for load balancing. Agents aim not only to support effective work of the sector, but also collaboration with other agents for maintenance of reliable work of all system. In terms of multiagent approach there is a set of different definitions of the term the agent, we represent the definition, the most suitable for our purposes: an agent is an autonomous entity, which acts upon an environment and directs its activities toward achieving some specified intentions [10].

The system consists of 4 interacting agents $\{A_i\}_{i=1}^4$, which are carrying out calculation of service parameters for 4 sectors. Agents interact with the environment, accepting the requests which arrived on service and using environment resources to serve the requests. Resources of the environment are described quantitatively and qualitatively (quantity – $\{c_i\}_{i=1}^4$ and skills – $\{r_i\}_{i=1}^4$ of employees).

Skills (roles) of employees are defined by types of requests which they can process. Generally, all employees are universal and can process any type of requests, but for reduction of the general time of requests implementation due to mechanization of often performed operations, all types of requests are classified on 4 categories and are transferred only to the relevant sector to which certain employees are assigned.

Each agent is characterized by the skills and a state. Skills $\{R_i\}_{i=1}^4$ of agents represent calculation of working parameters for each sector according to the queueing theory. For the considered sectors the queueing system types describing them are identical and their parameters pay off by the same rules, but generally, sectors can belong to various types of queueing systems.

The agent has information only on parameters of its sector, it obtains information on parameters of other sectors by means of exchange of messages with other agents. For the considered system the most important parameters of effective work are an average waiting time of requests in queue and queue length, in this work queue length we accept as the balancing parameter. Each sector has reference parameters of service calculated from a condition that waiting time of a request in queue shouldn't exceed 10 min for the sectors serving students and not to exceed 15 min for the sectors serving departments of educational process support. We will designate reference queue length for the sector i through L_i.

Depending on parameters of system the state of the agent is defined. The set of states S of agent is described by 5 names: standard loading – s_1, high loading – s_2, $S^1 = \{s_1, s_2\} \subseteq S$, passive state – s_p, state of transfer – s_t, state of reception – s_r, $S^2 = \{s_p, s_t, s_r\} \subseteq S$. Standard loading is described by the following ratio:

$$L_i^c \leq 1.2 \cdot Li, \tag{7}$$

where L_i^c – the current expected queue length calculated taking into account the current intensity of an input stream. Loading level, higher than the standard, referred as high loading.

The state of transfer is the state when an agent carrying out redirection of requests, state of reception is the state when an agent receives requests, the state when agent isn't occupied with redistribution will be considered as passive. The current state of the agent is described by a combination of states from 2 sets:

$$s^1 \times s^2 = \{s_1, s_2\} \times \{s_p, s_t, s_r\}$$

Agents who can communicate and react to messages from other agents we will call as friendly. In the considered system we will consider as friendly the agents who are in same zone (friendliness of type $1 - fr_1$, cost of unloading is 10%) and also the agents serving same type of clients (friendliness of type $2 - fr_2$, cost of unloading is 15%). The cost of unloading represents a share of requests serving time growth in the new sector. Requests serving time increases owing to low mechanization of actions according to requests from other sector.

We will consider data exchange between sectors of different types in various zones inadmissible as it will have the high cost of unloading because of functions low similarity in diagonal sectors.

Interaction of agents is described by a set of rules for sending messages and reactions to messages of each agent.

Rules of Interaction of Agents:
Each agent accepts a flow of requests on an entrance. The agent A_i can transfer each request in one of directions: D_{ii} – to serve in the sector; D_{ij}, $i,j = \overline{1,4}$ – to transfer to agent with friendliness of type 1 or type 2.

Serving of a request has the cost T_j – serving time of the request in system. Serving a request in other sectors increases this cost. Serving of the request in a section with friendliness of type 1 has cost $T_{j,norm} = T_j \cdot 10\%$; 10% – expenses of lack of mechanization; Serving of the request in a section with friendliness of type 2 has cost $T_{j,norm} = T_j \cdot 15\%$.

As the parameter presenting the state of the current agent to other agents we will accept the value based on queue length L_i. As this parameter has different values for the different queues, in order to avoid cases of constant redistribution in favor of systems with high value of the parameter, it is necessary to carry out normalization. The idea of normalization is in leading the normalized values to sizes comparable among themselves, so that they could be compared directly. The most widespread way of normalization which we will also use, it:

$$L_{i,norm} = \frac{L_i^c}{L_i}. \tag{8}$$

Considering existence of cost for transfer a request to other agents, as decision-making value will be used $L_{i,s} = L_{i,norm} \cdot 20\%$, where 20% - the indirect expenses concerning a sector overload additional categories of functions.

Agents Communication Algorithm:

(1) at a starting time an agent is in the state $s_1 \times s_p$; counters of the redirected requests are zero;
(2) at receipt of a request in state $s_1 \times s_p$, put the request in queue to the current sector, execute recalculation of system states; if the state $s_2 \times s_p$ is reached, pass to step 3;
(3) send the message of type 1 to friendly agents whose marker of an overload isn't established;

   – *message of type 1* – a message with information on overload state parameter;

(4) if receives the message of type 1 from the friendly agent, calculate the system state, send the agent message of type 2; compare the state to the parameter of the initiator of communication; if

$$L_{j,norm} > L_{i,s} \tag{9}$$

where $i \neq j$, A_i – current agent, A_j – initiator agent, pass to step 5;

- *message of type 2* – a message with information on the state parameter;

(5) pass into the state $s^1 \times s_r, s^1 \in S^1$, register the initiator;
(6) if receives the message of type 2 in state $s_2 \times s^2, s^2 \in S^2$, check the condition (9), if it is satisfied at least for 1 agent, pass into the state $s_2 \times s_t$, register recipient agent (−s), who will receive requests; pass to step 7; if the condition (9) isn't satisfied for any agent, send all friendly agents the message of type 3;

- *message of type 3* – a message about need of unloading;

(7) in state $s_2 \times s_t$, check the list of recipient agents, if the list is not empty, pass to step 8; if the list is empty, calculate the state of type s^1 and change the state of type s^2 to s_p;
(8) redirect the last request in the queue and information on its type to the last recipient agent in the list; register the redistribution; delete the recipient from the list; pass to step 8;
(9) if a request from an agent arrives in state $s^1 \times s_p, s^1 \in S^1$, reject the request and send to the system administrator the message of type 6;

- *message of type 6* – an error message with error details;

(10) if a request from an initiator agent arrives in state $s^1 \times s_r, s^1 \in S^1$, accept the request, delete the initiator from initiators list; check the list of initiators: if the list is empty, calculate the state of type s^1 and change the state of type s^2 to s_p;
(11) if a request arrives in state $s^1 \times s_r, s^1 \in S^1$ from non-initiator agent, reject the request, send to the system administrator the message of type 6;
(12) if in state $s^1 \times s_r, s^1 \in S^1$ is idle time more than 1 m, change the condition to $s^1 \times s_p, s^1 \in S^1$;
(13) if the transferred request was rejected by a recipient agent, put the request in queue to the current sector; send to the system administrator the message of type 6;
(14) if obtaining the message of type 3 from friendly agent, check the system state; if the state is $s_2 \times s_p$, then register the sender; check the list of agents whose marker of an overload is established; if not for all of friendly agents established the marker, then pass to step 3; if the marker established for all friendly agents, then send to the initiator agent message of type 5;

- *message of type 5* – a message about impossibility of unloading;

(15) if changing the state from $s_2 \times s_t$ to $s^1 \times s_p$, check the list of the senders of request for unloading; if the list is not empty, send all a message of type 4; compare the state to the parameter of the agent from the list; if condition (9) is satisfied, pass to step 5;

   – *message of type 4* – a message with state parameter value;

(16) if the message of type 5 arrives from an agent, register agent as a marked and register the time of receipt of the message;

(17) if after marking of the agent there passed 30 min, then remove marking;

(18) if the message of type 4 arrives from an agent, check the condition (9): if it is satisfied, pass into the state $s_2 \times s_t$, register recipient agent; pass to step 7.

4 Experimental Results of the Developed Model and Algorithm

For calculation of SSC queues parameters the basic data revealed during observation at traditional service in departments have been defined. One hour of the working day is accepted as a unit of time. On the basis of average values of λ_i (clients/hour) and μ_i (clients/hour), the necessary number of service channels c_i for each queue and the corresponding average queue length L_i (clients) and average waiting time W_i (hours) for the condition $W_i \leq 10$ min for consulting sectors and $W_i \leq 15$ min for administration sectors has been counted (Table 1). At the same time λ_i values are not constant and can change. Output parameters of queues are calculated according to average values of arrival stream parameters and don't consider peak loading.

Table 1. Parameters of SSC queues

No. of queue	c_i	λ_i	μ_i	ρ_i	L_i	W_i
1	12	61	5.6	0.9	6.5	0.1
2	4	13	4.2	0.77	1.8	0.14
3	8	59	8	0.92	9	0.15
4	4	20	6	0.83	3.3	0.16

We will check the work of algorithm and efficiency of service process on the basis of simulation.

The queueing systems of SSC can't provide effective service at essential changes of input parameters. The simulation (Fig. 1) represents the result of system functioning at increase in entrance load in two of 4 queues by 20%. The lengths of overloaded queues grow infinitely.

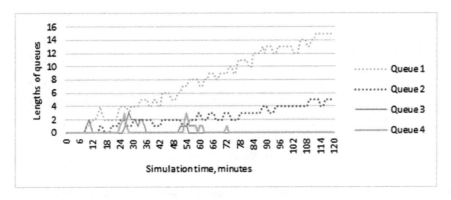

Fig. 1. Simulation of service process with constant overloading in some queues

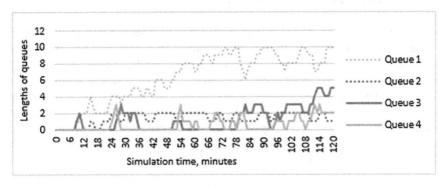

Fig. 2. The simulation result of load balancing according to the agents interaction algorithm

We can see in Fig. 2 that using the load balancing algorithm we can avoid infinitely growing queue length.

If there are periodic overloads in queues, despite gradual stabilization of length of queue, waiting time of the request in queue can go beyond admissible limits. Distribution of loading at a periodic overload in queues (Fig. 3) gives the chance to keep waiting time of requests in admissible limits.

Charts demonstrate that the load balancing using the proposed algorithm can become a solution of the problem of daily overloads in systems with parallel queues.

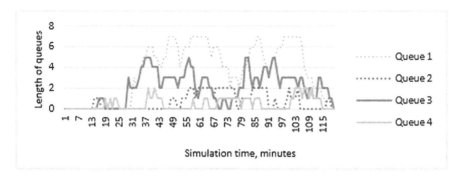

Fig. 3. The simulation result of load balancing according to the agents interaction algorithm in queues with periodic loading

5   Conclusion and Future Works

Thus, by means of methods of the queueing theory and agent-based load balancing many problems of planning, assessment and optimization of quality of SSC functioning can be solved. Agent-based approach allow to develop module system, which can be scaled easily: to add new sectors with new sets of functions and new nature of inter-action with other sectors. Now, after optimization of service process of educational process supporting zones, future work consists in optimization of further service routes of requests from these zones, thereby, optimization of processes in queueing networks. It will be also important to present results of work of the algorithm with use of real entrance data on system parameters.

References

1. Adan, I., Resing, J.: Queueing Systems. Eindhoven University of Technology (2015)
2. Yuan, X., Hwarng, H.B.: Managing a service system with social interactions: stability and chaos. Comput. Industr. Eng. **63**(4), 1178–1188 (2012). https://doi.org/10.1016/j.cie.2012.06.022
3. Jiang, Y., Li, Z.: Locality-sensitive task allocation and load balancing in networked multiagent systems: talent versus centrality. J. Parallel Distrib. Comput. **71**(6), 822–836 (2011). https://doi.org/10.1016/j.jpdc.2011.01.006
4. Down, D.G., Lewis, M.E.: Dynamic load balancing in parallel queueing systems: stability and optimal control. Eur. J. Oper. Res. **168**(2), 509–519 (2006). https://doi.org/10.1016/j.ejor.2004.04.041
5. Li, X., Mao, W., Zeng, D., Wang, F.-Y.: Agent-based social simulation and modeling in social computing. In: Yang, C.C., Chen, H., Chau, M., Chang, K., Lang, S.-D., Chen, P.S., Hsieh, R., Zeng, D., Wang, F.-Y., Carley, K., Mao, W., Zhan, J. (eds.) ISI 2008. LNCS, vol. 5075, pp. 401–412. Springer, Heidelberg (2008). https://doi.org/10.1007/978-3-540-69304-8_41

6. Klügl, F., Timpf, S.: Approaching interactions in agent-based modelling with an affordance perspective. In: El Fallah-Seghrouchni, A., Ricci, A., Son, T.C. (eds.) EMAS 2017. LNCS (LNAI), vol. 10738, pp. 21–37. Springer, Cham (2018). https://doi.org/10.1007/978-3-319-91899-0_2
7. Szymanezyk, O., Dickinson, P., Duckett, T.: Towards agent-based crowd simulation in airports using games technology. In: O'Shea, J., Nguyen, N.T., Crockett, K., Howlett, R.J., Jain, L.C. (eds.) KES-AMSTA 2011. LNCS (LNAI), vol. 6682, pp. 524–533. Springer, Heidelberg (2011). https://doi.org/10.1007/978-3-642-22000-5_54
8. Luo, J., Shi, Z., Wang, M., Huang, H.: Multi-agent cooperation: a description logic view. In: Lukose, D., Shi, Z. (eds.) PRIMA 2005. LNCS (LNAI), vol. 4078, pp. 365–379. Springer, Heidelberg (2009). https://doi.org/10.1007/978-3-642-03339-1_29
9. Monostori, L., Valckenaers, P., Dolgui, A., Panetto, H., Brdys, M., et al.: Cooperative control in production and logistics. Ann. Rev. Control **39**(1), 12–29 (2015). https://doi.org/10.1016/j.arcontrol.2015.03.001. Elsevier
10. Seel, N.M. (ed.): Encyclopedia of the Sciences of Learning. Springer, Boston (2012)

Text Processing and Information Retrieval

Handling Concept Drift and Feature Evolution in Textual Data Stream Using the Artificial Immune System

Amal Abid[✉], Salma Jamoussi[✉], and Abdelmajid Ben Hamadou

Multimedia InfoRmation Systems and Advanced Computing Laboratory, Digital
Research Center of Sfax, Sfax, Tunisia
abidamal90@gmail.com, salma.jammoussi@isims.usf.tn,
abdelmajid.benhamadou@isimsf.rnu.tn

Abstract. Data stream mining is an active research area that has attracted the
attention of many researchers in the machine learning community. Discovering
knowledge from large amounts of continuously generated data from online
services and real time applications constitute a challenging task for data analytics where robust and efficient online algorithms are required. This paper
presents a novel method for data stream mining. In particular, two main challenges of data stream processing are addressed, namely, concept drift and feature
evolution in textual data streams. To address these issues, the proposed method
uses the Artificial Immune System metaheuristic. AIS has powerful adapting
capabilities which make it robust even in changing environments. Our proposed
algorithm AIS-Clus has the ability to adapt its model to handle concept drift and
feature evolution for textual data streams. Experimental results have been performed on textual dataset where efficient and promising results are obtained.

Keywords: Data stream clustering · Concept drift · Feature evolution
Artificial Immune System metaheuristic

1 Introduction

Recently, real time applications such as real-time surveillance systems, telecommunication systems and social media services become ubiquitous in our daily life. Million of
thousands of data are generated every day. This fast and continuous flow of information
generated by such real time applications is known as data stream. Data stream can be
defined as an ordered sequence of data items that arrive continuously in high speed rate.
Data stream differs from static data which are offline processed and bounded in size.
The time-varying characteristic leads to a variation in the structure of data and thus data
distribution may change with time. This changing characteristic in the distribution of
data constitutes a challenging problem known as concept drift. Identifying and
detecting such changing character in the distribution of the data requires online algorithms that analyze data only once due to their unbounded size and their high incoming
rate.

Data stream mining is becoming the center of interest of many researchers that have
been focused on tuning and changing data mining techniques to respond to the problem

© Springer Nature Switzerland AG 2018
N. T. Nguyen et al. (Eds.): ICCCI 2018, LNAI 11055, pp. 363–372, 2018.
https://doi.org/10.1007/978-3-319-98443-8_33

of storing and processing this huge volume of data [9]. However, the learning algorithm should be persistent in differentiating between noise and change in the data distribution. Many textual data sources are offered by the world wide web and are considered as the most commonly shared and produced type of data between web users. In this vein, several studies are particularly dedicated to text stream mining fields such as social networks in order to analyze and discover useful information from textual streams. Text stream poses additional challenges to data stream processing due to its dynamic evolving nature and thus its unknown feature space. This is becoming the center of interest of many researchers that have been focused on tuning and changing data mining techniques to respond to the problem of feature evolution in huge volume of text. In this paper we aim to design a novel method for text data stream mining. For this reason, we suggest using meta-heuristics to solve the main issues of this problem. Nature has provided inspiration for the fields of computer science and engineering. Among bio-inspired algorithms, Artificial Immune System (AIS) has attracted a great deal of attention for the data mining community [12]. This method owns the property of easy adaptability to dynamic environments which is required by data stream learning algorithms.

This paper is organized as follows. Section 2 presents related work dealing with data stream mining for concept drift detection and feature evolving identification challenges. Section 3 describes a brief background of the AIS meta-heuristic. In Sect. 4, we present the proposed algorithm for text data stream mining. In Sect. 5, we discuss the experimental results and finally in Sect. 6 we conclude this paper.

2   Related Work

Mining text data streams is considered as a challenging problem requiring specific algorithms to handle the dynamic and scalable aspects of data. Text streams represent some additional difficulties to the data stream problem due to their huge feature space. The problem of feature evolving and the probability of concept drift are higher than concept drift in classical data stream. In addition, noise is usually present in text streams which can affect results. Specific algorithms are then required to manage all those problems. The concept drift phenomena happens over time due to the changing characteristics of our dynamic environment. It occurs when new observations appear while old ones become irrelevant. To cope with this issue, applications have to include a change detection strategy in their learning process in order to identify concept drift [14].

The framework proposed in [6] builds a classification model based on a slice of data instances in a given time frame. Initially, a classification phase is performed to classify arriving documents. Then for the labeling purpose, the most useful instances are selected using an active learning based approach. Relaying on those examples, the classifier is updated by the most recent concepts. In text streams, concept drift leads to a change in the distribution of data and thus these changing characteristic can be identified by the presence of novel features. This evolution can happen in two ways: (1) Adding novel attributes to the underlying data e.g. new words appear in the case of text data. Then old features may fade out and replaced by new ones. (2) New values appear for existing attributes, this case is particularly designed for numerical data. The feature evolving phenomena generally occurs where new features emerge as time flies. Due to the unbounded size of data stream, a dynamic feature vector is required as new features may appear over time. The work presented in [4] treats the main problem of text stream classification which is the dynamic feature space. At the first stage, a feature selection method is applied to select the N best features. In the second stage, an incremental learning algorithm is performed for the classification purpose. A fading factor is associated to each word to indicate its importance in the feature vector. As a new document arrives it is checked whether it contains any new words.

3   AIS Background

Many researchers have mimicked the intelligent behavior of natural behavior of individuals and have created intelligent systems based on what happens in our natural environment. Artificial immune system is a novel computational intelligence technique which simulates the immune features of biological immune system. To tackle real world problems, biological immune system has inspired computer scientists to simulate the concepts of natural immune system on real applications. This has led to the emergence of a new branch of computational intelligence known as the Artificial Immune Systems (AIS) meta-heuristic [3]. The AIS is mainly based on three theories: The Immune Network Theory [15] has been widely used for the data clustering. The negative selection algorithm is initially introduced in [11] as a learning algorithm to protect computers from malicious intruders such as virus. And finally the clonal selection principal used to perform machine learning and pattern recognition tasks [2]. However the most used model is the clonal selection principle. This model describes the process of simulation and proliferation of B-cells in order to defend a malicious attacker. AIS represents an efficient technique to cope with the evolving nature of data streams since it is inspired from Immune System that treats this kind of properties within foreign substances. Due to its powerful ability of learning, its decentralized property and its self adaptation in managing the dynamic environment, AIS was successfully applied in many clustering problems [10, 13]. AIS is a meta-heuristic that can be efficient to manipulating the data stream problems too. In [8] the authors propose a new clustering method (TECNO-STREAMS) based on the Artificial Immune System. This new method addresses the problem of mining evolving user profiles in noisy data sets. Initially, the K-means algorithm is performed to cluster the initial introduced population. Each time a new antigen is presented, the influence zone will decrease with the distance of the antigen. So, the more recent data will have higher influence zone.

4   Methodology

The work presented in this paper aims to cluster text data stream. This is mainly based on handling concept drift and feature evolving in textual data. To address those challenging issues the Artificial Immune System meta-heuristic was used to develop the AIS-Clus algorithm.

The process of the proposed framework is described as follow: For each incoming data X_{new} (antigen), a scoring function is calculated in order to examine the affinity values to each cluster (B-cell). Once determining the closest clusters, the AIS is involved to determine which cluster will absorb this novel invader. For this purpose we apply both the clonal selection principal and the negative selection mechanism. The first model, the clonal selection, consists in the production of antibodies to best match the new invader. Those antibodies are created from existing B-cells (clusters) to undergo a mutation process. This model seeks to handle concept drift and feature evolution challenges. The negative selection mechanism enables detecting noisy data. In fact, instances declared as outliers may belong to novel classes. To deal with the concept drift issue, a life cycle is associated to each term to indicate its contribution over time. By this fact, some terms will loose their importance as time flows so they will be eliminated and replaced by new appearing features. As a result of that, clusters which are inactive for a long period of time are removed and replaced by novel appearing classes. On the other hand, the feature evolving challenge is handled by considering the feature space dynamics. Whenever a new data instance arrives, the set of its novel features are included in the original feature space. In order to ensure greater importance of more recent concepts, a survive factor is assigned to each term. This strategy aims to manage the high dimensional size of feature vector by eliminating useless features and promoting novel ones.

Our proposed framework consists of two phases: (1) The static phase where clustering historical data is achieved. The first step to perform is the text preprocessing which aims to reduce the feature space vector. Then a clustering algorithm is used in order to obtain homogeneous clusters from static data. (2) The stream preprocessing phase: This takes as input the constructed clusters of the static phase. New data are divided into equal blocks. Each novel instance is declared either an outlier or belonging to an existing class. The constructed buffer of outliers is scanned periodically to decide whether novel classes appear. Text is characterized by its high dimensional nature and sparsity of data. A preprocessing step is crucial to reduce the huge size of textual data. Irrelevant data such as punctuation marks, non text characters, stop words should be removed. The set of remaining words is called the vocabulary of a document collection. To deal with concept drift detection and feature evolving issues, a robust feature selection method is required. We define a survive factor for each term to keep the most important words.

Each document is represented as a vector in m-dimensional space. More formally, a document D is represented as $(X_1, X_2, \ldots, X_m)$ where X_i represents the ith feature. In our proposed framework each feature X_i is defined as a tuple $(Word, SURV, Weight)$. Each tuple component is defined as follow: (1) A $Word$ is a string, survive factor $SURV \in [0, \infty[$ calculates the word period of life. $SURV$ refers to the age of a word as time flows and $Weight$, is a normalized value that indicates the importance of that term in a document D.

Initially all data points share the same feature vector and each instance will have a specific feature vector. The new incoming data is divided into equal sized chunks where each batch contains a fixed number N of instances. To cluster text data, we consider a number of tweets where each tweet represents an instance. So, we define an instance as a tuple $(ID, \overline{FeatureWords}, \overline{OutlierWords})$ where ID is a unique number to identify the instance, $\overline{FeatureWords}$ is a vector of features and $\overline{OutlierWords}$ is a vector of features which does not belong to the vector $\overline{FeatureWords}$.

Clusters are dynamic with time, some of them will disappear, others will grow and novel clusters appear. Another gap is that clusters are density shaped. Our algorithm consists in creating clusters from text streams. First the input data is scanned and the DBSCAN [1] clustering algorithm is performed to construct homogeneous groups of data. DBSCAN (Density-Based Spatial Clustering of Applications with Noise) is essentially based on two parameters: the neighborhoods parameter that defines a region of a given radius (eps) and the number of objects (MinPts) it contains. Density Based algorithms seem efficient in the context of text stream clustering since text data contains plenty of noise because of its huge number of features. Likewise, clustering data stream is characterized by its arbitrary shaped clusters and thus density based algorithms are efficient to handle this issue by detecting arbitrary shaped clusters. In the context of text mining, the cosine similarity is the most commonly used for high dimensional spaces. As we aim to cluster textual data, we are based on the cosine metric to calculate how much two documents are closer from each other.

After performing the clustering of available data and obtaining homogeneous clusters, summary information is associated to each cluster to provide a global description about it. The summarization technique plays an important role in the data stream mining and it is widely used to reduce the volume of manipulated information and the memory space. The summary statistics provide a global and sufficient description of a given cluster. So, a cluster Clus is represented as $Clus = (\overline{WordClus}, \overline{OutlierClus})$ where $\overline{WordClus}$ is a vector of features which are words occurring in points of the cluster. $\overline{OutlierClus}$ is a vector of terms. Given the set of points belonging to a cluster, $\overline{OutlierClus}$ is composed by features which are $\overline{OutlierWords}$ of the points of this cluster.

These summaries are used in order to calculate scores for novel arrived instances. The cluster having the strong cohesion with the novel data will absorb it. Otherwise, it is declared as outlier and will be used for farther analysis. The process of the proposed framework is described as follow:

Whenever a novel instance arrives, a score metric (defined as affinity function) will be calculated to quantify the compactness of this new point with each existing cluster. To calculate the affinity between a point P and a cluster C_i the following formula is used:

$$Affinity(P, C_i) = \alpha \times e^{-\gamma \times Cosine(P,C_i)} + \beta \times V \tag{1}$$

Where α and β are two weighting factors. In order to make a consideration to the similarity information, we use the exponential distribution. As we aim to obtain a small value of $Affinity(P, C_i)$, the exponential function $e^{-\gamma \times Cosine(P,C_i)}$ is low when the $Cosine(P, C_i)$ is high and vice versa.

V is an entropy based formula calculated as follow:

$$V = Entropy(B\_cell_{after}) - Entropy(B\_cell_{before}) \qquad (2)$$

B_cell_{after} is the cluster after adding the point P and B_cell_{before} is the cluster before it changes. If $V > 0$ then the added point disturbs the cluster characteristics otherwise the novel added point reduces the cluster entropy and then it fits with that cluster.

Given a random variable X, the entropy value $Entropy(X)$ measures the uncertainty of this variable. In our framework, we use the entropy measure in order to evaluate the orderliness of a given cluster before and after putting a novel point in it.

Suppose the vector $\overline{WordClus} = (X_1, \ldots, X_n)$ and $X_j (1 \leq j \leq n)$ is a feature word. $Entropy(C_i)$ of a cluster C_i is calculated as:

$$Entropy(C_i) = \sum_{j=1}^{n} H(X_j) = - \sum_{i=1}^{d} p_i \log_2(p_i) \qquad (3)$$

p_i defines the probability of an event $X_j = v_k$ where v_k belongs to the possible values of X_j. Whenever a foreign point enters into a cluster where most of its data are similar, it will disturb its homogeneity leading to an increase of the entropy value.

As we adopt features of AIS, the scoring formula represents the affinity value between B-cells and foreign invaders. Affinity measure evaluates the degree of interaction of an antibody-antigen.

After calculating the affinity value with all clusters (B-cells), they will undergo a selection step to be ready for cloning and mutation. The process of cloning consists in producing similar copies from the original clusters. Those copies will be used during the mutation process. The number of clones to make from each cluster depends on the affinity value between P and C_i.

$$NumberClones(C_i, P) = \frac{ComWord(C_i, P) + \rho}{2} \qquad (4)$$

The $ComWord(C_i, P)$ refers to the number of jointly features between novel point P and the cluster C_i. The number of common words may be too large, so we have divided this value by 2. Also in the case where the number of common words is too small, we have added a constant ρ to control the suitable number of clones.

In genetics, mutation consists in changing characteristics of different organisms. This operation allows clones to diversify their repertory and to evolve by changing their structure. In our algorithm, the mutation process aims to make existing clusters more similar to novel incoming data. The clonal selection principle produces cloned cells ready to undergo a mutation process. In our proposed method, the mutation process consists in adding a number of features to the $\overline{WordClus}$ vector. The mutation rate is the number of words to add to a given cluster. It is defined as:

$$WordsToAdd(C_i, P) = \eta \times ComWord(C_i, P) \qquad (5)$$

New added features are selected from $\overline{OutlierClus}$ vector according to their weights. η is a constant used to adjust the mutation rate parameter. To promote best features to be chosen, the roulette-wheel selection procedure is applied.

After performing the mutation process, clones evolve and change their structure to best fit introduced antigens. Closest clones to the antigen are selected to replace their parents. Clones which are far from antigen are removed.

5   Results

In order to test the efficiency of AIS-Clus, we adopt twitter data set as social media data is best modeled as data stream. Sanders corpus consisting of 5513 hand-classified tweets with four topics (Apple, Google, Microsoft, Twitter). After performing a pre-processing step, 2965 tweets are obtained for our clustering purpose.

Initially we have processed 1200 instances belonging to "Apple" and "Google" classes. For the streaming scenario, we consider some instances belonging to existing and other ones are considered belonging to novel classes which are "Microsoft" and "Twitter".

We carried a number of experiments in order to fix the appropriate parameters of AIS-Clus as follows: MinPts = 16, eps = 0.58, $\alpha = 0.6$, $\beta = 0.4$, $\rho = 2$ and $\eta = 2$.

Figure 1 illustrates the F_measure, the accuracy, the recall and the precision values obtained with AIS-Clus for the Sanders dataset.

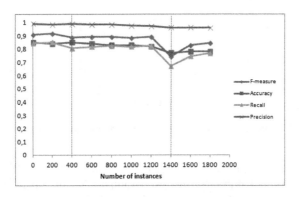

Fig. 1.   Results obtained on the Sanders dataset.

The two dashed lines indicate the presence of novel class. For example, for the novel arrived class "Microsoft", our algorithm detects it when 400 instances are processed. Likewise for "Twitter" which is detected at chunk 1400 and then novel class instances belonging to "Twitter" are added to it. However, there is a decrease of the recall curve at instance 1400. This is because our system delays a bit in detecting the novel class. So, when it appears, its recall value is relatively low. Then as time flows its recall value will increase.

To cope with concept drift in text data streams, we are based on a weighting term strategy. By simulating the biological metaphor of AIS, a term can be considered as a living organism having a life cycle where terms are born, they grow up and they die. Thus, the contribution of terms changes over time depending on their nourishment. Accordingly, by tracking the occurring changes in the life cycle of terms we can easily detect the concept drift. Besides, to track the evolving nature in text data (feature evolution), we consider the $\overline{WordClus}$ feature vector which defines hot terms inside the corresponding cluster. In fact, to cope with the high volume of text data, this feature vector is updated whenever its size exceeds a certain threshold. As a consequence, this method keeps the newer features, limit the memory and the run time of our algorithm.

The following experiment concerns the concept drift and feature evolution issues on the cluster "Apple" which is detected in the initial phase.

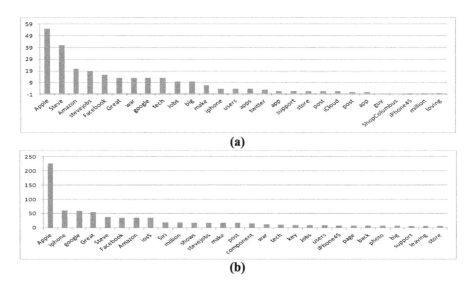

Fig. 2. Tracking concept drift and feature evolution on the Sanders dataset.

The Fig. 2(a) and (b) give the hot terms after updating the vector WordsClus of the cluster "Apple". As shown in these figures, the importance of terms is evaluated over time by considering cases where an update of the feature vector occurs Likewise some terms become more contributing such as "iphone" while other terms like "twitter" fade out and will disparate as time flows. Also, in Fig. 2(a) last words such as "ShopColumbus" and "iphoneS4" are considered as "drop terms" since they have negative survive factor. This negative value can be explained by the fact that these words are inactive for a long period of time. Thus they should be removed to promote novel appearing words. In addition, we can notice that novel features appear such as "ios5". This proves that our approach can effectively detect evolving features in text data streams. Another issue, which appears in data stream applications, is the presence of noise. In order to test the robustness of the proposed model we vary the noise rate and we observe the behavior of our algorithm.

Table 1 shows results of our experiments with different noise percentages. The F-measure is the harmonic mean of precision and recall, the error rate ERR calculates the total misclassification error and M_{new} calculates the rate of novel class instances misclassified as existing class.

Table 1. Noise variation on the Sanders dataset.

Noise variation	F-measure	ERR	M_{new}
0% noise	0.86	0.22	0.06
20% noise	0.85	0.25	0.063
30% noise	0.84	0.27	0.071

As mentioned in Table 1, here is a meaningless decrease in the F_measure value. Also, the presence of noise causes a meaningless increase in the ERR and M_{new} values. We can deduce that even when the noise level increases, our model can achieve very high clustering quality.

6 Conclusion

The revolution of technology and the emergence of real time applications have created a huge volume of data. Consequently, the necessity to adapt traditional data mining methods to cope with challenges posed by the data stream is became a crucial task. In this paper we have proposed a clustering algorithm for text data stream. AIS-Clus is an adaptive algorithm which uses the strength of the clonal selection theory to adapt clusters for the novel arriving streams to handle concept drift and feature evolving issues. Experimental results are performed on textual dataset and accurate results are obtained. AIS-Clus showed its effectiveness in handling the feature evolving issue in the context of textual data streams. As a future work, we aim to improve the robustness of our model by considering the semantic relation between features. Thereby, exploring the concept of ontology as future research can facilitate the attainment of this goal.

Acknowledgment. This paper was made possible by NPRP grant #9-175-033 from the Qatar National Fund (a member of Qatar Foundation). The statements made herein are solely the responsibility of the authors.

References

1. Dorigo, M., Colorni, A., Maniezzo, V.: A density-based algorithm for discovering clusters in large spatial databases with noise. In: Proceedings of 2nd International Conference on Knowledge Discovery and Data Mining, pp. 226–231 (1996)
2. De Castro, L.N., Von Zuben, F.J.: Learning and optimization using the clonal selection principle. IEEE Trans. Evol. Comput. **6**(3), 239–251 (2002)

3. John, H.: Adaptation in Natural and Artificial Systems: An Introductory Analysis with Applications to Biology, Control and Artificial Intelligence. MIT Press, Cambridge (1992)
4. Grigorios, T., Ioannis, K., Ioannis, V.: Dynamic feature space and incremental feature selection for the classification of textual data streams. In: International Workshop on Knowledge Discovery from Data Streams, ECML/PKDD-2006, p. 107. Springer (2006)
5. Norman, H., Packard, J., Doyne, F., Alan, S.: The immune system, adaptation and machine learning. Physica D **22**, 187–204 (1986)
6. Delany, S., Jane, L., Namee, B.: Handling concept drift in a text data stream constrained by high labelling cost. In: FLAIRS Conference. AAAI Press (2010)
7. Masud, M., Chen, Q., Khan, L., Aggarwal, C., Gao, J., Han, J., Srivastava, N., Oza, C.: Classification and adaptive novel class detection of feature-evolving data streams. IEEE Trans. Knowl. Data Eng. **25**(7), 1484–1497 (2013)
8. Nasraoui, O., Uribe, C., Gonzalez, F.: Tecno-streams: tracking evolving clusters in noisy data streams with a scalable immune system learning model. In: Proceedings of the Third IEEE International Conference on Data Mining, ICDM 2003, Washington, DC, p. 235. IEEE Computer Society (2003)
9. Sergio, R., Bartosz, K., Salvador, G., Michał, W., Francisco, H.: A survey on data preprocessing for data stream mining: current status and future directions. Neurocomputing **239**, 39–57 (2017)
10. Kuo, R.J., Chen, S., Cheng, W., Tsai, C.: Integration of artificial immune network and K-means for cluster analysis. Appl. Artif. Intell. **40**(3), 541–557 (2013)
11. Lawrence, A., Stephanie, F., Alan, S., Rajesh, C.: Self-nonself discrimination in a computer. In: Proceedings of the 1994 IEEE Symposium on Research in Security and Privacy, Los Alamitos. IEEE Computer Society (1994)
12. Jon, T., Thomas, K.: Artificial immune systems: using the immune system as inspiration for data mining. In: Data Mining: A Heuristic Approach, Chapter XI, pp. 209–230. Group Idea Publishing, September 2001
13. Yanmin, Z., Shuai, C., Tinggui, C.: K-means clustering method based on artificial immune system in scientific research project management in universities. Int. J. Comput. Sci. Math. **8**(2), 129–137 (2017)
14. Žliobaitè, I., Pechenizkiy, M., Gama, J.: An overview of concept drift applications. In: Japkowicz, N., Stefanowski, J. (eds.) Big Data Analysis: New Algorithms for a New Society. SBD, vol. 16, pp. 91–114. Springer, Cham (2016). https://doi.org/10.1007/978-3-319-26989-4_4
15. Jerne, N.: Towards a network theory of the immune system. Ann. Immunol. **125**, 373–389 (1974)

A Tweet Summarization Method Based on Maximal Association Rules

Huyen Trang Phan[1], Ngoc Thanh Nguyen[2], and Dosam Hwang[1(✉)]

[1] Department of Computer Engineering, Yeungnam University,
Gyeongsan, Republic of Korea
`huyentrangtin@gmail.com`, `dosamhwang@gmail.com`
[2] Faculty of Computer Science and Management,
Wroclaw University of Science and Technology, Wroclaw, Poland
`Ngoc-Thanh.Nguyen@pwr.edu.pl`

Abstract. A lot of information about different topics is posted by users on Twitter in just one second. People only want a way to get short, full, and accurate content which they are interested in receiving information. Tweet summarization to create that short text is a convenient solution to solve this problem. Many previous works were trying to solve the Tweet summarization problem. However, those researchers generated short texts based on the frequency of words in Tweet. They ignored word order in each Tweet. Moreover, they rarely considered the semantics of the words. This study tries to solve existing on above. The significant contribution of this study is to propose a new method to summary the semantics of the tweets based on mining the maximal association rules on a set of real data. The experiment results show that this proposal improves the accuracy of a summary, in comparison with other methods.

Keywords: Tweet summarization · Semantics summarization
Maximal association rule

1 Introduction

Summarization is the main component of natural language processing. Normally, summarization is applied to text data that are formal and complex, such as research papers. Recently, it has also been used for summarizing Tweets. In this study, Tweets are posted on Twitter by users and are chosen as the data on which to implement summarization. In 2006, Twitter began operations, and it has developed at a rapid rate. Twitter users post more than 400 million Tweets per day. Tweets are short text messages of 280 characters, which are created and shared by both users and data analysts. Tweet summarization is a method to generate short texts by combining the contents of Tweets on the same topic.

Twitter is used like a main communication channel, where users can post anything they think; where they can find their friends; where they can follow famous people they are interested in; where they can learn information about

© Springer Nature Switzerland AG 2018
N. T. Nguyen et al. (Eds.): ICCCI 2018, LNAI 11055, pp. 373–382, 2018.
https://doi.org/10.1007/978-3-319-98443-8_34

things that happen over a period of time. Besides the advantages, challenges include how users can capture information of interest in the fastest, most concise, and most complete way in other words, from among a huge number of Tweets, how to automatically generate short texts based on different topics that help users without losing a lot of time to get useful information. Therefore, we have to find a way to combine Tweets on the same topic, and then try to generate a short text corresponding to one topic. The summarization method is the best way to solve this problem. Tweet summarization has become significant, and it is applied in a variety of fields, including political, commercial, and scientific.

Some Tweet summarization methods have been proposed, and some systems have been developed [3,4,6–8,12]. However, the problem is that most of them focused on the frequency at which words appear. They ignored word order in each Tweet. Meanwhile, word order has a lot of influence on the semantics of the final summary. In other words, they rarely considered the semantics of the created short texts. That leads two significant problems which need to solve as follows: Assume that we have a set of tweets, $T = \{t_1, t_2, ..., t_n\}$ which are collected from many Twitter account. Let p_j be a topic is given by the user. The first problem needs to be solved that is from a large number of tweets like that, how to collect the tweets t_i belong to the topic p_j, in which t_i must have similarity semantic together. The second problem needs to be solved that is how to create a summary with rich semantics corresponding to the topic p_j.

In this study, a new method for summarizing Tweets is proposed by taking into account the following steps. First, gather all Tweets belonging to the selected topic. Next, remove Tweet outliers, which are Tweets that do not have the same semantics as the other Tweets in each topic. Then, maximal association rules are mined, and the Tweets are presented again based on the maximal association rules. Finally, calculate the weight of the Tweets based on the weight of maximal association rules, and the Tweets that fit are chosen to create a summary.

The rest of this paper is organized as follows. In the next section, related work on tweets summarization is investigated briefly. Section 3 explains the proposed method in detail. The experimental results and evaluations are shown in Sect. 4. Lastly, conclusions and future work are presented in Sect. 5.

2   Related Works

Tweet summarization is the method used to combine Tweets from many different sources into a short text. That text has to satisfy some criteria, such as being concise, full of information, and the reader must understand its meaning. Tweet summarization is a form of multi-text summarization, in which each text corresponds to one Tweet. Just like a text summary, Tweet summarization includes two main categories: abstractive and extractive. For an abstractive summary, the result is created by rewriting the Tweets. Meanwhile, with an extractive summary, the result is generated by selecting Tweets that have a high frequency of occurrence.

Most of the previous studies belong to the extractive summary category [3,4,6,8,12]. Chellal et al. [3] proposed a new model to create a real-time summary by choosing an incoming Tweet if it passes three filters for relevance, informativeness, and novelty. Therefore, the generated short summary has high coverage. Dutta et al. [4] offered a graph-based approach for summarizing Tweets. First, a graph is constructed based on the similarity among Tweets, which is measured using various features, including features based on WordNet synsets. Then, a detection technique is used on the graph to group similar Tweets. Finally, one Tweet is chosen from each cluster to be included in the summary. Hole and Takalikar [6] offered an approach to real-time summarization of Game Tournament, which analyzed the sentiment of the person posting a Tweet. It includes the following steps: detection of Tweets belonging to the topic as soon as they emerge, summarization of the Tweets related to the topic, and analyzing the feelings of fans of the game. Samuel and Sharma [8] provided a modified LexRank model to summarize Tweets on the basis of time. The results gave information about the time the event happened. This method improved the original method by using the temporal aspect of the Tweet. Therefore, a summary of the Tweets can easily be generated.

The previous studies basically summarized the Tweets into a brief text. However, the created text was mostly created by choosing Tweets that have a high appearance frequency. Meanwhile, the frequency of Tweets depends on the frequency of words in these Tweets. They rarely considered the order of the words. That leads to the lost semantics of the text, or the readers even do not understand the content of the text.

3   The maximal association rules-based Tweet summarization

This section discusses how to build Tweet summarization based on maximal association rules. The method focuses on finding a set of maximal association rules. Let $T = \{t_1, t_2, ..., t_n\}$ be a set of Tweets, let g_j be a cluster (corresponding to the topic, p_j, is given by the user) that contains Tweets t_i, let $H = \{H_1, H_2, ..., H_m\}$ be a set of association rules, and let $H_{max} = \{H_{max_1}, H_{max_2}, ..., H_{max_k}\}$ be a set of maximal association rules. The architecture of the proposed model is shown in Fig. 1. and includes the following stages: collect all Tweets, t_i, that belong to the cluster, g_j. Remove outliers that are not the same as the other Tweets in the group. Mine the maximal association rules, H_{max}. Present Tweets, t_i, again according to the maximal association rules. Choose the Tweets based on their weight to generate the final summary. The details of the proposed method are presented in the following subsections.

3.1   Collecting the Appropriate Tweets to Topic

In this subsection, a crawler collects Tweets from Twitter via the Twitter API. Tweets are preprocessed by removing elements like mentions, hyperlinks, hashtags, and symbols. This experiment does not consider Tweets that are less than

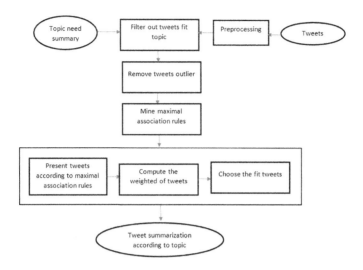

Fig. 1. The workflow of the proposed method

four tokens. Intuitively, a short Tweet rarely contains important content, nor does it include the context to be discovered. Therefore, it should be ignored. After the Tweets are gathered, we filter out the appropriate Tweets to the topic the users gave.

3.2   Removing Tweets Outliers

Tweets are informal data. They contain a lot of noise. Although Tweets are filtered based on the topic, they can also include some outliers. That affects the accuracy of the summary. This process is implemented by computing the semantic similarity among Tweets. Let *average* be a set of the average vectors for all words in every Tweet; let n be the number of elements in *average*; let k be the number of Tweets after removing outliers, and let α be a parameter used to indicate a threshold of similarity. Algorithm 1 shows the steps to remove outliers.

In this subsection, the words are converted to vectors by using Google's Word2vec[1]. It is a fantastic open source software package that is used to transfer words into a high-dimensional vector space.

3.3   Mining the maximal association rules

Let $I = \{i_1, i_2, ..., i_m\}$ be a set of items that contains *n_gram* of words, and let $T = \{t_1, t_2, ..., t_n\}$ be a set of n Tweets in which t_i contains a subset of items. $C = \{c_1, c_2, ..., c_k\}$ where c_l is a category. For example, if a is an item, then the category is a set t_i contains a.

---

[1] http://douglasduhaime.com/posts/clustering-semantic-vectors.html.

Algorithm 1. The removing Tweets outliers algorithm

Input: A set of tweets;
Output: A set of tweets after removing outliers;
1: $average = \varnothing$
2: **for** each tweet t_i in T **do**
3:   $Total = 0; count = 0;$
4:   **for** each word w_j in t_i **do**
5:     Convert w_j into a vector v_{w_j};
6:     $count = count + 1$
7:     $Total = Total + |v_{w_j}|$;
8:   **end for**;
9:   $average(t_i) = \dfrac{Total}{count}$;
10: **end for**
11: $average = \bigcup average(t_i)$
12: **for** i $= 1$ to n-1 **do**
13:   **for** j $= $ i+1 to n **do**
14:     $sim(average(t_i), average(t_j)) = 1 - cosine(average(t_i), average(t_j))$
15:   **end for**
16:   **if** $sim(average(t_i), average(t_j)) < \alpha$
17:       Remove $average(t_j)$
18: **end for**
19: **return** $T = \{t_i, i = (1, .., k), k <= n\}$;

Definition 1. *A maximal association rule for a Tweet is an association rule that is denoted $X \rightarrow_{max} Y$ and satisfies three criteria:*

(i) X is the largest in c_l which is denoted $c_{l\_max(X)}$
(ii) $sup(X) = \max(sup(X))$
(iii) $sup(Y) \geq 0$

The support and confidence of the maximal association rule are determined by the formulas following:

$$max\_sup_{(X \rightarrow_{max} Y)} = \frac{|X \rightarrow_{max} Y|}{n} \tag{1}$$

$$max\_conf_{(X \rightarrow_{max} Y)} = \frac{max\_sup_{(X \rightarrow_{max} Y)}}{|T(c_{(max(X))}, c_Y)|} \tag{2}$$

where X is as an antecedent and Y is as a consequent of the maximal association rule; $|X \rightarrow_{max} Y|$ is the number of Tweets that maximal support X and also support Y; $|T(c_{(max(X))}, c_Y)|$ is the number of Tweets in c_l that contain $c_{l\_max(X)}$ and at least $c_l(Y)$.

Mining the maximal association rules is the process for finding associations where max_sup is above some user-defined minimum max_sup, and the max_conf is above some user-defined minimum max_conf.

Mining the association rules is the main technique employed in this subsection. For each cluster, g_j, a set of the association rules, H, is found from a set of tweets, T. Then, we filter the maximal association rules [2] H_{max}. This manipulation is illustrated in the Algorithm 2.

Algorithm 2. The algorithm of maximal association rules mining

---

Input: - A set of Tweets;
        - minimum max_sup;
        - minimum max_conf;
Output: A set of maximal association rules;
1: Find all frequent word sets of L;
2: **for** each frequent word set called I **do**
3:    Determine all subset J of I;
4:    **for** each subset J of I **do**
5:        Determine all association rules of the form called $H: I - J \rightarrow I$;
6:    **end for**;
7: **end for**;
8: **for** $0 \leq i \leq k - 1$ **do**
9:    **for** $i \leq j \leq k$ **do**
10:      **if** $H_i \subseteq H_j$ and $sup(H_i) = sup(H_j)$: Remove H_i from H;
11:    **end for**;
12: **end for**;
13: **for** each H_i in H **do**
14:    **if** $(max\_sup_{H_i} \leq$ minimum $max\_sup)$ and $(max\_conf_{H_i} \leq$ minimum $max\_conf)$: Remove H_i from H;
15: **end for**;
16: $H_{max}:=H$;
17: **return** A set of maximal association rules is H_{max};

3.4 Creating Tweet summarization

In this subsection, we present the steps to generate a short text from the maximal association rules set. The problem needs to solve is how to create the final summary from the set of maximal association rules, H_{max}. The problem is solved by taking into account the steps follows:

Presenting Tweets According to the Maximal Association Rules: After mining the maximal association rules set, the Tweets are presented again via the maximal association rules as follows:

Definition 2. *The presentation of the Tweets using the maximal association rules is described as follows: If $X(H_{max_j}) \subseteq t_i$ with $i = \{1, .., n\}$ then:*

$$t_i = \bigcup_j^m H_{max_j} \tag{3}$$

where m is the number of maximal association rules that are existed in t_i.

Computing the Weighted of Tweets: The weighted Tweets are calculated based on the weight of the maximal association rules that the Tweet contains.

Definition 3. *The weight of a maximal association rule is computed by the following formula:*

$$w_{H_{max_j}} = \frac{|t|}{n} \tag{4}$$

where $|t|$ is the number of Tweets that contain the maximal association rule H_{max_j}; n is the number of Tweets.

Definition 4. *The weight of Tweet t_i is defined as follows:*

$$w_{t_i} = \sum_{j=1}^{m} w_{H_{max_j}} \tag{5}$$

where $w_{H_{max_j}}$ is the weight of the maximal association rule H_{max_j}, and m is the number of maximal association rules that are contained in t_i.

Choosing the Tweets to Generate the Short Text: When the short text is created, we need to be interested in problems such as the number of Tweets in the short text and the criteria that the selected Tweets must satisfy.

The number of Tweets needs to have in the summary which has to response:

$$\sum_{i=1}^{n} \alpha_i \leq N \tag{6}$$

where n is the number of Tweets; $\alpha_i = 1$ if the Tweet is chosen, but otherwise, $\alpha_i = 0$; N is the number of Tweets that people want to read in the summary.

The summary is created by using the Tweets that satisfy the target function.

Definition 5. *The target function of Tweet summarization based on maximal association rules is defined as follows:*

$$argmax \sum_{i=1}^{n} w_{t_i} \alpha_i \tag{7}$$

where, n is the number of tweets, t_i; if t_i is chosen, $\alpha_i = 1$, otherwise, $\alpha_i = 0$.

$$\alpha_i = \begin{cases} 1, if \ w_{t_i} is \ maximum \ and \ t_i \supseteq X(H_j) \\ 0, if \ otherwise \end{cases} \tag{8}$$

4   Experiments

4.1   Dataset

We gathered 145,000 raw Tweets from Twitter by using the available Streams API for Python. For each Twitter home page, we collected the maximum number of Tweets. After the Tweets were processed, we filtered 500 tweets belong to "President Obama" topic that we used to evaluate the performance of the model.

4.2   Evaluation

To evaluate the proposed model, we relied on the evaluation of the final results that the model obtained, that is, an evaluation of the summaries the model created. We assessed the model based on two criteria: information retrieval and similarity between the summaries of the baseline systems and the proposed model.

For information retrieval, we used a suite of automatically measured Recall-Oriented Understudy for Gisting Evaluation (ROUGE) metrics. They are widely used to compare similarities between summaries of another system and of the proposed model. The simplest ROUGE metric is the ROUGE_N metric [10].

$$ROUGE\_N = \frac{\sum_{s \in Summ_{ref}} \sum_{n\_grams \in s} match(n\_gram)}{\sum_{s \in Summ_{ref}} \sum_{n\_grams \in s} count(n\_gram)} \tag{9}$$

where $Summ_{ref}$ is the set of summaries of another system; $count(n\_gram)$ is the number of n_grams in summary by another system; $match(n\_gram)$ is the number of co-occurring n_grams between the summaries of another system and the proposed model, and s is a particular summary of another system.

For similarity between the summaries, we employed Precision, Recall, and F-score [1]. They are calculated as follows:

$$Precision = \frac{Summ_{ref} \cap Summ_{gen}}{Summ_{gen}} \tag{10}$$

$$Recall = \frac{Summ_{ref} \cap Summ_{gen}}{Summ_{ref}} \tag{11}$$

$$F - score = \frac{2 * Precision * Recall}{Precision + Recall} \tag{12}$$

In this study, we evaluated our proposed model by implementing it step by step as follows. We employed the Sumbasic method [11] to create the reference summaries. The first hypothesis summaries were generated by using the proposed method. The hybrid term frequency-inverse document frequency (TF-IDF) method [9] was used to give the second hypothesis summaries and is a feature-based approach. The third hypothesis summaries were created by using the LexRank method [5] which is a graph-based approach. Finally, we compared the results of the three approaches via the ROUGE_N metric, Precision, Recall, and F-score.

4.3   Results and Discussions

The proposal was subjected to experiments with the dataset described in Sect. 4.1. The final result that the proposed method generated was a semantic summary of the Tweets based on the available topic. We set $\alpha = 0.5$, $N = 10$, and then computed the ROUGE_N metric, Precision, Recall, and F-score. The accuracy of the result was compared to the other methods and is illustrated in Table 1 and Fig. 2.

Table 1. The average performance of three methods

Method	F-score	Precision	Recall
Proposed method	0.55	0.56	0.55
Hybrid TF-IDF	0.54	0.55	0.54
LexRank	0.51	0.52	0.50

In Table 1, we can see that the Precision, Recall, and F-score of the proposed method are higher than the baseline systems. For F-score, the smallest difference between the proposed method to other methods is 1.0%, and the most significant difference is 4.0%. For Precision, the distinct scope extends from 1.0% to 4.0%. And for Recall, the distinct scope extends from 1.0% to 5.0%. Thus, the similarity between the summarization by the proposed method and the summarization of the reference methods is better than in the summarization of the other methods.

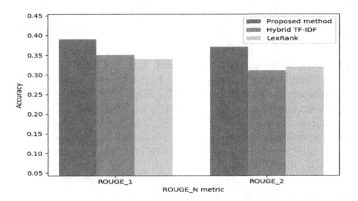

Fig. 2. The ROUGE_N metric of the proposed method and other methods

The accuracy for information retrieval of the methods was compared and is shown in Fig. 2. For ROUGE_1, the smallest difference between the proposed method to other methods is 4.0%, and the most significant difference is 5.0%. For ROUGE_2, the distinct scope extends from 5.0% to 6.0%. Thus, the information retrieval with our approach is better than the baseline methods.

According to the results analyzed above, although the differences between metrics of the proposed and the baseline methods are not significant, it also proves that our approach is more promising than others.

5   Conclusion and Future Work

This study introduced a new approach that is based on maximal association rules to summarize the content of Tweets. We conducted experiments based on a combination of machine learning and data mining for creating a short text from Tweet streams. The proposed method can improve the performance of Tweet summarization. The experiments demonstrated that this combination achieved better performance than other methods.

In future work, we plan to use the results of this study to construct a Tweet integration system. This integration system will provide a full process to combine the semantics of Tweets based on maximal association rules. In this system, Tweet summarization will be the final step to creating the result, in the hope that the performance from integrating Tweets will be improved.

Acknowledgment. This research was supported by Basic Science Research Program through the National Research Foundation of Korea (NRF) funded by the Ministry of Science, ICT&Future Planning (2017R1A2B4009410).

References

1. Aliguliyev, R.M.: A new sentence similarity measure and sentence based extractive technique for automatic text summarization. Expert Syst. Appl. **36**(4), 7764–7772 (2009)
2. Amir, A., Aumann, Y., Feldman, R., Fresko, M.: Maximal association rules: a tool for mining associations in text. J. Intell. Inf. Syst. **25**(3), 333–345 (2005)
3. Chellal, A., Boughanem, M., Dousset, B.: Multi-criterion real time tweet summarization based upon adaptive threshold. In: IEEE/WIC/ACM International Conference on Web Intelligence (WI), pp. 264–271. IEEE (2016)
4. Dutta, S., Ghatak, S., Roy, M., Ghosh, S., Das, A.K.: A graph based clustering technique for tweet summarization. In: 4th International Conference on Reliability, Infocom Technologies and Optimization (ICRITO) (Trends and Future Directions), pp. 1–6. IEEE (2015)
5. Erkan, G., Radev, D.R.: Lexrank: graph-based lexical centrality as salience in text summarization. J. Artif. Intell. Res. **22**, 457–479 (2004)
6. Hole, V., Takalikar, M.: Real time tweet summarization and sentiment analysis of game tournament. Int. J. Sci. Res. **4**(9), 1774–1780 (2013)
7. Menéndez, H.D., Plaza, L., Camacho, D.: Combining graph connectivity and genetic clustering to improve biomedical summarization. In: IEEE Congress on Evolutionary Computation (CEC), pp. 2740–2747. IEEE (2014)
8. Samuel, A., Sharma, D.K.: Modified lexrank for tweet summarization. Int. J. Rough Sets Data Anal. (IJRSDA) **3**(4), 79–90 (2016)
9. Sharifi, B., Hutton, M.A., Kalita, J.K.: Experiments in microblog summarization. In: IEEE Second International Conference on Social Computing (SocialCom), pp. 49–56. IEEE (2010)
10. Steinberger, J., Ježek, K.: Evaluation measures for text summarization. Comput. Inform. **28**(2), 251–275 (2012)
11. Vanderwende, L., Suzuki, H., Brockett, C., Nenkova, A.: Beyond sumbasic: task-focused summarization with sentence simplification and lexical expansion. Inf. Process. Manag. **43**(6), 1606–1618 (2007)
12. Wenerstrom, B., Kantardzic, M., Arabmakki, E., Hindi, M.: Multi-tweet summarization for flu outbreak detection. In: AAAI Fall Symposium: Information Retrieval and Knowledge Discovery in Biomedical Text (2012)

DBpedia and YAGO Based System for Answering Questions in Natural Language

Tomasz Boiński[(✉)], Julian Szymański, Bartłomiej Dudek, Paweł Zalewski, Szymon Dompke, and Maria Czarnecka

Department of Computer Architecture, Faculty of Electronics, Telecommunication and Informatics, Gdańsk University of Technology, Gdańsk, Poland
{tobo,julian.szymanski,bdudek,pzal,szydom,mczar}@eti.pg.edu.pl

Abstract. In this paper we propose a method for answering class 1 and class 2 questions (out of 5 classes defined by Moldovan for TREC conference) based on DBpedia and YAGO. Our method is based on generating dependency trees for the query. In the dependency tree we look for paths leading from the root to the named entity of interest. These paths (referenced further as fibers) are candidates for representation of actual user intention. The analysis of the question consists of three stages: query analysis, query breakdown and information retrieval. During these stages the entities of interest, their attributes and the question domain are detected and the question is converted into a SPARQL query against the DBpedia and YAGO databases. Improvements to the methods are presented and we discuss the quality of the modified solution. We present a system for evaluation of the implemented methods, showing that the methods are viable for use in real applications. We discuss the results and indicate future directions of the work.

1 Introduction

Work on automatic question answering systems started as early as the late sixties [26]. Over the years, the systems became more and more complex while often producing very good results [10,29]. Currently, most of the approaches are based on keywords where the answer is derived directly from the specified by the user keywords and can be either an explicit answer or (usually) the set of documents containing the keywords from the query (and potentially containing the answer). In the latter case the results obtained in this way are quite good; however, they require additional user verification and lookup within the documents provided.

The situation differs when questions are formulated in natural language. In this case there are no explicitly given keywords. The nature of the natural language can make the queries further ambiguous due to indirect subjects or lack of context. In this case many systems rely on identifying interrogative pronouns to detect the entity of interest. Such an approach however does not work well for common-sense knowledge questions like "How many legs does a dog have?".

© Springer Nature Switzerland AG 2018
N. T. Nguyen et al. (Eds.): ICCCI 2018, LNAI 11055, pp. 383–392, 2018.
https://doi.org/10.1007/978-3-319-98443-8_35

The questions formulated in natural language vary in difficulty depending on their complexity and ambiguity. Moldovan et al. [22] defined five classes of question difficulty. The first class consists of factoid questions, where the answer usually can be directly found in the database ("When did Beethoven died?"). The second class requires some knowledge about the question and the database structure. In this case the answer might not be syntactically close to the question ("Who is the spouse of Grover Cleveland?"). Class 3 of questions requires reasoning that is based on multiple, not always compatible, sources; class 4 of questions are interaction-based, and class 5 questions that require expert systems able of analogical reasoning.

In this paper we present a method for answering questions formulated in natural language. In our research we focus on the first two classes. The aim of the proposed method is to answer questions of class 1 and 2, providing, wherever possible, direct answers as found in DBpedia and YAGO databases. Most of the other solutions use their own, dedicated knowledge bases and either are complicated systems, or provide a list of potential answers for the user to choose from. The YAGO and DBpedia databases are vast and constantly improving, so we decided to base our solution on these two sources. The paper also aims at evaluating the solution against the well-known TREC question database.

The structure of the paper is as follows. Section 2 describes different approaches to question answering and our previous evaluation solution that served as a baseline for our approach. Section 3 presents in detail our approach, and shows improvements introduced to the algorithm. In Sect. 4 evaluation of the method is given. Finally in Sect. 5 we discuss the results and draw conclusions.

2 Existing Solutions

Over the years many interesting approaches to question answering emerged [5,21]. The first systems, like PRECISE [25] or BASEBALL [15], focused on providing natural language interfaces to databases, mapped user queries to SQL queries.

Further, question answering systems were proposed on a selected open domain. Here the approaches were not fine-tuned towards a specified domain, and needed to give answers to general questions. Many of the solutions were prepared during the TREC conference [33]. The best of those systems could answer as many as 70% of the questions from TREC database [32], ranging from simple factoid questions to complex, indirect questions. Some like LASSO [22], used deep lexical analysis of the question, to provide the answer using an iterative process. Others, like QRISTAL [17] or QALC [12], were based on semantic similarities and usually map the question into triples/queries. These approaches differ also in terms of complexity, from knowledge-rich systems [16] to simple systems like AskMSR [2].

Another group of systems are ontology-based solutions. Such systems take queries given in natural language and, based on the used ontology they return the answer from one or more knowledge bases that are compatible with that ontology.

Examples of such systems represent a very broad spectrum of solutions. Some, like SQUALL [11], SPARQL2NL [24] or ORAKEL [6] convert the question into a SPARQL query that is evaluated against the given knowledge base.

In recent years, one very complex state of the art system was created, namely IBM Watson [14]. The system was developed as part of the DeepQA Project, started in 2007. The first IBM Watson implementation was made using a cluster consisting of around 2500 CPUs, 15 TB of RAM and, without a connection to the Internet. The quality of the system was so good that it managed to win the Jeopardy TV show [13]. Its strength comes from multiple algorithms that cooperate to calculate the best answer. The drawback of this solution is its complexity and limited availability for the wider audience.

In our approach we aim at providing a Wikipedia-based solution for a question answering system. Wikipedia itself is not very formalized, but previous research shows that it can be formalized [28,30]. DBpedia [1] and YAGO [27] based solutions are viable for class 1 and class 2 questions. Other researchers also follow this route. Tahri et al. proposed a Support Vector Machines-based algorithm for a DBpedia-based question answering system [31]. QASYO [23] is a YAGO based system designed to use YAGO ontology to answer questions. Yahya et al. combined DBpedia and YAGO as sources to their approach and generated SPARQL queries based on the questions asked [34].

The aforementioned works, combined with our previous research, shows that such questions can be converted into formal SPARQL queries [4], which can be processed by the DBpedia and YAGO databases. The solution is based on the observation that dependency trees [7] of most of the queries had one or more paths leading from the root to the named entity of interest. These paths (so-called fibers) are candidates for representation of actual user intention. The goal of this step is to retrieve minimal fiber [4]. The question analysis consists of three stages: query analysis, query breakdown and information retrieval. The general architecture of the proposed solution was described in detail in our previous work [4]. In general the algorithm steps are as follows:

The first stage focuses on query retrieval and grammatical parsing. During this stage a Stanford NLP Parser [9] is used to detect structure of the query and convert it into an ordered tree [19]. Next step of this stage is concept retrieval where concepts are detected in the query, usually the entities and their properties. The sentence is tagged then using Penn-Treebank notation, and supplied with a list of detected entities and their properties.

The second stage aims at determining the entity and properties in question. During this stage we analyze dependencies within the query and perform minimization of the dependency tree. The created tree is then analyzed and fibers are detected within its branches. Finally the fibers detected are minimized.

The last stage, information retrieval, orders the fibers based on their score of relevancy and converts them to SPARQL queries which are used to retrieve the information from the database (YAGO and DBpedia).

The approach was tested using a subset of TREC-8[1] questions that were of class 1 and class 2 and gave precision of value 0.67 and recall equal to 0.36 resulting in F-measure equal 0.47. The specificity was 0.82 and the accuracy was 0.59 [4]. The results were not satisfactory, partially because of the algorithm's drawbacks and partially due to the quality of the databases, especially DBpedia. However, after a closer look many of the unanswered or wrongly answered questions had correct answers within the databases so we decided to extend our solution.

3   Extended Fiber Based Solution

Both the original and extended solutions are looking for answers to the queries based on paths in dependency trees representing the user's query. Each such tree usually has one or more paths leading from the root to a named entity of interest. These paths (so called fibers) are candidates for representation of actual user intention. The original evaluation implementation had several drawbacks limiting its quality. During recent works we identified and tried to eliminate these drawbacks which allowed us to improve the quality of the proposed methods. The solution allows one to submit queries and retrieve answers along with full description of the analysis process. We are not focused on the performance of the algorithms as we use online YAGO and DBpedia endpoints which significantly influence the performance.

The first problem we identified in the original solution was the metric used to match question elements to attribute names. The original method used the Levenhstein [18] metric which gave high similarity for words with matching parts. Sequences like 'death place' were thus matched to attributes like 'deathPlace' and 'birthPlace'. This resulted in additional, usually wrong answers, e.g. for question *What is death place of Mohammad Khaled Hossain?* it produced six answers: *Mount Everest, Nepal, 2013-05-21, 2013-5-20, Bangladesh, Munshiganj.* We decided to use the difflib Python module[2] for the matching which allowed us to eliminate the additional answers. After changes, the system gave only 2, proper answers: *Mount Everest, Nepal.*

We also extended the algorithm for attributes matching. In the original solution the query parts were matched to the label of the attributes found in the used databases. After a closer look it occurred that the actual name of the attribute is often almost a direct match (whereas the label can be misleading).

Further, we also included redirection in possible matches for query attributes. If the attribute could not be matched but low scored candidates included redirection we followed those. This way it is possible to match entities like alternative names of cities or countries (e.g. *Ulan Bator* and *Ulanbaatar*).

The original solution did not analyse synonyms when looking for entities or attributes matches. In the current implementation we use WordNet [20] as

---

[1] http://trec.nist.gov/data/qa/T8_QAdata/topics.qa_questions.txt.
[2] https://docs.python.org/2/library/difflib.html.

a backend for the nltk Python module[3] [3] for extending the list of attributes checked. For example the original program could not answer the question *Who is partner of Donald Trump?* as the databases do not have the attribute *partner* to identify the proper entity. The database contains however an attribute *spouse*. Using synonyms we can modify the question into *Who is spouse of Donald Trump?* and thus get the correct answer. We also use the same mechanism when an attribute or its synonym cannot be found in the databases. In this case we use the nltk module to find the closest words for missing attributes and use them to generate the answer. An example of a question requiring such actions is *What is decease place of Chopin?* as neither *decease place* nor its synonyms could be found in the databases for the entity *Chopin*. This way we are able to generate alternative questions that should have the same answer, and in this way reach the answer.

Attributes name matching sometimes gave misleading results, e.g. in the question *Where is birth place of Jimi Hendrix?* attribute *birth place*, due to different similarity measures used [8], was mapped to attributes *birthPlace* and *birthDate*. To bypass the limitation of similarity measures in our approach we try to detect interrogative pronouns within the question and their domain. For answer generation we analyse only those matched attributes that share the same domain as detected from the interrogative pronoun of the question. This increased the quality of the answers but introduced another problem. In some cases the attribute is not within the same domain as the interrogative pronoun but the answer is, e.g. in the question *Who is successor of George Washington?* the attribute *successor* is not the attribute of a person. However the answer to the question, like *John Adams*, is. Thus, after finding a potential answer the domain check must be performed to check whether the answer is from the same domain regardless of the attribute that led to the answer.

The same nltk module also allowed us to ask questions where entities attributes are represented as actions, e.g. *When died Jim Morrison?*. Based on the verbs in the question we generate nouns, which are in turn the most often used as attributes names within DBpedia and YAGO. In this case there is no *Jim Morrison* entity attribute called *died*. There is, however, a proper answer stored under *deathDate*, which, combined with the domain deduction described earlier is derived from noun *death*, which in turn was generated using the nltk module from the verb *died*. So the answer was 1971-07-03.

The biggest problem occurred for entities with multiple meanings. Many words or names in natural language can have more then one meaning, e.g. entity *Washington*[4] which can have high number of pages related to. Such entities are usually represented using disambiguation entities that have references to different meanings of the entity in question. We look through those referenced entities for a potential answer and then match the domain of the referenced entity, the attribute and the interrogative pronoun to present the best answer.

---

[3] https://www.nltk.org/.

[4] https://en.wikipedia.org/wiki/Washington.

This mechanism still needs further work as it generates additional, mostly wrong, answers. The list of alternative meanings can also be long.

4   Evaluation

We evaluated the solution using the aforementioned TREC-8 questions that belonged to class 1 and class 2. Out of the 40 questions 32 had answers in the DBpedia and YAGO databases, the remaining 8 did not. For all non-answerable questions the algorithm did not give any answers. Out of 32 answerable questions the algorithm answered correctly 20 questions. In 2 cases we got wrong answers and the remaining 10 questions were left unanswered. In 7 cases the algorithm gave additional, wrong answers. Those additional answers are the reason why the sum of true/false positives/negatives is higher than the number of questions used for the evaluation.

Summary in terms of answer correctness and performance evaluation using precision, recall and F-measure are presented in Tables 1 and 2 respectively.

Table 1. Answer correctness

True positives	True negatives	False positives	False negatives
20	8	9	12

Table 2. Performance scores

Precision	Recall	F-Measure
0.69	0.625	0.66

The updated algorithm achieved much higher recall (0.625 versus 0.36 in our evaluation implementation) while keeping precision at the same level (0.69 versus 0.67). The final F-measure increased from 0.47 to 0.66.

As can be seen in Table 3 there still were issues with some types of questions. In some cases the attributes detected within the query were matched incorrectly to entity attributes in the databases. In some cases this leads to providing completely wrong answers. It can be observed e.g. in question *Who is leading actor in 'The Godfather'?* where the attribute *leading actor* was marked as a synonym to *direct actor* and as a result matched to the attribute *director*. In other cases wrong match generated additional answers, like in question *What is location of Taj Mahal?* where word *location* had borderline similarity value of 80% with attribute *caption*. The same error can be observed in the question *What is population of Tucson?*. In this case the word *population* was incorrectly matched to attributes *populationAsOf, populationMetro, populationUrban, populationBlank*

Table 3. Examples of questions with wrong answers

Question	Answer in database	Correct answer returned	Wrong answer returned	Comment
What is location of US Declaration of Independence?	Yes	No	Yes	'Position' from 'location' and 'position' property holds a value for picture ('right')
Who is leading actor in 'The Godfather'?	Yes	No	Yes	'Direct actor' reasoned as synonym to 'leading actor', then associated with property 'director'
What is date of Battle of the Somme?	Yes	Yes	Yes	Wrong answers got from property 'seeAlso' ('see' was obtained as a synonym to word 'date')
What is location of Taj Mahal?	Yes	Yes	Yes	Wrong answers got from property 'caption' (which has 80% of similarity with word 'location')
What is population of Tucson?	Yes	Yes	Yes	Wrong answers from properties 'populationAsOf', 'populationMetro', 'populationUrban', 'populationBlank', 'populationEst' (too similar to 'population')
What is population of Ulan Bator?	Yes	Yes	Yes	Wrong answer from property 'populationAsOf' (too similar to 'population')
What is population of Ushuaia?	Yes	Yes	Yes	Wrong answers from property 'populationAsOf' (too similar to 'population')
What is produced by Peugeot?	Yes	Yes	Yes	Returned also answer '1739000' from property 'production'
Where was Washington born?	Yes	Yes	Yes	Problems with disambiguation

and *populationEst*. Such results were observed in almost all questions with additional, usually wrong, answers. In almost all cases however, the algorithm was able to give the correct answer alongside the wrong ones. This proves that the query analysis and entity detection process is correct. In further works we will focus on algorithms for better attribute matching.

In one case (*Who is Voyager manager?*) the question is formulated in a way that, even for humans, it is not easy to determine the correct answer without additional context information. We might assume the question asks for the manager of the Voyager space program, however the algorithm decided to match the ambiguous name *Voyager* to a TV show *Earth Star Voyager*[5] and gave as the answer the name of the show director (which from a technical point of view can be treated as correct). In this case the algorithm should be able to generate more potential answers from different domains. The system as a whole could than present those answers and ask the user to specify the question by selecting the entity domain.

5   Conclusion

Proposed solution can, in most cases, answer the questions formulated in natural language and is not domain-specific. The strength of the proposed idea lies also in utilization of widely available tools and databases. In most cases such a system is also able to give a direct answer to the question asked (e.g. the date of an event in question). With certain improvements, mainly in matching and disambiguation algorithms, the system might be able to answer questions belonging to class 3 type of questions (answering based on multiple sources). The proposed query analysis method may also be used to extract semantic data from text.

The advantage of the system is its speed and small hardware requirements. The system can be run on a standard PC and does not require any other special resources. Many other solutions (especially IBM Watson) requires dedicated and powerful hardware.

Most of the problems found with the method are related to insufficient quality of attributes matching. Such cases were observed during analysis of results given for almost all questions with additional, usually wrong, answers. The algorithm however, in most cases, was able to give the correct answer alongside the wrong ones. This proves that the query analysis and entity detection process is correct.

Most of the wrong answers can also be eliminated by formalization and harmonization of attributes used to describe concepts within the same and different domains in the databases itself. The DBpedia and YAGO databases still use very shallow and not well interlinked internal ontologies. Data linking occurs on different levels and is very domain dependent. Constant development and formalization of the databases used can thus have a positive impact on the results obtained by the algorithm, and in time should have a positive impact on the quality of the proposed method.

The problems described in the evaluation section of this paper touch a very difficult problem which is natural language disambiguation. Further improvements of the proposed solution should be thus focused on improving the matching algorithm and ability to better understand the context and domain of the questions asked. We believe, that after further improvements, the proposed method can be used as a base for a system able to answer questions formulated in natural language.

---

[5] https://en.wikipedia.org/wiki/Earth_Star_Voyager.

References

1. Auer, S., Bizer, C., Kobilarov, G., Lehmann, J., Cyganiak, R., Ives, Z.: DBpedia: a nucleus for a web of open data. In: Aberer, K., et al. (eds.) ASWC/ISWC -2007. LNCS, vol. 4825, pp. 722–735. Springer, Heidelberg (2007). https://doi.org/10. 1007/978-3-540-76298-0_52

2. Banko, M., Brill, E., Dumais, S., Lin, J.: AskMSR: question answering using the worldwide web. In: Proceedings of 2002 AAAI Spring Symposium on Mining Answers from Texts and Knowledge Bases, pp. 7–9 (2002)

3. Bird, S., Klein, E., Loper, E.: Natural Language Processing with Python. O'Reilly Media, Sebastopol (2009)

4. Boiński, T., Ambrożewicz, A.: DBpedia and YAGO as knowledge base for natural language based question answering—the evaluation. In: Gruca, A., Czachórski, T., Harezlak, K., Kozielski, S., Piotrowska, A. (eds.) ICMMI 2017. AISC, vol. 659, pp. 251–260. Springer, Cham (2018). https://doi.org/10.1007/978-3-319-67792-7_25

5. Bouziane, A., Bouchiha, D., Doumi, N., Malki, M.: Question answering systems: survey and trends. Procedia Comput. Sci. **73**, 366–375 (2015)

6. Cimiano, P., Haase, P., Heizmann, J.: Porting natural language interfaces between domains: an experimental user study with the ORAKEL system. In: Proceedings of the 12th International Conference on Intelligent User Interfaces, pp. 180–189. ACM (2007)

7. Culotta, A., Sorensen, J.: Dependency tree kernels for relation extraction. In: Proceedings of the 42nd Annual Meeting on Association for Computational Linguistics, p. 423. Association for Computational Linguistics (2004)

8. Czarnul, P., Rościszewski, P., Matuszek, M., Szymański, J.: Simulation of parallel similarity measure computations for large data sets. In: 2015 IEEE 2nd International Conference on Cybernetics (CYBCONF), pp. 472–477. IEEE (2015)

9. De Marneffe, M.C., MacCartney, B., Manning, C.D., et al.: Generating typed dependency parses from phrase structure parses. Proc. LREC. **6**, 449–454 (2006)

10. Duch, W., Szymański, J., Sarnatowicz, T.: Concept description vectors and the 20 question game. In: Kłopotek, M.A., Wierzchoń, S.T., Trojanowski, K. (eds.) Intelligent Information Processing and Web Mining. Advances in Soft Computing, vol. 31, pp. 41–50. Springer, Heidelberg (2005). https://doi.org/10.1007/3-540-32392-9_5

11. Ferré, S.: SQUALL: a controlled natural language as expressive as SPARQL 1.1. In: Métais, E., Meziane, F., Saraee, M., Sugumaran, V., Vadera, S. (eds.) NLDB 2013. LNCS, vol. 7934, pp. 114–125. Springer, Heidelberg (2013). https://doi.org/10.1007/978-3-642-38824-8_10

12. Ferret, O., Grau, B., Hurault-Plantet, M., Illouz, G., Monceaux, L., Robba, I., Vilnat, A.: Finding an answer based on the recognition of the question focus. In: TREC (2001)

13. Ferrucci, D., Levas, A., Bagchi, S., Gondek, D., Mueller, E.T.: Watson: beyond jeopardy!. Artif. Intell. **199**, 93–105 (2013)

14. Ferrucci, D.A.: IBM's Watson/DeepQA. In: ACM SIGARCH Computer Architecture News, vol. 39. ACM (2011)

15. Green, Jr, B.F., Wolf, A.K., Chomsky, C., Laughery, K.: BASEBALL: an automatic question-answerer. In: Papers Presented at the 9–11 May 1961, Western Joint IRE-AIEE-ACM Computer Conference, pp. 219–224. ACM (1961)

16. Harabagiu, S.M., et al.: FALCON: boosting knowledge for answer engines. In: TREC (2000)

17. Laurent, D., Séguéla, P., Nègre, S.: Cross lingual question answering using QRISTAL for CLEF 2006. In: Peters, C., Clough, P., Gey, F.C., Karlgren, J., Magnini, B., Oard, D.W., de Rijke, M., Stempfhuber, M. (eds.) CLEF 2006. LNCS, vol. 4730, pp. 339–350. Springer, Heidelberg (2007). https://doi.org/10.1007/978-3-540-74999-8_41

18. Lcvenshtcin, V.I.: Binary codes capable of correcting deletions, insertions, and reversals. Soviet Phys. Doklady **10**, 707–710 (1996)

19. Marcus, M.P., Marcinkiewicz, M.A., Santorini, B.: Building a large annotated corpus of English: The Penn Treebank. Comput. Linguist. **19**(2), 313–330 (1993)

20. Miller, G.A., Beckitch, R., Fellbaum, C., Gross, D., Miller, K.: Introduction to WordNet: An On-line Lexical Database. Princeton University Press, Cognitive Science Laboratory (1993)

21. Mishra, A., Jain, S.K.: A survey on question answering systems with classification. J. King Saud Univ. Comput. Inf. Sci. **28**(3), 345–361 (2016)

22. Moldovan, D.I., Harabagiu, S.M., Pasca, M., Mihalcea, R., Goodrum, R., Girju, R., Rus, V.: LASSO: A tool for surfing the answer net. TREC **8**, 65–73 (1999)

23. Moussa, A.M., Abdel-Kader, R.F.: QASYO: a question answering system for YAGO ontology. Int. J. Database Theor. Appl. **4**(2), 99–112 (2011)

24. Ngonga Ngomo, A.C., Bühmann, L., Unger, C., Lehmann, J., Gerber, D.: Sorry, I don't speak SPARQL: translating SPARQL queries into natural language. In: Proceedings of the 22nd International Conference on World Wide Web, pp. 977–988. ACM (2013)

25. Popescu, A.M., Etzioni, O., Kautz, H.: Towards a theory of natural language interfaces to databases. In: Proceedings of the 8th International Conference on Intelligent User Interfaces, pp. 149–157. ACM (2003)

26. Simmons, R.F.: Natural language question-answering systems: 1969. Commun. ACM **13**(1), 15–30 (1970)

27. Suchanek, F.M., Kasneci, G., Weikum, G.: Yago: a core of semantic knowledge. In: Proceedings of the 16th International Conference on World Wide Web, pp. 697–706. ACM (2007)

28. Szymański, J.: Self–organizing map representation for clustering wikipedia search results. In: Nguyen, N.T., Kim, C.-G., Janiak, A. (eds.) ACIIDS 2011, Part II. LNCS (LNAI), vol. 6592, pp. 140–149. Springer, Heidelberg (2011). https://doi.org/10.1007/978-3-642-20042-7_15

29. Szymański, J.: Words context analysis for improvement of information retrieval. In: Nguyen, N.-T., Hoang, K., Jędrzejowicz, P. (eds.) ICCCI 2012, Part I. LNCS (LNAI), vol. 7653, pp. 318–325. Springer, Heidelberg (2012). https://doi.org/10.1007/978-3-642-34630-9_33

30. Szymański, J., Duch, W.: Semantic memory knowledge acquisition through active dialogues. In: 2007 International Joint Conference on Neural Networks, IJCNN 2007, pp. 536–541. IEEE (2007)

31. Tahri, A., Tibermacine, O.: Dbpedia based factoid question answering system. Int. J. Web Seman. Technol. **4**(3), 23 (2013)

32. TREC: TREC 2008 (2008). http://trec.nist.gov/data/qa/T8_QAdata/topics.qa_questions.txt. Accessed 19 Nov 2017

33. Trec: Text REtrieval Conference (TREC) (2012). http://trec.nist.gov/. Accessed 12 Mar 2018

34. Yahya, M., Berberich, K., Elbassuoni, S., Weikum, G.: Robust question answering over the web of linked data. In: Proceedings of the 22nd ACM International Conference on Information & Knowledge Management, pp. 1107–1116. ACM (2013)

Bidirectional LSTM for Author Gender Identification

Bassem Bsir[(✉)] and Mounir Zrigui[(✉)]

LATICE Laboratory Research, Department of Computer Science,
University of Monastir, Monastir, Tunisia
Bsir.bassem@yahoo.fr, mounir.zrigui@fsm.rnu.tn

Abstract. Author profiling consists in inferring the authors' gender, age, native language, dialects or personality by examining his/her written text. This important task is a very active research area because of its utility in crime, marketing and business.

In this paper, we address the problem of gender identification by applying the Long Short-Term Memory neural network architecture. Which is a novel type of recurrent network architecture that implements an appropriate gradient-based learning algorithm to overcome the vanishing-gradient problem. Experimental results show that our composition outperformed the traditional machine learning methods on gender identification.

Keywords: LSTM neural network · Author profiling · Gender identification
Deep learning

1 Introduction

Authorship profiling is the task of inferring author's characteristics such as gender, age group, level of education, social class and cultural background. It has gained great importance, in the last few years, thanks to its new application and utility in marketing and legal linguistics. For instance, author profiling can be applied to discover cyber-criminal acts and detect sexual harassers or pedophiles [17, 25]. It can help also in analyzing the human behaviour and recognizing personality type [2]. Author profiling can be viewed as a special case of the text categorization (classification) problem. While dealing with such issue, typical text categorization usually focuses on classifying automatically the domain of the text (Sport, Politic, Economy, etc.) based on its content (topic-based) [18]. Nevertheless, it was shown that automated text categorization techniques could exploit combinations of simple lexical and syntactic features to infer the author's gender.

In fact, each individual has a distinct style of writing Twitter tweets, blogs and reviews. In these texts, the structure of sentence, vocabulary and way of representing thoughts vary from one person to another. Despite these differences, people belonging to similar group share certain aspects of writing. By "similar group", we refer to community made up of people having the same gender. The main challenge, in author profiling task, is to simplify the input sequence and keep the most important information reflecting the stylistic differences in texts between men and women [29].

© Springer Nature Switzerland AG 2018
N. T. Nguyen et al. (Eds.): ICCCI 2018, LNAI 11055, pp. 393–402, 2018.
https://doi.org/10.1007/978-3-319-98443-8_36

Over the past few years, Deep Learning (DL) architectures and algorithms have made impressive advances in fields such as image recognition [15], speech processing [27] and sentiment analysis [30]. Indeed, our idea in this paper is to use RNNs as discriminative binary classifiers to predict a characteristic label at every anonym text.

The aim of our approach is to investigate whether reasonable results for gender identification could be obtained by applying recently-emerged bi-directional neural network architectures, combined with word embed dings input representations. We conducted experiments on a dataset consisting of twitter texts presented by the PAN Lab at CLEF 2017 [26]. The simulation results show that our proposed model achieved significant improvement over those presented in the state-of-the-art in terms of accuracy.

2   State of the Art

Computational authorship profiling, attribution or identification methodology for both natural and computing languages require two main steps: The first one is the extraction of variables representing the author's style (authorial features) [4, 8, 12]. However, the second step consists in applying a statistical or machine learning algorithm to these variables in order to develop models that are capable of discriminating several authors. The most difficult task in these approaches is to define the variables and discovering the best classification algorithm.

Indeed, Sap et al. [30] derived a predictive lexica (words and weights) for age and gender identification using regression and classification models based on word usage in Facebook, blog and twitter data with associated demographic labels.

Säily [28] analyzed a part-of-speech tagged version of the Corpus of Early English Correspondence consisting of more than two million words written impersonal letters produced by about 660 writers of both genders. Although this finding could not be tested for statistical significance, authors confirmed, in their study, that males generally use more nouns and fewer pronouns than females. In fact, this result was statistically significant for every span of time between 1415 and 1681, except for one year.

Clauset et al. [10] used LIWC to analyze 46 million words produced by 11,609 participants. The studied texts include written texts composed as part of psychology experiments carried out by universities in the US, New Zealand and England. Full texts of fictional novels, essays written for university evaluation and transcribed free conversations from research interviews were investigated in these studies. The authors concluded that some of their LIWC variables were statistically significant and showed a gender effect.

We can also mention, for instance, the work of Wang et al. [32] who examined 10,000 short blog messages; each of which contains 15 words. They got 72.1% accuracy for gender prediction. Peersman et al. [25] presented an exploratory study in which they applied a text categorization approach for the prediction of age and gender on a corpus of chat texts collected from the Belgian social networking site Net log. They tried to identify the types of features that are most informative for a reliable prediction of age and gender on these difficult text types.

Bamman et al. [5] examined the most frequent words in a corpus of short messages written by more than 14,000 users of Twitter for gender patterns to predict the gender of anonymous text. The results they obtained confirmed that social network homophile is correlated to the use of same-gender language markers.

Pham et al. [26] relying on a corpus of 3524 Vietnamese Weblog pages of 73 bloggers and exploiting 298 features, they obtained 82.12 accuracy rate, for occupation, and 78.00 for location dimension by employing IBK [26].

Poulston et al. [1] used the genism Python library for LDA topic extraction with SVM classifiers. Their results proved that the topic models are useful in developing author's profiling systems.

Argamon et al. [3] analyzed an analogous sample taken from the BNC consisting of fiction and non-fiction documents. Their corpus includes 604 texts equally divided by genre and controlled for authorial origin for a total size of 25 million words. Their analysis consists in a frequency count of basic and most frequent function words, part-of-speech tags and part-of-speech two-grams and three-grams. The counts were processed by a machine-learning algorithm used to classify the texts according to the author's gender. They obtained an accuracy of 80%.

Martinc et al. [23] based on the corpus collected from Twitter text written by four different languages (Arabic, English, Portuguese and Spanish), [23] obtained 70.02 by using logistic regression by combining character, word POS n-grams, emoji's, sentiments, character flood in gland lists of words per variety in PAN 2017 competition [27].

González-Gallardo et al. [7] predicted the gender, age and personality traits of Twitter users. They accounted stylistic features represented by character N grams and POS N-grams to classify tweets. They applied Support Vector Machine (SVM) with a linear kernel called LinearSVC and obtained 83.46% for gender detection.

Deep learning-based approaches, such as the recursive neural network (RNNs) model [31] and the convolutional neural network model [19], have recently dominated the state-of-the-art in Natural Language Processing (NLP) researchers. Models, in this paradigm, can take advantage of general learning procedures relying on back-propagation, 'deep learning', a variety of efficient algorithms and some other tricks to further improve the training procedure.

Neural networks have recently demonstrated their great performance in many NLP tasks. For instance, Socher et al. [31], in 2016, introduced Recursive Neural Ten-sor Network and the Stanford Sentiment Treebank. In the same year, Yin et al. [34] investigated convolutional neural networks (CNNs) and basic RNNs for relation classification. They reported the higher performance of CNN, compared to RNN, and gave evidence that CNN and RNN provided complementary information.

Both Gokturk et al. [15] and Kalchbrenner et al. [20] proposed a convolutional neural network for dialogue act classification (sentence model and a discourse mod-el). Cho et al. [9] proposed a neural network model called RNN Encoder-Decoder consisting of two recurrent neural networks (RNN). One RNN encodes a sequence of symbols into a fixed-length vector representation, while the other decodes the representation into another sequence of symbols.

Zhou et al. [35] combined the strengths of CNN and long-short term memory (LSTM) architectures and proposed a novel model called C-LSTM for sentence representation and text classification. Their model utilizes CNN to extract a sequence of higher-level phrase representations and feeds into a long short-term memory recurrent neural network (LSTM) to obtain the sentence representation.

Zhou et al. [36] introduced Bidirectional Long Short-Term Memory Networks with Two Dimensional Max Pooling (BLSTM-2DPooling) to capture features on both the time-step dimension and the feature vector dimension. They first utilized Bidirectional Long Short-Term Memory Networks to transform the text into vectors.

Malmasi et al. [22] for the first time employed deep learning techniques for author profiling in 2016. They showed a considerable difference between traditional ma-chine learning models and deep learning models in the participant teams evaluated in the Third Workshop on NLP for Similar Languages, Varieties and Dialects. They attempted to narrow this gap using convolutional neural networks (CNN) as a first approach for author's profiling.

Kodiyan et al. [21] in their latest research in 2017 implemented a bi-directional Recurrent Neural Network with a Gated Recurrent Unit (GRU) combined with an Attention Mechanism for author profiling detection. In this work, researchers achieved an average accuracy of 75.31%, in gender classification, and 85.22% in language variety classification. Based on the same corpus, Miura et al. [24] presented model integrating word and character information with multiple conventional neural network layers. Their models marked joint accuracies of 64, 86% in the gender identification and the language classification, respectively in PAN 2017.

For Arabic texts, research works of authorship analysis performed on Arabic texts are not numerous due to two reasons: (i) the lack of reliable text collections to train and test automatically derived classifiers and (ii) the richness of Arabic morphology which makes part-of-speech tagging a more difficult task in Arabic than in other languages such as English where each part-of-speech is typically represented as a separate word. Estival et al. [13] collected Arabic and English e-mails written by 1033 English people and other e-mails written by 1030 Egyptian Arabic speakers. They studied several demographic and psychometric features for author's profiling. Bassem and Zrigui [6] a constructed an Arabic corpus taken from Facebook for age and gender detection. They used different techniques for classification combined with linguistic, stylistic and structural features to determine the author's gender and age. They obtained 71.52 accuracy to infer the author's gender and 53.08% for age detection.

3 The Proposed Approach

In the following section, we will present our system of determining the author's gender from Arabic corpus in details, as shown in Fig. 1. The main challenge in our system is to simplify the input sequence and keep the most important information how can infer the gender of author. For this reason, we applied a long short-term memory bidirectional RNN architecture combined with word embedding techniques.

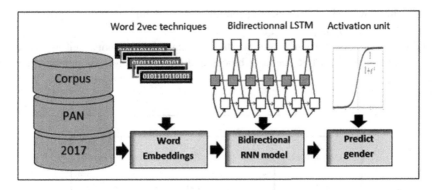

Fig. 1. General architecture of the proposed system.

3.1 Data

The dataset is part of author profiling task of PAN@CLEF 2017 [4], as shown in Table 1. It was collected from Twitter. For each tweet collections, Arabic texts are composed of tweets written by 2400 authors; 100 tweets per authors. For the Arabic language, four varieties were used in this corpus: Egypt, Gulf, Levantine, and Maghrebi.

Table 1. PAN CLEF 2017 training corpus statistic for gender detection.

Authors	Tweets	Language varieties	Genders
2400	240 k	Gulf, Levantine, Maghrebi, Egypt	1200 M; 1200 F

3.2 Word Embedding

In this layer, we transformed each word into embedded features. The Word2Vec model was formed by the corpus of Arabic Wikipedia[1] with 4 million tweets extracted in order to enrich the vocabulary list with words that do not exist in Wikipedia. For training, we used the skip-gram neural network model with a window of size 5 (1 center word + 2 words before and 2 words after), a minimum frequency of 15 and a dimension equal to 300.

3.3 Long Short-Term Memory Bidirectional RNN Model

LSTM model is an extension of the RNNs structure. It is a bi-directional recurrent neural network (BRNN) introduced by [16] who replaced the conventional neuron with a so-called memory cell controlled by input, output and forget gates in order to overcome the vanishing gradient problem of traditional RNNs. The BRNN can be trained without the limitation of using input information just up to a preset future

---

[1] Wikipedia, "WikimediaDownloads."https://dumps.wikimedia.org/arwiki/ 20170401/, 2017. [Online. Accessed 10 Apr 2017]

frame, which was simultaneously accomplished by being trained in positive and negative time directions [33, 35].

LSTM networks introduce a new structure called a memory cell where each memory cell is made of two memory blocks and an output layer, as demonstrated in Fig. 2.

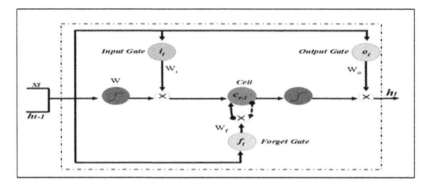

Fig. 2. A graphical representation of LSTM memory cells.

Each LSTM cell computes its internal state by applying the following iterative process and for multiple blocks; the calculations were randomly repeated for each block

$$i_t = \sigma(W_{hi}h_{t-1} + W_{xi}xt + W_{ci}c_{t-1} + b_i) \qquad (1)$$

$$f_t = \sigma(W_{hf}h_{t-1} + W_{xf}xt + W_{cf}c_{t-1} + b_f) \qquad (2)$$

$$o_t = \sigma(Woh_{t-1} + W_{xo}xt + Woc_t + b_o) \qquad (3)$$

$$c_t = f_t \odot c_{t-1} + i_t \odot \tanh(W_{xc}x_t + W_{hc}h_{t-1} + b_c) \qquad (4)$$

$$h_t = o_t \odot \tanh(c_t) \qquad (5)$$

where

i_t: Input gate. It shows the amount of new information that will be transmitted through the memory cell.

f_t: Forget gate is responsible for throwing away information from memory cell.

o_t: Output gate show how much information will be passed to expose to the next step.

c_t: Self-recurrent, which is equal to the standard RNN.

σ: The sigmoid function.

h_t: Final output.

$\odot$:   Denotes the element-wise vector product.

W: Matrices with different sub-scripts are parameter matrices.

b: The bias vector.

LSTM contains a hidden state and three gates: forget, input and an output gate. These gates use the previous output $ht - 1$, the current input xt and their output vectors. All gates were generated by applying the sigmoid function on the ensemble of input and the preceding hidden state.

- Forget Gate is used to decide what information would be thrown away from the cell state. It takes as input ht − 1and xt. It generate an outputs result $\in [0, 1]$. The value (1) means completely retain this, and 0 means completely forget this.
- Input Gate decides what information would be stored in the cell state. A sigmoid layer, called input gate layer, decides what information should be updated. After generating a new candidate by the tanh layer, combination was done by applying $\odot$.
- Output Gate decides the information that should be removed from the cell state, and the new information which must be added to the cell state and the cell state.

4   Experimentation

4.1   Performance Metric

In this section, we will evaluate the prediction accuracy of our method using the afore-mentioned corpora. We used three results obtained in PAN@CLEF2017 as a baseline method to assess our technique and show its efficiency. Those obtained by applying deep learning method in PAN@CLEF2017: the result of [21] based on using BRNN Gated Recurrent Unit and [24] relying on CNN architecture as well as the best results of gender identification obtained in PAN@CLEF2017 [27], as shown in Table 2.

Table 2. The three accuracy results of classifying gender for Arabic language on PAN 2017 testing.

Authors	Methods	Accuracy
Basile et al., 2017 [27]	SVM +tf-idf n-grams	80.06%
kodiyan et al., 2017 [21]	GRU	71.50%
Miura et al., 2017 [24]	CNN	76.44%

4.2   Activation Unit Type

Based on the BRNN-LSTM, we trained our model with 20% of dataset from PAN corpus in 100 epochs with four different activation units. The test set was imported for gender identification accuracy testing. The relationship between the accuracy rate and the different activation units is shown in Fig. 3.

Experiments have shown that with Sigmoid we obtained 78.4% accuracy to infer the gender of the author.

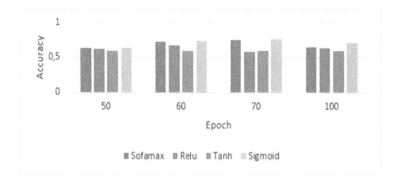

Fig. 3. Accuracy of gender identification under different activation units.

4.3   Evaluation

We used 10-fold cross-validation. The dataset was divided at the note level. We separated out 10% of the training set to form the validation set. This validation set was employed to evaluate our bi-directional RNN model. We also utilized a maximum of 100 epochs to train our model on an Intel core i7 machine with 16 GB memory. The accuracy result obtained by applying our method for gender identification is 79.23% for test data, as show Fig. 4. Obviously, our results are very encouraging for two reasons: our model outperformed deep learning methods and (ii) the gap between our accuracy is very minimal compared the best results obtained in PAN@CLEF 2017.

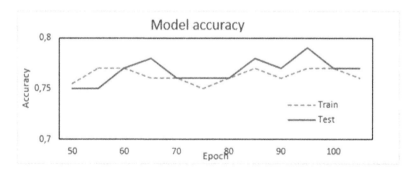

Fig. 4. Gender accuracy for train and test corpus.

In general, using BRNN-LSTM model with word embeddings provided an encouraging results and proved that bi-directional deep networks is crucial in Arabic author profiling, especially for gender identification task.

5   Conclusion

In this paper, we combined word embeddings and deep-learning long short-term memory bidirectional neural network to predict the authors' gender of Twitter texts. Our bi-directional recurrent neural networks model showed a good performance for gender identification. The obtained results were encouraging, since neural network models were efficiently applied to solve natural language processing problems (sentiment analysis, text classification, etc.).

As future works, we plan to extend to other attributes for Arabic author's identification such as language variety and personality features.

References

1. Poulston, A.: Stevenson, M., Bontcheva, K.: Topic models and n–gram language models for author profiling. In: Proceedings of CLEF 2015 Evaluation Labs (2015)
2. Alvarez-Carmona, M.A., et al.: INAOE's participation at PAN 2015: Author profiling task. In: Working Notes Papers of the CLEF (2015)
3. Argamon, S., Koppel, M., Fine, J., Shimoni, A.R.: Gender, genre, and writing style in formal written texts. Text-The Hague Then Amsterdam Then Berlin- 23(3), 321–346 (2003)
4. Aslam, T., Krsul, I., Spafford, E.H.: Use of a taxonomy of security faults (1996)
5. Bamman, D., Eisenstein, J., Schnoebelen, T.: Gender identity and lexical variation in social media. J. Socioling. 18(2), 135–160 (2014)
6. Bassem, B., Zrigui, M.: An empirical method for evaluation of author profiling framework. In: PACLIC 31. Cebu (2017)
7. González-Gallardo, C.E., Montes, A., Sierra, G., Núñez-Juárez, J.A., Salinas-López, A.J., Ek, J.: Tweets classification using corpus dependent tags, character and POS N-grams. In: Proceedings of CLEF 2015 Evaluation Labs (2015)
8. Chaski, C.E.: Who wrote it? Steps toward a science of authorship identification. Nat. Inst. Justice J. 233(233), 15–22 (1997)
9. Cho, K., et al.: Learning phrase representations using RNN encoder-decoder for statistical machine translation (2014). arXiv preprint arXiv:1406.1078
10. Clauset, A., Moore, C., Newman, M.E.: Hierarchical structure and the prediction of missing links in networks. Nature 453(7191), 98 (2008)
11. Collobert, R., et al.: Natural language processing (almost) from scratch. Journal of Machine Learning Research, 2493–2537, 12 Aug 2011
12. Ding, H., Samadzadeh, M.H.: Extraction of Java program fingerprints for software authorship identification. J. Syst. Softw. 72(1), 49–57 (2004)
13. Estival, D., et al.: Author Profiling for English and Arabic Emails (2008)
14. Gehring, W.J., et al.: A neural system for error detection and compensation. Psychol. Sci. 4(6), 385–390 (1993)
15. Gokturk, S.B., et al.: System and method for providing objectified image renderings using recognition information from images. U.S. Patent No. 9,430,719. 30 Aug 2016
16. Hochreiter, S., Schmidhuber, J.: LSTM can solve hard long time lag problems. In: Advances in neural information processing systems, pp. 473–479 (1997)
17. Inches, G., Crestani, F.: Overview of the international sexual predator identification competition at PAN-2012. In: CLEF (Online working notes/labs/workshop), vol. 30 (2012)

18. Joachims, T.: Text categorization with Support Vector Machines: learning with many relevant features. In: Nédellec, C., Rouveirol, C. (eds.) ECML 1998. LNCS, vol. 1398, pp. 137–142. Springer, Heidelberg (1998). https://doi.org/10.1007/BFb0026683

19. Kalchbrenner, N., Blunsom, P.: Recurrent continuous translation models. In: Proceedings of the 2013 Conference on Empirical Methods in Natural Language Processing (2013)

20. Kalchbrenner, N., Grefenstette, E., Blunsom, P.: A convolutional neural network for modelling sentences (2014). arXiv preprint arXiv:1404.2188

21. Kodiyan, D., et al.: Author profiling with bidirectional RNNs using attention with GRUs: notebook for PAN at CLEF 2017. In: CLEF 2017 Evaluation Labs and Workshop–Working Notes Papers, Dublin, Ireland. 11–14 September 2017

22. Malmasi, S., et al.: Discriminating between similar languages and arabic dialect identification: a report on the third dsl shared task. In: VarDial 3 (2016)

23. Martinc, M., Škrjanec, I., Zupan, K., Pollak, S.: Pan 2017: Author profiling gender and language variety prediction. CLEF (Working Notes) 2017. In: CEUR Workshop Proceedings 1866, CEUR-WS.org (2017)

24. Miura, Y., et al.: Author Profiling with Word + Character Neural Attention Network. CLEF (Working Notes) 2017. In: CEUR Workshop Proceedings 1866, CEUR-WS.org (2017)

25. Peersman, C., Daelemans, W., Van Vaerenbergh, L.: Predicting age and gender in online social networks. In: Proceedings of the 3rd International Workshop on Search and Mining User-Generated Contents, pp. 37–44. ACM (2011)

26. Pham, D.D., Tran, G.B., Pham, S.B.: Author profiling for Vietnamese blogs. In: International Conference on Asian Language Processing, 2009. IALP 2009, pp. 190–194. IEEE (2009)

27. Rangel, F., et al.: Overview of the 5th author profiling task at pan 2017: gender and language variety identification in twitter. In: Working Notes Papers of the CLEF (2017)

28. Säily, T.: Variation in morphological productivity in the BNC: Sociolinguistic and methodological considerations. Corpus Linguist. Linguist. Theory 7(1), 119–141 (2011)

29. Sallis, P.J., et al.: Identified: Software authorship analysis with case-based reasoning (1998)

30. Sap, M., Park, G., Eichstaedt, J., Kern, M., Stillwell, D., Kosinski, M., Schwartz, H.A.: Developing age and gender predictive lexica over social media. In: Proceedings of the 2014 Conference on Empirical Methods in Natural Language Processing (EMNLP), pp. 1146–1151 (2014)

31. Socher, R., et al. Recursive deep models for semantic compositionality over a sentiment treebank. In: Proceedings of the 2013 Conference on Empirical Methods in Natural Language Processing (2013)

32. Wang, P., Xu, J., Xu, B., Liu, C., Zhang, H., Wang, F., Hao, H.: Semantic clustering and convolutional neural network for short text categorization. In: Proceedings of the 53rd Annual Meeting of the Association for Computational Linguistics and the 7th International Joint Conference on Natural Language Processing (Volume 2: Short Papers). vol. 2, pp. 352–357 (2015)

33. Williams, J.D., Zweig, G.: End-to-end lstm-based dialog control optimized with supervised and reinforcement learning (2016). arXiv preprint arXiv:1606.01269

34. Yin, W., et al. Comparative study of cnn and rnn for natural language processing (2017). arXiv preprint arXiv:1702.01923

35. Zhou, C., Sun, C., Liu, Z., Lau, F.: A C-LSTM neural network for text classification (2015). arXiv preprint arXiv:1511.08630

36. Zhou, P., Qi, Z., Zheng, S., Xu, J., Bao, H., Xu, B.: Text classification improved by integrating bidirectional LSTM with two-dimensional max pooling (2016). arXiv preprint arXiv:1611.06639. 2016

A New Text Semi-supervised Multi-label Learning Model Based on Using the Label-Feature Relations

Quang-Thuy Ha[1], Thi-Ngan Pham[1,2(✉)], Van-Quang Nguyen[1,3],
Minh-Chau Nguyen[1], Thanh-Huyen Pham[1,4], and Tri-Thanh Nguyen[1]

[1] Vietnam National University, Hanoi (VNU), VNU-University of Engineering
and Technology (UET), No. 144, Xuan Thuy, Cau Giay, Hanoi, Vietnam
{thuyhq,15021766,ntthanh}@vnu.edu.vn
[2] The Vietnamese People's Police Academy, Hanoi, Vietnam
nganpt.di12@vnu.edu.vn
[3] Tohoku University, Sendai, Japan
quanguet@gmail.com
[4] Ha Long University, Ha Long, Quang Ninh, Vietnam
phamthanhhuyen@daihochalong.edu.vn

Abstract. Multi-label learning has become popular and omnipresent in many real-world problems, especially in text classification applications, in which an instance could belong to different classes simultaneously. Due to these label constraints, there are some challenges occurring in building multi-label data. Semi-supervised learning is one possible approach to exploit abundantly unlabeled data for enhancing the classification performance with a small labeled dataset. In this paper, we propose a solution to select the most influential label based on using the relations among the labels and features to a semi-supervised multi-label classification algorithm on texts. Experiments on two datasets of Vietnamese reviews and English emails of Enron show the positive effects of the proposal.

Keywords: Influential label · Label constraints · Multi-label learning
Semi-supervised multi-label learning · Label-feature relation

1 Introduction

There are two main approaches to multi-label classifications including the problem transformation approach and the adaptation approach [9, 15]. In the problem transformation approach, a multi-label classification task is transformed into a single-label one. After applying a certain single-label classification algorithm on domain data with different labels, a combination step of the classifiers' outputs is carried out to get the final results. Some popular algorithms, such as Binary Relevance, Classifier Chains, and Calibrated Label Ranking, belong to this approach. In the adaptation approach, popular learning techniques have been adapted directly for dealing with multi-label data. Some popular algorithms, such as ML-kNN, ML-DT, Rank-SVM, and CLM, follow this direction.

© Springer Nature Switzerland AG 2018
N. T. Nguyen et al. (Eds.): ICCCI 2018, LNAI 11055, pp. 403–413, 2018.
https://doi.org/10.1007/978-3-319-98443-8_37

Due to the complex label constraints occurring in multi-label dataset, the construction of a labeled dataset for multi-label modeling is difficult and time-consuming. Therefore, some semi-supervised multi-label classification algorithms have been proposed to exploit available large volumes of unlabeled data [3–5, 7, 8, 14]. Kong et al. [4] proposed a model of transductive multi-label learning by using label set propagation TRAnsductive Multi-label classification (TRAM) to assign a set of multiple labels to each instance. Zhan and Zhang proposed CO-training for INductive Semi-supervised multi-label learning (COINS) algorithm [8], which adapts the well-known co-training strategy which naturally works in the inductive setting for multi-label classifications. A semi-supervised multi-label classification algorithm has been proposed in [5] namely MULTI-label ClaSsification (MULTICS) based on adapting the semi-supervised multi-label classification algorithm TESC [10].

In MULTICS algorithm, the *most influential label* is determined based on the number of documents associated with that label. Most of previous text classification applications exploit label-feature relations (for example TF, TF.IDF, and topic [1, 2] models) which include a rich semantic meaning among labels. This paper proposed to use label-feature relations, which have more semantic meaning than the "greedy" solution in MULTICS, to determine the most influential label.

This paper's main contributions are (i) refining the phase to transform a multi-label classification problem into a single-label classification problem for TESC, (ii) proposing a solution to determine the most influential label based on using the label-feature matrix.

The rest of this article is organized as follows. In the next section, a new text semi-supervised multi-label learning model is described. Its predecessor MULTICS and the corresponding single-label algorithm TESC are introduced. Section 3 shows the experiments in multi-label classifications on Vietnamese reviews in the hotel domain and other datasets. Related works are introduced in Sect. 4. Conclusions are shown in the last section.

2   A New Text Semi-supervised Multi-label Learning Model

2.1   Problem Formulation

Let D be the set of documents in a considered domain. Let L be the set of labels $\{l_1, l_2, \ldots, l_q\}$. Let $\overline{D}$ and $\overline{D}^U$ be the collections of labeled and unlabeled documents, correspondingly. For each d in $\overline{D}$, *label(d)* denotes the set of labels assigned to d.

The task is to derive a multi-label classification function $f : D \rightarrow 2^L$ i.e., given a new unlabeled document $d \in D$, the function identifies a set of relevant labels $f(d) \subseteq L$.

2.2   The TESC Algorithm

TESC, an acronym of TExt classification using Semi-supervised Clustering, is a semi-supervised clustering method for single label classification proposed by Zhang et al. [10]. TESC can only perform on single label dataset in which each document can be associated with only a single label.

TESC consists of two phases: (1) clustering process aiming to find the set of partitions based on the labeled and unlabeled texts; (2) classification process aiming to assign a label for a new unseen text based on the trained clusters. In the clustering phase, TESC uses labeled texts to capture the silhouettes of text clusters. Next the unlabeled texts are added to the corresponding clusters to adjust their centroids. The set of clusters is used as the model for classification phase. In classification phase, given a new unlabeled text, the model returns the label of the nearest cluster to this text.

2.3   The MULTICS Algorithm

MULTICS [5], the abbreviation of MULTI-label ClaSsification, is a recursive adaptation multi-label classification algorithm based on TESC for multi-label classification. For each iteration, MULTICS is run on the training dataset D including a labeled dataset D^L and an unlabeled dataset D^U, i.e., $D = (D^L, D^U)$. Moreover, there exists a default label set L_1 of which all the data in D^L have. L_2 be the set of possible labels of which some data in D^L might have. MULTICS training phase consists of three steps:

Step 1, MULTICS finds "*the most influential label*", denoted by λ ($\lambda \notin L_1$, $\lambda \in L_2$), from D^L, which is the first label having the biggest number of occurrences. λ is used as the clue to create clusters, i.e., a simulative label set L^* is created including three macro labels λ_1, λ_2, and λ_3. The multi-label labeled dataset D^L is divided into three single-label labeled dataset D^{L1}, D^{L2}, and D^{L3}. The dataset D^{L1} includes documents associated with the same label set $\lambda_1 = L_1 \cup \{\lambda\}$. The dataset D^{L2} includes documents associated with the label set $\lambda_2 = L_1 \cup \{\lambda\}$ and at least another label. The dataset D^{L3} is the remaining part of D^L, i.e., $D^{L3} = (D^L \backslash D^{L1}) \backslash D^{L2}$, has macro label λ_3. It is clear that the label λ is not associated with any document in D^{L3}. Hence, the multi-label classification problem (D, L) is transformed into a single-label classification problem $(D, \{\lambda_1, \lambda_2, \lambda_3\})$.

Step 2, the TESC algorithm is called (denoted by TESC $(D, \{\lambda_1, \lambda_2, \lambda_3\}, C_{TESC})$) aiming to produce a set of the clusters C_{TESC} in D with respect to the set of three labels $\{\lambda_1, \lambda_2, \lambda_3\}$. Denote $D_1, D_2,$ and D_3 the set of the clusters related to the macro label λ_1, λ_2, and λ_3, correspondingly.

Step 3, it is clear that all the clusters in D_1 are assigned to the same label set $L_1 \cup \{\lambda\}$, and the set of the clusters in D_1 is included in the classification model. There exist two situations for D_2 (D_3). In the case that all the documents in D_2 (D_3) are associated with the same label set, then D_2 (D_3) is treated as D_1. In the other situation, a recursion of MULTICS is called on D_2 (D_3).

In classification phase, kNN is applied to find out the nearest cluster, and the label set of the found cluster is returned as the label set for the new data instance.

2.4 The Most Influential Label on Label-Feature Relations

In MULTICS algorithm, the label λ is determined by using the label-document relation, which is represented by the label-document matrix LD. Let $N = |D|$; K be the cardinality of the current label set; LD be a $K \times N$ matrix, in which $LD_{i,j}$ is assigned to 1 if j^{th}-document is assigned to the ith-label. Then the most influential label λ is the first row having the biggest sum of the row's values.

This paper chooses the most influential label λ base on an assumption that the label constraints in the feature space are better than data space. Therefore, instead of using the label-document matrix LD, this paper uses the label-feature matrix LF. Documents in the current training dataset are represented as M-size feature vectors, where M is the number of features. The document-feature matrix DF is defined based on N of M-size feature vectors, i.e., DF is an $N \times M$ matrix. The label-feature matrix LF is the result of multiplying the matrix LD with the matrix DF, i.e., $LF = LD \times DF$. A weighted social network is built on LF for identifying the most influential label based on social network graph. We compare several feature representations of TF, TF.IDF and TF.IDF in combination with hidden topics (LDA) for documents.

2.5 The New Version of MULTICS

The new version of MULTICS includes **MULTICSLearn2** procedure for learning a classification model, and **MULTICSClassify2** procedure for classification. The MULTICSLearn algorithm [5] is modified into below MULTICSLearn2.

Procedure MULTICSLearn2 (D^L, D^U, L_1, L_2, C)

Input:

    D^L: the current labeled dataset;

    D^U: the current unlabeled dataset;

    L_1: the set of current predefined labels for all documents in D^L. Initially, L_1 is an empty set.

    L_2: the set of current possible labels in L excluding L_1 which could be assigned for instances in D^L, in other words $L_2=L \backslash L_1$. Initially, L_2 is the set of all labels L.

Output:
    C: a global collection of labeled document clusters. Initially, C is an empty set, C will be updated after each iteration calling **MULTICSLearn2**.

Method:
    $D \leftarrow (D^L, D^U)$;

Step 1. Determine the most influential label

1. Construct the label-document matrix LD with K rows and N columns, where $K=|L_2|$, $N=|D|$, and $LD[i, j] = 1$ if the jth document in D has i^{th} label in L_2, otherwise $LD[i, j] = 0$;
2. Construct the document-feature matrix DF with N rows and M columns, where $DF[i, j]$ is the value of the jth feature in the data representation of the ith document in D;
3. Compute the label-feature matrix LF: $LF \leftarrow LD \times DF$;
4. Compute the label-label matrix LL: $LL \leftarrow LF \times LF^T$;
5. Construct a weighted social network $WN=(V, E, W)$, where V is L_2, E is $L_2 \times L_2$, and $W[i, j]$ is the scalar product of ith-row and j^{th}-row of LL matrix. Determine the most influential label λ is the first label having the biggest degree in the social network WN.

Step 2. Transform the multi-label classification task into single-label classification tasks and apply the TESC algorithm

6. Set $D_1 \leftarrow \{d \in D^L| \text{ label } (d) = L_1 \cup \{\lambda\}\}$;
    $\forall d \in D_1$: $d \leftarrow \lambda_1$ (λ_1 is the first dummy label);
7. Set $D_2 \leftarrow \{d \in D^L \backslash D_1| \text{ label } (d) \supset L_1 \cup \{\lambda\}\}$;
    $\forall d \in D_2$: $d \leftarrow \lambda_2$ (λ_2 is the second dummy label);
8. $D_3 \leftarrow D^L \backslash \{D_1 \cup D_2\}$; $\forall d \in D_3$: $d \leftarrow \lambda_3$ (λ_3 is the third dummy label);
9. Call TESC (D, $\{\lambda_1, \lambda_2, \lambda_3\}$, C_{TESC}).

Step 3. Add local resulted clusters into the global cluster set and call recursions

10. $C_1 \leftarrow \{c \in C_{TESC} | \text{ label}(c) = \lambda_1\}$; $D_1 \leftarrow \{d \in D | \exists c \in C_1: d \in c\}$;
    $\forall c \in C_1$: label$(c) \leftarrow L_1 \cup \{\lambda\}$; $\forall d \in D_1$: label$(d) \leftarrow L_1 \cup \{\lambda\}$; $C \leftarrow C \cup C_1$;
11. $C_2 \leftarrow \{c \in C_{TESC} | \text{ label}(c) = \lambda_2\}$; $D_2 \leftarrow \{d \in D | \exists c \in C_2: d \in c\}$;
    if $(\forall (d_1, d_2) \in DL_2 \times DL_2 : \text{ label}(d_1)=\text{label}(d_2) = \sigma)$ then
    begin $\forall c \in C_2$: label$(c) \leftarrow \sigma$; $\forall d \in D_2$: label$(d) \leftarrow \sigma$; $C \leftarrow C \cup C_2$ end
    else call MULTICSLearn2 ($D_2 \cap D^L$, $D_2 \cap D^U$, $L_1 \cup \{\lambda\}$, $L_2 \backslash \{\lambda\}$, C);
12. $C_3 \leftarrow \{c \in C_{TESC} | \text{ label}(c) = \lambda_3\}$; $D_3 \leftarrow \{d \in D | \exists c \in C_3: d \in c\}$;
    if $(\forall (d_1, d_2) \in DL_3 \times DL_3 : \text{ label}(d_1)=\text{label}(d_2) = \sigma)$ then
    begin $\forall c \in C_3$: label$(c) \leftarrow \sigma$; $\forall d \in D_3$: label$(d) \leftarrow \sigma$; $C \leftarrow C \cup C_3$ end
    else call MULTICSLearn2 ($D_3 \cap D^L$, $D_3 \cap D^U$, L_1, $L_2 \backslash \{\lambda\}$, C);

The training and classification procedures are described as follows

Input:

L: the set of all labels;
D^L: the input labeled dataset,
D^U: the input unlabeled dataset.

Output:

C: a global collection of labeled document clusters.

Method:

$C \leftarrow \varnothing$;
Call MULTICSLearn2 ($D^L, D^U, \varnothing, L, C$)

Procedure MULTICSClassify2 (*C, d*)

Input:

C: the labeled document clusters, $C=\{c: \text{label}(c) \subseteq L, \forall c \in C, \forall d \in c: \text{label}(d)=\text{label}(c)\}$.
u: a new unlabeled document.

Output:

The label set $\sigma \subseteq L$

Method:

$D =\cup_{c \in C} c$; // The set of all labeled documents in all clusters
$d' \leftarrow \text{kNN} (D, u)$; // d' is the nearest neighbor of u
$\sigma \leftarrow \text{label}(d')$

3 Experiments and Result Evaluations

3.1 The Dataset

This paper uses two databases to perform experiments. Because of performing the semi-supervised learning, this paper uses a sub-dataset of our dataset of 1524 Vietnamese hotel reviews [5] and a sub-dataset of Enron Email Dataset of 1702 emails[1]. The experimental dataset *Hotel1* includes 520 Vietnamese hotel reviews and the experimental dataset *Enron1* includes 520 Enron emails. Table 1 describes characteristics of two datasets used in experiments. In the table, the terms of |S|, Dim(S), L(S), F(S), LCard(S), Lden(S), DL(S), PDL(S) have the same meaning as those in [11].

Table 1. Characteristics of two datasets used in experiments

Dataset	|S|	Dim(S)	L(S)	F(S)	LCard(S)	Lden(S)	DL(S)	PDL(S)
Enron1	520	1000	53	text	3.658	0.070	323	0.621
Hotel1	520	1266	5	text	1.250	0.250	24	0.046

---

[1] http://waikato.github.io/meka/

On each experimental dataset, for investigating the contribution of the labeled and unlabeled documents, we generate four labeled datasets with very small size (N) of 20, 40, 60 and 80 documents, and for each labeled datasets, four unlabeled datasets of N, N + 20, N + 40, N + 60 of other documents are used. We use the test dataset of 300 documents from the remaining part of each experimental dataset.

3.2 Experimental Scenarios

In experiments, both methods to find the most influential label, i.e., using label-document relation (denoted by LD) in [5], and the proposed method using label-feature relation (denoted by LF), are implemented. Moreover, this study considers four popular features for text representation including Binary (B), Term frequency (TF), TFIDF and the combination of TFIDF and LDA.

Since the dataset of Enron is formatted in the binary representation, it is represented in Binary, and Binary in combination with LDA, with the different number of hidden topics. The hotel review dataset is represented in TF, TFIDF, and TFIDF in combination with LDA, with the different number of hidden topic. We set $k = 1$ for kNN in MULTICSClassify2.

In more details, the experiments using LDA features will be taken in three different numbers of hidden topics to investigate the effects of this LDA to the proposed method, where the number of hidden topics is 15, 20 and 25.

Besides, to illustrate the new MULTICS algorithm in different labeled and unlabeled datasets, we made four groups of combination (named A_i where $i = 1, 2, 3, 4$) of each labeled dataset with N instances, and four unlabeled datasets with N, $N + 20$, $N + 40$ and $N + 60$ instances, where N is 20, 40, 60 and 80 in different experiments.

Totally, one hundred forty-four experimental cases are done in this study.

3.3 Experimental Results and Discussions

The results of the proposed methods on the Hotel1 review dataset are shown in Table 2. In the case of macro-averaging AUC merit, the best performance is the experiment using LD method in TF representation. However, in other representation methods of TFIDF and TFIDF in combination with LDA in different number of hidden topics, the method of LF seems to be better than the method of LD in most of the experiments. The performances of the system are improved when the size of labeled dataset is increased.

The results of the proposed methods on the Enron1 dataset are shown in Table 3. In two merits of macro-averaging AUC and micro-averaging AUC, the method of LF is better than the method of LD in most of experiments. The results are quite low due to the fact that the number of labels (53) is much larger than the size of labeled dataset (which has 20, 40, 60 and 80 documents) in relation with the size of testing dataset (300 documents).

Table 2. The experimental results on the Vietnamese review dataset Hotel1, where No is the number of labeled reviews, Av is the average number

No	TF		TF.IDF		TF.IDF. LDA15		TF.IDF. LDA20		TF.IDF. LDA25	
	LD	LF	LD	LF	LD	LF	LD	LF	LD	LF
Macro-averaging AUC (%)										
20	56.43	56.43	50.28	50.28	50.27	50.27	50.27	50.27	50.28	50.28
40	62.07	60.20	54.87	54.87	53.89	54.65	54.06	54.81	54.71	54.87
60	59.85	59.85	55.27	55.40	55.31	55.43	55.31	55.44	55.31	55.43
80	61.90	62.05	57.35	57.33	57.61	57.61	57.22	57.22	57.40	57.40
Av	60.06	59.63	54.45	54.47	54.27	54.49	54.22	54.43	54.42	54.50
Micro-averaging AUC (%)										
20	57.93	57.93	57.93	59.85	57.93	59.85	57.93	59.85	57.93	59.85
40	66.13	64.77	64.16	62.94	64.16	62.89	64.16	62.94	64.16	62.94
60	67.29	67.32	67.52	64.07	67.52	64.12	67.52	64.12	67.52	64.12
80	66.62	66.75	67.14	65.41	67.14	65.53	67.14	65.14	67.14	65.41
Av	64.49	64.19	64.19	63.07	64.19	63.10	64.19	63.01	64.19	63.08

Table 3. The experimental results on the email dataset Enron1, where No is the number of labeled emails, Av is the average number

No	Binary		Binary. LDA15		Binary. LDA20		Binary. LDA25	
	LD	LF	LD	LF	LD	LF	LD	LF
Macro-averaging AUC (%)								
20	52.72	52.54	53.12	53.10	53.11	53.09	53.11	53.11
40	52.87	53.21	52.82	53.03	52.53	52.90	52.72	52.91
60	52.70	53.43	52.96	53.02	53.09	53.23	52.85	53.03
80	53.66	53.55	53.25	53.22	53.78	53.52	53.35	53.15
Av	52.99	53.18	53.04	53.09	53.13	53.19	53.01	53.05
Micro-averaging AUC (%)								
20	64.84	66.85	65.34	65.46	65.42	65.54	65.42	65.51
40	61.62	61.69	55.00	58.96	55.32	58.55	55.78	58.56
60	60.28	61.78	57.72	58.61	57.99	59.25	57.24	58.72
80	62.23	62.70	59.31	59.88	59.96	60.40	59.01	60.16
Av	62.24	63.26	59.34	60.73	59.67	60.94	59.36	60.74

3.4 A Comparing MULTICS with Multi-label Supervised Learning Algorithms

A comparison of the proposed multi-label semi-supervised algorithm with some multi-label supervised learning semi-supervised algorithms, ten experimental cases of the proposed algorithm on the email dataset Enron [11] have been done. The remaining dataset D included 1402 data is divided into five subsets $D1$, $D2$, $D3$, $D4$, and $D5$. The sizes of $D1$, $D2$, $D3$, and $D4$ are 280, the size of $D5$ is 282. For each experiment, the union of two of the five subsets are chosen as the labeled dataset while the rest is chosen as the unlabeled one. Table 4 shows the experimental results by proposed semi-supervised learning method on the ten experimental cases (Exp. Case) and the average of ten results in Micro-averaging F1 (Micro_F1) and Macro-averaging AUC (AUC_-Macro) measures [9].

Table 4. Experimental results on the email dataset **Enron** in Micro-averaging F1 (Micro_F1) and Macro-averaging AUC (AUC_Macro) measures.

Exp. Case	Micro_F1 (%)		AUC_Macro (%)	
	LD	LF	LD	LF
#1	32.5459	34.7558	51.7145	51.6538
#2	34.8380	34.0329	52.1214	51.9761
#3	37.5332	36.6851	54.0866	53.2677
#4	32.8852	36.9295	52.0851	52.5600
#5	33.7913	36.9856	52.9042	53.3916
#6	35.9213	36.2865	53.6970	54.2096
#7	35.0482	33.4965	53.0291	52.8677
#8	32.1041	36.4706	53.8063	53.8270
#9	36.5011	34.7418	53.3392	53.3932
#10	31.7888	35.7849	53.3810	54.2419
Average	**34.29571**	**35.61692**	**53.01644**	**53.13886**

In comparison of the proposed method with the Multi-label kNN algorithm family on Micro-averaging F1 [16] (from 33.1% to 49.5%), with BR and RAKEL on Macro-averaging AUC [11] (57.9% and 59.6%, correspondingly), the results by proposed semi-supervised learning method is acceptable. Moreover, the output data of MULTICS with a high confidence may be used as new labeled data in the semi-supervised algorithm EM.

4   Related Works

Zhan and Zhang [8] focused on a semi-supervised learning approach for multi-label classifications. Label constraints were considered based on investigating the feature space. They proposed a co-training method named COINS (CO-training for INductive Semi-supervised multi-label learning), which includes two parts of the set of features. Two parts of feature space for two classification models are chosen based on a diversity maximization. The highlight of COINS is the supervision information to be communicated to the classification models, which are chosen based on pairwise ranking predictions. On the macro-averaging AUC measure, COINS is shown to have a higher performance than that of ECC, SMSE, TRAM, and iMLCU. Our paper considers label constraints based on using the label-feature matrix.

In [13], Zhang et al. also proposed solutions for label constraints based on investigating label-feature relations for multi-label classifications. The authors proposed a multi-label learning approach named MLFE (Multi-label Learning with Feature-induced labeling information Enrichment). The authors argued that the underlying structure of feature space is characterized by reconstructing relationships among training examples; then by calling MFLE, the structural information in the feature space is leveraged to enrich the labeling information of multi-label examples. Three steps of structural information discovery, labeling information enrichment, and predictive model induction are three main phases in MLFE. A weighted directed graph is constructed where the vertex set corresponds to the set of training instances, the set of directed edges connect any pair of instances with nonzero weight, and the weight matrix encodes the relationships among all training examples. With four evaluation measures (Ranking loss, Average precision, Macro-averaging F1, Micro-averaging F1), experiments on seven datasets (cal500, emotions, medical, llog, msra, image, scene, yeast) showed that MLFE excelled in many cases. In this paper, some document representations are used for the label-feature matrix, which is used for determining the most influential label.

5   Conclusions

This work provided a solution to determine the most influential label based on the label-feature matrix in the multi-label classification algorithm MULTICS [5], i.e., an upgraded version of MULTICS was described clearly. Experimental results showed that the upgraded MULTICS is useful, particularly with the ability to work with a very limited number of labeled examples. Moreover, the upgraded MULTICS can be used not only for text domains but also for any domains.

In this work, MULTICS was applied experimentally for only two datasets of our Vietnamese hotel review dataset and the Enron email dataset. More experiments on other datasets should be done. Social networks of labels [6] based on the label-feature matrix should be investigated for determining the most influential label.

References

1. Blei, D.M., Ng, A.Y., Jordan, M.I.: Latent dirichlet allocation. J. Mach. Learn. Res. **3**, 993–1022 (2003)
2. Blei, D.M.: Probabilistic topic models. Commun. ACM **55**(4), 77–84 (2012)
3. Guo, Y., Schuurmans, D.: Semi-supervised multi-label classification: a simultaneous large-margin, subspace learning approach. In: Flach, Peter A., De Bie, T., Cristianini, N. (eds.) ECML PKDD 2012. LNCS (LNAI), vol. 7524, pp. 355–370. Springer, Heidelberg (2012). https://doi.org/10.1007/978-3-642-33486-3_23
4. Kong, X., Ng, M.K., Zhou, Z.-H.: Transductive multilabel learning via label set propagation. IEEE Trans. Knowl. Data Eng. **25**(3), 704–719 (2013)
5. Pham, T.-N., Nguyen, V.-Q., Tran, V.-H., Nguyen, T.-T., Ha, Q.-T.: A semi-supervised multi-label classification framework with feature reduction and enrichment. J. Inf. Telecommun. **1**(2), 141–154 (2017)

6. Szymański, P., Kajdanowicz, T.: Is a data-driven approach still better than random choice with naive Bayes classifiers? In: Nguyen, N.T., Tojo, S., Nguyen, L.M., Trawiński, B. (eds.) ACIIDS 2017. LNCS (LNAI), vol. 10191, pp. 792–801. Springer, Cham (2017). https://doi. org/10.1007/978-3-319-54472-4_74
7. Wang, B., Tsotsos, J.: Dynamic label propagation for semi-supervised multi-class multi-label classification. Pattern Recognit. **52**, 75–84 (2016)
8. Zhan, W., Zhang, M.-L.: Inductive semi-supervised multi-label learning with co-training. In: KDD 2017, pp. 1305–1314 (2017)
9. Zhang, M.-L., Zhou, Z.-H.: A Review on multi-label learning algorithms. IEEE Trans. Knowl. Data Eng. **26**(8), 1819–1837 (2014)
10. Zhang, W., Tang, X., Yoshida, T.: TESC: an approach to text classification using semi-supervised clustering. Knowl. Based Syst. **75**, 152–160 (2015)
11. Zhang, M.-L., Lei, W.: LIFT: multi-label learning with label-specific features. IEEE Trans. Pattern Anal. Mach. Intell. **37**(1), 107–120 (2015)
12. Zhang, M.-L., Li, Y.-K., Liu, X.-Y., Geng, X.: Binary relevance for multi-label learning: an overview. Front. Comput. Sci. **12**(2), 191–202 (2018)
13. Zhang, Q.-W., Zhong, Y., Zhang, M.-L.: Feature-induced labeling information enrichment for multi-label learning. In: AAAI-18 (2018, in press)
14. Zhao, F., Guo, Y.: Semi-supervised multi-label learning with incomplete labels. In: IJCAI 2015, pp. 4062–4068 (2015)
15. Zhou, Z.-H., Zhang, M.-L.: Multi-label Learning. Encyclopedia of Machine Learning and Data Mining, pp. 875–881. Springer, Boston (2017). https://doi.org/10.1007/978-1-4899-7687-1
16. Reyes, O., Morell, C., Ventura, S.: Effective lazy learning algorithm based on a data gravitation model for multi-label learning. Inf. Sci. **340–341**, 159–174 (2016)

Sensor Networks and Internet of Things

A DC Programming Approach
for Worst-Case Secrecy Rate
Maximization Problem

Phuong Anh Nguyen[(✉)] and Hoai An Le Thi

Department Computer Science and Application, LGIPM,
University of Lorraine, 3 rue Augustin Fresnel, BP 45112,
57073 Metz Cedex 03, France
{phuong-anh.nguyen,hoai-an.le-thi}@univ-lorraine.fr

Abstract. This paper is concerned with the problem of secure transmission for amplify-and-forward multi-antenna relay systems in the presence of multiple eavesdroppers. Specifically, spatial beamforming and artificial noise broadcasting are chosen as the strategy for secure transmission with robustness against imperfect channel state information of the intended receiver and the eavesdroppers. In such a scenario, the objective is to maximize the worst-case secrecy rate while guaranteeing the transmit power constraint at the relay and the norm-bounded channel uncertainty. We reformulate the problem as a general DC (Difference-of-Convex functions) program (i.e. minimizing a DC function under DC constraints) and develop a very inexpensive DCA based algorithm for solving it. Numerical results illustrate the effectiveness of the proposed algorithm and its superiority versus the existing approach.

Keywords: DC programming · DCA · Physical layer security
Artificial noise · Beamforming · Relay networks

1 Introduction

Wireless networks are more vulnerable than wired networks due to their broadcast nature. Traditionally, information transmission is secured by encryption techniques on the application layer. This solution has disadvantages and could be completely broken with the rapid development of computational devices [1]. In recent years, physical layer security has emerged as an alternative or a complement to cryptosystems [2].

Physical layer security was first studied from an information-theoretic perspective of discrete memoryless channels by Wyner [3]. It proved that secure communication can be guaranteed only if the channel condition of legitimate receiver is better than that of the eavesdroppers. In 1978, Cheong and Hellman generalized Wyner's results [4], where the secrecy capacity was shown to be the difference between the capacity of the main channel and that of the eavesdropper channels.

© Springer Nature Switzerland AG 2018
N. T. Nguyen et al. (Eds.): ICCCI 2018, LNAI 11055, pp. 417–425, 2018.
https://doi.org/10.1007/978-3-319-98443-8_38

If the secrecy capacity downs to zero, the wireless channels become unreliable [5]. Fortunately, this can be overcome by taking advantage of multiple antenna beamforming techniques [6–8]. Recently, Goel and Negi [9] proposed an efficient beamforming strategy where artificial noise signal is added to the signal for confusing the eavesdroppers. The secrecy rate performance in the case of joint beamforming and artificial noise can be improved as compared to the conventional beamforming design [10].

So far, the main challenge of joint beamforming and artificial noise approaches comes from imperfect channel state information (CSI) assumption of legitimate receiver and eavesdroppers. The challenge was handled by [11], in which the problem was modeled as maximizing the worst-case secrecy of the system with multiple eavesdroppers under relay power constraint. It is a nonconvex optimization problem for which [11] proposed the so called SCA algorithm based on the semidefinite relaxation and the successive convex approximation technique.

It is well known that (see e.g. [18]) SCA is a special version of DCA, a power approach in nonconvex programming framework. The main feature of DCA is its flexibility in the choice of DC decomposition, and there are as many DCA as there are DC decompositions. Exploiting the nice effect of DC decomposition, we propose in this paper a new DCA scheme via a suitable DC decomposition for solving the same problem considered in [11]. It turns out that the convex subproblems in our DCA are simpler than the one in SCA. Numerical experiments show the efficiency and the superiority of the proposed DCA versus SCA in terms of both quality and rapidity.

The rest of this paper is organized as follows. The considered secrecy rate maximization problem is stated in Sect. 2 while the solution method based on DC programming and DCA is presented in Sect. 3. Numerical experiments are reported in Sect. 4, and finally, Sect. 5 concludes the paper.

For beginning, let us introduce some notations. Throughout this paper, boldface upper-case and lower-case letters denote matrices and vectors, respectively. We use $(.)^T, (.)^H$ and $\text{Tr}(.)$ to stand for the transpose, Hermitian and trace of a matrix, respectively. $\mathbf{I}_K$ denotes $K \times K$ identity matrix; $\otimes$ denotes the Kronecker product; $\| \cdot \|$ represents the Frobenius norm; $\succeq$ means the property of semidefinite; $\log(.)$ is taken to the base of 2; $x \sim \mathcal{CN}(\mu, \Omega)$ is a random vector following a complex circular Gaussian distribution with mean μ and covariance Ω.

2 Robust Joint Beamforming and Artificial Noise Design

In this section, let us briefly restate the problem formulated in [11].

A relay network comprises a transmitter, a receiver in presence of M eavesdroppers through a K-antenna relay. Especially, the CSI knowledge at the receiver and the eavesdroppers are imperfect. The system consists of two hops.

In the first hop, information is transmitted with the power P_s. The channel coefficient from source to relay is denoted by $\mathbf{f} \in \mathbb{C}^K$. In the second hop, the relay forwards the source information to receiver by a beamforming matrix

$\mathbf{W} \in \mathbb{C}^{K \times K}$ and then superimposes the artificial noise signal $\mathbf{v}_r \sim \mathcal{CN}(\mathbf{0}, \boldsymbol{\Phi})$ with $\boldsymbol{\Phi} \succeq \mathbf{0}$. The relay power is constrained by sum power P. The relay is assumed to have perfect CSI from source to the relay, but only knows imperfect CSI of legitimate receiver and eavesdroppers. Let $\mathbf{h} \in \mathbb{C}^{\mathbb{K}}$ and $\mathbf{g}_i \in \mathbb{C}^K$ be the channel coefficients from the relay to receiver and the ith eavesdropper, respectively as

$$\mathbf{h} = \overline{\mathbf{h}} + \Delta\mathbf{h}, \ \|\Delta\mathbf{h}\| \leqslant \epsilon_{\mathbf{h}}, \ \mathbf{g}_i = \overline{\mathbf{g}}_i + \Delta\mathbf{g}_i, \ \|\Delta\mathbf{g}_i\| \leqslant \epsilon_{\mathbf{g}_i}, \ i = 1, ..., M,$$

where $\overline{\mathbf{h}}$ and $\overline{\mathbf{g}}_i$ are the estimates of CSI at relay; $\epsilon_{\mathbf{h}}$ and $\epsilon_{\mathbf{g}_i}$ are the coefficient errors.

According to [11], the worst-case secrecy rate maximization (WCSRM) problem is reformulated as

$$\max_{\mathbf{F}, \mathbf{E}, \mathbf{X}, \boldsymbol{\Phi}, R_d, R_e, x_i, y_i, \gamma_i^l, \gamma_i^u} R_d - R_e, \tag{1}$$

$$\text{s.t. } \text{Tr}(\mathbf{F}) + \text{Tr}(\mathbf{X}) + \text{Tr}(\boldsymbol{\Phi}) \leqslant P, \tag{2}$$

$$\mathbf{F} = P_s \mathbf{f}^T \otimes \mathbf{I}_K \mathbf{X}(\mathbf{f}^T \otimes \mathbf{I}_K)^H, \tag{3}$$

$$\mathbf{E} = \sum_{i=1}^{k} \mathbf{E}_i \mathbf{X} \mathbf{E}_i^H + \boldsymbol{\Phi}, \tag{4}$$

$$\begin{bmatrix} -\gamma_0^u \mathbf{I}_K - \mathbf{F} & -\mathbf{F}\overline{\mathbf{h}} \\ -\overline{\mathbf{h}}^H \mathbf{F} & -\overline{\mathbf{h}}^H \mathbf{F}\overline{\mathbf{h}} + x_0 + \gamma_0^u \epsilon_{\mathbf{h}}^2 \end{bmatrix} \preceq 0, \tag{5}$$

$$\begin{bmatrix} -\gamma_0^l \mathbf{I}_K + \mathbf{E} & \mathbf{E}\overline{\mathbf{h}} \\ -\overline{\mathbf{h}}^H \mathbf{E} & 1 + \overline{\mathbf{h}}^H \mathbf{E}\overline{\mathbf{h}} - y_0 + \gamma_0^l \epsilon_{\mathbf{h}}^2 \end{bmatrix} \preceq 0, \tag{6}$$

$$\begin{bmatrix} -\gamma_i^u \mathbf{I}_K + \mathbf{F} & \mathbf{F}\overline{\mathbf{g}}_i \\ \overline{\mathbf{g}}_i^H \mathbf{F} & \overline{\mathbf{g}}_i^H \mathbf{F}\overline{\mathbf{g}}_i - x_i + \gamma_i^u \epsilon_{\mathbf{g}_i}^2 \end{bmatrix} \preceq 0, \tag{7}$$

$$\begin{bmatrix} -\gamma_i^l \mathbf{I}_K - \mathbf{E} & -\mathbf{E}\overline{\mathbf{g}}_i \\ \overline{\mathbf{g}}_i^H \mathbf{E} & -1 - \overline{\mathbf{g}}_i^H \mathbf{E}\overline{\mathbf{g}}_i + y_i + \gamma_i^l \epsilon_{\mathbf{g}_i}^2 \end{bmatrix} \preceq 0, \tag{8}$$

$$\mathbf{F} \succeq 0, \ \mathbf{E} \succeq 0, \ \mathbf{X} \succeq 0, \ \boldsymbol{\Phi} \succeq 0, \gamma_i^l \geq 0, \gamma_i^u \geq 0, i = 0, ..., M. \tag{9}$$

$$\frac{x_0}{y_0} \geqslant 2^{2R_d} - 1, \tag{10}$$

$$\frac{x_i}{y_i} \leqslant 2^{2R_e} - 1, \ i = 1, ..., M. \tag{11}$$

where $\mathbf{E}_i = \mathbf{e}_i^T \otimes \mathbf{I}_M$, $\mathbf{e}_i$ is a unit vector with ith entry being one.

In the following section, we investigate DC programming and DCA in the face of the nonconvex constraints (10), (11). DCA is an efficient method for coping with large-scale nonconvex problems.

3   Solution Method Based on DC Programming and DCA

Before developing a DCA based algorithm for solving (1), we give a short introduction of DC programming and DCA.

3.1   DC programming and DCA

DC programming and DCA [13–15,17,18] constitute the backbone of smooth/nonsmooth nonconvex programming and global optimization. They address the problem of minimizing a function f which is a difference of lower semicontinuous proper convex functions g, h on the whole space $\mathbb{R}^n$, namely

$$\alpha = \inf\{f(x) := g(x) - h(x) : x \in \mathbb{R}^n\}, \ (P_{dc}).$$

Such a function f is called DC function, and $g - h$, DC decomposition of f while g and h are DC components of f. The convex constraint $x \in C$ can be incorporated in the objective function of (P_{dc}) by adding the indicator function of C ($\chi_C(x) = 0$ if $x \in C$, $+\infty$ otherwise) to the first DC component g of the DC objective function f.

   The principle idea of DCA is quite simple: at the iteration k, DCA replaces h by its affine minorization (taking $y^k \in \partial h(x^k)$) and solves the resulting convex program

$$x^{k+1} \in \arg\min_{x \in \mathbb{R}^n}\{g(x) - h(x^k) - \langle x - x^k, y^k\rangle\}.$$

   The extension of DC programming and DCA was investigated for solving general DC programs with DC constraints [16] as follows:

$$\begin{aligned}
&\min_x f_0(x) \\
&\text{s.t. } f_i(x) \leqslant 0, \ \forall i = 1, ..., m, \ x \in C,
\end{aligned} \tag{12}$$

where $C \subseteq \mathbb{R}^n$ is a nonempty closed convex set; $f_i : \mathbb{R}^n \to \mathbb{R}, \ \forall i = 0, 1, .., m$, are DC functions, i.e., $f_i(x) = g_i(x) - h_i(x)$ with f_i and h_i being convex functions. It is apparent that this class of nonconvex programs is the most general in DC programming and as a consequence it is more challenging to deal with than standard DC programs. Two approaches for general DC programs were proposed in [14,15] to overcome the difficulty caused by the nonconvexity of the constraints. Both approaches are built on the main idea of the philosophy of DC programming and DCA, that is approximating (12) by a sequence of convex programs. The former was based on penalty techniques in DC programming while the latter was relied on the convex inner approximation method. Because we use the idea of the second approach to solve the problem mentioned in this article, we presented herein its main scheme.

   General DCA scheme consists of linearizing the concave part of DC decompositions of all DC objective function and DC constraints by $y_i^k \in \partial h_i(x^k)$, $\forall i = 1, ..., m$ and computing x^{k+1} by solving the following convex problem

$$\begin{aligned}
&\min_x g_0(x) - \langle y_0^k, x\rangle \\
&\text{s.t. } g_i(x) - h_i(x^k) - \langle y_i^k, x - x^k\rangle \leqslant 0, \ \forall i = 1, ..., m, \ x \in C.
\end{aligned} \tag{13}$$

3.2 DC Programming and DCA for the Problem (1)

In this subsection, we reformulate the nonconvex constraints (10) and (11) as DC constraints and propose an approach based on DC programming and DCA to handle the problem (1).

It is easy to see that the inequality constraints (10d), (10f) in [11] hold with equalities at the optimal solution $(\mathbf{W}^*, \boldsymbol{\Phi}^*, y_i^*)$. Thus,

$$\max_{\|\Delta\mathbf{h}\|\leqslant\epsilon_\mathbf{h}} 1+(\overline{\mathbf{h}}+\Delta\mathbf{h})^H\mathbf{W}^*(\mathbf{W}^*)^H(\overline{\mathbf{h}}+\Delta\mathbf{h})+(\overline{\mathbf{h}}+\Delta\mathbf{h})^H\boldsymbol{\Phi}^*(\overline{\mathbf{h}}+\Delta\mathbf{h}) = y_0^*, \quad (14)$$

$$\min_{\|\Delta g_i\|\leqslant\epsilon_{g_i}} 1+(\overline{\mathbf{g}}_i+\Delta\mathbf{g}_i)^H\mathbf{W}^*(\mathbf{W}^*)^H(\overline{\mathbf{g}}_i+\Delta\mathbf{g}_i)+(\overline{\mathbf{g}}_i+\Delta\mathbf{g}_i)^H\boldsymbol{\Phi}^*(\overline{\mathbf{g}}_i+\Delta\mathbf{g}_i) = y_i^*,$$
$$(15)$$

for $i = 1, ..., M$. Since the left side of (14), (15) are positive, we add to the problem (1) the constraints $y_i \in \mathbb{R}^+$, $i = 0, ..., M$.

For convenience, we denote $\mathbf{z}_0 = (x_0, y_0, R_d)$, $\mathbf{z}_i = (x_i, y_i, R_e)$ with $i = 1, ..., M$.

The nonconvex constraint (10) is expressed as a DC constraint

$$F_0(\mathbf{z}_0) = G_0(\mathbf{z}_0) - H_0(\mathbf{z}_0) \leqslant 0,$$

where

$$G_0(\mathbf{z}_0) = \frac{1}{2}\rho_0(x_0^2 + y_0^2 + R_d^2),$$

$$H_0(\mathbf{z}_0) = \frac{1}{2}\rho_0(x_0^2 + y_0^2 + R_d^2) + x_0 + y_0 - y_0 2^{2R_d}$$

are convex functions for ρ_0 being determined in Proposition 1. Similarly, the nonconvex constrainst (11) are reformulated as DC constraints

$$F_i(\mathbf{z}_i) = G_i(\mathbf{z}_i) - H_i(\mathbf{z}_i) \leqslant 0, \; i = 1, ..., M,$$

where

$$G_i(\mathbf{z}_i) = \frac{1}{2}\rho_i(x_i^2 + y_i^2 + R_e^2),$$

$$H_i(\mathbf{z}_i) = \frac{1}{2}\rho_i(x_i^2 + y_i^2 + R_e^2) - x_i - y_i + y_i 2^{2R_e}$$

are convex functions for ρ_i being determined in Proposition 1.

Proposition 1. *Denote* $D_0 = P(\|\overline{\mathbf{h}}\| + \epsilon_\mathbf{h})^2 + 1$ *and* $D_i = P(\|\overline{\mathbf{g}}_i\| + \epsilon_{g_i})^2 + 1$ *for* $i = 1, ..., M$. *The following statements hold*

(a) *If* $\rho_0 \geqslant 2(\ln 2)D_0 + 2(\ln 2)^2 D_0^2$, *then both* $G_0(\mathbf{z}_0), H_0(\mathbf{z}_0)$ *are convex.*
(b) *If* $\rho_i \geqslant 2(\ln 2)D_0 + 2(\ln 2)^2 D_0 D_i$, *then both* $G_i(\mathbf{z}_i), H_i(\mathbf{z}_i)$ *are convex for* $i = 1, ..., M$.

From the Proposition 1, we can choose $\rho_0 = 2(\ln 2)D_0 + 2(\ln 2)^2 D_0^2$ and $\rho_i = 2(\ln 2)D_0 + 2(\ln 2)^2 D_0 D_i$, $i = 1, ...M$. Hence, we obtain the following DC formulation of the problem (1)

$$\min_{\mathbf{F},\mathbf{E},\mathbf{X},\boldsymbol{\Phi},R_d,R_e,x_i,y_i,\gamma_i^l,\gamma_i^u} R_e - R_d, \qquad (16)$$

$$\text{s.t.} \qquad (2) - (9),$$
$$G_i(\mathbf{z}_i) - H_i(\mathbf{z}_i) \leqslant 0, \quad i = 0, ..., M.$$

Following the general DCA scheme, a DCA based algorithm for solving the problem (16) can be described in Algorithm 1.

Algorithm 1. DCA based algorithm for solving the problem (16)

Initialization: Let $(x_i^0, y_i^0, R_e^0, R_d^0)$ be an initial point, $i = 0, ...M, 0 \leftarrow k$.
repeat
    1. For $i = 1, ..., M$, compute

$$\nabla H_0(\mathbf{z}_0^k) = (\rho_0 x_0^k + 1, \ \rho_0 y_0^k + 1 - 2^{2R_d^k}, \ \rho_0 R_d^k - 2\ln 2 y_0^k 2^{2R_d^k}),$$
$$\nabla H_i(\mathbf{z}_i^k) = (\rho_i x_i^k - 1, \ \rho_i y_i^k - 1 + 2^{2R_e^k}, \ \rho_i R_e^k + 2\ln 2 y_i^k 2^{2R_e^k}),$$

    2. Compute $(x_i^{k+1}, y_i^{k+1}, R_e^{k+1}, R_d^{k+1})$, $i = 0, ..., M$ by solving the problem

$$\min_{\mathbf{F},\mathbf{E},\mathbf{X},\boldsymbol{\Phi},R_d,R_e,x_i,y_i,\gamma_i^l,\gamma_i^u} R_e - R_d, \qquad (17)$$

    s.t. $\qquad (2) - (9),$

$$G_i(\mathbf{z}_i) - H_i(\mathbf{z}_i^k) - \langle \nabla H_i(\mathbf{z}_i^k), \mathbf{z}_i - \mathbf{z}_i^k \rangle \leqslant 0,$$

    3. $k \leftarrow k + 1$,
until stopping criterion.

---

4   Numerical Experiments

The numerical experiments aim to evaluate the performance of the proposed algorithm and to compare it with the existing method, SCA [11].

All algorithms were implemented in Matlab R2016b, and executed on a PC Intel® Core™2 Quad (2.83 GHz) of 4 GB RAM with the same datasets and initial point. The CVX toolbox [19] is used to solve the subproblem (17). All of the channel coefficients are independently generated following $\mathcal{CN}(0, 1)$. The number of antennas at relay is 4. The transmit power at the source is $P_s = 20\,\text{dB}$. We assume the coefficient error $\epsilon = \epsilon_h = \epsilon_{g_i}$, $i = 1, ..., M$. The stopping criterion for two algorithms is $|R_d^k - R_e^k - R_d^{k-1} + R_e^{k-1}| < 10^{-3}$.

Table 1. System secrecy rate (SSR) versus the number of eavesdroppers

M	DCA			SCA		
	SSR-AVER	SSR-Best	CPU(s)	SSR-AVER	SSR-Best	CPU(s)
10	**2.2624**	**3.8316**	**23**	2.0684	2.7052	71
50	**1.7543**	**2.4981**	**85**	1.6543	2.1846	476
70	**1.6094**	**2.1346**	**115**	1.6043	2.1034	700
90	**1.5887**	**2.1307**	**137**	1.5804	2.0989	957

We are interested in the effect of the number of eavesdroppers, of the relay power and of the coefficient error in the systems. For each value of these parameters the algorithms are tested on 10 independent channel relizations and the average values of the secrecy rate (SSR-AVER) and the best value (SSR-Best) are reported in Table 1 (when $P = 20$ dB, $\epsilon = 0.1$, with different number of eavesdroppers), Table 2 (when $M = 50$, $\epsilon = 0.1$ and the relay power varies) and Table 3 (when $M = 50, P = 20$ dB and the coefficient error varies).

Table 2. System secrecy rate (SSR) versus the relay power

$P(dB)$	DCA			SCA		
	SSR-AVER	SSR-Best	CPU(s)	SSR-AVER	SSR-Best	CPU(s)
5	**0.9420**	**1.4889**	**65**	0.9419	1.4867	346
10	**1.2930**	**1.8879**	**80**	1.2814	1.8377	448
15	**1.5249**	**2.1786**	**84**	1.4897	2.0433	446
20	**1.7382**	**2.4981**	**84**	1.6374	2.1850	473

Table 3. System secrecy rate (SSR) versus the coefficient error

ϵ	DCA			SCA		
	SSR-AVER	SSR-Best	CPU(s)	SSR-AVER	SSR-Best	CPU(s)
0.01	**1.9358**	**2.5569**	**110**	1.9063	2.3527	550
0.1	**1.8115**	**2.4981**	**84**	1.7023	2.1850	473
0.2	**1.4735**	**2.2263**	**87**	1.3953	1.9347	421
0.3	**1.1469**	**1.9323**	**95**	1.0464	1.6426	414

It is seen from Table 1 that the larger the number of eavesdroppers is, the smaller the worst-case secrecy rate is. In terms of computation time, we see that DCA is faster than SCA - the ratio of gain of DCA versus SCA varies from 3 times to 7 times. In Table 2, as we expected, increasing the relay power can yield a higher secrecy rate. Table 3 shows the sensitivity of the secrecy rate to the CSI error.

5   Conclusions

We have investigated DC programming and DCA to solve the secrecy rate optimization problem under the relay power constraint in the presence of imperfect CSI of legitimate receiver and eavesdroppers. Exploiting the nice effect of DC decomposition, we have proposed suitable DC decompositions for the objective function and the DC constraints that results in a very inexpensive DCA scheme. Numerical results have confirmed the superiority of DCA compared to SCA. Works on DCA for other optimization models in Physical layer security are in progress.

References

1. Kumar, S., Paar, C., Pelzl, J., Pfeiffer, G., Ruppp, A., Schimmler, M.: How to break DES for euro 8,980. In: Workshop on Special purpose Hardware for Attacking Cryptographic Systems – SHARCS 2006, Colgne, Germany (2006)
2. Li, Q., Song, H., Huang, K.: Achieving secure transmission with equivalent multiplicative noise in MISO wiretap channels. IEEE Commun. Lett. **17**(5), 892–895 (2013)
3. Wyner, A.D.: The wire-tap channel. Bell Syst. Tech. J. **54**(8), 1355–1387 (1975)
4. Leung-Yan-Cheong, S., Hellman, M.E.: The Gaussian wiretap channel. IEEE Trans. Inf. Theory **24**(7), 451–456 (1978)
5. Zou, Y., Wang, X., Shen, W.: Eavesdropping attack in collaborative wireless networks: security protocols and intercept behavior. In: Proceedings of the 17th IEEE International Conference on Computer Supported Cooperative Work in Design, Whistler, Canada, pp. 704–709 (2013)
6. Zhang, J., Gursoy, M.: Collaborative relay beamforming for secrecy. In: Proceedings of the IEEE Wireless Communication Symposium (ICC), Cape Town, South Africa, pp. 1–5 (2010)
7. Jeong, C., Kim, I., Dong, K.: Joint secure beamforming design at the source and the relay for an amplify-and-forward MIMO untrusted relay system. IEEE Trans. Signal Process. **60**(1), 310–325 (2012)
8. Mukherjee, A., Swindlehurst, A.: Robust beamforming for security in MIMO wiretap channels with imperfect CSI. IEEE Trans. Signal Process. **59**(1), 351–361 (2011)
9. Goel, S., Negi, R.: Guaranteeing secrecy using artificial noise. IEEE. Trans. Wireless Commun. **7**(6), 2180–2189 (2008)
10. Li, Q., Yang, Y., Ma, W.K., Lin, M., Ge, J., Lin, J.: Robust cooperative beamforming and artificial noise design for physical-layer secrecy in AF multi-antenna multi-relay networks. IEEE Trans. Signal Process. **63**(1), 206–220 (2015)
11. Zhang, C., Gao, H., Liu, H., Lv, T.: Robust beamforming and jamming for secure AF relay networks with multiple eavesdroppers. In: IEEE Military Communications Conference (MILCOM), pp. 495–500 (2014)
12. Zhang, L., Jin, L., Luo, W., Tang, Y., Yu, D.: Robust joint beamforming and artificial noise design for amplify-and-forward multi-antenna relay systems. In: IEEE Speech and Signal Processing (ICASSP), pp. 1732–1736 (2015)
13. Pham Dinh, T., Le Thi, H.A.: Convex analysis approach to DC programming: theory, algorithms and applications. Acta Mathematica Vietnamica **22**, 289–355 (1997)

14. Pham Dinh, T., Le Thi, H.A.: DC optimization algorithms for solving the trust region subproblem. SIAM J. Optim. **8**, 476–505 (1998)
15. Pham Dinh, T., Le Thi, H.A.: The DC (difference of convex functions) programming and DCA revisited with DC models of real world non convex optimization problems. Ann. Oper. Res. **133**, 23–46 (2005)
16. Le Thi, H.A., Huynh, V.N., Pham Dinh, T.: DC Programming and DCA for General DC Programs. In: van Do T., Thi H., Nguyen N. (eds) Advanced Computational Methods for Knowledge Engineering. Advances in Intelligent Systems and Computing, pp. 15–35 (2014)
17. Pham Dinh, T., Le Thi, H.A.: Recent advances in DC programming and DCA. Trans. Comput. Intell. **13**, 1–37 (2014)
18. Le Thi, H.A., Pham Dinh, T.: DC programming and DCA: thirty years of developments. Math. Program. **169**(1), 5–68 (2018). Special Issue: DC Programming - Theory, Algorithms and Applications
19. Grant, M., Boyd, S: CVX: Matlab software for disciplined convex programming, version 2.0 (2012). http://cvxr.com/cvx

System for Detailed Monitoring of Dog's Vital Functions

David Sec[1], Jan Matyska[1], Blanka Klimova[1], Richard Cimler[2(✉)],
Jitka Kuhnova[2], and Filip Studnicka[2]

[1] Faculty of Informatics and Management, University of Hradec Kralove,
Hradec Kralove, Czech Republic
david.sec@uhk.cz
[2] Faculty of Science, University of Hradec Kralove, Hradec Kralove, Czech Republic
richard.cimler@uhk.cz

Abstract. Epilepsy is the most common neurological disorder, affecting 0.6% to 0.75% of dogs. However, it is quite difficult to recognize the start of epileptic seizures. Consequently, the purpose of this article is to explore the devices that may be able to help detect epileptic seizures in dogs, including a discussion of their benefits and limitations.

We have designed a new solution because there are no suitable commercial devices or systems that can detect epileptic seizures in dogs, neither is there a solution for a potential detailed analysis of a dog's vital functions. The vision for the future research is to use the data obtained from the created system for monitoring epileptic seizure in dogs.

Several commercial sensors have been compared to determine their ability to monitor vital functions, focused on possible monitoring of epileptic seizures. Our system consists of a wearable sensor, a base station running Windows IoT, and a cloud server. This system enables us to monitor breath frequency and heartbeat, which might be used to detect an epileptic seizure.

Keywords: Epileptic seizures · Detection · Sensors · Heart beat
Breath frequency · Dogs

1 Introduction

Epilepsy is the most common neurological disorder, affecting 0.6% to 0.75% of dogs [16]. Generally, epilepsy is a lifelong disease. A seizure occurs when there is abnormal electrical activity in the brain, which leads to sudden but short-lived changes in a dog's behavior and/or movement. Some dogs may be more predisposed to epilepsy than others and their prevalence may be higher than others [1]. A dog is diagnosed with epilepsy if it has at least two unprovoked epileptic seizures more than 24 h apart [1]. These seizures can last for a few seconds or they can last for a couple of minutes [13]. In most dogs, recurrent

© Springer Nature Switzerland AG 2018
N. T. Nguyen et al. (Eds.): ICCCI 2018, LNAI 11055, pp. 426–435, 2018.
https://doi.org/10.1007/978-3-319-98443-8_39

seizures have no identifiable underlying cause [12]. The key characteristics of an epileptic seizure are as follows: a loss of voluntary control, often seen with convulsions (jerking or shaking movements and muscle twitching); irregular attacks that start and finish very suddenly; and, attacks that appear very similar each time and which have a repetitive clinical pattern [1].

It is quite difficult to recognize the beginnings of seizure attacks in dogs. Current research talks about triggers such as stress, tiredness, food or medication. Epilepsy in dogs cannot be cured, it can only be maintained by antiepileptic drug (AED) therapy, which has almost no side-effects. Therefore, detecting epileptic seizures before they break out is of considerable importance because it can help both owners and their dogs to prevent distress, injury, or even death among affected dogs. Consequently, it can contribute to the social, emotional, and physical improvement of the dog's quality of life.

The purpose of this article is to explore the range of available devices and to design a system that may help to monitor the dog's vital functions. In future research, we aim to detect epileptic seizures in dogs, which may help to improve the lives of both the dog and their owner.

The rest of this paper is structured as follows. The second section will review the current commercial solutions and different types of sensors. The third section describes the system's requirements. The fourth section presents a design solution, which consists of a three-layer system. The final section will analyse the data obtained from measurements taken of a dog.

2   Current Sensor Solutions

We have studied several systems for monitoring vital functions in humans and dogs, with a focus on the detection of epileptic seizures. Our literature review shows that there are currently several systems that can detect epileptic seizures in human beings; however, there are none for dogs [14]. Several commercial solutions are available, which consist of sensors and a gateway to aggregate and process the data [2,8,10]. Although they are not designed for pets, some sensors are available that can be used in our system. In addition, we will create our own algorithm to detect epileptic seizures [3–7,9].

2.1   Commercial Solutions for Monitoring Vital Functions in Dogs

The following section will review the systems that are able to monitor a dog's vital functions. These systems are able to monitor different vital functions, although none of them is able to monitor epileptic seizures.

PetPace Smart Collar [10]. One of the existing solutions is a PetPace Smart Collar. Although this manufacturer actually sells two different variants with different types of functions, the hardware is the same for both. The architecture of this system is divided into three layers. The first, lowest layer contains a smart sensor attached to the dog's collar. This sensor provides actual information about the dog's activity to the middle layer. This middle layer is a small base station, which communicates with a sensor on a collar and makes aggregations of this data. The collected data are sent to the highest, cloud layer on the Internet using a standard RJ-45 Ethernet 10/100 BASE-T interface, which allows the user to access the data using a web interface or a mobile application. How the sensor connects with the base station is not specified; however, it is stated that the transmission is ongoing within a range of 1000 ft. This indicates that a proprietary radio transmission or WiFi connection is being used.

The sensor on the collar contains an accelerometer and a built-in temperature sensor, which records activities such as running, sleeping, and the number of steps taken. From these values, it is also calculates the dog's Health Index, Activity Index, and Burned Calories. However, the fundamental problem is that the base station aggregates the data into sequences that are then stored on the cloud. There are several differences between the Basic and Pro variants. The first basic version is intended for home use. It aggregates the measured data after 30 min and then sends them for evaluation to the cloud service. In the Pro version, the user can select aggregation intervals from 2, 15, or 30 min. Compared to the first version, it also offers the possibility to monitor multiple collars using one base station. It also offers Intense monitoring, where data are aggregated every 2 min. In the Pro version, it is possible to select the sampling frequencies and set custom trigger values, which are used for alarm reporting.

PetPace is also the only device which promises to monitor epileptic seizures. However, this function is actually part of an ongoing research study, which is currently able to monitor heartbeat, activity, breath and temperature changes.

FitBark 2 [2]. The second device that is designed to monitor a dog's activity is FitBark 2. This device is very similar to the PetPace collar. Although Fit-Bark also has a three layer architecture and data aggregations, it does differ from PetPace. The first difference is that FitBark use Bluetooth Low Energy to make the connection with a base station. This technology promises lower energy consumption during data transmission and it extends the lifetime of the battery that is contained in the collar. Any smartphone, with Android or iOS, can be used as a base station. FitBark provides features such as monitoring the dog's physical activity and giving information about the total number of steps taken, the distance walked, or the calories burned.

Scollar [8]. The third device, Scollar, is another dog tracking solution that provides information about the dog's activity and which offers tracking functions. This system contains a base station and it includes a WiFi connection to the Internet. This base station can, as in the previous solution, be supplied by a smartphone. Data collection and transfer is the same as in the previous devices. The main difference is that Scollar is the only device that provides a GPS sensor and dog tracking service. The collar also contains a temperature sensor and integrated battery, with an estimated lifetime from 45 up to 60 days. However, this system cannot monitor any of the dog's vital functions and it cannot be used for seizure measurement.

2.2   Sensors for Monitoring Activity and Vital Data

None of these commercial solutions are suitable for monitoring detailed vital functions and none of them are able to detect epileptic seizures in dogs.

The data provided by these companies are not able to be used in a detailed study of the dog's vital functions because of its generality. It is not possible to get the raw data, only the aggregated data is available.

In the next section, we will focus on the various sensors that are currently available on the market. These sensors will be evaluated with respect to usability for monitoring vital functions and to detect epileptic seizures in dogs.

Microsoft Band [5]. Microsoft's Band is a smart wristband device that is used for human fitness and activity monitoring. This wristband contains an accelerometer with sampling frequencies up to 62 Hz. The user can access this data using Microsoft SDK. Unfortunately, it is not possible to read the sensor data at a sufficient frequency for measurement purposes. In 2017, Microsoft withdrew the SDK for its Band bracelet.

Oblu [6]. Oblu are a small startup company that is developing a universal sensing platform for IoT. This development kit contains an accelerometer and gyroscope, with sampling frequencies up to 1000 Hz. This board also contains many GPIO pins that are used to attach other sensors. This board communicates using Bluetooth Low Energy, version 4.1. This board does not contain an integrated battery; therefore, power needs to be supplied to the board needs from an eternal power source or battery pack. The main disadvantage of this board is that is has a power consumption of about 100 mA, which would place high demands on the capacity of the battery and would also increase the weight of the device. Another weaknesses of this solution is that it as a higher price, which starts at $ 112. Finally, this startup company may be unable to provide long-term support and equipment production.

Flora [3]. Flora is a representation of many Arduino compatible devices. This board only contains a micro-controller and programming circuits, without using any batteries and sensors. All of the sensors need to be connected to GPIO around this circuit. This board is particularly impressive in its small dimensions and it is designed for use in portable electronics. However, the ATMEGA32U4-AU chip that it uses does not provide sufficient power for the required sampling frequencies.

GENEActiv Wireless [7]. This company offers just four variations of bracelets. All variants include an accelerometer and a gyroscope, but only the

GENEActiv Wireless version is equipped with wireless data transmission via a Bluetooth interface to make the wireless data transfer from the bracelet. The built-in three-axis gyroscope allows readings up to 1000 Hz. The only limitation of this device is its high cost, the limited capabilities of developing its own SW and, above all, the very low battery life, of up to 8 h, due to the required frame rate and transfer technology limits.

Xiaoimi MiBand 2 [4]. This is the cheapest variant of all of the tested devices. This fitness tracker contains an accelerometer and notification LEDs. With the original manufacturer's firmware, this device only provides aggregated data. It could, however, be modified to provide realtime data. During testing, the data reading frequency was very low (around 10 Hz) which is too low to measure any health activity.

MbientLab MetaWear [9]. This is a smart sensor module that contains an accelerometer, gyroscope barometer, thermometer, and special mathematical preprocessor to combine data from each sensor integrated on the board. This device can measure with a frequency of up to 1.5 kHz. However, it has a real-time data stream that allows it to send upto 100 samples per second. This value is the minimal threshold for successful heartbeat data analysis. This module also has an integrated 100 mAh battery, with an estimated lifetime from 2 to 4 days (Table 1).

Table 1. Overview of individual sensor parameters.

	Microsoft BAND	Oblu	Flora	GENEActiv	MiBand2	MetaWear
Price [$]	35	112	20	264	35	82
Accelerometer	YES	YES	External	YES	YES	YES
Gyroscope	YES	YES	External	YES	NO	YES
Sample rate [Hz]	62	150		200	10	100
Battery cap. [mAh]	Li-ion 200	—	—	Lipol	70	100
Estimated lifetime	5–10	1	—	0.3	15–30	2–4
Dimension [mm]	40.0	40.0	44.0	43.0	40.3	35.0
	19.0	29.7	2.0	40.0	15.7	29.0
	8.7	21.6		13.0	10.5	10.0
Communication	BLE	BT	External	RT	BLE	BLE
Available SDK	NO	YES	YES	NO	NO	ANO
Sustainability	NO	NO	NO	YES	YES	YES

3   Measurement Requirements

High precision beat-to-beat heart monitoring is possible using a hollow ECG (Holter) and breathing can be monitored using a chest strap. However, both of these methods are very limiting for a dog and long-term portable realization is also very difficult. Another method of measurement is to use a ballistocardiographic signal to sense heart function. This signal is the result of mechanical micro-movements induced by cardiac activity. This signal can be measured by extremely sensitive accelerometers or gyroscopes with a high signal-to-noise ratio. It is also possible to analyze other activities, such as the dog's breathing and physical activity.

The method based on the articles by [15, 17] examines changes, especially changes in sleep, where a number of seizures occurs. When the dog is asleep, the sensors also monitor the lowest degree of background noise (where noise is also a dog's movement, for example). The ballistocardiography method is especially effective during sleep and it will allow us to examine in detail the moments preceding seizures. With this information, it is possible to predict seizures, even in a waking state.

To measure all of these values with the necessary precision, we need to collect all data with minimal sampling frequencies. These frequencies are defined by the frequency of measured value itself. The Nyquist–Shannon sampling theorem [18] defines this sample rate as twice or more of the original sampled signal. Heart rate is the most important value to monitor. Typical heart frequencies are below 120 bps (beats per second). The minimal sampling frequency is 5 Hz. However, this sample rate will still only be an average value (e.g., the number of beats per minute–the average in 60 s). Although this is sufficient for fitness monitoring in healthy dogs, it is completely unusable for epileptic detection. To predict the accuracy of individual heartbeat measurements, which is also called beat-to-beat precision, we need to be able to detect every single heartbeat with sufficient precision (about 10 ms). Because the harmonic frequencies in the mechanical signal from the heart contain waves of the order of magnitude of 20 ms, we have to reach a scanning frequency at least a resolution of 10 ms; hence, the equal sampling frequency is 100 Hz.

Many methods are available to extract the heart and breath rate from a measured signal, which starts with simple frequency analysis methods and includes complex signal processing and filtering algorithms. Because of the need for a highly accurate determination of heart rate variability, we decided to use methods based on the following publications [15, 17]. Autocorrelation of the measured signal is used to extract variability. Although this method is partly reminiscent of pattern recognition, it only focuses on the time base of the signal and it neglects minor variations, which is ideal for our case. An analogous method for extracting breathing information can be used but it is necessary to clean the input signal with appropriate frequency filters from the components related to the cardiac function.

4   Provided Solution

As was mentioned previously, there is currently no suitable solution available on the market. Consequently, we have designed our system using sensors from MetaWear, which are provided by the MbientLab company. Our system consist of a MetaWear sensor, a gateway, and cloud storage—see Fig. 1.

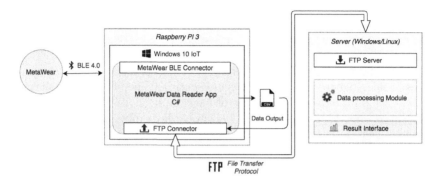

Fig. 1. System overview

For the initial phase of testing, Windows IoT running on a Raspberry Pi3 prototyping board has been used as a data collecting central hub. All of the devices in range are automatically connected immediately after an initial handshake with this hub and they then start to send data. All of the measured data are saved into a file on a SD card and every 10,000 samples these data are sent to an evaluation server via an integrated FTP connection. This system uses Windows 10 IoT, which has some limitations. The most serious deficiency is that it is impossible to terminate the connection with a BT device, which was discovered during the development of this device. (Microsoft's official documentation is "not yet implemented" [11].) Another major problem is the inability to establish a direct FTP connection for the file upload. Therefore, it was necessary to make a connection through WebSocket as an alternative. In addition, the integrated BT chip on the Raspberry Pi prototyping board has an effective signal range of just a few meters. The BT and WiFi network is operated by the same integrated circuit chip; therefore, it is almost impossible to send measured data over WiFi in realtime and all of the data has to be sent by a wired Ethernet interface. To start recording, it is enough to connect the gateway to an Ethernet network and plug in a 5 V power adapter. The rest of the connection process between the network and device starts automatically after device boot.

These limitations are major reasons to develop a similar application based on another platform. As the most suitable solution, Android 8 has been chosen. The SDK for this system is more frequently updated due to its expansion. Even so, Android has its flaws. The most basic flaws include the frequency of sending and receiving BT LE communication packets. When using Android 5.x, it is possible to set the shortest time between packets to about 7.5 ms, while on newer versions it is only about 15 ms due to internal battery saver politics. This also partially limits the speed of data recording from the sensor module. This solution can provide a more stable connection between the phone and BT sensor but it requires more interaction with the user.

5   Evaluation of the Results

To validate its functionality, the developed system was first tested on a healthy dog. The dog was a 3 year old German Shepperd, with a sensor attached to his collar. Measurements were provided in an open air real environment. The gateway was located at the dog's shed and it was powered by a 10000 mAh power bank. The data was saved to a SD card, where it was downloaded manually and uploaded to FTP due to the absence of a wired Internet connection in the shed.

Several hours of recording have been processed to evaluate the measured data. The typical patterns of measured and all of the calculated variables are shown in the following figures. All of the data has been processed using Matlab software.

Figure 2 demonstrates data output from a three-axis gyroscope. These data are not edited. The black arrows mark some of the individual strokes of the heart. From the measured waveforms, it is possible to determine individual strokes of the heart. These are formed by certain waves, which are also called ballistocardiographic waves, that are induced by the stroke of the heart and the movement of blood in the large arteries. We are particularly interested in the time distance of individual heart strokes. Because of their statistics, it might be possible to impose an onset epileptic seizure. For this particular dog, it is interesting to observe a certain type of arrhythmia where the distances of individual heartbeats periodically change (they grow and then fall, see Fig. 2). This is called sinus respiratory arrhythmia, which is caused by breathing and occurs especially in

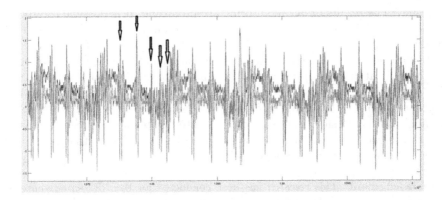

Fig. 2. Typical measured signal with three axis gyroscope—raw data from a sensor detecting heartbeats

young individuals. Arrhythmia is natural and it does not pose any health risks for this dog.

The last Fig. 3 shows a breathing curve. This curve is obtained by filtering raw data using a low pass filter with defined parameters. From the measured curve, it is possible to obtain the breathing frequency with an accuracy of individual breaths using autocorrelation.

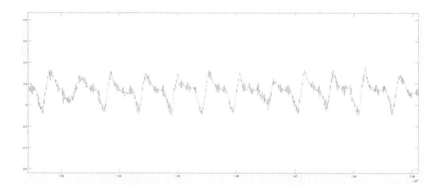

Fig. 3. Computed breathing curve

6   Conclusion

The results of this study reveal that there is a lack of systems and sensors to help detect epileptic seizures in dogs. However, research shows that a dog's epileptic seizures are comparable with those of human beings and they can be monitored. Several sensors are available for monitoring activity, which might be used for epilepsy monitoring. We have used the MetaWear solution in our system to monitor a dog's vital functions, focusing on heartbeat and breath frequency. The results of our monitoring are promising. It is possible to process the measured data and monitor heartbeat and breath frequency. The device that we have designed is able to monitor both heartbeat and breath frequency.

In our research, we have made an effort to move all of the calculations directly to the base station in realtime without sending all of the data onto to the server. The next issue is to extend the communications options, such as using GSM or an external notification module to notify the user about a seizure via SMS or sending Android Push Notification directly into the phone and regular reports via email. Simultaneously, we are now monitoring several dogs with epilepsy and collecting data about their epileptic seizures.

Acknowledgment. This research has been conducted for the NF2 s.r.o. company. Permission to publish the results from NF2 s.r.o and the support of the Specific Research Project at FIM UHK are gratefully acknowledged. Financial support from the Faculty of Science University of Hradec Králové is also gratefully acknowledged.

References

1. Epilepsy in dogs, December 2017. https://www.thekennelclub.org.uk/health/for-owners/epilepsy/
2. Fitbark home, November 2017. https://www.fitbark.com/
3. Flora accelerometer, December 2017. https://learn.adafruit.com/flora-accelerometer
4. Mi band, December 2017. http://www.mi.com/en/miband2/
5. Microsoft health, December 2017. http://developer.microsoftband.com/bandSDK
6. Oblu iot, December 2017. https://www.oblu.io/
7. Professional wearables & accelerometer research watches, December 2017. https://www.activinsights.com/
8. Scollar the smart pet collar, December 2017. https://www.scollar.com/
9. Sensors for r&d, December 2017. https://mbientlab.com/
10. Smart-sensing collar, December 2017. https://petpace.com/smart-sensing-collar/
11. Windows devices bluetooth, November 2017. https://docs.microsoft.com/en-us/uwp/api/windows.devices.bluetooth
12. Chandler, K.: Canine epilepsy: what can we learn from human seizure disorders? Vet. J. **172**(2), 207–217 (2006)
13. Grath, J.M.: Epilepsy in dogs: causes and symptoms, December 2017. https://www.petfinder.com/dogs/dog-health/epilepsy-in-dogs-causes/
14. Jory, C., Shankar, R., Coker, D., McLean, B., Hanna, J., Newman, C.: Safe and sound? a systematic literature review of seizure detection methods for personal use. Seizure-Eur. J. Epilepsy **36**, 4–15 (2016)
15. Nakano, M., Konishi, T., Izumi, S., Kawaguchi, H., Yoshimoto, M.: Instantaneous heart rate detection using short-time autocorrelation for wearable healthcare systems. In: Engineering in Medicine and Biology Society (EMBC), 2012 Annual International Conference of the IEEE, pp. 6703–6706. IEEE (2012)
16. Packer, R., Volk, H.A.: Epilepsy beyond seizures: a review of the impact of epilepsy and its comorbidities on health-related quality of life in dogs. Vet. Rec. **177**(12), 306–315 (2015)
17. Piotrowski, Z., Różanowski, K.: Robust algorithm for heart rate (HR) detection and heart rate variability (HRV) estimation. Acta Phys. Pol. A **118**(1), 131–135 (2010)
18. Shannon, C.E.: Communication in the presence of noise. Proc. IRE **37**(1), 10–21 (1949)

Driver Supervisor System with Telegram Bot Platform

Emir Husni$^{(\boxtimes)}$ and Faisal Hasibuan

School of Electrical Engineering and Informatics, Institut Teknologi Bandung,
Bandung, Indonesia
ehusni@lskk.ee.itb.ac.id, faicanhas@gmail.com

Abstract. The biggest factor causes traffic accidents is human errors. Bad driving behavior could increase the risk factor of accidents. Besides, it would lead to damage the vehicle quickly. The bad behavior could come from the less awareness of the driver or the lack of knowledge in the manner of the good and safe driving behavior. Therefore, in this research, a system is developed to know the driver's behavior while he/she is driving, especially for car rental agencies. The system will work as a supervisor which notify the driver and a car rental administrator when the driver makes a mistake based on desired rules and tell him/her what should be done. The driver will receive supervisions through text messages over a Telegram Channel and voicemails over a speaker. Utilizing vehicle diagnostic data which are retrieved directly from the vehicle using the On-Board Diagnostic (OBD) II unit, the driver's behavior can be analyzed in real time. OBD II is connected with Raspberry Pi 2 and then integrated with Telegram Bot platform, adopting Internet of Things (IoT) technology. The results of this research show the complete driving log based on selected parameters and supervision messages received by the driver related to him/her mistakes. The field test generated 7098 lines of data in a log file, which is 1145 of them exceed the limit. At least 71 supervision messages received by the driver within limit of 4 messages per minute. The calculated bandwidth used is 68.8 bytes per minute per message and it is recommended to use at least 12.1 MB monthly data plan to accommodate 175,200 supervision messages monthly. Stable Internet connection is required more than high speed Internet to keep the real-time connection alive.

Keywords: Internet of Things (IoT)
On-Board Diagnostic (OBD) II · Raspberry Pi 2 · Telegram Bot
Driving behavior · Driving supervisor

1 Introduction

Road traffic injury is one of the biggest causes of death worldwide [1]. In Indonesia, more than 40% road users died by 4 wheeled or more vehicles, including drivers and passengers, followed by motorcycle for more than 35% [2]. Human

© Springer Nature Switzerland AG 2018
N. T. Nguyen et al. (Eds.): ICCCI 2018, LNAI 11055, pp. 436–444, 2018.
https://doi.org/10.1007/978-3-319-98443-8_40

error is claimed to be the main factor causing the accident to occur. Traffic accidents often caused by lack of awareness and lack of driver's knowledge in driving. However, engine failure because of engine overheat may trigger the accident, even harmful for the driver itself [3]. The ethical driver behavior is a preventive action for road safety users.

Internet of Things (IoT) is a concept of computation where objects become smart, programmable, connected to the Internet, and able to interact with humans [4,5]. Firstly introduced by Kevin Ashton in 1999 [6], IoT is also known as pervasive computing, ubiquitous computing, and ambient intelligence.

The result of this research is the development of system that connect a car and humans, which become the supervisor for the driver. Using the On-Board Diagnostics (OBD II) [7], data from the car will be analyzed on Raspberry Pi 2. The system then tells the driver and the administrator of the vehicle about the driver's mistake through Telegram Bot Platform. These data are based on various parameters such as speed, engine rpm, engine load, and engine coolant temperature.

Many researchers applied the OBD II for the remote monitoring and reporting with embedded system [8,9], and some researchers improved it to integrated it into cloud computing technology [10,11]. Some researchers did researches related to the driver's behavior, such as predicting dangerous driving moments [12], reviewing some approaches made to observe the driving behavior and driver [13], and detecting and identifying of the abnormal driving behavior [14].

This paper is organized as follows. In Sect. 2, the system components are given. In Sect. 3, the architecture and the way of work of the system is presented. The experimental results of the implemented system based on the real roadway test are also presented in Sect. 4. Finally, conclusions of this paper is given in the last section.

2   System Components

2.1   On-Board Diagnostic (OBD) II

Engine control system has a brain built in Powertrain Control Module (PCM) [15], also known as Engine Control Unit (ECU). The OBD II is a system to monitor advanced vehicle, built in directly into vehicle PCM [7]. The OBD II was Firstly applied in 1994–1996, The OBD II is the next generation of the first generation launched in 1988. The original purpose of applying OBD was to fulfill emission target regulated by Environmental Protection Agency (EPA), which could decrease vehicle emission up to 90%.

In the OBD II, if there are problems like malfunction of a vehicle component, then the system will tell about the problems through the problem code called Diagnostic Trouble Code (DTC). This code contains of Five alphanumeric digits and officially referred to international standard SAE J2012 [16].

To connect with the car's OBD, an OBD II scanner like ELM327 v1.5 HH OBD Advanced Bluetooth Scanner is needed. It is plugged into Diagnostic Link Connector (DLC) using the 16-pin OBD connector, usually located under dashboard or steering wheel. The OBD II protocols are supported by the DLC which is depended on OBD standards issued by vehicle manufacturers. Some of the existing OBD protocols are SAE J1850 PWM, SAE J1850 VPW, and ISO9141-2.

2.2 Bluetooth

Founded in 1994, the Bluetooth® technology is understood as a wireless alternative for data cable, the way of data exchange is done by using radio transmission [17]. Nowadays, there are generally several kinds of Bluetooth technologies like the Bluetooth Basic Rate/Enhanced Data Rate (BR/EDR), the Bluetooth Low Energy (BLE), and the Bluetooth High Speed (BHS).

The ELM327 v1.5 HH OBD Advanced Bluetooth Scanner transmits the OBD data from the car wirelessly through the Bluetooth connection. A Bluetooth dongle is connected into Raspberry Pi 2 so, the Raspberry Pi is able to connect with Bluetooth devices. Therefore, the OBD data will be transmitted from OBD scanner to Raspberry Pi 2 using the Bluetooth transmission.

2.3 Raspberry Pi 2

Raspberry Pi 2, the second generation of Raspberry Pi family, is a credit card sized computer. Despite the tiny size, Raspberry Pi 2 is powerful enough to run some CPU intensive tasks but with low consume power. It is because Raspberry Pi 2 is powered by 900 MHz quad-core ARM Cortex-A7 CPU, RAM 1 GB, and Micro SD card slot to install the OS as well as for internal storage. Equipped with 4 USB ports, so we can add external devices such as USB Bluetooth dongle and USB 3G Modem to empower Raspberry Pi with outside world.

The data retrieved from the OBD scanner will be processed and analyzed in Raspberry Pi 2. The processed data will be recorded into log files whereas the analyzed data will trigger the supervision action if the driver makes the mistake. A supervision message will be delivered through customized Telegram app based on Telegram Bot platform. Besides, the Raspberry Pi 2 is also connected with a portable speaker to make the supervision messages delivered loudly.

2.4 Cellular Data

Mobile devices can communicate each other using Internet Protocol (IP). If WiFi connection is not affordable, then one way to connect to the internet wirelessly is using cellular data. Cellular data use the Node B in 3G networks.

To make Raspberry Pi 2 connected to the internet using cellular data, it is needed to connect the Pi with an USB 3G modem. The modem will have at least one IP address given by cellular carrier when it is initiating a connection to the carrier network. The IP address will change dynamically as well as the mobile device leaves a carrier coverage area to another carrier coverage area, or in other words, connected with a different node.

2.5   Telegram Bot Platform

The Telegram is a cloud-based mobile and desktop messaging app with a focus on security and speed [18]. This app comes from Russia and was founded by Pavel Durov and Nikolai Durov. The Telegram uses 2 layers of security encryption: server-client encryption and client-client encryption. They are based on AES 256-bit symmetric encryption, RSA 2048-bit encryption, and Diffie-Hellman's secure key exchange. As a result, the Telegram combines security, reliability, and speed in any networks.

The Telegram also has API Bot, a platform for developers that allows anyone to easily build tools devoted to Telegram. Bot is a third-party application running inside Telegram. Bot is not human. Although, bot can interact with humans with some commands, messages, or inline requests.

The supervision program code is integrated with Telegram Bot API. The Bot API will be connected to Telegram Cloud in real time, so that it will consumed some cellular data even there are no messages sent. The supervision messages will be received by the users on their Telegram App. Telegram Channel is specifically used to accommodate drivers and car rental manager to stay connected during driving.

3   The Architecture and the Way of Works of the System

The idea of the driver supervisor system is as follows. Firstly, the OBD II scanner is plugged into DLC, then it connects to Raspberry Pi 2 using Bluetooth connection, and transmits the OBD data. The OBD data, which is the PID code, are then interpreted with an open library to be understood by humans. Afterwards, the data is processed and modified in such a way to obtain final data. This final data consist of 4 parameters: speed (km/h), engine rpm (rpm), engine load (%), and engine coolant temperature (°C). These data are then processed and analyzed with the given rules in each parameter so that the behavior of the driver can be known.

If the system detects the driver's mistake from the result of the analysis, then it will send supervision messages to the users through Telegram App. It also gives voicemails for the driver. The messages contain the mistake done by the driver and give recommendations to the driver about what should be done. The car rental administrator can send broadcast messages through Telegram Channel to give information or additional instructions to the drivers.

Besides, log files are also provided to record diagnostic data in every single second and store it into a flash storage for the whole driving data. These log files are very useful for further analysis if there are internet connectivity issues while driving, resulting in a lot of data loss error. The architecture of the system can be seen in Fig. 1.

There are rules given for each parameter. If the driver breaks the limit of the rules, then this will trigger the program to send supervision messages, both of texts or audio. The rules are implemented with if statement within the program supervisi.py and categorized as follow (Table 1).

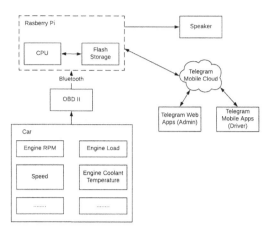

Fig. 1. System architecture

Table 1. Values of parameter limitation rule

No	Parameter	Operator	Value	Unit
1	Speed	<=	80	km/h
2	Engine RPM	<=	2500	rpm
3	Engine load	<=	60	%
4	Engine coolant temperature	<=	90	°C

To make Raspberry Pi 2 connected to the Telegram App, it is needed to make Telegram Bot API in Raspberry Pi. A Python framework called Telepot is used to do this. But first, a Bot account is setup using BotFather until Token given. This Token is used to authenticate and authorize the Bot to connect using the API. Then, Telegram Channel is connected the Bot, the drivers, and the manager. Finally, the code is inserted to do telematics analysis including supervision messages delivery in the end of code and make sure the code is always loop.

4 Experimental Results

In this section, the roadway test result is given. In this test, a road vehicle test was conducted. Tests was carried out from Rancaekek to Lembang areas, Bandung, Indonesia. This journey was taking up to 2 h, began around 16.00 until 18.00 local time. Road terrain varies, from long straight way in the high way until uphill road in the mountains. The car used in this test is Mitsubishi Colt T 120 SS 2007, which is an OBD II compliant car.

A log file was successfully created with 7098 lines of data, where there are 1145 lines of data and 1272 times broke the parameter limit values. Supervisor messages had been received 71 times as stated below (Table 2).

Table 2. Number of supervisor messages

No	Category	Amount
1	Speed	1
2	Engine RPM	24
3	Sudden acceleration	46
4	Engine coolant temperature	0
	Total	71

The supervisor messages successfully sent both in Telegram texts and audio format through the speaker. The messages received by the driver are shown in Table 3.

Table 3. Supervisor messages

No.	Error category	Supervisor messages	
		Text message	Audio
1	Speed	Your speed is more than 80 km/h Reduce your speed immediately!	speedsound.wav
2	Engine RPM	Engine RPM exceeding 2500 rpm Reduce stepping on the accelerator, change the gear higher!	rpmsound.wav
3	Sudden acceleration	Do not step on the accelerator suddenly! Step on the accelerator slowly!	acceleratorsound.wav

The audio files were made using Audacity version 2.1.2 on MacBook Air 2015, 1.6 GHz dual-core Intel Core i5 with 3 MB shared L3 cache, 4 GB of 1600 MHz LPDDR3 onboard memory, with macOS Sierra. A voice-over is used with some additional sound effects and audio editing. Table 4 the complete list of audio files were made. There is also one additional sound as welcome greeting and driving supervisor instruction.

Since internet quota is typically limited by cellular data providers, it is important to make sure that the supervisor messages are using large bandwidth. The audio files are stored and played internally in the Raspberry Pi, so it does not take any internet bandwidth. Using a tool called the Bandwidthd, running via background service in the Raspberry Pi, it measured data traffic of supervisor message through Telegram Bot platform. See Table 5 and Fig. 2 for more details.

From the measurement, the total bandwidth consumed is 446.5 KB with 214 KB of outgoing data traffic and 232.5 KB of incoming data traffic. As many

Table 4. List of sound files for supervision

No.	Category	File name	Duration (s)	Size (MB)
1	Speed	speedsound.wav	11	2
2	Engine RPM	rpmsound.wav	14	2.5
3	Sudden acceleration	acceleratorsound.wav	9	1.6
4	Engine coolant temperature	coolanttempsound.wav	13	2.2
5	Welcome greeting and supervisor instruction	WelcomeGreeting.wav	47	8.4

Table 5. Bandwidth

IP address	Total bandwidth (KB)	Total sent (KB)	Total received (KB)	HTTP (KB)	TCP (KB)	UDP (KB)
192.168.8.100	446.5	214.0	232.5	243.0	243.0	203.5

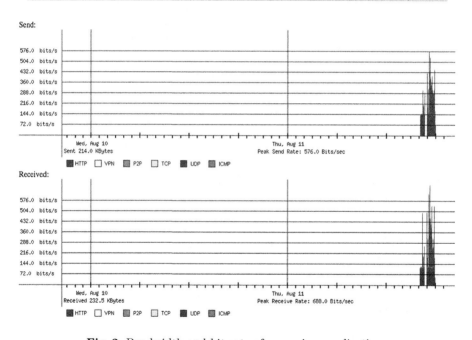

Fig. 2. Bandwidth and bit rate of supervisor application

as 243 KB data traffic was used for delivering messages, while the remaining 203.5 KB was consumed to keep the real-time connection to Telegram Bot server alive. The peak send rate is only 576.0 bits/sec, while the peak receive rate is only 688.0 bits/sec.

The average amount of bandwidth consumed per message of supervision in this test is equal to 68.8 bytes per minute per message. The limitation is given to maximize supervisor messages received in 1 min, which is 4 messages. There will be 175,200 messages maximum monthly. Therefore, the system needs the minimum bandwidth at least 12.1 MB monthly to work properly.

5 Conclusion

This system supervises the driving behavior of the driver. If the driver does a bad behavior related to given parameters, the driver supervisor will give an alert to the driver. The OBD data has been successfully retrieved and processed to give the 4 parameters: speed, engine rpm, engine load and engine coolant temperature. The analytic program has been successfully analyzed the data to identify the driver's mistake. The experiment result show that the supervisor messages have been successfully received by the driver and car rental administrator when the driver made a mistake. The car rental administrator now is able to find out the driver's behavior. It will help the car rental administrator to determine the rate of each driver as a reference for the next rental. As the bandwidth consumption is small, this supervisor system can be implemented extensively with any options of data plan.

References

1. World Health Organization: The top 10 causes of death, Fact sheet No. 310 (2014). http://www.who.int/mediacentre/factsheets/fs310/en/
2. World Health Organization: Country Profiles, In Global Status Report On Road Safety 2015, 1211 Geneva 27, Management of Noncommunicable Diseases, Disablitiy, Violence and Injury Prevention (NVI), p. 148 (2015)
3. RACQ: The causes of engine overheating and its consequences can be many and varied. https://www.racq.com.au/cars-and-driving/cars/owning-and-maintaining-a-car/car-maintenance/engine-overheats
4. IEEE: Internet of Things Community, IEEE (2016). https://www.ieee.org/member ship-catalog/productdetail/showProductDetailPage.html?product=CMYIOT736
5. Cisco: Internet of Things (IoT). http://www.cisco.com/c/en/us/solutions/internet-of-things/overview.html
6. IoT Council: The Internet of Things (2016). http://www.theinternetofthings.eu/what-is-the-internet-of-things
7. NAPA Institute of Automotive Technology: Introduction to OBD II, in OBD II and Second Generation Scan Tools, NAPA Institute of Automotive Technology (1998)
8. Lin, C.E., Shiao, Y.-S., Li, C.-C., Yang, S.-H., Lin, S.-H., Lin, C.-Y.: Real-time remote onboard diagnostics using embedded GPRS surveillance technology. IEEE Trans. Veh. Technol. 56(3), 1108–1118 (2007)
9. Ceuca, E., Tulbure, A., Taut, A., Pop, O., Farkas, I.: Embedded system for remote monitoring of OBD Bus. In: 36th International Spring Seminar on Electronics Technology, pp. 305–308 (2013)

10. Jhou, J.-S., Chen, S.-H., Tsay, W.-D., Lai, M.-C.: The implementation of OBD-II vehicle diagnosis system integrated with cloud computation technology. In: 2013 Second International Conference on Robot, Vision and Signal Processing, pp. 9–12 (2013)
11. Amarasinghe, M., Kottegoda, S., Arachchi, A.L., Muramudalige, S., Bandara, H.M.N.D., Azeez, A.: Cloud-based driver monitoring and vehicle diagnostic with OBD2 telematics. In: 2015 IEEE International Conference on Electro/Information Technology (EIT). IEEE (2015)
12. Fao, C.Y., Wu, B.Y., Wang, J.M., Chen, S.W.: Dangerous driving event prediction on expressways using fuzzy attributed map matching. In: Proceeding of the Ninth International Conference on Machine Learning and Cybernetics. IEEE (2010)
13. Kang, H.B.: Various approaches for driver and driving behaviour monitoring: a review. In: 2013 IEEE International Conference on Computer Vision Workshops. IEEE (2013)
14. Chen, Z., Yu, J., Zhu, Y., Chen, Y., Li, M.: Abnormal driving behaviors detection and identification using smartphone sensors. In: 2015 12th Annual IEEE International Conference on Sensing, Communication, and Networking (SECON). IEEE (2015)
15. Duffy, J.E.: Modern Automotive Technology. Goodheart-Willcox Company (2003)
16. SAE Vehicle E/E System Diagnostics Standards Committee, SAE J2012: Diagnostic Trouble Code Definitions, Society of Automotive Engineers (2002)
17. Bluetooth SIG, Inc.: Bluetooth, The Story Behind Bluetooth Technology. https://www.bluetooth.com/what-is-bluetooth-technology/bluetooth
18. Telegram.org, Home: Telegram, a new era of messaging, Telegram Messenger. https://telegram.org

Multi-agent Base Evacuation Support System Using MANET

Shohei Taga[1](✉), Tomofumi Matsuzawa[2], Munehiro Takimoto[2],
and Yasushi Kambayashi[1]

[1] Department of Computer and Information Engineering, Nippon Institute
of Technology, 4-1 Gakuendai, Miyashiro-machi, Minamisaitama-gun,
Saitama 345-8510, Japan
ac1l15313@gmail.com, yasushi@nit.ac.jp
[2] Department of Information Sciences, Tokyo University of Science,
2641 Yamazaki, Noda 278-8510, Japan
{t-matsu,mune}@cs.is.noda.tus.ac.jp

Abstract. In this paper, we propose an evacuation support system that provides evacuation routes in the case of a disaster, and verify the usefulness of the system. In recent years, with the development of communication and portable device technologies, people can collect and disperse information using the Internet regardless of time and place. Current popular wireless communication infrastructure is supported by a series of base stations and one communication equipment in such a base station handles a lot of communication. Therefore, when problems occur at an equipment in such a communication base station, it may be difficult, even if possible, for the smartphones to use the Internet. In fact, in the 2011 off the Pacific coast of Tohoku Earthquake in Japan, we have observed a large-scale communication failure due to corruption of the communication equipment and traffic congestion. Paralyzed communication infrastructure made it difficult for people to collect information about the conditions of transportation and safety information about family and friends using smartphones. Our proposed system address this problem by using multiple kinds of mobile agents as well as static agents on smartphones that use a mobile ad hoc network (MANET). The proposed system collects information by mobile agents as well as diffuses information by mobile agents so that the system provides an optimal evacuation route for each user in a dynamically changing disaster environment.

Keywords: Mobile ad hoc network · Mobile agent · Multi agent
Contingency plan · Risk management

1 Introduction

In this paper, we propose an evacuation support system that provides evacuation routes in the case of a disaster, and verify the usefulness of the system. In recent years, with the development of communication and portable device technologies, people can collect and spread information using the Internet regardless of time and place. Current popular wireless communication infrastructure is supported by a series of base stations

© Springer Nature Switzerland AG 2018
N. T. Nguyen et al. (Eds.): ICCCI 2018, LNAI 11055, pp. 445–454, 2018.
https://doi.org/10.1007/978-3-319-98443-8_41

and one communication equipment in such a base station handles a lot of communication. Therefore, when problems occur at an equipment in such a communication base station, it may be difficult, even if possible, for the smartphones to use the Internet. In fact, in the 2011 off the Pacific coast of Tohoku Earthquake in Japan, we have observed a large-scale communication failure due to corruption of the communication equipment and traffic congestion. Paralyzed communication infrastructure made it difficult for people to collect information about the conditions of transportation and safety information about family and friends using smartphones.

Our proposed system addresses this problem of communication infrastructure by constructing mobile ad hoc network (MANET) by wireless communication between users' smartphones. Then users can share information in such a network. Since MANET is a network constructed with only portable devices, it is possible to avoid problems due to failure of communication infrastructure. However, the ever changing topology of MANET makes stable communication extremely difficult. Therefore, we propose an information sharing method using mobile software agents. A mobile agent is a program with mobility, it has a feature of perceiving environment and deciding behavior. Our proposed system constructs optimal evacuation routes by using such mobile agents to share and collect information necessary for evacuation.

In previous studies, we proposed basic configuration of the evacuation support system and verified its feasibility by simulators [1, 2]. In this study, we have expanded the features of the system and verify its usefulness with a simulator namely NS-3. In the previous paper, we have only employed mobile agents for diffusing information about the circumstances. The evacuation users, therefore, have received information not related to their potential evacuation routes, while having not receiving crucial information for their evacuation (i.e. potential dangers on their evacuation routes). In order to ameliorate the system, in this paper, we introduce yet another kind of mobile agents that actively collect information closely related to the users' potential evacuation routes.

NS-3 is a discrete-event network simulator which targeted primarily for research and educational use [3]. The previous simulators were crude and cannot reflect the real MANET situation. On the other hand, NS-3 models the communication environments and processes in a form close to the real world, and makes it possible to perform various verification experiments. Using NS-3 makes the simulation results of ad hoc network feasible.

The structure of the balance of this paper is as follows: the second section describes related works. The third section describes the details the proposed system. The fourth section describes the numerical experiments and discusses the results, and the fifth section discusses the future works and concludes the discussion.

2   Related Works

Avilés et al. proposed an evacuation support system using MANET and Ant Colony Optimization (ACO) for in-door environments [4]. In the study, they implemented ACO by using mobile agents so that the ACO algorithm can take the movement trajectory and speed of evacuees in consideration. Then the evacuation support system constructs an optimal evacuation route for each user.

Likewise, Ohta et al. studied on evacuation support methods using ACO and MANET [5]. In the study, they pointed out a problem such as a conventional ACO may include dangerous locations when it constructs evacuation routes. This problem is caused by the dynamic nature of the disaster environments such as conflagration or tsunami. In order to mitigate this problem, they proposed an improved ACO-based evacuation support system that equips deodorant pheromone which erases ACO pheromone traces when dangerous locations are found. Goto et al. applied this proposed method to real data of tsunami damage [6]. They showed practical results by using the data of Rikuzentakata city which suffered great damage due to the 2011 off the Pacific coast of Tohoku Earthquake. They verified the method of Ohta et al. is feasible based on the real data.

Kambayashi et al. proposed and implemented a system that collects safety information of evacuees using mobile agents and MANET [7]. In the study, they proposed a method to reduce load on transmission by combining multiple mobile agents into one. Nishiyama et al. proposed communication system using portable devices that switch between MANET and Delay Tolerant Networking (DTN) according to the communication situation [8]. DTN is a method for coping with a network environment where maintaining stable communication connection is hard to achieve. When communication is disconnected, portable devices accumulate data, and then transmitted when communication is resumed. Their proposed system apply MANET when there are many portable devices in the surroundings, and apply DTN when there are not much. With such a method, they achieved to cope with various network environments.

3 Agent Base Evacuation Support System

In this section, we describe our proposed system in detail. The proposed system aims to provide an optimal evacuation route for each user (hereafter we call the evacuation user). Since the proposed system maintains the map information of the evacuation area, it is possible to calculate the shortest route to the destination (i.e. safe place). However, at the time of a disaster, there is a possibility of occurrence of a point unsuitable for evacuation (hereafter we call dangerous point) due to fire, building collapse, or inundation. Since nobody knows these points before the occurrence of a disaster, it is necessary to collect the information during evacuation. When an evacuation user finds a dangerous point, he or she inputs the position information to the system. Then the proposed system constructs a new evacuation route avoiding this dangerous point, and provides it to the user. At the same time, the system diffuses the information about the dangerous point and new route to other users' smartphones. As a result, evacuation users other than the discoverer can know the dangerous point and avoid it in advance. In order to realize this function, we use multiple mobile agents.

A multi-agent system is a system that consists of multiple agents and achieves tasks by their cooperative operations. The agents can be categorized into two types: mobile agents and static agents. A mobile agent is generated when it is needed and executes a task through migrating among communication sites including smartphones. Every mobile agent has a unique identifier. A static agent resides on communication site including, of course, a smartphone. Unlike mobile agents, static agent has no unique identifier. We describe the details of each agent we use in the proposed system below.

3.1 Static Agents

Information Agent. Information Agent is a static agent residing on a smartphone that interacts with mobile agents and constructs evacuation routes. When requested from the system, it creates mobile agents.

The information agent processes the request in the following order. (i) It generates the requested mobile agent. (ii) It acquires the information necessary for the generated mobile agent from the node management agent and passes it to the mobile agent. (iii) It stores the mobile agent in a queue. It periodically checks the queue, and dispatches the mobile agents to the neighboring smartphones. When a mobile agent comes from another smartphone, the information agent receives information held by the mobile agent. Then it passes the requested information to the arrived mobile agent and store it in the queue in the same way as the above step (iii). The information agent records the unique identifier of the mobile agent that visited the smartphone as well as it created in a list called *visitor list*. The information agent requests the visitor list of other smartphones when it communicates with them. It then passes the received visitor list to the mobile agent that needs it in the queue. The mobile agent decides the next destination from this visitor list.

The information agent constructs evacuation routes based on the information it initially has, and the information collected from the visited mobile agents. The evacuation route is the shortest route to the destination avoiding dangerous points that are currently known by the information agent. The evacuation route is determined based on the Dijkstra's algorithm. The Dijkstra's algorithm is an algorithm for solving the shortest path problem between two nodes in a graph, and was proposed by Edgar Dijkstra in 1959 [9]. In the proposed system, the graph consists of intersection as the nodes, and the distances between intersections as the edge weights. The information agent constructs an initial evacuation route at the system startup time. After that, when a mobile agent arrives and let the information agent know a dangerous point exists on the current evacuation route, the information agent reconstructs the evacuation route.

Node Management Agent. Node Management Agent is another static agent residing on a smartphone for managing the information on the smartphone. The node management agent stores the dangerous point information collected from the visited mobile agents in the information table. At this time, if the same information already exists in the information table, the node management agent delete older information. Also, if the information agent requests information about dangerous points, the node management agent passes the requested information.

3.2   Mobile Agents

Information Diffusion Agent. Information Diffusion Agent is a mobile agent that diffuses the information of the dangerous point found by the evacuation user to neighboring smartphones.

When an evacuation user finds a dangerous point and input its information into the system, the information agent generate the information diffusion agent. The information agent passes the coordinates of the discovered point to the generated information diffusion agent. And then, the information diffusion agent waits until a link with another smartphone is established. After that the information diffusion agent act as follows: (i) When communication links with other smartphones are established, the information diffusion agent copies itself by the number of linked smartphones and moves to each smartphone. However, if the information diffusion agent finds that it has already visited the smartphone (i.e. the smartphone's visitor list has its identifier), it does not move but commit suicide. (ii) After the movement, it passes its own information to the information agent on the destination, and be into standby state until the next communication link being established. It repeats this process a constant number of hops. When the information diffusion agent copies itself, it copies not only its own information but also its own unique identifier. Therefore, it does not move to the smartphone that its own copy has visited. This mechanism prevents a smartphone from receiving the same information multiple times.

Information Collecting Agent. Information Collecting Agent is another mobile agent that collects information about the events on the evacuation route and returns to the original smartphone. The information diffusion agent diffuses the information discovered by the evacuation user around the discovery point, but there is no guarantee that this information can be conveyed to all evacuation users who need it. In order to solve this problem, we propose the information collecting agent that actively collects information diffused by the information diffusion agent as shown in Fig. 1.

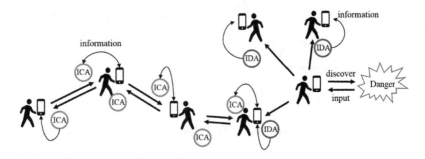

Fig. 1. Information diffusing and information collecting.

The information collecting agent returns to the original smartphone after the information collecting process, but there is a problem with the return method. In the situation of the proposed system being used, it is difficult to return through the movement history of the agent in reverse order due to the disappearance of the smartphones it has visited. They may move out of wireless communication range or their batteries may be exhausted. For this reason, the proposed system predicts the current location of the original smartphone based on the evacuation route, moving speed and elapsed time of the original smartphone.

The information agent generates the information collecting agent at regular time intervals. Then, the information agent passes the current evacuation route information, the moving speed of the user, and the life time of the information collecting agent, to the generated information collecting agent. And this information collecting agent waits until a communication link with another smartphone is established. The information collecting agent has two states, the collecting state and the return state, and the behavior changes depending on the state. Initially it is in the collecting state. The collecting state is a state of collecting information and acts as follows. (i) The information collecting agent moves to the smartphone located close to the current evacuation route of the user and furthest from the user. Though, of course, it excludes the smartphone that it has already visited. (ii) It acquires information from the destination smartphone. The information collecting agent repeats this process during the collecting state. If the information collecting agent acquires information that tells there is a dangerous points on the evacuation route, or exceeded half of the its own life time, or cannot find the next migration candidate, the information collecting agent becomes in the return state. The return state is a state of returning to the original smartphone and acts as follows. (i) The information collecting agent calculates the predicted current position of the original smartphone based on the information it has; i.e. evacuation route information of the original smartphone and the moving speed of the evacuation user of the smartphone, and the elapsed time since it was generated as shown in Fig. 2. (ii) It moves to the smartphone closest to the calculated expected current position.

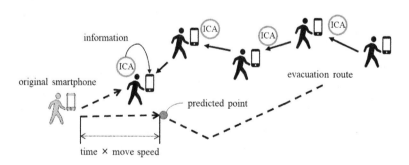

Fig. 2. Predict the position of original smartphone.

Unlike the collecting state, in the return state, it is not excludes the smartphone it has visited. It repeats this process until returning to the original smartphone. When the information collecting agent return to the original smartphone, it passes the collected information to the information agent on the smartphone and disappears. If the information collecting agent cannot returns even exceed the own life time, or if it cannot find the original smartphone in the vicinity of the calculated predicted current position, it commits suicide.

One of the disadvantages of the information collecting agent is that, since it is generated from each smartphone, the network load tends to increase. In order to mitigate this problem, the proposed system controls the generation of the information collecting agents by broadcasting messages to the neighboring smartphones that request to stop generating the information collecting agent for certain period. This message contains the address and the evacuation route of the sender, and the life time of the information collecting agent. The smartphones that receive this message stop generating information collecting agents if its own evacuation route is the same as described one in the message. If the own evacuation route is different from what described in the message, it ignores and discards the message. When the information collecting agent returns to the sender smartphone of the message, this smartphone broadcasts messages that permit generating information collecting agents and the collected information to the neighboring smartphones. The smartphones that have stopped generating information collecting agent resume generating of the agents when they receive this message. The smartphones also resume generating information collecting agents when the stopping period in the message elapses.

4 Numerical Experiment

This section describes the numerical experiment of the proposed system. We verified by simulation in situations that people use the proposed system at the disaster area. We used NS-3 for simulation. NS-3 is a discrete-event network simulator which is open for research and educational use [3]. The model of the communication environments and processes in a form close to the real world, and the simulator makes it possible to perform various verification experiments. In this experiment, we verified how far evacuees can avoid dangerous points and how many agents are transmitted when using the proposed system.

4.1 Experimental Conditions

We created the simulation map that represents the evacuation area (Fig. 3). This simulation map is a model of the real map which about 1 km^2. Evacuation users are randomly placed and move toward assigned destination (safe area). There are four destinations (depicted as green zones), and randomly assigned to each evacuation user. The moving speed of the evacuation user is set to 1 m/s. The evacuation user knows the map and the destination of the evacuation area and moves through the route constructed by the Dijkstra's algorithm. However, the evacuation user does not know dangerous points in advance, and they will know for the first time by actually touching to the dangerous points or notified by other evacuation users through mobile agents. When an evacuation user knows the new dangerous point, the information agent in his or her smartphone reconstructs the evacuation route avoiding this dangerous point. Ten dangerous points are randomly allocated from all intersections. The communication distance of the smartphone is 50 m. The evacuation user who arrives at the destination

Fig. 3. Simulation map (Color figure online)

terminates the communication. The simulation finishes when all the evacuation users arrive at the designated destinations.

In this verification experiment, we measured (1) the number of the touches to the dangerous points of all evacuees and (2) the number of agent transmissions of all the smartphones. In (1), we divided the experiment into two cases. One is the case that people use the proposed system. In this case, the information diffusion agent diffuse discovered information, and the information collecting agent collects information on evacuation route. The number of hops of an information diffusion agent is 2, and the life time and the generation interval of an information collecting agent is set to 50 s. The other case is that people do not use the proposed system. In this case, people spreads the discovered information by simple flooding. The number of hops is not limited, and if the link with another smartphone is established at the time of transmission, it is diffused any number of times. However, as with the proposed system, the same information is never transmitted to the same smartphone. In both cases, we further divided the cases that the number of evacuees 100, 200, 300, 400, 500. We have carried out all the cases 50 times each, and the average value was taken as the results.

4.2   Result and Discussion

Figure 4 shows the results of the number of the touches to the dangerous points. In this figure, "proposed system" is the case of using the proposed system and "flooding" is the case of not using it. In all the results, we found that the case of using proposed system suppressed the number of touches to the dangerous points than the case of not using it.

The result of the number of agent transmissions was as shown in Fig. 5. We found that as the number of evacuation users increases, the number of transmissions also increases. Also, the average number of transmissions per evacuation user increased according to the number of people. We have considered that this is due to as the number of the agent generation source smartphones increases, the number of transmission of the relay smartphones increases. In this experiment, the maximum number

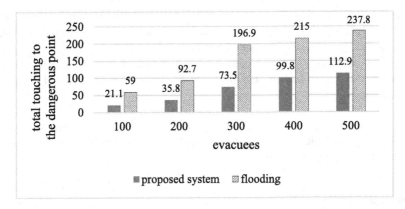

Fig. 4. The number of the touches to the dangerous points.

of evacuation users was set at 500 people and the number of agent transmissions is tolerable, but in a real situation, the number of people may increase further. In such a case, the load of the relay smartphone may increases to an extent that cannot be ignored. We must construct yet another mechanism to suppress the number of generating mobile agents.

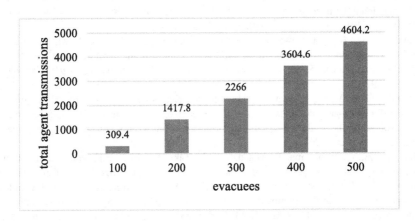

Fig. 5. The number of agent transmissions

5   Conclusion

In this paper, we proposed an evacuation support system that enables information sharing under environments where people cannot use the Internet communication due to the disaster, and verified this system.

As a future work, it is necessary to address the increase in the load of relay smartphones. Since the information collecting agent moves along the evacuation route of the evacuation user, they frequently move in a busy street such as main streets and in front of evacuation centers. As a result, the network load increases locally. Therefore, it is necessary to develop a mechanism for controlling the flow amount of the agent in such a place. In addition, we also need to extend the simulation. Although we used a two-dimensional map in the simulation, a three-dimensional map is needed to consider tsunami damage.

Acknowledgements. This work is partially supported by Japan Society for Promotion of Science (JSPS), with the basic research program (C) (No. 17K01304 and 17K01342), Grant-in-Aid for Scientific Research (KAKENHI).

References

1. Taga, S., Matsuzawa, T., Takimoto, M., Kambayashi, Y.: Multi-agent approach for return route support system simulation. In: Proceeding of the Eighth International Conference on Agents and Artificial Intelligence, vol. 1, pp. 269–274 (2016)
2. Taga, S., Matsuzawa, T., Takimoto, M., Kambayashi, Y.: Multi-agent approach for evacuation support system. In: Proceeding of the Ninth International Conference on Agents and Artificial Intelligence, vol. 1, pp. 220–227 (2017)
3. NS-3 Homepage. https://www.nsnam.org. Accessed 22 Mar 2017
4. Avilés, A., Takimoto, M., Kambayashi, Y.: Distributed evacuation route planning using mobile agents. In: Nguyen, N.T., Kowalczyk, R., Fred, A., Joaquim, F. (eds.) Transactions on Computational Collective Intelligence XVII. LNCS, vol. 8790, pp. 128–144. Springer, Heidelberg (2014). https://doi.org/10.1007/978-3-662-44994-3_7
5. Ohta, A., Goto, H., Matsuzawa, T., Takimoto, M., Kambayashi, Y., Takeda, M.: An improved evacuation guidance system based on ant colony optimization. Intelligent and Evolutionary Systems. PALO, vol. 5, pp. 15–27. Springer, Cham (2016). https://doi.org/10.1007/978-3-319-27000-5_2
6. Goto, H., Ohta, A., Matsuzawa, T., Takimoto, M., Kambayashi, Y., Takeda, M.: A guidance system for wide-area complex disaster evacuation based on ant colony optimization. In: Proceedings of the Eighth International Conference on Agents and Artificial Intelligence, vol. 1, pp. 262–268 (2016)
7. Kambayashi, Y., Nishiyama, T., Matsuzawa, T., Takimoto, M.: An implementation of an ad hoc mobile multi-agent system for a safety information. In: Grzech, A., Borzemski, L., Świątek, J., Wilimowska, Z. (eds.) Information Systems Architecture and Technology: Proceedings of 36th International Conference on Information Systems Architecture and Technology – ISAT 2015 – Part II. AISC, vol. 430, pp. 201–213. Springer, Cham (2016). https://doi.org/10.1007/978-3-319-28561-0_16
8. Nishiyama, H., Ito, M., Kato, N.: Relay-by-smartphone: realizing multi-hop device-to-device communications. IEEE Commun. Mag. **52**(4), 56–65 (2014)
9. Dijkstra, E.W.: A note on two problems in connexion with graphs. Numerische Mathematik **1**, 269–271 (1959)

Analysis of Software Routing Solution Based on Mini PC Platform for IoT

Josef Horalek and Vladimir Sobeslav[✉]

Faculty of Informatics and Management, University of Hradec Kralove,
Hradec Kralove, Czech Republic
{josef.horalek,vladimir.sobeslav}@uhk.cz

Abstract. The following article presents results of the research aimed at the possibility of software routing implemented into a Mini PC platform in IoT area. The article presents the basic principles of Quagga and Bird software routers. For every software router, its routing architecture in connection to the operating system is presented, including a draft of complex architecture of the software routing solution. In the research, emphasis was put on usability of the given platforms in software routing on the Mini PC platform. An analysis of Quagga and Bird platforms' respective system resource requirements with the focus mainly on the CPU usage during routing on Humming-Board Gate and Raspberry Pi – Model B platforms, is also included. Furthermore, an analysis of the routing effectiveness using the given testing topology during the simultaneous use of both hardware solutions during the implementation of static and dynamic routing. Acquired results are presented using box graphs, with the course of long-term behavior during individual routings being depicted using 2nd period of moving average, which offers a relevant idea about the course of the routing and both routing daemons.

Keywords: Quagga · BIRD · Raspberry PI · HummingBoard
IoT

1 Introduction

The reader of this article is probably already familiar with the term of Internet of Things (IoT). For completeness, one of the possible definitions will be mentioned. In the reference [1], Intel Corporation microprocessor manufacturer describes IoT as evolution of mobile, home, and embedded applications that are connected to the Internet, integrating computing capabilities and using data analytics to extract valuable information. Therefore, billions of devices will be connected to the Internet, with more of them to follow. Such devices can become an intelligent system of systems, sharing data over the cloud storage. All of these change our daily activities, lives, and our world in countless ways. The future smart world can be viewed e.g. in a model released by Libelium Comunicaciones Distribuidas company [2], one of the most prominent developer centers in the smart world area.

This concept is taken into consideration by the most prominent organizations in the world as well. From among the academic institutions, let us mention e.g. Massachusetts

© Springer Nature Switzerland AG 2018
N. T. Nguyen et al. (Eds.): ICCCI 2018, LNAI 11055, pp. 455–466, 2018.
https://doi.org/10.1007/978-3-319-98443-8_42

Institute of Technology, which, according to the reference [3], asserts that there are going to be 50 billion devices connected to the Internet by 2020. Networking hardware manufacturer Cisco Systems, Inc. estimates that by that year, 46% of all the communication will be realized solely among the devices (M2M) [4]. Furthermore, IBM claims that approximately 13 billion devices are already connected and it is expected to rise to 29 billion by 2020 [5].

Gartner, Inc., which is a company dealing with research and consulting regarding IT, stated in January 2017 in the source [6] that by the end of that year there were going to be 8.4 billion devices connected to IoT. According to that research, the number of devices increased by 31% compared to 2016. The enormous rise in the use of IoT is, according to this organization, going to grow further and by 2020 reach 20.4 billion devices. The total price of the individual devices and services in 2017 is estimated to be 2 trillion USD. However, the prognoses regarding the development of IoT may be, it is apparent that it is a trend with huge potential.

1.1   Research Goal

The following article is a loose continuation of the software routing platforms published in [7]. The aim of the research was a comparative analysis of routing performance and efficiency of selected routing platforms over Linux core operating systems. The obtained results proved the possibility of efficient use of the software routers in various areas, including smart IoT networks. An important characteristic of the architecture and use of IoT networks is that whether they be used for intelligent house state monitoring [8], embedded appliances [9], or smart grid applications [10], the common feature is their event-based architecture based on the concept of real-time processes and operations. To meet this requirement, it is necessary to have segmented, routed, and secure network. The IoT network does not necessarily need to be based on a deterministic or syntactic model, but it can focus on the context of an event itself and be able to quickly respond to it. At the same time, it is not suitable to use high-performance hardware routers for the IoT networks in the aforementioned examples, and using platforms designed primarily for IoT world is recommended. With these affordable computing systems, it is possible to realize software routing in order to meet the basic aims and logic of modern IoT networks. To confirm this assertion was the goal of the research that had been performed earlier. This research focused on monitoring the workload of CPU caused by the processes of selected routing daemons and routing speed while using static routing, distance-vector routing (RIP protocol), and link-state routing (OSPF) over IPv4 protocol.

2   Introduction to Software Routers

A software router is a routing platform, sometimes also called routing daemon, designated for the most of GNU/Linux OSs. Article [11] describes software routers as user applications running outside of the OS core. The main purpose of software routers is primarily management of OS routing tables. It means that an OS by itself can resend packets between target devices as it has a simple routing table, however it does not understand routing protocols and other more complicated network parameters, which is managed by the aforementioned routing daemons. Therefore, these tools allow the administrator to define individual routing protocols, according to which the operating system makes decisions later [12].

As there are a plenty of possible routing solutions, the decision as to which routing daemons to use depends on various factors. In our research, we have decided to use a comparative analysis of such routing platforms that are universally distributed under GNU/GPL open-source license.

The first daemon is Quagga, which is a standard in the area of routing under GNU/Linux OS. This daemon can pride itself with vast history. It is universally one of the best-known and most-used GNU/GPL routing projects of its kind.

The other routing platform we have surveyed is BIRD, a relatively young Czech routing platform, which not only is very ambitious, but also has several successes in outperforming its aforementioned better-known competitor.

2.1 Quagga

Quagga platform is a daemon designated for Unix-like OSs [13, 14] that offers routing services for TCP/IP networks. The creator of this project, Kunihiro Ishiguro, initially developed Quagga as a fork of the GNU Zebra project. Besides static routing, Quagga is also capable of implementation of RIPv1, RIPv2, RIPng, OSPFv2, OSPFv3, and BGPv4+ routing protocols, with the last incorporating addressing for multicast and IPv6. The daemon also externally supports IS-IS routing protocol both in IPv4 and IPv6 version. Among others, Quagga newly supports Optimized Link State Routing (OLSR) protocol via wireless routing plugin, Multiprotocol Label Switching (MPLS) Label Distribution Protocol (LDP), as a derivative of LDP designated for OpenBSD OS with Bidirectional Forwarding Detection (BFD). From the listing of supported routing protocols, it is apparent that this daemon supports both their IPv4 and IPv6 versions [13]. Thanks to this reference we also know that OSPFv3 and IS-IS routing protocols is at the moment in so-called beta phase, and contains certain imperfections in the implementation of OSPFv3 and IS-IS in their respective IPv6 versions.

Among significant advantages of this platform belongs mainly its wide user community, owing to which plenty of various and relatively easily-accessible information regarding this platform exists.

2.2 BIRD

Bird Internet Routing Daemon (BIRD) [15] is a relatively new Czech routing platform designated for all Unix-like OSs. The beginning of the projects can be dated back to 1999, however, shortly after the project was put on hold and it was not before the verge of 2008 and 2009 when it was renewed by CZ.NIC labs. The daemon was developed as a school project at the Faculty of Mathematics and Physics of Charles University in Prague. Similarly to Quagga [3], BIRD started to support MPLS. The authors of the platform state that the BIRD project is designated mainly for fully dynamic routing. The platform is being developed primarily for OS GNU/Linux, however, it is also compatible with other Unix-like OSs. Analogously to Quagga, BIRD is being distributed under GNU/GPL license. BIRD supports both static and dynamic routing via BGP, RIP, OSPF, BFD routing protocols, and recently also via Babel routing protocol. BIRD, similarly to Quagga, supports both IPv4 and IPv6 routing. Similarly to Quagga platform, this routing platform is capable of displaying (although not changing) the settings via command line (CLI). Its syntax is completely different, and it differs from the one being used by the devices by Cisco Systems, Inc. (which is being imitated by the developers of Quagga software router). From among BIRD's advantages we can mention programmable filters, well organized configuration files, extensive documentation, low memory and CPU requirements, well organized code, and support of more independent routing tables [15].

3   Proposed Solution Model

In this chapter, we are going to introduce our designed and implemented solution of a software router on Mini PC for IoT platform. The first step while establishing an IoT router is the choice of a proper hardware platform. It needs to have at least two network interfaces or a possibility to add these interfaces to the hardware platform. As suitable platforms for SW router realization, Raspberry Pi (Raspberry Pi (1) Model B) and HummingBoard i2 Gate were used. Both of these have only one network port at disposal, and therefore it was necessary to expand the network ports by additional Ethernet interfaces. To do so, an external network adapter by Edimax was utilized. This adapter conforms to IEEE 802.3/802.3u standards and has USB 2.0 interface at disposal (including backwards compatibility with USB 1.1). It supports the function of auto-negotiation at speed of 10/100 Mbps. Inside the adapter, a miniature data buffer SRAM 20 kB RX a 20 kB TX is integrated [6]. It also supports low power consumption while in idle mode or under low workload (600 mW during full workload, 46 mW during remote wake-up and 0.7 mW without remote wake-up). Additionally, it does not require an external power adapter and, above all, supports GNU/Linux OS.

Next step was to choose a suitable control system that would perform individual network tasks, manage drivers, manage routing tables, and secure the functioning of specific routing protocols.

For HummingBoard hardware platform, choice of suitable operating system was based on the in-depth analysis [17], according to which the only OSs officially supported by SolidRun company are Android OS, Debian OS, and Yocto OS. Android is being massively used in touchscreen smartphones. On the other hand, Yocto OS is developed primarily for employment in embedded systems [18]. Unlike the two previous options, Debian GNU/Linux is a highly universal OS and because of that, it was used for the implementation of our software router. As for the OS for Raspberry Pi 1 Model B, we have chosen the only manufacturer-supported OS GNU/Linux distribution, which was Raspbian OS [19].

3.1   Architecture Solutions

After choosing suitable HW and SW, control platforms were chosen. Via these platforms, router administrator is going to run the whole routing process. In the research, we have decided not only to implement the aforementioned solution, but also to perform a reciprocal analysis. We have specifically used two different HW platforms with two partially different OSs, although we have used two identical control platforms for Quagga [20] and BIRD [21] SW daemons. The comparison of the principal properties of the two implemented routing platforms is presented in Table 1.

Table 1. Comparison of routing platform properties

		BIRD	QUAGGA
IP protocol	IPv4	Yes	Yes
	IPv6	Yes	Yes
Static routing		Yes	Yes
Dynamic routing protocols	RIPv1	Yes	Yes
	RIPv2	Yes	Yes
	RIPng	No	Yes
	OSPFv2	Yes	Yes
	OSPFv3	Yes	Yes (Beta version)
	IS-IS	No	Yes (Beta version)
	BGPv4	Yes	Yes
	OLSR	No	Yes (External)
	Babel	Yes	No
License		GNU/GPL	GNU/GPL
Debugging option		Complex	Complex
Support for Linux		From kernel 2.0	from kernel 2.0
Multicast		No	No
Interactive CLI		Yes (just for listing information)	Yes (IOS-like syntax)
Other interesting features and protocols		MPLS, BFD	MPLS, BFD

As for the functioning of the router, the platforms we have used are the fundamental element of our solution. An illustration of complex architecture of the solution is depicted in Fig. 1.

The overall architecture comprises of three principal parts. They are HW, OS core (depicted as KERNELMODE), and environment of OS user applications (depicted as KERNELMODE). The whole "networking" process must be processed by the OS first as the majority of network operations is asynchronous. Therefore, it is necessary to identify, order, and process individual messages before passing them to a specific process. The process of sending messages between the network interface and the user application is administered by the OS functioning in KERNEL MODE. Besides, the OS must monitor the execution of the processes based on their "network behavior". Additionally, all the problems regarding the routing or IP addresses are as well implemented in the OS core [22].

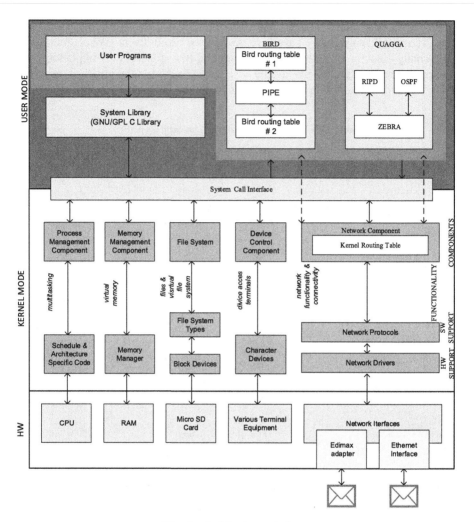

Fig. 1. Architecture solutions

If we are now going to follow the path of an incoming message, we will first arrive into network interface (in our case either Ethernet port integrated on Mini PC motherboard, or Edimax network adapter). As the network port is purely HW issue, it is necessary to firstly process the incoming message (at the moment in the form of bits) and convert it to a form accepted by higher layers. The conversion is performed by the (network) drivers. Name of the specific network driver used GNU/Linux OS can be generally found using #lspci command.

After translating the message and collating the recipient's address with own address, it is determined whether the message is designated for our machine, or for someone else. If the message is designated for an application in USER MODE, the

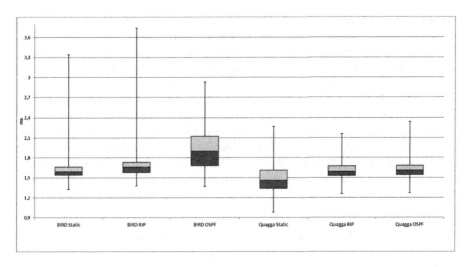

Fig. 2. Comprehensive data measurement

router starts to behave similarly to a destination station – it starts to process and "unpack" the packet.

However, if the message is designated for another station, the router first checks whether forwarding is on. If it is off, the message is immediately discarded. If forwarding is on, the router starts to perform the routing based on the routing protocol defined in Network Protocols module (Fig. 3).

This module cannot normally function on its own as it needs access to the OS core routing table (KERNEL ROUTING TABLE located in NETWORK COMPONENTS). The OS routing table contains the information about where the given message is to be forwarded. This content can be determined via the routing daemons and it can be done so either statically (static routing), or dynamically (dynamic routing). The SW routing daemons, however, unlike all the aforementioned components, work in USER MODE, and can be therefore also operated by the device administrator. It is, therefore, a fundamental control element that indirectly defines the contents of the OS core routing table, specifying the used routing protocol (via NETWORK PROTOCOLS) via SYSTEM CALL INTERFACE located between USER MODE and KERNEL MODE. The OS core subsequently sets a new address of the recipient based on the configuration and returns the packet to the drivers, which in turn send it to specific interface. This way, a new message leaves our router.

4   Comprehensive Data Measurement

To evaluate the efficiency and functionality of the proposed software routing solution, practical measurements were performed in order to allow for a comparative analysis of the routing speed. The values below stand for the time in milliseconds, which it takes before the destination. In order to measure these values SW routers implemented on

Raspberry PI and HummingBoard platforms were used. We consecutively used both BIRD and Quagga routing platforms, on which we realized static and dynamic routing via RIP and OSPF routing protocols. For every measurement we tried to procure mutually homogenous environment. We always sent exactly 500 packets. All the measured and statistically processed values are depicted in Fig. 5.

From Fig. 5 we can easily and intuitively discern individual minimum, average, and maximum response delays for every routing variant we measured. Furthermore, the values of individual quartiles (Q1, Median, and Q3) can also be easily identified. Besides the graphically represented data above, it is also appropriate to mention specific statistical calculations and values directly derived from our measurements (average response time, variation range (R), modus, dispersion (Var(x)), obliquity (G1), and sharpness (G2).

Table 2. Statistical evaluation of measured data

	BIRD			QUAGGA		
	Static	RIP	OSPF	Static	RIP	OSPF
Min	1,335	1,388	1,373	0,980	1,263	1,274
AVG	1,607	1,681	1,906	1,489	1,608	1,618
Max	3,347	3,745	2,938	2,279	2,165	2,341
Q1	1,540	1,580	1,680	1,340	1,530	1,540
Median	1,590	1,660	1,900	1,460	1,590	1,610
Q3	1,660	1,730	2,123	1,610	1,673	1,680
MOD	1,590	1,670	2,160	1,510	1,550	1,540
R	2,012	2,357	1,565	1,299	0,902	1,067
Var(x)	0,019	0,034	0,066	0,042	0,016	0,016
G1	0,013	0,031	0,004	0,005	0,001	0,001
G2	0,022	0,052	0,011	0,007	0,001	0,001

From the information from Fig. 3 and Table 2, it can be inferred that the data obtained during the test have character of normal distribution and the measured values do not encompass notable extremes, staying close to the middle values. Behavior of both tested software routers was stable and usable during the implementation of all three routing protocols.

From the viewpoint of the routing platforms themselves, Quagga daemon generally attained lower values of all the parameters during the implementation of the three routing types, i.e. it was "faster" and more stable while performing the tests.

Routing efficiency while employing static routing and dynamic routing protocols is only illustrational as they are significantly influenced by the simple topology used. It can be, however, stated that the software routing platforms do not affect the logic and efficiency of the routing and its algorithms.

Behavior of the routing platforms during the implementation of static routing and RIP and OSPF dynamic routing protocols is depicted in the following figures. If employing smoothing methods while running diagnostics of the behavior, the measured data is the most accurately reflected in 2nd period of moving average, which is defined by Eq. (1) and depicted in Figs. 4, 5, and 6.

$$\overline{y_t} = \frac{y_{t-m} + y_t + y_{t+m}}{2m+1} \tag{1}$$

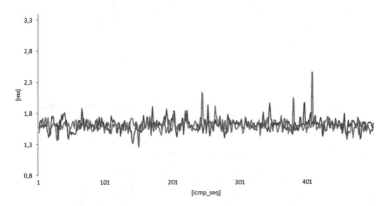

Fig. 3. 2nd period of moving average (Bird Static - red and QuaggaStatic - blue) (Color figure online)

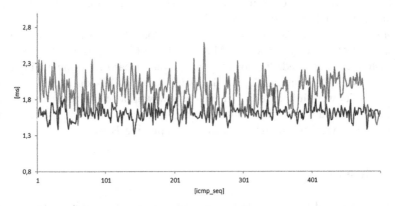

Fig. 4. 2nd period of moving average (Bird OSPF - red and QuaggaOSPF - blue) (Color figure online)

4.1 System Requirements for Routing Daemons

During the comparative analysis we also focused on comparison of the workload that the individual daemons allocate when running on the CPU of given routers. In order to simulate the workload, ICMP packets were used. Those were generated on all the network elements depicted in Fig. 2 the way that every time at one specific moment, every element in the computer network was sending ICMP packets to all other stations

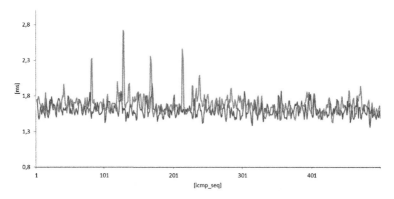

Fig. 5. 2nd period of moving average (Bird RIP - red Quagga RIP - blue) (Color figure online)

Table 3. CPU usage

	QUAGGA			BIRD		
	Static	RIP	OSPF	Static	RIP	OSPF
HummingBoard Gate	1,34%	1,16%	0,85%	0,21%	0,32%	0,17%
Raspberry Pi - Model B	1,63%	1,86%	1,25%	0,35%	0,40%	0,30%

in the topology. In doing so, during the highest-workload measurement we performed, the values listed in Table 3 were obtained.

From the measured data it can be observed that the maximum workload was reached using RIP routing protocol on Mini Raspberry PI while using Quagga software router. On the other hand, the lowest values were measured using OSPF routing protocol on Mini PC HummingBoard Gate while using BIRD software router. However, the results were influenced by relatively low workload and short routing table. Despite that, it is apparent that while employing Quagga software router the CPU workload was several times higher than when employing BIRD.

5   Conclusion

The article presented the results of the research aimed at the use of software routing implemented in Mini PC architecture for IoT networks. Two software router platforms were selected – Quagga and BIRD. Additionally, two hardware platforms were chosen – Humming-Board Gate and Raspberry Pi - Model B – that served as the base for the implementation of software routing daemons. The architecture to meet the routing requirements of the software daemons and their cooperation with Mini PC hardware layer and software platforms was designed.

During the comparative analysis, workload which the given routing daemon allocates from their respective router's CPU while running was measured. From the measured data it is apparent that the maximum workload was reached using RIP routing protocol on Mini Raspberry PI while using Quagga software router. On the other hand, the lowest values were measured using OSPF routing protocol on Mini PC HummingBoard Gate while using BIRD software router.

From the viewpoint of the routing platforms themselves, Quagga daemon generally attained lower values of all the parameters during the implementation of the three routing types, i.e. it was "faster" and more stable and without any relevant extreme values.

The fundamental result of the research is that software based routing platforms do not influence the logic and efficiency of the routing and its algorithms in no significant way. From this, it can be inferred that during their implementation in Mini PC architecture, software based routing platforms are highly efficient and can be employed in IoT networks to a full extent in order to secure the efficiency of the IoT architecture, security, and administration.

Acknowledgement. This work and the contribution were supported by project of specific science "Computer networks for cloud, distributed systems, and Internet of Things", Faculty of Informatics and Management, University of Hradec Kralove, Czech Republic. We would like also thank to students Lubos Mercl and Pavel Blazek.

References

1. Intel IoT: What Does The Internet of Things Mean? In: Youtube. Intel, Santa Clara (2014). https://www.youtube.com/watch?v=Q3ur8wzzhBU. Accessed 29 Oct 2017
2. Libelium Smart World: Libelium Comunicaciones Distribuidas S.L (2017). http://www.libelium.com/libelium-smart-world-infographic-smart-cities-internet-of-things/. Accessed 14 Jan 2018
3. Internet of Things: Roadmap to a Connected World. MIT: xPRO. MIT, Cambridge (2016). https://mitxpro.mit.edu/courses/course-v1:MITProfessionalX+IOTx+2016_T1/about. Accessed 29 Oct 2017
4. Internet of Things (IoT): The IoT links objects to the Internet, enabling data and insights never available before. Cisco: Internet of Things (IoT). Cisco Systems, San Jose (2017). https://www.cisco.com/c/en/us/solutions/internet-ofthings/overview.html# ~ stickynav=1. Accessed 29 Oct 2017
5. IBM: Watson Internet of Things (IoT). IBM, New York (2017). https://www.ibm.com/internet-of-things/. Accessed 29 Oct 2017
6. Gartner Says 8.4 Billion Connected "Things" Will Be in Use in 2017: Up 31 Percent from 2016. Gartner, Egham (2017). https://www.gartner.com/newsroom/id/3598917. Accessed 29 Oct 2017
7. Horalek, J., Matyska, J., Sobeslav, V.: Comparative analysis of software routers on Linux. In: IEEE 15th International Symposium on Computational Intelligence and Informatics (CINTI), pp. 295–300. IEEE (2014). https://doi.org/10.1109/cinti.2014.7028693. Accessed 12 Jan 2018. ISBN 978-1-4799-5338-7
8. Kelly, S.D.T., Suryadevara, N.K., Mukhopadhyay, S.C.: Towards the implementation of IoT for environmental condition monitoring in homes. IEEE Sens. J. **13**(10), 3846–3853 (2013). https://doi.org/10.1109/jsen.2013.2263379. Accessed 12 Jan 2018. ISSN 1530-437x

9. Yashiro, T., Kobayashi, S., Koshizuka, N., Sakamura, K.: An Internet of Things (IoT) architecture for embedded appliances. In: IEEE Region 10 Humanitarian Technology Conference, pp. 314–319. IEEE (2013). https://doi.org/10.1109/r10-htc.2013.6669062. Accessed 12 Jan 2018. ISBN 978-1-4673-5963-4

10. Yun, M., Yuxin, B.: Research on the architecture and key technology of Internet of Things (IoT) applied on smart grid. In: International Conference on Advances in Energy Engineering, pp. 69–72. IEEE (2010). https://doi.org/10.1109/icaee.2010.5557611. Accessed 12 Jan 2018. ISBN 978-1-4244-7831-6

11. Argyraki, K., Ratnasamy, S., Baset, S., et al.: Can software routers scale? In: Proceedings of the ACM Workshop on Programmable Routers for Extensible Services of Tomorrow, PRESTO 2008, p. 21. ACM Press, New York (2008). https://doi.org/10.1145/1397718. 1397724. Accessed 18 Dec 2017. ISBN 9781605581811

12. Mark, B.L., Zhang, S., Mcgeer, R., Brassil, J., Sharma, P., Yalagandula, P.: Performance of an adaptive routing overlay under dynamic link impairments. In: IEEE Military Communications Conference, MILCOM 2007, pp. 1–7. IEEE (2007). https://doi.org/10.1109/milcom.2007.4455100. Accessed 18 Dec 2017. ISBN 978-1-4244-1512-0

13. Horalek, J., Matyska, J., Sobeslav, V.: Comparative analysis of software routers on Linux. In: IEEE 15th International Symposium on Computational Intelligence and Informatics (CINTI), pp. 295–300. IEEE (2014). https://doi.org/10.1109/cinti.2014.7028693. Accessed 18 Dec 2017. ISBN 978-1-4799-5338-7

14. Quagga Software Routing Suite: NonGNU (2013). http://www.nongnu.org/quagga/index. html. Accessed 27 Oct 2017

15. Filip, O.: The BIRD Internet Routing Daemon Project. CZ.NIC, Praha (2004). http://bird. network.cz/. Accessed 26 Oct 2017

16. USB 2.0 Fast Ethernet Adapter: Edimax.com. Edimax, New Taipei City (2014). http://www. edimax.com/edimax/mw/cufiles/files/download/datasheet/transfer/USB/EU-4208/EU-4208_ v2_Datasheet.pdf. Accessed 26 Oct 2017

17. Official IMX6 Distributions: Wiki.solid-run.com. SolidRun, Yokne'am Illit (2016). https:// wiki.solid-run.com/doku.php?id=products:imx6:software:os:officiall. Accessed 26 Oct 2017

18. Introducing the Yocto Project: Yoctoproject.org. Linux Foundation, San Francisco (2016). https://www.yoctoproject.org/tools-resources/videos/introducing-yocto-project. Accessed 26 Oct 2017

19. Raspberry Pi: Raspberry Pi Foundation, Cambridge (2017). https://www.raspberrypi.org/. Accessed 26 Oct 2017

20. Quagga Routing Suite: Quagga Routing Software Suite, GPL licensed (2017). http://www. nongnu.org/quagga/. Accessed 10 Nov 2017

21. The BIRD Internet Routing Daemon: Ondrej Filip (2013). http://bird.network.cz/. Accessed 10 Nov 2017

22. Linux Kernel and Its Architecture: Knowstuffs (2012). https://knowstuffs.wordpress.com/ 2012/06/11/linux-kernel-andarchitecture/. Accessed 26 Oct 2017

Data Mining Methods and Applications

SVM Parameter Optimization Using Swarm Intelligence for Learning from Big Data

Yongquan Xie[1], Yi Lu Murphey[1]($\boxtimes$)(iD), and Dev S. Kochhar[2]

[1] University of Michigan-Dearborn, Dearborn, MI 48128, USA
{yongquan,yilu}@umich.edu
[2] Ford Motor Company, Dearborn, MI 48126, USA
dkochhar@ford.com

Abstract. Support vector machine (SVM) is one of the most successful machine learning algorithms to solve practical pattern classification problems. The selection of the kernel function and its parameter plays a vital role on the results. Radius basis function (RBF) is a prevalently used kernel. For an RBF-SVM, two parameters, c and γ, control the SVM performance. In this paper, we present a SVM parameter learning algorithm, DL&BA, effective for learning from big data. The DL&BA algorithm has two stages. At the first stage, we use a distributed learning (DL) to search for a region which promises optimal parameter pairs. At the second stage, a swarm intelligent optimization algorithm - the Bees Algorithm (BA) is used to search for an optimal pair of c and γ. We applied the DL&BA algorithm to solving an important automotive safety problem, driver fatigue detection, which involves a large amount of real-world driving data. Our experimental results show that DL&BA is not only computational efficient but also effective in finding an optimal pair of c and γ.

Keywords: Support vector machine · Bees Algorithm
Fatigue detection

1 Introduction

SVM is a supervised learning method popularly used for classification and regression. SVM was initially introduced by Vapnik in 1990s [2], and has been successfully applied to a wide range of applications including vehicle fault diagnostics [5], driver state classifications [17], housing price prediction [3], etc. In constructing an SVM model, an important step is to select proper model parameters associated with selected kernel function. A popular kernel function used in a SVM model is the Gaussian Radial Basis Function (RBF). For a RBF-SVM, the two parameters control the performance of the SVM are c and γ.

This work is supported in part by research grants from Michigan Institute of Data Science (MIDAS), ZF-TRW, and Ford Motor Company.

N. T. Nguyen et al. (Eds.): ICCCI 2018, LNAI 11055, pp. 469–478, 2018.
https://doi.org/10.1007/978-3-319-98443-8_43

The penalty parameter c determines the trade-off between the minimization of the fitting error and the maximization of the between-class margin. The parameter γ determines the bandwidth of RBF. Section 2 presents a brief discussion of the existing techniques for parameter selection for SVMs.

In this paper we present an intelligent parameter search algorithm, Distributed Learning and Bees Algorithm (DL&BA), that is designed to select an optimal parameter pair of c and γ based on a large training data. The DL&BA algorithm has two stages. At the first stage, we use a distributed learning (DL) to search for a region which promises optimal parameter pairs. At the second stage, a swarm intelligent optimization algorithm - the Bees Algorithm (BA) is used to search for an optimal pair of c and γ. By utilizing a swam-based intelligent search technique, the DL&BA algorithm has the capability of finding an optimal solution located either within or outside the original searching regions, on or between the search grids. We applied the DL&BA algorithm to solving an important automotive safety problem, driver fatigue detection, which involves a large amount of real-world driving data. Our experimental results show that DL&BA is not only computational efficient but also effective in finding an optimal pair of c and γ. More details will be given in Sect. 3. Throughout the paper, $[c, \gamma]$ is used to represent a parameter pair.

2   Preliminaries

The distribution of optimal values for $[c, \gamma]$ could be a multimodal surface in a domain [4]. The grid-search using cross-validation is the most straight forward and commonly used approach to identify a reasonable parameter pair. But for training data set of large size, Grid-search shows very costly in computation due to exhaustive search strategy. Therefore, advanced methods have been consistently putting forward such as Hybrid Genetic Algorithm (GA) [12], Particle Swarm Optimization (PSO) [6], Ant Colony Optimization (ACO) [19], and Bat Algorithm [11] to find proper $[c, \gamma]$. It should be noted that these algorithms can only run properly when they are configured appropriately. For example, GA requires population size, crossover rate and mutation rate to be set and PSO requires the configuration of population size and two learning factors. The choice of these parameters can have a large impact on optimization performance, and there is no universal criterion available for the configuration. In addition, large amount of existing works do not investigate how to configure these parameters. The DL&BA, however, aims to optimize SVM parameters selection process at a controllable computational complexity. Although it also has several configurable parameters, the settings of them is simple, and our experiments show the ultimate performance is not sensitive to a variation of their values.

The Bees Algorithm (BA) is a swarm intelligent optimization tool originally proposed by Pham et al. [9] in 2005. It is elaborated in [8] as well. Inspired by foraging behavior of honey bees in nature, it uses a population of artificial bees to discover optimal solutions. In the natural food foraging process, a population of honeybees is divided into two main types: scouts and foragers. Food collection begins by scouts being sent to explore randomly areas surrounding the hive to find food sources. Scouts that have discovered sources providing high-quality nectar or pollen, return to the hive and perform a "Waggle Dance" to advertise the locations of the food sources they found. Foragers in the hive watch this waggle dance and join the dancers in harvesting the food source. After performing the waggle dance, the scouts return to the food source for a new harvest. In BA, the artificial bees detect and gather their food using two distinct search methods, namely, neighborhood search (or local search) and global search. As in nature, local search is carried out by foragers primarily to exploit profitable food sources, while global search is performed by scouts to locate potential food sources. This algorithm has been improved and applied to solve many problems in both numerical optimization and practical applications [14–16, 20].

3   The Proposed DL&BA Algorithm

The DL&BA algorithm consists of two stages, (i) distributed learning on a large training data, and followed by (ii) Bees Algorithm to exploit significant pairs generated by stage one, and conducts parallel random search in global area. The first stage is designed for finding significant parameter pairs through distributed learning, so as to reduce computational cost. The second stage uses the significant pairs obtained at the first stage as initial seeds. Details are described below.

At the first stage, a training data set Ω^{tr} is randomly partitioned into n_c sets, Ω_i^{tr}, $i = 1, 2, ..., n_c$, each of the partitions is used as training data for coarse grid-search. Let $n_{c\gamma}$ to be the number of $[c, \ \gamma]$ over the grids. The parameter pairs occurring in $S^c = \cap_{i=1}^{n_c} S_i^c$ are considered globally significant. Algorithm 1 summarized the computational steps of this stage.

Algorithm 1. First stage of DL&BA

Input: Ω^{tr}
1: Divide Ω^{tr} into n_c subsets Ω_i^{tr}.
2: **for** $i = 1$ to n_c **do**
3:     Perform grid-search on Ω_i^{tr} and obtain c-v accuracy of the SVM trained by each pair on grid.
4:     Select $k_1 \cdot n_{c\gamma}$ pairs that have the highest accuracy and put into S_i^c
5: **end for**
6: Obtain $S^c = \cap_{i=1}^{n_c} S_i^c$
Output: S^c

The second stage is presented in Algorithm 2. The entire training data Ω^{tr} is used by the Bees Algorithm to evaluate parameter pairs. The Bees Algorithm performs a combination of exploitation on the existing significant solutions by using a neighborhood search, and exploration on the solution space by global random search. The integration of the two types of search techniques enhances the algorithm's ability to avoid local optima and locate global optimum. The original Bees Algorithm begins by initializing a population of scout bees with random solutions. However, in DL&BA, the population of the scouts is initialized by $S^f \subset S^c$, satisfying $|S^f| = k_2 \cdot |S^c|$ ($|\cdot|$ gives the size of the set) and S^f are top-performed pairs on Ω^{tr} evaluated in a cross-validation process. S^f is then used by the Bees Algorithm as initial seeds to exploit through neighborhood search. Equation 1 formulates the generation of artificial forager bees denoted by $[c_i^{ngh}, \gamma_i^{ngh}]$ in neighborhood,

$$\begin{cases} c_i^{ngh} = c + r_1 \times c_{step} \\ \gamma_i^{ngh} = \gamma + r_2 \times \gamma_{step} \end{cases} \tag{1}$$

where c and γ represent a significant parameter pair in S^f, c_i^{ngh} and γ_i^{ngh} is a pair generated in the neighborhood of $[c, \gamma]$, i indexes the number of pairs generated in neighborhood of a significant pair, r_1 and r_2 are random values uniformly distributed in $(-1, 1)$. Each pairs generated in a neighborhood is used to train an SVM using cross-validation on data Ω^{tr}. In an iteration of neighborhood search around of a pair, greedy selection is implemented in the way that only the pair of the best performance is kept for the next iteration.

Algorithm 2. Second stage of DL&BA

Input: S^c, Ω^{tr}
 1: Obtain c-v accuracy of each pair in S^c on Ω^{tr}, sort them in descending order, put the top $k_2 \cdot |S^c|$ pairs in S^f.
 2: **while** have not met stopping criteria **do**
 3:     **for** each $[c, \gamma]$ in S^f **do**
 4:         Generate pairs in neighborhood using Equation 1.
 5:         Train SVM on Ω^{tr} with pairs generated in neighborhood and obtain c-v accuracy.
 6:         Perform greedy-selection.
 7:     **end for**
 8:     **for** $i = 1$ to $(1 - k_2) \cdot |S^c|$ **do**
 9:         Generate pairs for global search using Equation 2.
10:         Obtain c-v accuracy on Ω^{tr} of generated pairs.
11:     **end for**
12:     Sort all pairs (neighborhood & gloabl search) according to c-v accuracy in descending order, and update S^f
13: **end while**
14: $[c*, \gamma*] =$ the first one of the ranked pairs
Output: $[c*, \gamma*]$

The number of global random pair for exploratory search is $(1 - k_2) \cdot |S^c|$. Equation 2 formularize the generation of random pairs,

$$\begin{cases} c_i^{glb} = c_{bdmax} + r_3, \; or \; c_i^{glb} = c_{bdmin} - r_3 \\ \gamma_i^{glb} = \gamma_{bdmax} + r_4, \; or \; \gamma_i^{glb} = \gamma_{bdmin} - r_4 \end{cases} \tag{2}$$

where $c_{bdmax}, c_{bdmin}, \gamma_{bdmax}, \gamma_{bdmin}$ are the boundaries of the significant pairs in S^c, c_i^{glb} and γ_i^{glb} is a pair generated for global search, i indexes the number of pairs for global search, r_3 and r_4 are random values following Gaussian distribution with zero mean and standard derivation of c_{step} and γ_{step}, respectively. As well, the performance of a pair generated is evaluated by the c-v accuracy of the SVM trained by Ω^{tr} using this pair.

In each iteration, the population of the Bees Algorithm is sorted in descending order according to the c-v accuracy, only the top $k_2 \cdot |S^c|$ pairs are selected for neighborhood search in the next iteration. The stopping criteria are set to be (i) the maximum iteration, and (ii) the highest c-v accuracy of the entire bees population is not improved for the consecutive lim iterations (which is called the limit of stagnation in the original Bees Algorithm). Meeting either of the two conditions will terminate the algorithm.

The computational complexity is analyzed as follows. For a given $[c, \gamma]$ pair and training data set Ω^{tr}, the computational complexity of obtaining an SVM model with RBF kernel is $O(dn_s^2)$ [1], where $n_s = |\Omega^{tr}|$, and d the sample dimension. To optimize $[c, \gamma]$, the grid-search method must evaluate $n_{c\gamma}$ pairs of $[c, \gamma]$ in search domain. Therefore the computational complexity of finding S_i^c in the first stage from one partition is $n_{c\gamma} \cdot O(dn_s^2)$. To simplify the analysis by omitting insignificant items, the overall complexity of DL&BA can be deduced to $p \cdot O[d(\beta \cdot n_s)^2] + q \cdot O[d(n_s)^2]$, where $\beta = \dfrac{1}{n_c} \ll 1$, p and q denote the number of evaluations has to be made to rank parameter pairs in the first stage and second stage, respectively. It is noticeable that for even larger training data set, increasing n_c will reduce the algorithm's computation cost.

4 Case Study: Driver Fatigue Detection

In this section, we present our research in applying the DL&BA to building an SVM system for driver fatigue detection from a large training data set. More details about this application can be found in exiting studies [7,10,13,18] and will not be repeated here.

We collected real-world driving data that contained 29 trips taken by 20 drivers with ages ranging from 30 to 50 traveling on highways and local streets in the U.S. Each trip is about 50 min long. Each data sample at time instance t is labeled with either "1" as alert (non-fatigue) or "0" as fatigue by human experts. Each trip contains video captured by a camera mounted inside the vehicle to capture the front view of the car. We use three signals extracted from video data for driver fatigue detection, lane position (L_p), lane heading (L_h), and lateral distance ($D_{lateral}$). [13] is a previous study, where more detailed descriptions can be found.

After pre-processing procedures, the data collected, denoted as Ω, has about 608k samples. It is partitioned randomly into Ω^{tr} and Ω^{te}, which have 405k and 203k samples respectively. In our previous studies, we set $c = 2^{-5}, 2^{-4}, \cdots, 2^{12}$ and $\gamma = 2^{-10}, 2^{-14}, \cdots, 2^7$. We found promising parameters often appear within the region $c = [2^2, 2^7]$ and $\gamma = [2, 2^6]$. So in this study, we expand the each region to $c = (0, 200]$ and $\gamma = (0, 90]$ with linear c_{step} and γ_{step}.

Influence of k_1 and n_c. These two variables are used in the first stage. k_1 controls the number of significant parameter pairs in S^c. Small k_1 value would lead to very small $|S^c|$ that may not have sufficient number of parameter pairs for the second stage to learn from. Large k_1 value would generate large S^c that may require too much computation time. n_c determines the number of partitions Ω_i^{tr} that Ω^{tr} is divided into. Small n_c would generate large partition size $|\Omega_i^{tr}|$ that can make computational cost of the first stage too high. Large n_c may give too many small partitions that are not representative to the entire training data. We investigate the influence of k_1 and n_c by fixing one value and increasing the other one from small to large. Figure 1a and b show the c-v accuracy obtained by $[c*, \gamma*]$ generated by DL&BA when k_1 and n_c vary respectively. An obvious rise in c-v accuracy can be observed when k_1 or n_c grows at the beginning. However, the c-v accuracy is not sensitive to further growth of them. Also, it can be observed that, with the $[c, \gamma]$ found by DL&BA, the c-v accuracy can achieve the exhaustive grid-search (as it obtains the most convincing pair). Figure 2a and b show that larger k_1 and smaller n_c result in higher computational cost. Therefore, for the data of current scale, setting k_1 to 10%–12%, and n_c to 10–20 is suggested.

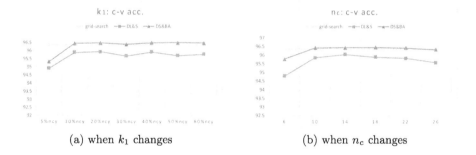

(a) when k_1 changes   (b) when n_c changes

Fig. 1. Comparison of c-v accuracy as k_1 and n_c changes

Influence of k_2 and n_{ngh}. k_2 determines the number of pairs in S^f to perform neighborhood search $(k_2 \cdot |S^c|)$ and the number of pairs used in the global random search $((1-k_2) \cdot |S^c|)$. Therefore, k_2 controls the equilibrium of exploitation and exploration ability of the Bees Algorithm. n_{ngh} specifies the number of pairs generated in the neighbor of a pair. Large n_{ngh} allows the algorithm to search more intensively around a significant pair, but at the same time elevates the computation cost. Figure 3a shows that increasing k_2 can initially enhance the c-v accuracy, which then reaches a plateau. The computational cost, however, is elevated along with the growth of k_2. Similarly in Fig. 3b, the variation of n_{ngh} shows that more extensive search in a neighborhood leads to higher computational cost. It is recommended to use $k_2 = 70\%$, and $n_{ngh} = 3$.

Influence of c_{step} and γ_{step}. c_{step} and γ_{step} determines search resolution in the first stage. Generally, using lower resolution can reduce the time complexity,

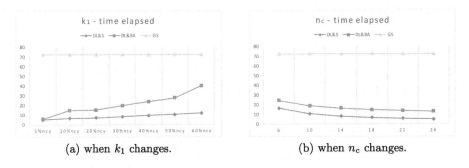

(a) when k_1 changes.   (b) when n_c changes.

Fig. 2. Comparison of computational time as k_1 and n_c changes.

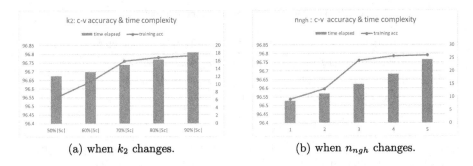

(a) when k_2 changes.   (b) when n_{ngh} changes.

Fig. 3. Comparison of c-v accuracy and computational time as k_2 and n_{ngh} changes

but at the same time increase the likelihood of missing a good parameter pair. Since the second stage of DL&BA incorporates a heuristic search method, the DL&BA as a whole should be able to capture optimal parameter pair even with lower search resolution at the beginning. Figure 4a and b show how the growth of c_{step} and γ_{step} can respectively affect the c-v accuracy on Ω^{tr}. The vertical axis on the left represents accuracy (for the curve), and the axis on the right represent $n_{c\gamma}$ - total number of $[c, \gamma]$ to be evaluated for one partition. It is obvious that computational cost of the first stage drops in exponential scale (the larger $n_{c\gamma}$ is, the more complex the first stage is) when c_{step} or γ_{step} grows. However, the DL&BA's ability of obtaining a pair that produces high accuracy is not degraded. For example, with c_{step} growing from 5 to 40, or γ_{step} growing from 5 to 35, the accuracy obtained by the algorithm remains stable (only slight drop of 0.2%).

Performance on Testing Data. The performances of the DL&BA are evaluated using testing data set Ω^{te}. The algorithm is configured as follows: $k_1 = 10\%$ and $n_c = 18$, $k_2 = 70\%|S^c|$, and $n_{ngh} = 3$. The stopping criteria are (i) maximal iteration $= 15$, and (ii) $lim = 3$. Table 1 presents the performances of three parameter search algorithm. Grid-search (a popular exhaustive search method), DL&S, and DL&BA. It can be seen that (i) the DL&BA is able to find an

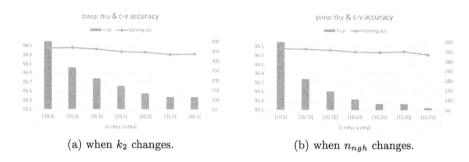

(a) when k_2 changes.   (b) when n_{ngh} changes.

Fig. 4. Comparison of c-v accuracy and complexity as c_{step} and γ_{step} changes

Table 1. Algorithms' performance comparison

Algorithm	$[c^*, \gamma^*]$	Tr. acc.	Te. acc.	Time elapsed
Grid-search	$[120, 25]$	96.36%	90.21%	70.8 h
DL&S [13]	$[110, 25]$	95.89%	90.05%	6.7 h
DL&BA	$[113.373, 27.069]$	96.45%	90.29%	12.6 h

optimal parameter pair outside of the pre-defined parameter domain and off the pre-defined grids; (ii) DL&BA gave the highest c-v accuracy on Ω^{tr} and testing accuracy on Ω^{te}; (iii) both DL&BA and DL&S show significant reduction in computational cost. The DL&BA and DL&S algorithm have exactly the same search method at the first stage. But the second stage of DL&BA involves iterations of parameter search within and outside the region boundaries and between grid points, and evaluation of the SVMs generated by these parameter pairs on the entire training data Ω^{tr}. Therefore, the computational cost of DL&BA is higher than DL&S. However, in comparison with the grid-search, the DL&BA algorithm used only 18% of the computational time taken by the grid-search algorithm.

We also conducted an experiment to evaluate the performance of purely second stage with random initialization on Ω^{tr}. On the same platform, it took more than 3 days to converge to a solution that performs equivalently to DL&BA. Therefore, only using the BA to search without considering initialization is not recommended (even simple grid-search is more efficient than purely BA search).

5   Conclusion

In this paper, we present a heuristic parameter search algorithm DL&BA for building a RBF-SVM effectively as well as efficiently from big training data. The DL&BA algorithm consists of two stages. At the first stage, a distributed grid-search method is used to a set of significant parameters $([c, \gamma])$. The Bees Algorithm uses the significant parameter pairs generated by the first stage as seed parameters to find optimal parameter pairs through both the neighborhoods and global search procedures. We applied the DL&BA algorithm to solving a driver fatigue detection problem, which has a training data of large size (608k) collected from real-world driving trips. We showed in this study that for a data set of size at scale around 10^5, it is appropriate to use the following parameter settings in the DL&BA algorithm: k_1 to 10%–20%, n_c to 10–20, k_2 to 70% (due to the nature of BA), and n_{ngh} to 3 based on two considerations: (1) efficient search time and effective search results. We demonstrated through experiment results that the parameter values in these ranges generate the SVMs that give similar c-v accuracy. Our experimental results show that the DL&BA can achieve the same level in c-v and testing accuracy as exhaustive search but at lower time cost. Another significant contribution of DL&BA is that, due to the incorporation of the BA, it has the capability of searching not only at cross points of grids defined by search steps of c and γ, but also find good parameters located inside or outside the bounding areas of significant parameter pairs set generated by the first stage of the DL&BA algorithm, which implies that it could potentially find optimal parameter pairs outside the original grid search region. DL&BA also has the capability to scale up to big data by adjusting its control parameters, such as increasing the n_c in the first stage, and/or reducing the population in BA when training data set becomes bigger. These will be further investigated in our future research.

References

1. Burges, C.J.: A tutorial on support vector machines for pattern recognition. Data Min. Knowl. Discov. **2**(2), 121–167 (1998)
2. Drucker, H., Wu, D., Vapnik, V.N.: Support vector machines for spam categorization. IEEE Trans. Neural Netw. **10**(5), 1048–1054 (1999)
3. Gu, J., Zhu, M., Jiang, L.: Housing price forecasting based on genetic algorithm and support vector machine. Expert Syst. Appl. **38**(4), 3383–3386 (2011)
4. Hsu, C.W., Chang, C.C., Lin, C.J., et al.: A practical guide to support vector classification (2003)
5. Jegadeeshwaran, R., Sugumaran, V.: Fault diagnosis of automobile hydraulic brake system using statistical features and support vector machines. Mech. Syst. Signal Process. **52**, 436–446 (2015)
6. Lin, S.W., Ying, K.C., Chen, S.C., Lee, Z.J.: Particle swarm optimization for parameter determination and feature selection of support vector machines. Expert Syst. Appl. **35**(4), 1817–1824 (2008)
7. Mandal, B., Li, L., Wang, G.S., Lin, J.: Towards detection of bus driver fatigue based on robust visual analysis of eye state. IEEE Trans. Intell. Transp. Syst. **18**(3), 545–557 (2017)
8. Pham, D., Ghanbarzadeh, A., Koc, E., Otri, S., Rahim, S.: The bees algorithm - a novel tool for complex optimisation. In: Intelligent Production Machines and Systems-2nd I*PROMS Virtual International Conference, 3–14 July 2006. sn (2011)
9. Pham, D., Ghanbarzadeh, A., Koc, E., Otri, S., Rahim, S., Zaidi, M.: The bees algorithm. Technical note. Manufacturing Engineering Centre, Cardiff University, UK, pp. 1–57 (2005)

10. da Silveira, T.L., Kozakevicius, A.J., Rodrigues, C.R.: Automated drowsiness detection through wavelet packet analysis of a single EEG channel. Expert Syst. Appl. **55**, 559–565 (2016)
11. Tharwat, A., Hassanien, A.E., Elnaghi, B.E.: A BA-based algorithm for parameter optimization of support vector machine. Pattern Recognit. Lett. (2016)
12. Wu, C.H., Tzeng, G.H., Lin, R.H.: A novel hybrid genetic algorithm for kernel function and parameter optimization in support vector regression. Expert Syst. Appl. **36**(3), 4725–4735 (2009)
13. Xie, Y., Bian, C., Murphey, Y.L., Kochhar, D.S.: An SVM parameter learning algorithm scalable on large data size for driver fatigue detection. In: IEEE Symposium Series on Computational Intelligence. IEEE Computational Intelligence Society (2017)
14. Xie, Y., Zhou, Z., Pham, D.T., Liu, Q., Xu, W., Ji, C., Lou, P., Tian, S.: A forager adjustment strategy used by the bees algorithm for solving optimization problems in cloud manufacturing. In: ASME 2015 International Manufacturing Science and Engineering Conference, p. V002T04A014. American Society of Mechanical Engineers (2015)
15. Xie, Y., Zhou, Z., Pham, D.T., Xu, W., Ji, C.: A multiuser manufacturing resource service composition method based on the bees algorithm. Computational Intell. Neurosci. **2015**, 12 (2015)
16. Xu, W., Tian, S., Liu, Q., Xie, Y., Zhou, Z., Pham, D.T.: An improved discrete bees algorithm for correlation-aware service aggregation optimization in cloud manufacturing. Int. J. Adv. Manuf. Technol. **84**(1–4), 17–28 (2016)
17. Yeo, M.V., Li, X., Shen, K., Wilder-Smith, E.P.: Can SVM be used for automatic EEG detection of drowsiness during car driving? Saf. Sci. **47**(1), 115–124 (2009)
18. Zhang, W., Murphey, Y.L., Wang, T., Xu, Q.: Driver yawning detection based on deep convolutional neural learning and robust nose tracking. In: International Joint Conference on Neural Networks (IJCNN), pp. 1–8. IEEE (2015)
19. Zhang, X., Chen, X., He, Z.: An ACO-based algorithm for parameter optimization of support vector machines. Expert Syst. Appl. **37**(9), 6618–6628 (2010)
20. Zhou, Z., Xie, Y., Pham, D., Kamsani, S., Castellani, M.: Bees algorithm for multimodal function optimisation. Proc. Instit. Mech. Eng. Part C J. Mech. Eng. Sci. **230**(5), 867–884 (2016)

A CNN Model with Data Imbalance Handling for Course-Level Student Prediction Based on Forum Texts

Phuc Hua Gia Nguyen[(✉)] and Chau Thi Ngoc Vo

Faculty of Computer Science and Engineering, Ho Chi Minh City University of Technology, Vietnam National University, Ho Chi Minh City, Vietnam
{1412961, chauvtn}@hcmut.edu.vn

Abstract. Nowadays teaching and learning activities in a course are greatly supported by information technologies. Forums are among information technologies utilized in a course to encourage students to communicate with lecturers more outside a traditional class. Free-styled textual posts in those communications express the problems that the students are facing as well as the interest and activeness of the students with respect to each topic of a course. Exploiting such textual data in a course forum for course-level student prediction is considered in our work. Due to hierarchical structures in course forum texts, we propose a method in this paper which combines a deep convolutional neural network (CNN) and an adopted and adapted loss function for more correct recognitions of instances of the minority class which includes students with failure. A CNN model with data imbalance handling is a novel method appropriate for the course-level student prediction task. Indeed, through an empirical evaluation, our method has been confirmed to be an effective solution. Compared to other methods such as C4.5, Support Vector Machines, and Long-Short Term Memory networks, the proposed method can provide higher Accuracy, Precision, Recall, and F-measure on average for early predictions of the students with either success or failure in two different real courses. Such better predictions can help both students and lecturers beware of students' study and support them in time for ultimate success in a course.

Keywords: Convolutional neural network · Mean squared error
Mean squared false error · Data imbalance · Student prediction

1 Introduction

Nowadays, using a forum in a course is popular for teaching/learning activities. It is very important for distance courses where interactions between students and lecturers take place mainly. Nevertheless, it is also helpful for traditional courses to enhance out-class interactions. A lot of contents have been generated in those forums and become a valuable source for knowledge discovery in the education domain. Indeed, many related works in [6–9, 12, 19–23] have utilized the forum discussions for discovered knowledge about students and their study. In this paper, our work defines a solution to a course-level student prediction task based on forum texts.

© Springer Nature Switzerland AG 2018
N. T. Nguyen et al. (Eds.): ICCCI 2018, LNAI 11055, pp. 479–490, 2018.
https://doi.org/10.1007/978-3-319-98443-8_44

In the educational data mining area, there are a large number of the existing works such as [2, 4–9, 11, 12, 14, 16, 18–23]. Several approaches have been proposed with supervised learning methods [2, 12], semi-supervised learning ones [11, 14, 18], and recently deep learning ones [22]. Compared to these works, our work exploits forum texts for student predictions while [2, 4, 11, 12, 14, 16, 18] did not. In addition, deep learning is one of the most recent methods considered in this area. Only [22] and ours have investigated deep learning-based solutions. In [22], a deep fully connected feed-forward neural network was defined for drop-out predictions in massive open online courses (MOOCs). Different from [22], a deep convolutional neural network (CNN) is proposed in our work with data imbalance handling in traditional courses using forums. Not only [22] but also many other works based on forum texts for knowledge discovery has not yet addressed the data imbalance issue in their tasks.

Regarding data imbalance handling, we have studied several existing solutions in deep learning. Many of them have been defined with promising results. However, most of them such as [3, 10] have investigated the methods and examined those on image data. As an exception, [17] has introduced an algorithm-level solution for deep neural networks with new loss functions and examined them on both image and textual data with many imbalance ratios. Two new loss functions are mean false error (MFE) and mean squared false error (MSFE) which is an improved version of the previous MFE one. Therefore, in our work, the MSFE loss function is adopted and adapted to our CNN for the task. Compared to [12] where a cost matrix must have been pre-defined manually for data imbalance handling, our work deals with this data imbalance issue automatically in the learning process.

Defining a CNN with data imbalance handling, our work is the first one providing students and lecturers with predictions of students' final study status after the end of a course based on their texts posted in a course forum. Via experimental results on real data from two courses of the same subject in two contiguous academic years, the deep learning method in our solution has higher accuracy than other traditional and deep learning methods. It can achieve more than 95% correct predictions. In addition, it can greatly improve the correct recognitions of instances in the minority class with the MSFE loss function compared to those with the mean squared error (MSE) and cost-based strategies. With such results, we believe that the outcome of our prediction task can be used to support both students and lecturers to forecast the students' study performance after each course. Warning messages can be sent to them as soon as possible before or by midterm break so that they can have enough time to prepare more suitable learning and teaching activities for in-trouble students predicted with failure.

2 A Course-Level Student Prediction Task

Different from our previous works in [14, 18], this work aims at a solution to a course-level student prediction task. The task is to predict a final study status of a student at the course level. A final study status is either pass or fail a course. Early prediction is expected before the end of a course so that both student and lecturer can beware of the student's current study performance and prepare more for an ultimate success in the course.

To perform this task, each student is represented in a computational form. A vector space model is used. Instead of examining the study performance of each student for constructing a prediction model, our work considers the discussions and feedbacks of students. Those are collected from student's posts in a course forum. Texts from each student are transformed into a vector in a p-dimensional space in the preprocessing phase. For a student in the previous courses in the past, his/her status is known. Their data are then used in the model building phase. For a student in the current course, his/her status is going to be predicted in the predicting phase.

Using the extracted and transformed data, a supervised learning process is conducted. Due to textual data characteristics, our work is developed with deep learning. In addition, the number of course-failing students is less than the number of course-succeeding students. Data imbalance needs to be considered. A convolutional neural network model with a loss function for data imbalance handling is thus defined and proposed as our task solution in the next section.

3   The Proposed Method

3.1   Method Description

In this subsection, our method is sketched in Fig. 1 with three main phases as briefly mentioned in Sect. 2: *(1) Preprocessing, (2) Model building*, and *(3) Predicting*.

In the *Preprocessing* phase, raw texts are gathered from a course forum using a crawler. Each message in the raw texts is timestamped. Their spellings, short forms, and tokenization are studied. Those actions are related to the performing of a text preprocessor. Its output includes preprocessed texts which are then processed by a *word2vec* model [13] to obtain vectors in a vector space model. A *skip-gram* model has been built and used in our work. As a result, each student is represented as a vector, reflecting contents and interactions which have been created in the course forum.

In the *Model building* phase, a training data set is generated from the vectors of the known students with their course final status in the past. If the student fails the course, one (1) is used to present his/her course final status. Otherwise, zero (0) is used. A deep convolutional neural network is then defined with a loss function different from the traditional mean squared error (MSE) function to handle data imbalance. A prediction model is built with special consideration on predicting the instances of the minority class instead of treating all the instances equally. The resulting model is then evaluated to ensure its effectiveness. The output of this phase is a validated model.

In the *Predicting* phase, a validated model is used to predict a course final status of a current student. It receives only the vector representing a current student and then output 0 for predicting the success in the course and 1 for predicting the failure in the course of the corresponding student. In our work, such predictions are made as soon as possible before the end of the course. Our study uses the midterm break as a critical point in time for early predictions at the course level.

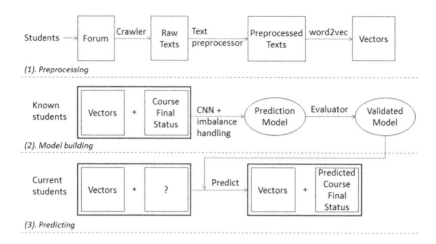

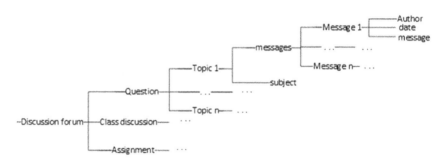

Fig. 1. Description of the proposed method.

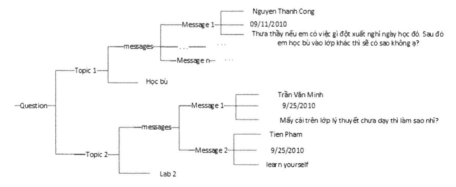

Fig. 2. Data hierarchy in a course forum.

Fig. 3. Sample data based on the previous data hierarchy in a course forum.

3.2 From Input Data to Vectors in the Computational Form

In a course forum, aka discussion forum, there are many different threads. Each thread has many topics discussed by both students and lecturers via corresponding messages. Each message includes the author information, timestamps, and textual content. So, we explore textual data in a course forum in a data hierarchy shown in Fig. 2. Sample data for each message is given in Fig. 3.

For each student, we gather his/her posts together in sequence based on the timestamp of each post. They are then tokenized and transformed into a p-dimensional vector in a vector space model. Although flattened as a sentence, their hierarchical structure is inherently captured in the texts along the time axis.

For vector modeling of our textual data, a *skip-gram* model is built using all the texts in a course forum. It is chosen due to its semantic support stated in [13]. Each token in all the post texts of each student s_j is represented by a vector v_i drawn from the resulting *skip-gram* model. These vectors are then placed next to each other to be a larger vector representing all the post texts of each student.

In order to make all the vectors s_j representing students have the same size, i.e. the same dimensionality, the maximum length *max_length* of the largest vector is selected as a standard dimensionality. Zeros padding is then used to align other smaller vectors with this standard dimensionality. As compared to the vector representation of forum texts in [15], data hierarchy is embedded in our vector representation. This makes our vectors high dimensional and detailed with not only a content of each message but also its temporal existence in a sequence of posts. As a result, it is formally written as:

$$s_j = v_1 \| v_2 \| \ldots \| v_i \| \ldots \| v_{max\_length} \tag{1}$$

where $\|$ is the concatenation operator defined in [15].

3.3 A Prediction Model: A CNN Model with Data Imbalance Handling

Popular for computer vision and image classification, convolutional neural networks (CNN) are deep neural networks with special discriminative deep architecture composed of convolutional layers, pooling layers, and fully connected layers in addition to input and output layers. Previously described, a convolutional neural network is built for our prediction model. This model type is chosen so that the textual data of each student in a course forum can be examined with multiple abstraction layers in a deeper manner, leading to more discrimination between instances of different classes.

Like any neural network, CNNs require thorough network design and tuned learning processes for particular domains and problems. Thus, in our work, a simple CNN model is developed with reference to the CNN model proposed in [15] for textual data classification in a vector space. More complex models can be designed and tuned for the task if the task effectiveness is not high enough. Our CNN is designed with the layers: INPUT-CONVOLUTIONAL-POOLING-FULLY CONNECTED-OUTPUT.

- The INPUT layer: There are *max_length* × *output_vector_size* nodes where *max_length* is the maximum number of tokens in textual data of each student and *output_vector_size* is the size of an output vector from the *skip-gram* model.

- The CONVOLUTIONAL layer: There are many filters, aka convolution kernels (128 in our model) with this layer for convolutions using a given kernel size (20 in our model). These filters are randomly initialized and then updated along the learning process. The ReLU activation function is also applied right after this layer to make a transformation on the output of this layer and obtain activation maps after local contextual information in the input has been extracted.
- The POOLING layer: there are many windows with a given size (2 in our model) using a give stride (2 in our model). Many kinds of windows can be used: max, min, average. MAX-POOLING is used in our model to select the maximal value from the output of the convolutional layer. This layer aims to combine the local contextual information from the previous layer to generate a global feature vector including relevant features subject to the next fully connected layer.
- The FULLY CONNECTED layer: there are full connections between the output of the POOLING layer and the nodes of this layer (100 in our model). This fully connected layer now performs the learning process for a classification model.
- The OUTPUT layer: as binary classification is supported, there is one node in this layer.

For the learning process of this CNN, the backpropagation algorithm is also applied adaptively with the standard mean squared error (MSE) loss function. Nevertheless, data imbalance is addressed in our task while MSE treats all the instances equally. This leads to correct predictions for most of the instances of the majority class and few of the instances of the minority class. In order to overcome the data imbalance issue, we adopt and adapt the mean squared false error (MSFE) loss function proposed in [17] to the learning process of this CNN. Their details are given below.

The standard MSE loss function of our CNN model is defined as:

$$l_{MSE} = \frac{1}{M} \sum_{i=1..M} \frac{1}{2} \left(d^{(i)} - y^{(i)} \right)^2 \tag{2}$$

Where: M is the total number of instances in the training data set, $d^{(i)}$ represents the desired class of the i-th instance, and $y^{(i)}$ represents the corresponding predicted class.

The MSFE loss function adopted from [17] and adapted to our CNN is defined as:

$$l_{MSFE} = FPE^2 + FNE^2 = \frac{1}{2} \left((FPE + FNE)^2 + (FPE - FNE)^2 \right) \tag{3}$$

Where:

$$FPE = \frac{1}{N} \sum_{i=1..N} \frac{1}{2} \left(d^{(i)} - y^{(i)} \right)^2 \tag{4}$$

$$FNE = \frac{1}{P} \sum_{i=1..P} \frac{1}{2} \left(d^{(i)} - y^{(i)} \right)^2 \tag{5}$$

Where: FPE is mean false positive error, FNE is mean false negative error, N is the number of instances in the negative class (i.e. majority class), and P is the number of instances in the positive class (i.e. minority class).

According to Eq. (3), the MSFE loss function reaches the minimum value when FPE equals FNE and both approach zero. This leads to the balanced accuracy for both classes. At that time, the number of correct predictions for instances of the minority class can be kept high, at least as high as that of the majority class. Compared to MSE, MSFE can balance correct predictions of instances of both classes. So, data imbalance can be tackled well with our CNN for the student prediction task.

In [17], MSFE has been proposed with deep neural networks on both image and textual data. Therefore, our work is the first one adapting this loss function to convolutional neural networks on textual data. Other imbalance handling methods have also been introduced so far. However, most of them like [3, 10] considered image data.

3.4   Method Characteristics

Previously described, our work also takes advantage of forum texts in a specific course in order to forecast a course final status of each student as soon as possible. As input texts in our work are inherently hierarchical, many hidden layers of abstraction exist in data. Deep convolutional neural networks are suitable for not only feature learning in our data but also making predictions. This is why our solution is based on the method proposed in [15]. Nevertheless, data imbalance handling is considered in our work for the educational data mining context while not in [15]. In addition, the loss function in [17] is adopted and adapted to a deep convolutional neural network in our work for data imbalance handling. In [17], only deep neural networks have been examined with this loss function on imbalanced non-educational data sets. Bringing them together, our solution has been defined originally for course-level student prediction by exploring only their forum textual data.

Compared to the related works in the educational data mining area, our work is the first one conducting the course-level student prediction task based on forum texts. This choice makes our work more practical in practical use because the from-course-to-course differences in teaching and learning activities can be avoided even for the same subject. For example, this year a lecturer gives many quizzes to the students and they are used in predictions; however, next year, another lecturer might use project-based activities and assessment instead of quizzes in predictions. At that moment, student characteristics observed are different from year to year. As a result, student representation and the model are thus redefined and rebuilt. By contrast, using our approach, predictions can be made for any course deployed with a discussion forum.

Nevertheless, forum discussions are optional for students. Some of them might have no post in the course forum. Therefore, it is supposed that most of our students join the forum and are active. Combining predictions in this work with those based on study performance and required teaching and learning activities is our future work.

4 Empirical Evaluation

4.1 Settings of Our Empirical Study

For an empirical evaluation of our method, we conduct several experiments in this study with real data from the Principle of Programming Languages subject at Faculty of Computer Science and Engineering, Ho Chi Minh City University of Technology, Vietnam National University at Ho Chi Minh City, Vietnam [1]. This subject is a requisite one for the third-year students in the Computer Science program. There are two courses of this subject in this empirical study. The first one took place in 2010 and the second one in 2011. We name the 2010 and 2011 data sets corresponding to these two courses. There are about 150 students active in forum discussions. The total number of students, the number of students with status "Pass", and the number of students with status "Fail" in each data set are recorded. The averaged number of posts for each student is calculated. Details are given in Table 1.

From Table 1, it can be seen that very high imbalance exists in our real data sets. Our minority class is a positive one while our majority class is a negative one. In addition, the successful students have more posts than the failed students on average.

Table 1. Data descriptions

Data set	Total		Pass		Fail		Averaged Post#	
	Student#	%	Student#	%	Student#	%	Pass	Fail
2010	151	100	134	88.74	17	11.26	14.28	8.76
2011	153	100	141	92.16	12	7.84	10.96	5.67

Using the aforementioned data sets, our evaluation examines the answers of the two following questions regarding the effectiveness of the proposed method:

- Question 1: Is a CNN model more appropriate than the other ones?
- Question 2: Does data imbalance handling improve the correct recognition of instances of the minority class?

For comparison, C4.5, Support Vector Machines (SVM), and Long-Short Term Memory (LSTM) networks are used with default settings. Two different loss functions (mean squared error (MSE) and mean squared false error (MSFE)) are associated with LSTM networks and CNN models. The cost learning strategy in [10] is also considered with CNN models. The implementations of these machine learning algorithms are based on the *sklearn* and *keras* libraries using Python.

In addition, only the *word2vec*-based vector representation is used although it has been examined in other vector representations with binary and *tf-idf* weights. All the above listed methods are performed with this *word2vec*-based vector representation. For this representation, a *skip-gram* model is built using the *gensim* library with *min_count* = 5 for the minimum word frequency and output vector *size* = 400.

In order to measure how effective a method is, Accuracy (%), Precision, Recall, and F-measure are used. We use Accuracy in percent in the range [0, 100] to check generally correct predictions of instances of any class while Precision, Recall, and F-measure in the range [0, 1] to check correct predictions of only instances of the minority class. Their higher values indicate that a method is more effective.

4.2 Experimental Results and Discussions

In Tables 2 and 3, our experimental results are displayed. The best results for each measure are shown in bold. The MSE or MSFE loss function is also specified along with LSTM and CNN models. The cost approach is examined with CNN models. Generally speaking, our method with CNN and MSFE can provide the best Accuracy, Recall, and F-measure and the second best Precision. This is understandable because the MSFE loss function favors instances of the positive class and thus, tends to give predictions a positive class, leading to an increasing of correct positive instance recognitions and misrecognitions of negative instances.

As for an answer to Question 1, our method outperforms the others with respect to all evaluation measures on both data sets. This performance shows the more effectiveness of CNN in comparison with those of C4.5, SVM, and LSTM. SVM and LSTM (MSE) are effective when classifying instances of the negative class. However, they fail to predict instances of the positive class. By contrast, C4.5 is not so effective but able to recognize some instances of the positive class. Different from these models, CNN (MSE) and CNN (MSFE) have better correct predictions for all the instances. Matching data characteristics and model characteristics, more effectiveness can be reached for this task. Multiple abstraction layers in our forum texts captured for each student can have been processed well by a deep CNN model.

Table 2. Experimental results on the 2010 data set

	C4.5	SVM	LSTM (MSE)	LSTM (MSFE)	CNN (cost)	CNN (MSE)	CNN (MSFE)
Accuracy (%)	78.38	88.40	88.40	52.13	95.27	90.13	**96.67**
Precision	0.11	0.00	0.00	0.17	**1.00**	**1.00**	0.92
Recall	0.13	0.00	0.00	**0.77**	0.57	0.20	**0.77**
F-measure	0.12	0.00	0.00	0.27	0.73	0.33	**0.84**

Table 3. Experimental results on the 2011 data set

	C4.5	SVM	LSTM (MSE)	LSTM (MSFE)	CNN (cost)	CNN (MSE)	CNN (MSFE)
Accuracy (%)	84.59	92.09	92.09	83.99	94.97	93.92	**96.93**
Precision	0.12	0.00	0.00	0.17	**1.00**	**1.00**	0.79
Recall	0.13	0.00	0.00	0.28	0.44	0.29	**0.86**
F-measure	0.12	0.00	0.00	0.21	0.61	0.45	**0.82**

Regarding Question 2, in both 2010 and 2011 cases, CNN (MSFE) is always more effective than CNN (MSE). This confirms the more goodness of the MSFE loss function than that of the MSE one on textual data with the CNN model. In addition, CNN (MSFE) outperforms CNN (cost) although CNN (cost) has higher Precision values. This is because the cost strategy in [10] was studied for image data while the MSFE one for textual data additionally and thus, more suitable for our problem. Consequently, our data imbalance handling with the MSFE loss function can improve the correct recognition of instances of the minority class. Indeed, it can make resultant predictions more accurate in general and also specifically for instances of the positive minority class. It has great improvement in Recall of the minority class from 0.2 to 0.77 for the 2010 data set and from 0.29 to 0.86 for the 2011 data set. Meanwhile, it sacrifices just a bit Precision for such better Recall.

In short, our combination of a convolutional neural network and a loss function is an appropriate method. It is regarded as a novel solution to the course-level student prediction task based on textual data in a course forum. More correct predictions can help identifying the students with predicted "Fail" status before or at latest midterm break of a course. They can be then warned for their current study performance and thus, supported in time so that the rest of time of a course is enough for them to study more and reach an ultimate success in the course. Such a result presents our significant educational data mining contribution to the education domain in practice.

5   Conclusions

In this paper, we propose a novel method to the course-level student prediction task where timestamped textual data in a course forum are exploited. Our method is a deep convolutional neural network with an adopted loss function that can favor the predictions of instances in a minority class. As a result, our method can provide more accurate predictions than the others. In addition, it can recognize more instances of the minority class correctly. Experimental results on two real data sets have shown its higher Accuracy and Precision, Recall, and F-measure of the minority class on average. The adopted loss function, MSFE, is also a better choice in our method with more significant improvements on Recall.

Nonetheless, in each course, there are also students with no posts. Some of them succeed in the course and some don't. Thus, other characteristics of the students are going to be considered more for the task. In addition, improvement in Recall is expected to be further examined with more experiments on data from many other course forums. Last but not least, web application development is going to be completed for warning both students and lecturers using the prediction results of this work.

Acknowledgments. This research is funded by Vietnam National University - Ho Chi Minh City (VNU-HCM), Vietnam, under grant number C2017-20-18.

References

1. Academic Affairs Office: Ho Chi Minh City University of Technology, Vietnam. http://www.aao.hcmut.edu.vn. Accessed 25 Mar 2018
2. Bayer, J., Bydzovska, H., Geryk, J., Obsivac, T., Popelinsky, L.: Predicting drop-out from social behaviour of students. In: Proceedings of the 5th International Conference on Educational Data Mining, pp. 103–109 (2012)
3. Buda, M., Maki, A., Mazurowski, M.A.: A systematic study of the class imbalance problem in convolutional neural networks. arXiv preprint arXiv:1710.05381 (2017)
4. Chaturvedi, S., Goldwasser, D., Daumé III, H.: Predicting instructor's intervention in MOOC forums. In: Proceedings of the 52nd Annual Meeting of the Association for Computational Linguistics, USA, pp. 1501–1511 (2014)
5. Crossley, S.A., Dascalu, M., McNamara, D.S., Baker, R., Trausan-Matu, S.: Predicting success in massive open online courses (MOOCs) using cohesion network analysis. In: Proceedings of CSCL 2017, pp. 103–110 (2017)
6. Ezen-Can, A., Boyer, K.E., Kellogg, S., Booth, S.: Unsupervised modeling for understanding MOOC discussion forums: a learning analytics approach. In: Proceedings of LAK 2015, pp. 1–5. ACM, USA (2015)
7. He, J., Bailey, J., Rubinstein, B.I.P., Zhang, R.: Identifying at-risk students in massive open online courses. In: Proceedings of the 29th AAAI Conference on Artificial Intelligence, pp. 1749–1755 (2015)
8. Li, Y., Fu, C., Zhang, Y.: When and who at risk? Call back at these critical points. In: Proceedings of the 10th International Conference on Educational Data Mining, pp. 168–173 (2017)
9. Khalil, M., Ebner, M.: Learning analytics in MOOCs: can data improve students retention and learning? In: Proceedings of EdMedia 2016, Canada, pp. 1–8 (2016)
10. Khan, S.H., Hayat, M., Bennamoun, M., Sohel, F., Togneri, R.: Cost-sensitive learning of deep feature representations from imbalanced data. IEEE Trans. Neural Netw. Learn. Syst. **PP**(99), 1–15 (2017)
11. Kostopoulos, G., Kotsiantis, S., Pintelas, P.: Estimating student dropout in distance higher education using semi-supervised techniques. In: Proceedings of the 19th Panhellenic Conference on Informatics, pp. 38–43 (2015)
12. Márquez-Vera, C., Cano, A., Romero, C., Ventura, S.: Predicting student failure at school using genetic programming and different data mining approaches with high dimensional and imbalanced data. Appl. Intell. **38**, 315–330 (2013)
13. Mikolov, T., Chen, K., Corrado, G., Dean, J.: Efficient estimation of word representations in vector space. In: The Workshop Proceedings of the International Conference on Learning Representations (2013)
14. Nguyen, D.H., Vo, T.N.C., Nguyen, H.P.: Combining transfer learning and co-training for student classification in an academic credit system. In: Proceedings of the 2016 IEEE RIVF International Conference on Computing & Communication Technologies, Research, Innovation, and Vision for the Future, pp. 55–60. IEEE, Vietnam (2016)
15. Pomponiu, V., Thing, V.L.L.: A deep convolutional neural network for anomalous online forum incident classification. In: Roychoudhury, A., Liu, Y. (eds.) A Systems Approach to Cyber Security, pp. 57–69. IOS Press (2017)
16. Sharkey, M., Sanders, R.: A process for predicting MOOC attrition. In: Proceedings of the 2014 Conference on Empirical Methods in Natural Language Processing (EMNLP), Qatar, pp. 50–54 (2014)

17. Wang, S., Liu, W., Wu, J., Cao, L., Meng, Q., Kennedy, P.J.: Training deep neural networks on imbalanced data sets. In: Proceedings of the 2016 International Joint Conference on Neural Networks (IJCNN), pp. 4368–4374 (2016)

18. Chau, V.T.N., Phung, N.H.: A random forest-based self-training algorithm for study status prediction at the program level: *minSemi-RF*. In: Sombattheera, C., Stolzenburg, F., Lin, F., Nayak, A. (eds.) MIWAI 2016. LNCS (LNAI), vol. 10053, pp. 219–230. Springer, Cham (2016). https://doi.org/10.1007/978-3-319-49397-8_19

19. Wang, F., Chen, L.: A nonlinear state space model for identifying at-risk students in open online courses. In: Proceedings of the 9th International Conference on Educational Data Mining, pp. 527–532 (2016)

20. Wen, M., Yang, D., Rosé, C.P.: Sentiment analysis in MOOC discussion forums: what does it tell us? In: Proceedings of the 8th International Conference on Educational Data Mining, pp. 1–8 (2014)

21. Whitehill, J., Mohan, K., Seaton, D., Rosen, Y., Tingley, D.: MOOC dropout prediction: how to measure accuracy? In: Proceedings of L@S 2017, pp. 1–4. ACM, USA (2017)

22. Whitehill, J., Mohan, K., Seaton, D., Rosen, Y., Tingley, D.: Delving deeper into MOOC student dropout prediction. arXiv:1702.06404v1 [cs.AI], pp. 1–9, 21 February 2017

23. Wong, J.-S., Pursel, B., Divinsky, A., Jansen, B.J.: An analysis of MOOC discussion forum interactions from the most active users. In: Agarwal, N., Xu, K., Osgood, N. (eds.) SBP 2015. LNCS, vol. 9021, pp. 452–457. Springer, Cham (2015). https://doi.org/10.1007/978-3-319-16268-3_58

Purity and Out of Bag Confidence Metrics for Random Forest Weighting

Mandlenkosi Victor Gwetu[✉], Serestina Viriri, and Jules-Raymond Tapamo

University of KwaZulu-Natal, Private Bag X54001, Durban 4000, South Africa
gwetum@ukzn.ac.za

Abstract. Random Forests are an ensemble classification technique that employs a committee of diverse decision trees to make predictive decisions based on training set observations. In the conventional RF algorithm, individual tree decisions are aggregated with equal weighting to arrive at a majority vote. Recent initiatives have found merit in the use of leaf node purity and out of bag sets for estimating the probability of an individual tree's classification accuracy on unseen instances. This study proposes the concepts of Purity Gap Gain (PGG) and Relative Tree Confidence (RTC) as new ways of rating a decision tree's classification competence and ultimately influencing the quality of the resulting ensemble decision. PGG extends the idea of leaf node purity by taking into account the rate of purity convergence and the depth at which it takes place. RTC is a comprehensive score which takes into account the confidence with which a tree makes both correct and incorrect out of bag classifications. Statistical tests based on UCI datasets demonstrate the significant relationship between a RF's strength and the relative confidence of its decision trees. When applied to RFs with high strength, the proposed weighting methods demonstrate classification accuracy results that are predominantly comparable but at times superior to conventional approaches.

Keywords: Random Forest · Weighting · Confidence · Purity

1 Introduction

Ensemble classification methods are premised on a training phase which generates several individual base classifiers and a testing phase which aggregates their predictions into a consolidated result [8]. The most common examples include boosting [12], bagging [4] and Random Forests (RFs) [5], all of which are based on decision trees. In boosting, trees are trained in succession and emphasis is placed on instances that are misclassified by previous classifiers; the ensemble is therefore a committee of evolving trees that increasingly adapt towards learning how to correctly predict the most difficult training instances. In bagging, trees can be trained concurrently and emphasis is placed on creating bootstrapped samples; the ensemble is therefore a committee of trees which have random

© Springer Nature Switzerland AG 2018
N. T. Nguyen et al. (Eds.): ICCCI 2018, LNAI 11055, pp. 491–502, 2018.
https://doi.org/10.1007/978-3-319-98443-8_45

similarities and differences. RFs complement bagging by introducing an extra level of stochastic behavior in the selection of features involved in node split point optimization. As a result, RFs are known to (1) have reliable classification performance that is comparable to Support Vector Machines (SVMs), (2) be tolerant of noise and robust to over fitting, and (3) be fast and simple to use since they only rely on 2 main parameters [8].

In the typical RF algorithm, N instances are drawn from the training set with repetition. Due to this bootstrapped sampling, there is on average a third of the training set not used by each tree [11]. This excluded collection of instances is known as the Out Of Bag (OOB) set, which is useful for determining feature importance as well as RF internal estimates for error, strength and correlation; in the context of the training set [5]. When investigation is carried out from a statistical perspective, these internal estimates are sufficient for demonstrating the effect of changes to RF properties such as the number of features used in splitting. In fact, Brieman argues that the OOB set eliminates the need for a separate test set since there is empirical evidence that the OOB set is equivalent to a test set of the same size as the training set [5]. On the other hand, conventional machine learning experiments tend to require dedicated training and testing sets to allow for objective and repeatable comparison amongst various classification methods. When this approach is taken, the OOB set can then be used to improve RFs in an unbiased manner since it is separate from the test set. In previous literature, this improvement has been proven to be significant when weighting mechanisms for aggregating the RF classification are based on OOB performance of individual base trees [2,11,14,15].

The simplest form of aggregating a RFs base tree predictions is through simple majority voting, where each base tree has an equal weight [5,9]. Veto voting [6,13] on the other hand, allows one base classifier to veto the decisions of other members of the ensemble such that it determines the final verdict. In weighted voting, each base classifier's prediction is weighted by its OOB error rate, which is an indication of the confidence placed on it for classifying previously unseen instances [1]. In dynamic integration, the test instance is used to influence either the set of base classifiers that will predict its class [11] or the weight applied to each member of the ensemble's prediction [15]. Since each base classifier will tend to be competent at classifying certain types of test instances but unreliable for others, this approach allows the prediction of each test case to be significantly influenced by classifiers which are most familiar with similar cases. Although dynamic integration has been proven effective, it has the drawback of dependence on cross validation to determine the level of similarity between test and training set instance pairs.

This study aims to contribute towards the improved performance of RFs firstly, by proposing a new metric for quantifying a leaf node's confidence in classifying instances. We note that node purity, which is typically used to measure tree classification confidence [9,17], does not take into account useful information such as the length of the path to the leaf node and consistency in the quality of its split criteria. Our proposed metric, Purity Gap Gain (PGG) is influenced by the J48 aggregate confidence measure used by Monteith and Martinez [9]. Their measure is a hybrid of 6 metrics which include purity, information gain along the tree path and OOB error, amongst others. PGG's uniqueness is justified by the fact that it is agnostic of the OOB set and it is a single calculation which favors leaf nodes, whose path from the root is shorter and exhibits rapid convergence in purity. Inspired by the boosting technique [12], we also propose a new weighting scheme, Relative Tree Confidence (RTC), which rewards correct classifications but penalizes errors on the OOB set. This work is part of a larger study to design RF internal measures that work at tree level. The ultimate objective is to formulate tree based measures which evaluate classification strength and diversity such that RF committees can be restricted to base classifiers meeting predetermined threshold values of these measures. As a first step, we seek to establish whether RTC has an association with RF strength. If this is the case, then limiting a RF to trees with high RTC scores should theoretically yield one with greater strength since it has been empirically proven that strength influences the upper bound of generalization error [3,5].

This paper comprises of 6 sections; the remaining sections are organized as follows. Section 2 contains theoretical background on RFs with particular emphasis on using the OOB set to calculate strength. Section 3 outlines our proposed new metrics for improving RFs while Sect. 4 describes the datasets, experimental protocol and performance measures used to evaluate the proposed methods. The results of the study are presented and discussed in Sect. 5 and future work is previewed in Sect. 6.

2 Random Forests

This section outlines the notation adopted to represent the various components of a RF and introduces some of its internal measures. As a supervised machine learning algorithm, RFs receive a training set X, of N instances, each of which can be represented by $X(i) = (\mathbf{x}, y)$ where $1 \leq i \leq N$ and $\mathbf{x} = (x_1, x_2, \ldots, x_A)$, is a vector of A feature values. Each instance $\mathbf{x}$ has a corresponding label y such that $1 \leq y \leq C$, from the set of C possible class values. A set of samples $\{\Theta_{k0}, 1 \leq k \leq T\}$ is drawn with replacement from the training set and used to create a RF classifier $h(x)$ made up of T corresponding decision tree classifiers $h_1(\mathbf{x}), h_2(\mathbf{x}), \ldots, h_T(\mathbf{x})$. The RF algorithm can therefore be said to accept X as input and return $h(x)$. During the classification stage, $h_k(x) = 0$ denotes a correct prediction whereas $h_k(x) = y$ specifies that x has been incorrectly labelled as a member of class y.

The sample Θ_{k0} is used to create the root node[1] of tree k, which is built by recursively splitting nodes until terminating conditions are met. We represent any node at depth $d \geq 0$ in the tree and its size by Θ_{kd} and $|\Theta_{kd}|$ respectively. Let $\Phi(\Theta_{kd}, j)$ represent the set of all possible ways of partitioning node Θ_{kd} into B branches based on attribute j and $\phi \in \Phi(\Theta_{kd}, j)$ be one such partition.

---

[1] The terms sample and node are used interchangeably in this text since they are closely related.

ϕ^* is the adopted partition as it is the most optimal in terms of decreasing the impurity of child nodes. When a partition is enforced, $\phi(z), 1 \leq z \leq B$ is the resulting child node of Θ_{kd} at depth $d+1$. For the sake of simplicity we represent $\phi(z)$ as Θ_{kd+1}. The probability of landing on branch z from Θ_{kd} when splitting based on ϕ is represented by $p(z)$. The probability of class y in node Θ_{kd} and the probability of class y in branch z of Θ_{kd} are represented by $p(y)$ and $p(y|z)$ respectively.

2.1   Gini Index

Individual decision trees within a RF, use simple rules to recursively partition a sample Θ_{kd} in an attempt to produce pure child nodes with instances belonging to the same class. Although the Gini Index (GI) [11] was the original measure chosen by Brieman [5] for measuring the level of decrease in impurity along a RF decision tree path, it is still widely used due to its simplicity and effectiveness [7,16]. Given that a partition $\phi \in \Phi(\Theta_{kd}, j)$ splits Θ_{kd} into B child nodes $(\phi(z))$, the resulting decrease in impurity is calculated as follows:

$$GI(\phi) = -\sum_{y=1}^{C} p(y)^2 + \sum_{z=1}^{B} p(z) \sum_{y=1}^{C} p(y|z)^2. \tag{1}$$

The optimal partition ϕ^* for splitting a node Θ_{kd} on attribute j is deduced as follows:

$$\phi^* = \underset{\phi \in \Phi(\Theta_{kd}, j)}{\operatorname{argmax}} GI(\phi). \tag{2}$$

2.2   Random Forest Strength

Let $P_o(\mathbf{x}, y)$ be the proportion of votes allocated to class y when an OOB instance $(\mathbf{x}, y)$ is classified by the RF trees for which $\mathbf{x} \notin \Theta_{k0}$:

$$P_o(\mathbf{x}, y) = \frac{\sum_{k=1}^{T} I(h_k(\mathbf{x}) = y; (\mathbf{x}, y) \in O_k)}{\sum_{k=1}^{T} I((\mathbf{x}, y) \in O_k)}, \tag{3}$$

where O_k is the OOB set of classifier $h_k(\mathbf{x})$ and $I(c)$ is an indicator function which returns 1 if the condition c is true and 0 otherwise. The margin on an OOB instance reflects the extent to which correct votes exceed votes for any other class and is represented by:

$$mr(\mathbf{x}, y) = P_o(\mathbf{x}, 0) - \max_{y=1}^{C} P_o(\mathbf{x}, y). \tag{4}$$

The strength of a RF is an internal measure of the average expected margin over the whole training set [11]:

$$s(X) = \frac{1}{N} \sum_{i=1}^{N} mr(X(i)). \tag{5}$$

3 Weighted Voting

With the preliminaries established, this chapter begins the search towards tree confidence based voting. We propose 2 new weighting methods (PGG and RTC) based on the concept of node purity [17], which we define as follows:

$$\hat{p}(\Theta_{kd}) = \max_{y=1}^{C} p(y). \tag{6}$$

3.1 Purity Gap Gain

The assumption behind PGG is that the root node Θ_{k0} of a decision tree has a low purity and the goal of its induction is to recursively perform splitting until we obtain leaf nodes with a high purity. We specifically seek the maximum purity of 1 and the quicker and/or more aggressively this occurs, the better. Given that Θ_{kd} is a child node of Θ_{kd-1} within an induced tree h_k, the PGG at a node can be defined recursively as follows:

$$PGG(\Theta_{kd}) = \begin{cases} \hat{p}(\Theta_{kd}), & \text{if } d = 0 \\ PGG(\Theta_{kd-1}), & \text{if } \hat{p}(\Theta_{kd-1}) \geq \hat{p}(\Theta_{kd}) \\ \frac{PGG(\Theta_{kd-1}) + \frac{\hat{p}(\Theta_{kd}) - \hat{p}(\Theta_{kd-1})}{1 - \hat{p}(\Theta_{kd-1})}}{d}, & \text{if } \Theta_{kd} \text{ is a leaf node} \\ PGG(\Theta_{kd-1}) + \frac{\hat{p}(\Theta_{kd}) - \hat{p}(\Theta_{kd-1})}{1 - \hat{p}(\Theta_{kd-1})}, & \text{otherwise} \end{cases} \tag{7}$$

At each node along the path from the root to a leaf, we calculate the gap between the level of purity in the parent node and the maximum possible purity. The local PGG at the current child node is a fractional value that reflects how much of this gap the split manages to cover. The leaf nodes record the average local PGG along their path from the root; non-leaf nodes simply store the cumulative sum of local PGGs along their path from the root. This cumulative sum only takes positive local PGGs into account. Negative values are penalized by retaining the cumulative local PGG value at the parent node whilst updating the path depth. This essentially yields a lower cumulative sum over a longer path when compared with the hypothetical scenario in which all local PGGs were instead positive. The PGG stored at the leaf nodes is the ultimate PGG value that we use to capture the confidence a tree has in its prediction.

When a tree h_k is presented with a test case x, it iteratively navigates this instance along a path from the root to the leaf node that ultimately determines its predicted class $(h_k(x))$. The PGG at this leaf node will be relatively high if the path from the root is short but records only high local PGGs, thus reflecting high speed and confidence in this classification. Conversely, long paths from the root coupled with low local PGGs, result in low PGGs at the leaf nodes; we label this as a prediction lacking confidence since it does not reflect decisiveness.

3.2 Relative Tree Confidence

The RTC weight is our cornerstone metric for modeling strength at tree level since it takes OOB performance into account like the RF strength does, albeit under different constraints. RTC is calculated by simply aggregating weighted classifications $h_k(x), x \in O_k$ whilst simultaneously penalizing incorrect predictions by negating their corresponding weights. To avoid non-positive aggregates we scale the weights from the range $[-1,1]$ to $[0,1]$. The objective of RTC is to capture the overall confidence that a tree exhibits in its OOB classifications whilst simultaneously noting the accuracy of these predictions. A tree which gets each of its OOB predictions correct with a confidence level of 1 will also achieve a RTC of 1. Conversely a tree which gets each of its predictions wrong with maximum confidence achieves a RTC of 0. We refer to such a case as demonstrating negative confidence and consider this worse that having low or no confidence over all predictions. We initially intended for RTC to specifically use PGG for determining tree prediction confidence but subsequently also used the purity of the classifying leaf node for experimental comparison. While the purity based approach towards modelling prediction confidence is simple and efficient, the PGG approach offers a more complex but intuitive alternative. The comparison between the two approaches effectively seeks to find out whether the computational overhead of PGG can be justified.

4 Experimental Protocol

The experiments used in this study are designed to achieve 2 objectives: (1) to assess the reliability of RTC as a means of measuring tree strength and (2) to provide a preliminary evaluation of the effectiveness of PGG and RTC as tree weighting metrics in RFs. All experiments were conducted using our own C++ implementation of dichotomous ($B = 2$) RFs that use the following settings [5,11]:

- Each RF is made out of $T = 100$ trees.
- The number of randomly selected attributes at each node of a tree is set to $\sqrt{A}$.
- The size of each bootstrapped sample Φ_{k0} is N, the size of X.
- Node splitting is stopped when any of the following conditions is met: $|\Theta_{kd}| \leq 5$, $\hat{p}(\Phi_{kd}) = 1$, or a depth $d = 30$ is reached.

Experiments were conducted on a Linux server[2] and took a combined run time of 1 h 35 min over all data sets.

4.1 Experiment A

To achieve the first objective, we introduce the concepts of RF sorting and subforests. We ensure that the trees in a RF store tree confidence outputs from OOB evaluations and these details are subsequently used as a basis for sorting the trees in ascending order. The sorted RF is then divided into 2 sub-forests at the middle such that all the trees in the top half store lower confidence values

---

[2] 64 cores and 694 GB of RAM; the actual CPU cores are Intel(R) Xeon(R) running at 2.70 GHz.

than those in the bottom half. We then calculate the strength of the combined forest, top sub-forest and bottom sub-forest, as defined in Eq. (5). Given a data set, we build 30 RFs and then generate sorted top and bottom sub-forests for each. We use the paired T-test to determine whether the mean difference between the two sub-forest strengths is zero. We formally define our hypotheses as follows:

- The null hypothesis (H_0) assumes that the true mean difference (μ_d) between the strengths of the top and bottom sub-forests is zero.
- The alternative hypothesis (H_1) assumes that μ_d is non zero.

If RTC is a reliable means of modelling RF strength at tree level, the experimental results should demonstrate sufficient evidence for us to reject (H_0) in favor of (H_1). Since the top and bottom sub-forest contain the weakest and strongest trees respectively, we do not expect their average strengths to be equal.

4.2   Experiment B

Given a data set, we generate 30 random training and test sets using 90% and 10% split ratios respectively in each case. Experiment A uses the 30 random training sets to build a set of corresponding RFs. To achieve the second objective, we select the RF with the highest strength from each data set and test its classification accuracy using its corresponding test set. Table 1 shows the adopted codes and brief descriptions of the various weighted voting schemes used for classification. The same RF classifier and test set are used in each case; the only difference is the weighting scheme applied.

   We are specifically interested in observing the performance of schemes 1 and 3 (which are the main contributions of these study) in comparison to other existing schemes. Our ultimate goal is to use tree confidence measures to automatically populate a RF with only strong trees in the hope of improving classification accuracy. In this experiment, we are interested in observing the comparative performance of our schemes in a context where a high RF strength has been deliberately ensured. Since the internal metrics rate each RF classifier in Experiment B as being relatively stronger than those from other training set splits, we thus artificially create such a context. The results of Experiment B should however not be seen as a definitive characterization of the voting schemes since the context of their application is controlled and the evaluation is only performed once.

4.3   Data Sets

The effectiveness of the methods proposed in this study is tested using 10 data sets from the UCI repository [10]. These data sets are drawn from Robnik [11] and Breiman's [5] studies on RFs; we exclude data sets with nominal attributes and missing values as these properties are beyond the scope of the present study. An additional constraint is enforced to exclude data sets with more than 3000 instances, for computational reasons. The characteristics of the chosen data sets

Table 1. Weighting schemes

Scheme	Code	Description
1	PUR	Purity of the classifying leaf node
2	PGG	PGG of the classifying leaf node
3	RTC-pur	RTC using purity as a confidence measure
4	RTC-pgg	RTC using PGG as a confidence measure
5	OOB	1 − OOB error rate
6	HYB	A hybrid weight calculated by combining schemes 1–5 above
7	EQU	Equal weighting

Table 2. UCI datasets

Dataset	N^*	M	C	Dataset	N^*	M	C
bupa	345	6	2	iris	150	4	3
ecoli	336	7	8	segmentation	2310	19	7
german-numeric	1000	24	2	sonar	208	60	2
glass	214	9	7	vehicle	846	28	4
ionosphere	351	34	2	yeast	1484	8	10

are summarized in Table 2, which reveals the diversity of the problems represented, in terms of data set size $(N^*)^3$, number of features (A) and number of classes (C).

Each of the chosen data sets is preprocessed by converting all feature values to floating points and class labels to integers, then saved in a tab delimited format.

5   Results and Discussion

Table 3 shows that for both the RTC-pur and RTC-pgg sorted RFs, the top sub-forest consistently exhibited a lower strength than the bottom sub-forest. Moreover the paired t-test shows that the difference between the means of the 2 sub-forest strengths is statistically significant, hence we reject (H_0) in favor of (H_1). We therefore infer that the RTC metric does indeed reflect RF strength at tree level. Other interesting observations include the fact that the bottom sub-forest of the RTC-pur sorted RF consistently shows higher strength than its counterpart in the RTC-pgg sorted RF. This shows that the purity confidence measure complements the RTC metric better than the PGG measure when its given trees are more accurate. The converse is however true for the top sub-forests where the PGG measure seems to achieve relatively higher strength from

---

[3] This value is larger than N, the size of a training set, X which is drawn from a data set.

less accurate trees. In both the RTC-pur and RTC-pgg sorted RFs, the bottom sub-forests show greater strength than the full RF, indicating great potential in the idea of carefully choosing the trees that constitute a RF. There is however a strong correlation between all forest (full, top and bottom) strengths over the different data sets, with the iris and yeast collections consistently achieving highest and lowest strengths respectively.

Table 3. Mean RF strengths

Dataset	Full RF	RTC-pur sorted		RTC-pgg sorted	
		Top	Bottom	Top	Bottom
bupa	0.18	0.12	0.25	0.12	0.24
ecoli	0.51	0.45	0.55	0.47	0.54
german-numeric	0.33	0.30	0.36	0.31	0.35
glass	0.30	0.24	0.35	0.24	0.35
ionosphere	0.65	0.60	0.69	0.62	0.68
iris	0.77	0.74	0.81	0.75	0.80
segmentation	0.50	0.44	0.56	0.45	0.54
sonar	0.33	0.25	0.41	0.25	0.41
vehicle	0.36	0.33	0.39	0.33	0.38
yeast	0.11	0.07	0.14	0.08	0.13

The results of Experiment B are captured in Table 4 which generally reflects great similarity amongst the considered weighting schemes; maximum or minimum accuracies for each data set are highlighted for emphasis. In all data sets, the OOB based schemes all achieved the exact same accuracy. This perhaps reveals why some studies [6] prefer to simply use the equal vote approach since it is less computationally intensive but can yield results that are equivalent to OOB based weighting schemes. We again note that the outcome could have been different if a comprehensive comparison (with multiple repeats and considering various split ratios) had been conducted. In this case, we are not able to convincingly confirm the findings of previous literature [2, 11, 14, 15] which attest to the superiority of OOB based weighting schemes, since these schemes outperformed equal voting in only one data set. It is interesting to note that the non-OOB based weighting schemes demonstrate greater diversity in classification accuracy, with the PGG scheme achieving maximum accuracy in 3 instances. There is certainly potential in PGG as a RF weighting scheme since diversity coupled with strength is desirable [11] in RFs. The secret to unleashing the full potential of RFs however seems to lie within acquiring a deeper understanding of the peculiarities associated with the different data sets since performance is closely related with specific data sets.

Table 5 shows the distribution of the various weighting schemes over all data sets and RFs. Although each scheme has its own unique profile, PGG

Table 4. RF accuracies

Data set	Weighting Scheme						
	PUR	PGG	RTC-pur	RTC-pgg	OOB	HYB	EQU
bupa	0.65	0.65	0.65	0.65	0.65	0.65	0.65
ecoli	0.76	0.79	0.79	0.79	0.79	0.79	0.76
german-numeric	0.66	0.66	0.65	0.65	0.65	0.65	0.65
glass	0.67	0.67	0.67	0.67	0.67	0.67	0.67
ionosphere	0.86	0.86	0.86	0.86	0.86	0.86	0.86
iris	0.87	0.87	0.87	0.87	0.87	0.87	0.87
segmentation	0.76	0.76	0.81	0.81	0.81	0.81	0.81
sonar	0.75	0.75	0.75	0.75	0.75	0.75	0.75
vehicle	0.69	0.70	0.69	0.69	0.69	0.69	0.69
yeast	0.52	0.54	0.52	0.52	0.52	0.53	0.52

Table 5. Weighting scheme value distributions

Weighting scheme	Minimum	Average	Maximum
PUR	0.20	0.93	1.00
PGG	0.01	0.39	1.00
RTC-pur	0.44	0.71	1.00
RTC-pgg	0.49	0.61	0.90
OOB	0.00	0.29	0.69
HYB	0.32	0.67	0.98

demonstrates the widest range of values that virtually spans the entire range of possible values. The PGG also showcases the mean value closest to 0.5 when compared against other schemes. This shows that there were trees that exhibited both extremely negative and positive confidence while the majority generally showed either indecisiveness or no confidence. Although the distributions of RTC-pur and RTC-pgg are generally similar, the former records a higher average than the latter while the latter has a higher minimum value than the former. This could be the explanation to our earlier observation in which RTC-pgg has higher strength in bottom sub-forests while the RTC-pur has higher strength in top sub-forests.

6   Conclusion

This study has successfully explored the idea of modelling RF strength at tree level, with the RTC based metrics showing great potential for use as tree selectors in RFs. We have also proposed a novel RF weighting scheme: PGG which is based on the purity metric but incorporates additional information that takes into account how quickly and/or aggressively a decision tree arrives at a classification. It has demonstrated performance that is in most cases comparable with existing methods but is at times superior. However, much work still needs to be done to unlock the full potential of RFs which still seem to be greatly influenced by the peculiarities of the different data sets.

Future work will focus on extending this study to modelling another important RF internal metric: correlation, at tree level. This will enable us to detect trees which demonstrate diversity in their classification tendencies. Since strength and diversity are empirically related to generalization error, combining these 2 measures at tree level has great potential. Other opportunities for extending this work include the characterization of tree classification successes and failures. Although this study has demonstrated the storage of quantitative data in trees, there is opportunity to store more useful information such as average profiles of commonly misclassified OOB instances. This will allow us to not only report on accuracy rates but elaborate on the common classification mistakes that decision trees and RFs make. There is also need for a more comprehensive comparison of weighting schemes in order to investigate the effect of factors such as multiple repeats and various split ratios.

References

1. Bauer, E., Kohavi, R.: An empirical comparison of voting classification algorithms: bagging, boosting, and variants. Mach. Learn. **36**(1–2), 105–139 (1999)
2. Bernard, S., Adam, S., Heutte, L.: Dynamic random forests. Pattern Recogn. Lett. **33**(12), 1580–1586 (2012)
3. Bernard, S., Heutte, L., Adam, S.: A study of strength and correlation in random forests. In: Huang, D.-S., McGinnity, M., Heutte, L., Zhang, X.-P. (eds.) ICIC 2010. CCIS, vol. 93, pp. 186–191. Springer, Heidelberg (2010). https://doi.org/10.1007/978-3-642-14831-6_25
4. Breiman, L.: Bagging predictors. Mach. Learn. **24**(2), 123–140 (1996)
5. Breiman, L.: Random forests. Mach. Learn. **45**(1), 5–32 (2001)
6. Fawagreh, K., Gaber, M.M., Elyan, E.: Random forests: from early developments to recent advancements. Syst. Sci. Control Eng. Open Access J. **2**(1), 602–609 (2014)
7. Hu, C., Chen, Y., Hu, L., Peng, X.: A novel random forests based class incremental learning method for activity recognition. Pattern Recogn. **78**, 277–290 (2018)
8. Liaw, A., Wiener, M., et al.: Classification and regression by random forest. R News **2**(3), 18–22 (2002)
9. Monteith, K., Martinez, T.: Using multiple measures to predict confidence in instance classification. In: The 2010 International Joint Conference on Neural Networks (IJCNN), pp. 1–8. IEEE (2010)
10. Newman, C.B.D., Merz, C.: UCI repository of machine learning databases (1998). http://www.ics.uci.edu/~mlearn/MLRepository.html
11. Robnik-Šikonja, M.: Improving random forests. In: Boulicaut, J.-F., Esposito, F., Giannotti, F., Pedreschi, D. (eds.) ECML 2004. LNCS (LNAI), vol. 3201, pp. 359–370. Springer, Heidelberg (2004). https://doi.org/10.1007/978-3-540-30115-8_34
12. Schapire, R.E., Freund, Y., Bartlett, P., Lee, W.S.: Boosting the margin: a new explanation for the effectiveness of voting methods. Ann. Stat. **26**, 1651–1686 (1998)
13. Shahzad, R.K., Fatima, M., Lavesson, N., Boldt, M.: Consensus decision making in random forests. In: Pardalos, P., Pavone, M., Farinella, G.M., Cutello, V. (eds.) MOD 2015. LNCS, vol. 9432, pp. 347–358. Springer, Cham (2015). https://doi.org/10.1007/978-3-319-27926-8_31

14. Tripoliti, E.E., Fotiadis, D.I., Argyropoulou, M.: An automated supervised method for the diagnosis of alzheimer's disease based on fMRI data using weighted voting schemes. In: IEEE International Workshop on Imaging Systems and Techniques, IST 2008, pp. 340–345. IEEE (2008)

15. Tsymbal, A., Pechenizkiy, M., Cunningham, P.: Dynamic integration with random forests. In: Fürnkranz, J., Scheffer, T., Spiliopoulou, M. (eds.) ECML 2006. LNCS (LNAI), vol. 4212, pp. 801–808. Springer, Heidelberg (2006). https://doi.org/10.1007/11871842_82

16. Wang, Y., Xia, S.T.: Unifying attribute splitting criteria of decision trees by Tsallis entropy. In: 2017 IEEE International Conference on Acoustics, Speech and Signal Processing (ICASSP), pp. 2507–2511. IEEE (2017)

17. Witten, I.H., Frank, E., Hall, M.A., Pal, C.J.: Data Mining: Practical Machine Learning Tools and Techniques. Morgan Kaufmann, San Francisco (2016)

A New Computational Method for Solving Fully Fuzzy Nonlinear Systems

Raheleh Jafari[1]([✉]), Sina Razvarz[2], and Alexander Gegov[3]

[1] Centre for Artificial Intelligence Research (CAIR),
University of Agder, Grimstad, Norway
jafari3339@yahoo.com
[2] Departamento de Control Automático CINVESTAV-IPN
(National Polytechnic Institute), Mexico City, Mexico
[3] School of Computing, University of Portsmouth, Buckingham Building,
Portsmouth PO13HE, UK

Abstract. Predicting the solution of complex systems is a significant challenge. Complexity is caused mainly by uncertainty and nonlinearity. The nonlinear nature of many complex systems leaves uncertainty irreducible in many cases.

In this work, a novel iterative strategy based on the feedback neural network is recommended to obtain the approximated solutions of the fully fuzzy nonlinear system (FFNS). In order to obtain the estimated solutions, a gradient descent algorithm is suggested for training the feedback neural network. An example is laid down in order to demonstrate the high accuracy of this suggested technique.

Keywords: Approximate solution · Complex system
Feedback neural network · Gradient descent algorithm · Simulation

1 Introduction

Neural network has drawn attention since it is viewed as to be successful in various applications [1–10]. Fuzzy neural network is coordinated with numerable characteristics expressed as learning capability, generalization and nonlinear mapping [11]. The standard neural network is viewed as to be approximator [12,13]. A new learning algorithm for training the fuzzy neural networks by taking into consideration of triangular fuzzy weights is suggested in [14]. The fuzzy delta learning rule in order to train the fuzzy neural network is proposed in [15]. Obtaining the solution of fuzzy problem by utilizing neural networks is investigated in [16]. The approximated solution of fully fuzzy matrix equation by using neural networks is studied in [17]. A static neural network is suggested in [18] in order to obtain the approximate solution of fuzzy polynomials. A dynamic neural network is utilized in order to extract the estimated solution of dual fuzzy polynomials in [19]. New methods for fuzzy identification and modeling are proposed in [20–22]. The state space model of a linear system is extended to fuzzy

© Springer Nature Switzerland AG 2018
N. T. Nguyen et al. (Eds.): ICCCI 2018, LNAI 11055, pp. 503–512, 2018.
https://doi.org/10.1007/978-3-319-98443-8_46

case in [23]. The homotopy method for solving fuzzy nonlinear system is suggested in [24]. A new algorithm for dynamical nonsingleton fuzzy control system is proposed in [25]. The Adomian methodology for solving fuzzy system of linear equations is suggested in [26]. The homotopy technique in order to obtain the solution of a system of fuzzy nonlinear equations is suggested in [27].

It is worthwhile to mention that no study has been initiated for solving FFNS by using neural networks. For this purpose, in this work a new method based on neural network is suggested in order to obtain the Z-number solutions of FFNS. The Z-number weights are adjusted by utilizing a learning algorithm which is based on the gradient descent method. The simulation results shows that the new proposed technique is effective in extracting the Z-number solutions of FFNS.

This work is organized as follows: In Sect. 2 some basic definitions are laid down. The structure of the feedback neural network for obtaining the Z-number solutions of FFNS is illustrated in Sect. 3. In Sect. 4 an example with application is given in order to show the efficiency of the proposed method. The conclusion is given in Sect. 5.

2 Preliminaries

Prior to the introduction of the FFNS, some important definitions associated with fuzzy numbers and Z-numbers are given in this section.

Definition 1. A fuzzy number B is a function $B \in E : \Re \to [0, 1]$, in such a manner that, (1) B is normal, (there exist $a_0 \in \Re$ in such a manner that $B(a_0) = 1$; 2) B is convex, $B(\varrho a + (1 - \varrho)\tau) \geq \min\{B(a), B(\tau)\}$, $\forall a, \tau \in \Re, \forall \varrho \in [0, 1]$; (3) B is upper semi-continuous on $\Re$, i.e., $B(a) \leq B(a_0) + \varepsilon$, $\forall a \in N(a_0)$, $\forall a_0 \in \Re$, $\forall \varepsilon > 0$, $N(a_0)$ is a neighborhood; (4) The set $B^+ = \{a \in \Re, B(a) > 0\}$ is compact.

The popular membership functions for fuzzy numbers are the triangular function

$$\mu_B = F(\lambda_1, \lambda_2, \lambda_3) = \begin{cases} \frac{a - \lambda_1}{\lambda_2 - \lambda_1} & \lambda_1 \leq a \leq \lambda_2 \\ \frac{\lambda_3 - \zeta}{\lambda_3 - \lambda_2} & \lambda_2 \leq a \leq \lambda_3 \end{cases} \tag{1}$$

otherwise $\mu_B = 0$, and trapezoidal function

$$\mu_B = F(\lambda_1, \lambda_2, \lambda_3, \lambda_4) = \begin{cases} \frac{a - \lambda_1}{\lambda_2 - \lambda_1} & \lambda_1 \leq a \leq \lambda_2 \\ \frac{\lambda_4 - a}{\lambda_4 - \lambda_3} & \lambda_3 \leq a \leq \lambda_4 \\ 1 & \lambda_2 \leq a \leq \lambda_3 \end{cases} \tag{2}$$

otherwise $\mu_B = 0$.

Definition 2. A Z-number has two components $Z = [B(a), \tilde{p}]$. The primary component $B(a)$ is defined as restriction on a real-valued uncertain variable a. The secondary component $\tilde{p}$ is defined as a measure of reliability of B.

The probability measure is

$$\tilde{P} = \int_{\Re} \mu_B(a)\tilde{p}(a)da \tag{3}$$

where $\tilde{p}$ is denoted as the probability density of a, also $\Re$ is denoted as the restriction on $\tilde{p}$. For discrete Z-numbers

$$\tilde{P}(B) = \sum_{i=1}^{n} \mu_B(a_i)\tilde{p}(a_i) \tag{4}$$

Definition 3. The α-level of the fuzzy number B is illustrated as [28],

$$[B]^\alpha = \{a \in \Re : B(a) \geq \alpha\} \tag{5}$$

where $0 < \alpha \leq 1$, $B \in E$.

Since $\alpha \in [0, 1]$, $[B]^\alpha$ is bounded which is defined as $\underline{B}^\alpha \leq [B]^\alpha \leq \overline{B}^\alpha$. The α-level of B between $\underline{B}^\alpha$ and $\overline{B}^\alpha$ is stated as

$$[B]^\alpha = \left(\underline{B}^\alpha, \overline{B}^\alpha\right) \tag{6}$$

Definition 4. The Z-numbers have three elementary operations; $\oplus, \ominus,$ and $\odot$ which are named as addition, subtraction, and multiplication.

Suppose $Z_1 = (B_1, \tilde{p}_1)$ and $Z_2 = (B_2, \tilde{p}_2)$ be two discrete Z-numbers expressing the uncertain variables a_1 and a_2, $\sum_{\iota=1}^{n} \tilde{p}_1(a_{1\iota}) = 1$, $\sum_{\iota=1}^{n} \tilde{p}_2(a_{2\iota}) = 1$. The operations are displayed as

$$Z_{12} = Z_1 * Z_2 = (B_1 * B_2, \tilde{p}_1 * \tilde{p}_2) \tag{7}$$

where $* \in \{\oplus, \ominus, \odot\}$.

We have

$$[B_1 \oplus B_2]^\alpha = [B_1]^\alpha + [B_2]^\alpha = [\underline{B}_1^\alpha + \underline{B}_2^\alpha, \overline{B}_1^\alpha + \overline{B}_2^\alpha] \tag{8}$$

$$[B_1 \ominus B_2]^\alpha = [B_1]^r - [B_2]^\alpha = [\underline{B}_1^\alpha - \underline{B}_2^\alpha, \overline{B}_1^\alpha - \overline{B}_2^\alpha] \tag{9}$$

$$[B_1 \odot B_2]^\alpha = \left(\begin{array}{c} \min\{\underline{B}_1^\alpha \underline{B}_2^\alpha, \underline{B}_1^\alpha \overline{B}_2^\alpha, \overline{B}_1^\alpha \underline{B}_2^\alpha, \overline{B}_1^\alpha \overline{B}_2^\alpha\} \\ \max\{\underline{B}_1^\alpha \underline{B}_2^\alpha, \underline{B}_1^\alpha \overline{B}_2^\alpha, \overline{B}_1^\alpha \underline{B}_2^\alpha, \overline{B}_1^\alpha \overline{B}_2^\alpha\} \end{array} \right) \tag{10}$$

For all $\tilde{p}_1 * \tilde{p}_2$ operations, we have

$$\tilde{p}_1 * \tilde{p}_2 = \sum_i \tilde{p}_1(a_{1,i})\tilde{p}_2(a_{2,(n-i)}) = \tilde{p}_{12}(a) \tag{11}$$

Definition 5. Absolute value of a triangular fuzzy number $B(a) = F(\lambda_1, \lambda_2, \lambda_3)$ is defined as

$$|B(a)| = |\lambda_1| + |\lambda_2| + |\lambda_3| \tag{12}$$

3   Numerical Solution of Fully Fuzzy Nonlinear System with Feedback Neural Network

In this section a new method based on feedback neural network is proposed in order to obtain the numerical solutions of FFNS.

3.1 Fully Fuzzy Nonlinear Systems

The FFNS is stated as below

$$\begin{cases} S_{11} \odot p \odot q \oplus ... \oplus S_{1n} \odot p^n \odot q^n = G_1 \\ S_{21} \odot p \odot q \oplus ... \oplus S_{2n} \odot p^n \odot q^n = G_2 \end{cases} \tag{13}$$

where $S_{1j}, S_{2j}, p, q, G_1, G_2$ belong to Z-number set (for $j = 1, ..., n$). In order to obtain estimated solutions, a feedback neural network which is equivalent to Eq. (13) is suggested. The proposed network is shown in Fig. 1.

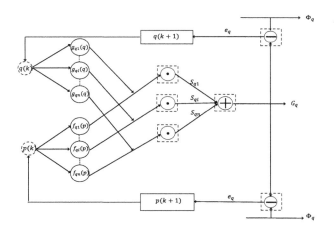

Fig. 1. Fully fuzzy nonlinear system in the form of feedback neural network.

3.2 Computation of Fuzzy Output

We propose a four layer feedback neural network where, the α-level sets of the Z-number parameters S_{qj} are nonnegative, i.e., $0 \leq \underline{S}_{qj}^{\alpha} \leq \overline{S}_{qj}^{\alpha}$ where $j = 1, ..., n$ and $q = 1, 2$. We have

- Input units

$$\begin{aligned} [p]^{\alpha} &= (\underline{p}^{\alpha}, \overline{p}^{\alpha}) \\ [q]^{\alpha} &= (\underline{q}^{\alpha}, \overline{q}^{\alpha}) \end{aligned} \tag{14}$$

- First hidden units

$$\begin{aligned} [u_j]^{\alpha} &= (\underline{u}_j^{\alpha}, \overline{u}_j^{\alpha}), \quad j = 1, ..., n \\ [v_j]^{\alpha} &= (\underline{v}_j^{\alpha}, \overline{v}_j^{\alpha}), \quad j = 1, ..., n \end{aligned} \tag{15}$$

where

$$\underline{u}_j^{\alpha} = \begin{cases} \underline{p}^{\alpha}(\underline{p}^{\alpha})^{j-1}, \underline{p}^{\alpha} \geq 0 \\ \overline{p}^{\alpha}(\overline{p}^{\alpha})^{j-1}, \underline{p}^{\alpha} < 0, \text{ j is even} \\ \underline{p}^{\alpha}(\underline{p}^{\alpha})^{j-1}, \underline{p}^{\alpha} < 0, \text{ j is odd} \end{cases} \tag{16}$$

$$\overline{u}_j^\alpha = \begin{cases} \overline{p}^\alpha(\overline{p}^\alpha)^{j-1}, \ \underline{p}^\alpha \geq 0 \\ \underline{p}^\alpha(\underline{p}^\alpha)^{j-1}, \ \overline{p}^\alpha < 0, \ j \text{ is even} \\ \overline{p}^\alpha(\overline{p}^\alpha)^{j-1}, \ \overline{p}^\alpha < 0, \ j \text{ is odd} \end{cases} \tag{17}$$

$$\underline{v}_j^\alpha = \begin{cases} \underline{q}^\alpha(\underline{q}^\alpha)^{j-1}, \ \underline{q}^\alpha \geq 0 \\ \overline{q}^\alpha(\overline{q}^\alpha)^{j-1}, \ \underline{q}^\alpha < 0, \ j \text{ is even} \\ \underline{q}^\alpha(\underline{q}^\alpha)^{j-1}, \ \underline{q}^\alpha < 0, \ j \text{ is odd} \end{cases} \tag{18}$$

$$\overline{v}_j^\alpha = \begin{cases} \overline{q}^\alpha(\overline{q}^\alpha)^{j-1}, \ \overline{q}^\alpha \geq 0 \\ \underline{q}^\alpha(\underline{q}^\alpha)^{j-1}, \ \overline{q}^\alpha < 0, \ j \text{ is even} \\ \overline{q}^\alpha(\overline{q}^\alpha)^{j-1}, \ \overline{q}^\alpha < 0, \ j \text{ is odd} \end{cases} \tag{19}$$

- Second hidden units

$$[o_j]^\alpha = (\underline{o}_j^\alpha, \overline{o}_j^\alpha), \quad j = 1, ..., n \tag{20}$$

where

$$\underline{o}_j^\alpha = \begin{cases} \underline{u}_j^\alpha \underline{v}_j^\alpha, \ \underline{u}_j^\alpha \geq 0, \ \underline{v}_j^\alpha \geq 0 \\ \underline{u}_j^\alpha \overline{v}_j^\alpha, \ \underline{u}_j^\alpha < 0, \ \underline{v}_j^\alpha \geq 0 \\ \overline{u}_j^\alpha \overline{v}_j^\alpha, \ \underline{u}_j^\alpha < 0, \ \underline{v}_j^\alpha < 0 \\ \overline{u}_j^\alpha \underline{v}_j^\alpha, \ \underline{u}_j^\alpha \geq 0, \ \underline{v}_j^\alpha < 0 \end{cases} \tag{21}$$

and

$$\overline{o}_j^\alpha = \begin{cases} \overline{u}_j^\alpha \overline{v}_j^\alpha, \ \overline{u}_j^\alpha \geq 0, \ \overline{v}_j^\alpha \geq 0 \\ \overline{u}_j^\alpha \underline{v}_j^\alpha, \ \overline{u}_j^\alpha < 0, \ \overline{v}_j^\alpha \geq 0 \\ \underline{u}_j^\alpha \overline{v}_j^\alpha, \ \overline{u}_j^\alpha \geq 0, \ \overline{v}_j^\alpha < 0 \\ \underline{u}_j^\alpha \underline{v}_j^\alpha, \ \overline{u}_j^\alpha < 0, \ \overline{v}_j^\alpha < 0 \end{cases} \tag{22}$$

- Output unit

$$[\Phi_q]^\alpha = (\underline{\Phi}_q^\alpha, \overline{\Phi}_q^\alpha), \quad q = 1, 2 \tag{23}$$

where

$$[\Phi_q]^\alpha = \left(\begin{matrix} \sum_{j \in M} \underline{S}_{qj}^\alpha \underline{o}_j^\alpha + \sum_{j \in N} \overline{S}_{qj}^\alpha \underline{o}_j^\alpha, \\ \sum_{j \in C} \overline{S}_{qj}^\alpha \overline{o}_j^\alpha + \sum_{j \in D} \underline{S}_{qj}^\alpha \overline{o}_j^\alpha \end{matrix} \right) \tag{24}$$

where $M = \{j|\ \underline{o}_j^\alpha \geq 0\}$, $N = \{j|\ \underline{o}_j^\alpha < 0\}$, $C = \{j|\ \overline{o}_j^\alpha \geq 0\}$, $D = \{j|\ \overline{o}_j^\alpha < 0\}$.

A cost function for α-level sets of the Z-number output Φ_q and the corresponding target output G_q is defined as below

$$\begin{aligned} e_q &= \underline{e}_q^\alpha + \overline{e}_q^\alpha \\ \underline{e}_q^\alpha &= \tfrac{1}{2}(\underline{G}_q^\alpha - \underline{\Phi}_q^\alpha)^2 \\ \overline{e}_q^\alpha &= \tfrac{1}{2}(\overline{G}_q^\alpha - \overline{\Phi}_q^\alpha)^2 \end{aligned} \tag{25}$$

3.3 Learning Algorithm

Z-number quantities $p = ((p^1, p^2, p^3, p^4), p)$ as well as $q = ((q^1, q^2, q^3, q^4), p)$ are initialized at random Z-numbers. For Z-number variable p adjust rule is stated as below [14, 29],

$$\begin{aligned} p^r(k+1) &= p^r(k) \oplus \Delta p^r(k), \quad r = 1, 2, 3, 4 \\ \Delta p^r(k) &= -\eta \frac{\partial e_q}{\partial p^r} \oplus \gamma \Delta p^r(k-1) \end{aligned} \tag{26}$$

where η is the learning rate and γ is the momentum term constant. $\frac{\partial e_q}{\partial p^r}$ is computed as below

$$\frac{\partial e_q}{\partial p^r} = \frac{\partial \underline{e}_q^\alpha}{\partial p^r} + \frac{\partial \overline{e}_q^\alpha}{\partial p^r}$$

(27)

Therefore,

$$\frac{\partial \underline{e}_q^\alpha}{\partial p^r} = \frac{\partial \underline{e}_q^\alpha}{\partial \underline{\Phi}_q^\alpha} \frac{\partial \underline{\Phi}_q^\alpha}{\partial p^r} = -(\underline{G}_q^\alpha - \underline{\Phi}_q^\alpha) \frac{\partial \underline{\Phi}_q^\alpha}{\partial p^r}$$

(28)

where

$$\frac{\partial \underline{\Phi}_q^\alpha}{\partial p^r} = \sum_{j \in M} \frac{\partial \underline{\Phi}_q^\alpha}{\partial \underline{o}_j^\alpha} \frac{\partial \underline{o}_j^\alpha}{\partial \underline{u}_j^\alpha} \frac{\partial \underline{u}_j^\alpha}{\partial (\underline{p}^\alpha)^j} \frac{\partial (\underline{p}^\alpha)^j}{\partial p^r}$$
$$+ \sum_{j \in N} \frac{\partial \underline{\Phi}_q^\alpha}{\partial \underline{o}_j^\alpha} \frac{\partial \underline{o}_j^\alpha}{\partial \underline{u}_j^\alpha} \frac{\partial \underline{u}_j^\alpha}{\partial (\underline{p}^\alpha)^j} \frac{\partial (\underline{p}^\alpha)^j}{\partial p^r}$$

(29)

and

$$\frac{\partial \overline{e}_q^\alpha}{\partial p^r} = \frac{\partial \overline{e}_q^\alpha}{\partial \overline{\Phi}_q^\alpha} \frac{\partial \overline{\Phi}_q^\alpha}{\partial p^r} = -(\overline{G}_q^\alpha - \overline{\Phi}_q^\alpha) \frac{\partial \overline{\Phi}_q^\alpha}{\partial p^r}$$

(30)

where

$$\frac{\partial \overline{\Phi}_q^\alpha}{\partial p^r} = \sum_{j \in C} \frac{\partial \overline{\Phi}_q^\alpha}{\partial \overline{o}_j^\alpha} \frac{\partial \overline{o}_j^\alpha}{\partial \overline{u}_j^\alpha} \frac{\partial \overline{u}_j^\alpha}{\partial (\overline{p}^\alpha)^j} \frac{\partial (\overline{p}^\alpha)^j}{\partial p^r}$$
$$+ \sum_{j \in D} \frac{\partial \overline{\Phi}_q^\alpha}{\partial \overline{o}_j^\alpha} \frac{\partial \overline{o}_j^\alpha}{\partial \overline{u}_j^\alpha} \frac{\partial \overline{u}_j^\alpha}{\partial (\overline{p}^\alpha)^j} \frac{\partial (\overline{p}^\alpha)^j}{\partial p^r}$$

(31)

In above relations the derivatives $\frac{\partial (\underline{p}^\alpha)^j}{\partial p^r}$ and $\frac{\partial (\overline{p}^\alpha)^j}{\partial p^r}$ can be written as below

$$\frac{\partial (\underline{p}^\alpha)^j}{\partial p^r} = \frac{\partial (\underline{p}^\alpha)^j}{\partial \underline{p}^\alpha} \frac{\partial \underline{p}^\alpha}{\partial p^r}$$
$$= \begin{cases} j(\underline{p}^\alpha)^{j-1} \begin{cases} 1-\alpha, & r=1 \\ \alpha, & r=2 \\ 0, & r=3 \\ 0, & r=4 \end{cases}, & \underline{p}^\alpha \geq 0 \\ j(\overline{p}^\alpha)^{j-1} \begin{cases} 0, & r=1 \\ \alpha, & r=2 \\ 1-\alpha, & r=3 \\ 0, & r=4 \end{cases}, & \underline{p}^\alpha < 0, \quad j \text{ is even} \\ j(\underline{p}^\alpha)^{j-1} \begin{cases} 0, & r=1 \\ \alpha, & r=2 \\ 1-\alpha, & r=3 \\ 0, & r=4 \end{cases}, & \underline{p}^\alpha < 0, \quad j \text{ is odd} \end{cases}$$

(32)

and

$$\frac{\partial(\overline{p}^\alpha)^j}{\partial p^r} = \frac{\partial(\overline{p}^\alpha)^j}{\partial \overline{p}^\alpha}\frac{\partial \overline{p}^\alpha}{\partial p^r}$$

$$= \begin{cases} j(\overline{p}^\alpha)^{j-1} \begin{cases} 1-\alpha, r = 1 \\ \alpha, \quad r = 2 \\ 0, \quad r = 3 \\ 0, \quad r = 4 \end{cases} & \overline{p}^\alpha \geq 0 \\[2em] j(\underline{p}^\alpha)^{j-1} \begin{cases} 0, \quad r = 1 \\ \alpha, \quad r = 2 \\ 1-\alpha, r = 3 \\ 0, \quad r = 4 \end{cases} & \overline{p}^\alpha < 0, \quad \text{j is even} \\[2em] j(\overline{p}^\alpha)^{j-1} \begin{cases} 0, \quad r = 1 \\ \alpha, \quad r = 2 \\ 1-\alpha, r = 3 \\ 0, \quad r = 4 \end{cases} & \overline{p}^\alpha < 0, \quad \text{j is odd} \end{cases} \tag{33}$$

The connection weights p_j are updated as follows

$$p_j(k+1) = (p(k+1))^j, \quad j = 2, ..., n \tag{34}$$

We can adjust the Z-number parameter q like p.

4   Applications

In this section, a real example is used to demonstrate how to apply feedback neural network in order to find the solutions of FFNS.

Example 1. Consider the series circuit consisting a voltage source and two bulbs, see Fig. 2. The power of bulbs is defined as a function of V and I. $P_1(v, i) = VI$ is power of first bulb and $P_1(v, i) = V^2 I^2$ is the power of second bulb, where V is the Voltage and I is the current of bulbs having Z-number amount. The total power equation for circuit is defined as follow

$$\begin{cases} ((7, 9, 11), p(0.7, 0.81, 0.9)) \odot V \odot I \oplus ((2, 5, 7), p(0.7, 0.8, 0.9)) \odot V^2 \odot I^2 \\ = ((29, 1201, 6011), p(0.7, 0.8, 0.91)) \\ ((9, 10, 12), p(0.7, 0.8, 0.9)) \odot V \odot I \oplus ((4, 6, 9), p(0.7, 0.85, 0.9)) \odot V^2 \odot I^2 \\ = ((55, 1501, 8001), p(0.75, 0.8, 0.9)) \end{cases}$$

$V = ((-5, -3, -2), p(0.8, 0.9, 1))$ and $I = ((-4, -3, -1), p(0.8, 0.9, 1))$ are the exact solutions. The neural network shown in Fig. 1 is utilized in order to estimate the solutions V and I. The maximum learning rate of neural network is $\eta = 0.001$. The neural network starts from $V(0) = ((-8, -6, -5), p(0.7, 0.8, 0.9))$ and $I(0) = ((-7, -6, -4), p(0.75, 0.8, 0.9))$. The approximation results are demonstrated in Table 1. The error between the approximate solution and the exact solution is demonstrated in Fig. 3.

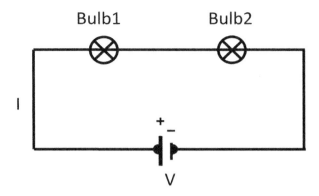

Fig. 2. Series circuit

Table 1. Neural network approximation for the solutions

k	$V(k)$	$I(k)$
1	$((-7.8876, -5.9203, -4.9099),$ $p(0.7, 0.8, 0.9))$	$((-6.8909, -5.9392, -3.9151),$ $p(0.7, 0.8, 0.9))$
2	$((-7.5038, -5.6027, -4.5932),$ $p(0.7, 0.8, 0.9))$	$((-6.5488, -5.6312, -3.6009),$ $p(0.7, 0.8, 0.9))$
3	$((-7.2192, -5.3202, -4.2984),$ $p(0.7, 0.8, 0.9))$	$((-6.2472, -5.3533, -3.2783),$ $p(0.7, 0.8, 0.9))$
$\vdots$	$\vdots$	$\vdots$
97	$((-5.0107, -3.0113, -2.0127),$ $p(0.7, 0.8, 0.9))$	$((-4.0112, -3.0106, -1.0121),$ $p(0.7, 0.8, 0.9))$
98	$((-5.0073, -3.0086, -2.0096),$ $p(0.7, 0.8, 0.9))$	$((-4.0089, -3.0078, -1.0093),$ $p(0.7, 0.8, 0.9))$
99	$((-5.0043, -3.0055, -2.0062),$ $p(0.7, 0.8, 0.9))$	$((-4.0056, -3.0049, -1.0061),$ $p(0.7, 0.8, 0.9))$

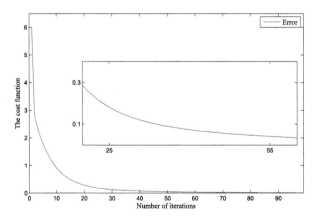

Fig. 3. The error between the approximate solution and the exact solution

5   Conclusion

In this work, a new method based on the feedback neural network is proposed in order to approximate the solutions of FFNS. In order to obtain the approximate solutions of FFNS a learning algorithm based on the gradient descent method is applied. An example is laid down in order to demonstrate the high accuracy of this suggested technique. The future work is the application of the mentioned methodology for FFNS equations on the basis of Z-numbers.

References

1. Jafari, R., Yu, W.: Fuzzy control for uncertainty nonlinear systems with dual fuzzy equations. J. Intell. Fuzzy. Syst. **29**, 1229–1240 (2015)
2. Jafari, R., Yu, W., Li, X.: Fuzzy differential equation for nonlinear system modeling with Bernstein neural networks. IEEE Access. (2017). https://doi.org/10.1109/ACCESS.2017.2647920
3. Jafari, R., Yu, W.: Uncertainty nonlinear systems modeling with fuzzy equations. Math. Probl. Eng. **2017** (2017). https://doi.org/10.1155/2017/8594738
4. Jafari, R., Yu, W.: Uncertainty nonlinear systems modeling with fuzzy equations. In: Proceedings of the 16th IEEE International Conference on Information Reuse and Integration, San Francisco, Calif, USA, pp. 182–188, August 2015
5. Jafari, R., Yu, W., Li, X.: Numerical solution of fuzzy equations with Z-numbers using neural networks. Intell. Autom. Soft Comput. **1**, 1–7 (2017)
6. Jafari, R., Yu, W., Li, X., Razvarz, S.: Numerical solution of fuzzy differential equations with Z-numbers using Bernstein neural networks. Int. J. Comput. Intell. Syst. **10**, 1226–1237 (2017)
7. Razvarz, S., Jafari, R., Yu, W.: Numerical solution of fuzzy differential equations with Z-numbers using fuzzy Sumudu Transforms. Adv. Sci. Technol. Eng. Syst. J. **3**, 66–75 (2018)
8. Razvarz, S., Jafari, R., Granmo, O.-Ch., Gegov, A.: Solution of dual fuzzy equations using a new iterative method. In: Nguyen, N.T., Hoang, D.H., Hong, T.-P., Pham, H., Trawiński, B. (eds.) ACIIDS 2018. LNCS (LNAI), vol. 10752, pp. 245–255. Springer, Cham (2018). https://doi.org/10.1007/978-3-319-75420-8_23
9. Pratama, M., Lu, J., Anavatti, S.G., Lughofer, E., Lim, C.P.: An incremental meta-cognitive-based scaffolding fuzzy neural network. Neurocomputing **171**, 89–105 (2016)
10. Pratama, M., Lu, J., Lughofer, E., Zhang, G., Er, M.J.: Incremental learning of concept drift using evolving type-2 recurrent fuzzy neural network. IEEE Trans. Fuzzy Syst. **25**, 1175–1192 (2017)
11. Abiyev, R.H., Kaynak, O.: Fuzzy wavelet neural networks for identification and control of dynamic plantsa novel structure and acomparative study. IEEE Trans. Ind. Electron. **55**, 3133–3140 (2008)
12. Hornik, K.: Approximation capabilities of multilayer feed-forward networks. Neural Netw. **4**, 251–257 (1991)
13. Scarselli, F., Tsoi, A.C.: Universal approximation using feedforward neural networks: a survey of some existing methods, and some new results. Neural Netw. **11**, 15–37 (1998)
14. Ishibuchi, H., Kwon, K., Tanaka, H.: A learning of fuzzy neural networks with triangular fuzzy weghts. Fuzzy Sets Syst. **71**, 277–293 (1995)

15. Hayashi, Y., Buckley, J.J., Czogala, E.: Fuzzy neural network with fuzzy signals and weights. Int. J. Intell. Syst. **8**, 527–537 (1993)
16. Buckley, J.J., Eslami, E.: Neural net solutions to fuzzy problems: the quadratic equation. Fuzzy Sets Syst. **86**, 289–298 (1997)
17. Mosleh, M.: Evaluation of fully fuzzy matrix equations by fuzzy neural network. Appl. Math. Model. **37**, 6364–6376 (2013)
18. Abbasbandy, S., Otadi, M.: Numerical solution of fuzzy polynomials by fuzzy neural network. Appl. Math. Comput. **181**, 1084–1089 (2006)
19. Jafarian, A., Jafari, R.: Approximate solutions of dual fuzzy polynomials by feedback neural networks. J. Soft Comput. Appl. (2012). https://doi.org/10.5899/2012/jsca-00005
20. Takagi, T., Sugeno, M.: Identification of systems and its applications to modeling and control. IEEE Trans. Man Cybern. **15**, 116–132 (1985)
21. Lughofer, E., Cernuda, C., Kindermann, S., Pratama, M.: Generalized smart evolving fuzzy systems. Evolv. Syst. **6**, 269–292 (2015)
22. Angelov, P.P., Filev, D.: An approach to online identification of Takagi-Sugeno fuzzy models. IEEE Trans. Syst. Man Cybern. Part B Cybern. **34**, 484–498 (2004)
23. Chen, G., Pham, T.T., Weiss, J.J.: Modeling of control systems. IEEE Aerosp. Electron. Syst. **31**, 414–428 (1995)
24. Abbasbandy, S., Ezzati, R.: Homotopy method for solving fuzzy nonlinear equations. Appl. Sci. **8**, 1–7 (2006)
25. Mouzouris, G.C., Mendel, J.M.: Dynamic non-singleton logic systems for nonlinear modeling. IEEE Trans. Syst. **5**, 199–208 (1997)
26. Allahviranloo, T.: The Adomian decomposition method for fuzzy system of linear equations. Appl. Math. Comput. **163**, 553–563 (2005)
27. Paripour, M., Zarei, E., Shahsavaran, A.: Numerical solution for a system of fuzzy nonlinear equations. J. Fuzzy Set Valued Anal. **2014**, 1–10 (2014)
28. Zadeh, L.A.: Toward a generalized theory of uncertainty (GTU) an outline. Inform. Sci. **172**, 1–40 (2005)
29. Alefeld, G., Herzberger, J.: Introduction to Interval Computations. Academic Press, New York (1983)

Facial Expression Recognition: A Survey on Local Binary and Local Directional Patterns

Kennedy Chengeta and Serestina Viriri[✉]

School of Mathematics, Statistics and Computer Science,
University of KwaZulu-Natal, Durban, South Africa
{216073421,viriris}@ukzn.ac.za

Abstract. Automated facial and emotional recognition has been extensively applied in computer science, medical neuroscience, law enforcement and crowd monitoring. The study evaluates use of popular feature descriptors, Local Binary Pattern (LBP) and Local Directional Pattern (LDP) variants in facial expression recognition feature extraction. It then classifies results of the local facial features of major emotional states, namely neutral, anger, fear, extraction and expression identification using a combined ratio of classifiers called Voting Classifer. Databases used in the experiments involved Cohn-Kanade Database and the Googleset datasets and the expression classification rate of around 99.13% was achieved. The proposed solution included a hybrid of Local Directional Pattern (LBP), Local Directional Pattern (LDP) as the feature extraction algorithms and weighted ensemble of classifiers called voting classifier classification algorithm.

1 Introduction

Facial expression identification automation has benefited fields like medicine, security, accounting, education and computing [3]. Widespread use of automated facial expression analysis has been witnessed in identity management systems, crowd surveillance, crime control and security control systems [3]. The human face's emotions guides a person's gender, attractiveness, age, personality and other characteristics as well. Automated facial recognizers have been applied in intelligent networks and transportation systems for drowsiness detection as well in the internet of things. For the deaf, sign languages use facial expressions to do grammar encoding in communication. Indications for sickness, stress, deception and happiness are derived through micro expressions since they leak behavior control. Facial expressions are either voluntary or intentional and in some cases non-voluntary or natural responses to stimuli [1,12]. Based on Darwin's emotions theory, fear arises when one opens the eyes or raises eyebrows [12]. Anger manifests when one has open eyes and raised nostrils [5]. One will be showing contempt by looking the other way. Surprise is shown by opening both the mouth and eyes wide [5]. AlchemyAPI, Emotiva, IBM Watson as well Windows

© Springer Nature Switzerland AG 2018
N. T. Nguyen et al. (Eds.): ICCCI 2018, LNAI 11055, pp. 513–522, 2018.
https://doi.org/10.1007/978-3-319-98443-8_47

Azure are some of the major facial emotional recognition industrial APIs used in sentiment analysis and crowd expression detection.

2   Literature Review-Facial Expressions

Key expressions include sadness, joy, neutral, surprise, fear and anger. Facial expression recognition algorithms research in personalised gaming has been successfully researched [5,13]. Chronic pain identification has been successfully identified using facial expression analysers in hospitals and places of work [1,12]. Automated smile identification was also used in hospitals with focus on chronic patients and young babies using real time embedded solutions [3]. Facial expressions for different emotions have been extracted using various feature extraction algorithms together with various supervised and unsupervised learning classifiers. Local binary patterns and local directional patterns, both local algorithms have been used successfully to identify all the different expressions [8,9] with holistic algorithms like Principal component analysis achieving great accuracy [14]. Facial expression feature extraction is done using namely geometric and appearance based feature extraction. The former method computes the locations and shape data using various facial parts example like nose, mouth, eye, distance between nose and the eye region. These facial parts are then combined into a feature vector which represents the geometry of the face [6]. The feature vector classification accuracy is affected by the dimensional properties and redundancy of some of the features. Appearance based algorithms represent a face as multiple raw intensity images where a given image is made up of multi-dimensional vectors. Statistical algorithms are used to extract feature space from the images. Some of the features in the given feature vectors are redundant taking for instance non-discriminating features which bring complexity and reduced recognition accuracy rates [1,2]. The research expands facial expression recognition work using LBP and LDP and voting classifier [12].

2.1   Holistic and Local Algorithms

Holistic and local features play a great role in facial expression analysis [4,7]. Holistic algorithms will extract the entier facial image whereas the local algorithms consider local facial subregions like nose, mouse and ears. The latter is position and face orientation dependent as well. Popular holistic algorithms include PCA (principal component analysis), discrete cosine transform (DCT) [4] and Fisher faces as well. Holistic algorithms are good because they is no information loss even after concatenation since they focus on specific points of interest which are not impacted by dimensional reduction [6]. Local features are extracted by algorithms such as Local Gabor Feature Vectors (LGFV) [6], Local Binary Patterns (LBP) and Local Directional Patterns.

2.2   Local Binary Patterns (LBP)

Local Binary Patterns divide an image into local subregions for feature extraction and or texture analysis in the field of facial recognition. This assists in overcoming facial recognition challenges like occlusion or non-rigidness in facial images [6,12]. The localisation of feature extraction allows classification in both coloured or gray level images with no variance [8,13]. The localised feature vectors are derived from feature extraction where each facial subregion is concatenated into a single histogram of feature vectors which is classified by traditional machine learning classifiers like k-nearest neighbour, random forest and voting classifiers [8,12]. Local Binary Pattern (LBP) [7,8] variants have been used in facial and emotion recognition and have exhibited invariancy and robustness to illumination changes and computational simplicity [8]. Various LBP variants were proposed as LBP extensions and modifications were done to improve facial recognition accuracy [7,9].

2.3 LBP Variants

Various LBP variants have been proposed and used in facial expression recognition. LBP variants included symmetric (CS-LBP) which is a modified LBP algorithm, Rotated Local Binary Patterns (RLBP), Uniform local binary patterns (ULBP) as well as LBPNet which is applied LBP on Convolutional Neural Networks (CNN) [7,12]. LBPs have been combined with Gabor features in facial expression application succesfully [7,9]. Rotated Local Binary patterns which consider LBP signs of differences when calculating final descriptors have also been applied in emotional recognition feature extractions [9]. Symmetric (CS-LBP), an enhanced LBP algorithm with longer feature vectors, higher dimensions and more robustness than the normal LBP has also been applied in facial expression feature extraction experiments with more success on flater images [7,9,12]. Enhancements have been applied to the LBP using the SIFT descriptors [1,2].

2.4 LBP Variants with Convolutional Neural Networks

Deep learning allows for high level data representation by going through several conv (processing) layers and allows for dual feature extraction and classification. Convolutional neural networks (CNN) have been implemented with success in facial expression recognition [10]. Yu and Zhang used the EmotiW idataset with 5 convolutional layers to recognised facial expressions [10]. Since CNN does both it has achieved better results than traditional feature extraction algorithms like LBP and LDP. Binary Patterns with Encoded Convolutional Neural Networks were successfully applied for texture recognition and remote sensing [10]. LBP-Net also applied LBP and Convolutional Neural Network (CNN) deep learning topology and replaced deep learning training kernels with Local Binary Pattern computer vision descriptors [11]. This was found to be less expensive approach and achieved with little data [11].

2.5 Local Directional Patterns

Local directional pattern algorithms use compass masks to encode directional facial local image components. The image is split into subregions and the feature extractions are grouped into one histogram to classify. More prominent regions get picked up in generating the more pronounced regions [7,9]. Different experiments were done where the descriptors accuracy where measured given the noise, different lighting conditions and varying time conditions and the LDP descriptors exhibited robust resistance to noise [7,9]. The top k-directional bits are given the value of 1 and the remaining bits take the value of zero. The image was then split into small regions where histograms were combined into an LDPv descriptor. Encoding is done using edge detection algorithms on more prominent edges based on the Kirsch algorithm which applies image convolution [7,12]. Edge responses are invariant to noise and non-monotonic illumination changes [9,12].

3   Methods and Techniques

3.1   Facial Expression Databases

Facial images were retrieved from the Googleset database and CK+ dataset. The dataset images depicted different facial expressions reflecting various emotions. The given emotions ranged from fear, sadness, happiness/joy, disgust to neutral [11]. The study participants performed several action units and facial displays and the images changed from neutral to peak through seven categories [15]. The google facial dataset was chosen due to its small dataset structure and the CK+ due to its large dataset and different races. The CK+ dataset has around hundred individuals of American, Asian and Latin origin [15]. The Cohn-Kanade (CK) AU-coded expression dataset included over 100 students aged in their teenage years and early adulthood [12]. The facial Google dataset was also used. Around an eighth of the dataset subjects were African American.

3.2   Preprocessing Face Alignment, Normalization and Dimensional Reduction

Noise was eliminated by re-sizing the data so that all images had same dimensions. The images were initially converted into grey level images. Kirsch masking at 45° rotation was used to extract edge responses in eight different directions. During normalization, the images went through standardization using size, pose and illumination as key parameters in relation to the image. Dimensional reduction was implemented using Principal component analysis (PCA) algorithm.

3.3   Feature Extraction

Feature extraction for local features of the images into the histogram was done based on Local Binary Patterns and Local Directional Patterns. For the 2 databases, Voting classifiers were then applied on the trained data. Accuracy and performance of the algorithms was measured on the data. A given LBP M, N operator is represented mathematically as follows

$$LBP_{(M,N)}(m_x, n_x) = \sum_{V-1}^{V=0} q(p_x - p_c)2^V. \tag{1}$$

For the LBP equation, the neighbourhood being depicted as M-bit binary resulting in V distinct values for a specific LBP code. The gray level is in the form of 2 V-bin unique LBP codes. For the Local Directional Number Pattern [12]. The facial expression images were subdivided into subregions where LDPx histograms were retrieved and combined into a single descriptor [12].

$$LDP_x(\sigma) = \sum_K^{r=0} \sum_L^{r=0} f(LDP_q(o, u), \sigma). \tag{2}$$

3.4 Classification

For both Local Binary Patterns (LBP) and Local Directional Patterns (LDP), the feature vectors were used for classification by machine learning classifiers and the trained dataset was also used as input into the deep learning neural network. Various machine learning classifiers, namely random forest, neural networks, k-nearest neighbor, 4.5 decision tree classifiers and support vector machines were examined. They were also combined in the form of a combined classifier called Voting Classifier. EnsembleVoteClassifier is a meta-classifier that combines various machine learning classifiers for classification via majority, weighted or probability voting [7]. The Ensemble classifier eclf is a combination of clf1, kNearest Neighbour with weight 0.35, clf2 representing Support Vector Machine with weight 0.2 and classifier clf3 with weight 0.45. The modal eclf classifier accuracy is then taken as the accuracy. We applied cross validation to ensure there is reduced overfitting.

$$eclf = EC(clfs = [clf1, clf2, clf3], weights = [0.35, 0.2, 0.45]) \tag{3}$$

EC in the above equation represents the EnsembleClassifier with 3 classifiers. The facial images were then saved in a MySQL Database in the form of feature vector forms for facial expressions. The expression classes identified included neutral, happiness, anger, fear, disgust and sadness (sorrow).

3.5 LBP Convolutional Neural Network Expression Recognition

Both the CK+ and Googleset databases were processed using 5 CNN convolutionary layers after the feature vectors had been already been extracted. The google python 2.7 tensorflow library was used in this process and accuracy results for each emotion measured.

4 Facial Recognition and Expression Results

The experiment included firstly finding the modal ensemble classifier which was then used in Local Binary Pattern feature vector classification and this is explained in Sect. 4.1. Several experiments were then run to compare facial expression accuracy firstly using Local Binary pattern variants with the modal ensemble classifier (Sect. 4.2). The same data was also run against a CNN variant Local Binary Pattern. The facial recognition and expression classification results were then compared respectively in Sects. 4.2 to 4.5. The 2 resulting algorithms were then compared in terms of accuracy, cost, time and simplicity in the conclusion Sect. 4.6.

4.1 Finding Modal Ensemble Voting Classifier

The classification used cross-validation based on a training/test ratio of 6 to 4 ratio on the image classification algorithms. The classification results are detailed in diagram in Fig. 1. The voting classifier combination for kNN, support vector machines (SVM), random forest (RF) and C4.5 decision tree (DT) with various ratios of the different classifiers [12]. Around 100 variations of the classifiers were executed with different classifier ratios (classes) to find the modal combination. The table in Fig. 1. shows 4 samples of randomly picked 100 ratio combinations executed with a 5:2:1:2 ratio, 3:1:3:4 ratio, 6:1:2:1 ratio and 4:1:4:1 ratio which was class 9, 34, 45 and 25 respectively. The modal accuracy also shown, class 25 was the one where kNN had high ratio which is 7:1:1:1 and this gave an accuracy for 99.13% for the CK+ database and 99.23% for the Google data-set. The combinations where decision trees were dominant showed worst performance in the tests as shown by class 45. The choice of the ensemble classifier was based on the LDP and ELBP (8, 2) feature extraction combination (Fig. 2).

classifier	class 9	class 34	class 45	class 25	class 100
kNearest Neighbour	x_1 * 0.5	x_1 * 0.3	x_1 * 0.2	x_1 * 0.7	x_1 * 0.4
Support Vector Machines	x_2 * 0.2	x_2 * 0.1	x_2 * 0.1	x_2 * 0.1	x_2 * 0.1
Random Forest	x_3 * 0.2	x_3 * 0.3	x_3 * 0.1	x_3 * 0.1	x_3 * 0.4
c4.5 Decision Tree	x_4 * 0.1	x_4 * 0.3	x_4 * 0.6	x_4 * 0.1	x_4 * 0.1
Accuracy(CK+ Data)	0.79	0.67	0.23	0.9923	0.86
Accuracy(Googleset)	0.72	0.64	0.26	0.9913	0.81

Fig. 1. Voting meta classifier

4.2 Facial Recognition Classification Results

The algorithms used in the study included LBP variants namely symmetric CS-LBP, local ternary LTP, the rotated RLBP, enhanced LBP, LBP with CNN and a combined algorithm with LDP and ELBP algorithm. Various classifiers were used namely k-Nearest neighbor, Support Vector Machine, Random Forest (RF) and a Voting Classifier. The modal accuracy, class 25 was the one where kNN had high ratio which is 7:1:1:1 and this gave an accuracy for 99.13% for the CK+ database and 99.23% for the Google data-set. The combinations where decision trees were dominant showed worst performance in the tests as shown by class 45. The choice of the ensemble classifier was based on the LDP and ELBP (8, 2) feature extraction combination.

GoogleSet Data	kNN+	Support Vector Machine	RF	Voting Classifier	Ave Time(s)	CK+ Data	kNN+	Support Vector Machine	RF	Voting Classifier
$LBP_{8,2}$	93.09%	97.12%	97.16%	97%	52.31s	$LBP_{8,2}$	92%	97%	97.6%	97%
$LBP_{16,2}$	94.26%	98.03%	97.28%	95.99%	44.23s	$LBP_{16,2}$	93%	98%	97%	96%
$CS\text{-}LBP_{8,2}$	93.51%	97.98%	97.33%	97.86%	51.87s	$CS\text{-}LBP_{8,2}$	93%	96%	96%	98.56%
$CS\text{-}LBP_{16,2}$	94.05%	96.92%	96.08%	98.31%	53.45s	$CS\text{-}LBP_{16,2}$	92.5%	97.2%	96.08%	97.1
$ELBP_{16,2}$	91.3%	91.02%	96.29%	97.09%	55s	$ELBP_{8,2}$	89%	87.2%	96%	96.9%
$ELBP_{16,2}$	88.21%	87.12%	95.75%	96.91%	52s	$ELBP_{16,2}$	85%	85%	96.5%	96.1%
$LTP_{8,2}$	89.71%	96.31 %	96.77%	96.74%	51s	$LTP_{8,2}$	85.7%	95.1 %	97%	95.54%
$LTP_{16,2}$	88.3%	96.44 %	97.42%	97.24%	54s	$LTP_{16,2}$	86.27%	95 %	97%	96.33%
$RLBP_{8,2}$	86.1%	97.01 %	96.8%	97.48%	53s	$RLBP_{8,2}$	85.1%	95.4 %	96.8%	96.61%
$RLBP_{16,2}$	84.3%	97.21 %	97.09%	97.64%	51s	$RLBP_{16,2}$	85.3%	95.21 %	96%	97.66%
$LDP+ELBP_{8,2}$	94.18%	97.81%	96.92%	99.13%	55.23s	$LDP+ELBP_{8,2}$	93.1%	98.2%	97.88%	99.23%
$LDP+ELBP_{16,2}$	94.78%	97.99%	97.02%	99.03%	55.6s	$LDP+ELBP_{16,2}$	94.3%	97.45%	98.39%	99.27%
$LBP+CNN(99.63\%)$					1620s	$LBP+CNN(99.77\%)$				

Fig. 2. Classifier for CK+ and Googleset dataset

4.3 Facial Expression Results: Small Datasets-Googleset

The facial expression results were based on the following emotions namely anger, disgust, fear, happy or joyness, sadness, neutral, surprise plus contempt. The results are summarised in the following namely Figs. 3 and 4 based on the Googleset data. The google set data was a small dataset and experiments were carried out from 100 images up to 1500 images though the best results picked were for 75 images as well as for 181 images all achieving an average well beyond 99%. Due to the small dataset the 75 images were only able to pick 4 emotions hence the reason the 3 emotions were all showing zero accuracy. For the Googleset large dataset test with the best accuracy a 1500 images were trained and using a two thirds classification ratio, 1000 images were classified into the 7 emotions.

	precision	recall	f1-score	support	Confusion Matrix		
anger	1	1	1	16	anger	[[16, 0, 0, 0, 0, 0, 0]	
disgust	0.98	1	0.99	44	disgust	0, 44, 0, 0, 0, 0, 0]	
fear	1	0.92	0.96	12	fear	0, 1, 11, 0, 0, 0, 0]	
happy	1	1	1	3	happy	0, 0, 0, 3, 0, 0, 0]	
disgust	0	0	0	0	disgust	0, 0 0, 0, 0, 0, 0]	
fear	0	0	0	0	fear	0, 0 0 0, 0, 0, 0]	
happy	0	0	0	0	happy	0, 0, 0, 0 0, 0, 0]]	
avg/total	0.99	0.99	0.99	75			

Fig. 3. Google set dataset facial expression recognition small dataset-75 images

	precision	recall	f1-score	support	Confusion Matrix		
anger	1	1	1	16	anger	[16, 0, 0, 0, 0, 0, 0],	
disgust	1	1	1	44	disgust	0, 44, 0, 0, 0, 0, 0],	
fear	0.92	1	0.96	12	fear	0, 0, 12, 0, 0, 0, 0],	
happy	1	1	1	39	happy	0, 0, 0, 39, 0, 0, 0],	
neutral	1	0.94	0.97	18	neutral	0, 0, 0, 0, 17, 0, 1],	
sadness	1	1	1	23	sadness	0, 0, 0, 0, 0, 23, 0],	
surprise	0.97	0.97	0.97	29	surprise	0, 0, 1, 0, 0, 0, 28]	
avg / total	0.99	0.99	0.99	181			

Fig. 4. Google set dataset facial expression recognition-181 images

4.4   Facial Expression Results for Large Datasets-CK+

CK+ database experiments involved tests with a small dataset of 500 up to a range of 4000 images as well. The accuracy in both scenarios was impressive averaging 99%+ in the larger dataset and 98% in the smaller dataset. Only 6 emotions were measured. As indicated in the overall classification table the Voting Classifier showed supreme accuracy compared to the individual classifiers. Sadness was the best accuracy emotion whilst neutral showed few individuals who were incorrectly classified (Figs. 5 and 6).

	precision	recall	f1-score	support		Confusion Matrix					
anger	0.98	1	0.99	317	anger	[[317,	0,	0,	0,	0,	0],
disgust	0.98	0.99	0.99	214	disgust	[2,	212,	0,	0,	0,	0],
fear	1	0.99	0.99	230	fear	[2,	0,	228,	0,	0,	0],
happy	0.99	0.99	0.99	276	happy	[1,	3,	0,	272,	0,	0],
neutral	1	0.99	1	214	neutral	[0,	1,	0,	1,	212,	0],
sadness	1	0.98	0.99	149	sadness	[0,	0,	1,	2,	0,	146]]
avg/total	0.99	0.99	0.99	1400							

Fig. 5. CK+ dataset facial expression recognition large dataset (1400 images)

	precision	recall	f1-score	support		Confusion Matrix					
anger	0.99	1	0.99	70	anger	[70,	0,	0,	0,	0,	0]
disgust	1	1	1	44	disgust	[0,	44,	0,	0,	0,	0]
fear	1	1	1	32	fear	[0,	0,	32,	0,	0,	0]
happy	1	1	1	40	happy	[0,	0,	0,	40,	0,	0]
neutral	1	1	1	36	neutral	[0,	0,	0,	0,	36,	0]
sadness	1	0.96	0.98	28	sadness	[1,	0,	0,	0,	0,	27]
avg/total	1	1	1	250							

Fig. 6. CK+ dataset facial expression recognition small dataset (250 images)

4.5   Results Analysis

LDP+ELBP feature extraction classified by a voting classifier gave accuracy of 99.23% and 99.13 on the Google-Set and CK+ databases which was the modal accuracy. This was much improved accuracy compared to other comparative LBP variant algorithms. The tests results from the experiments included runs of around a century of images, medium sized runs of a thousand to 2000 images and huge datasets beyond 5000 images for the CK+ database. The results did support the theory that large or small datasets have no impact on accuracy on LBP algorithms [12] with voting classifiers and will give high accuracy irregardless. The LDP algorithm, an edge detection algorithm's smoothening of the edges capability based on the Kirsch algorithm aided in removing edge noises from the grey level images. The google-set data also had impressive accuracy when classified by an ensemble classifer with accuracy of 99.23% being the modal classification result. The classifier with better results was the Voting Classifier and it showed consistent results for low number of images from around 100 and also for medium range of images which is around 1000 as well as for huge number of images ranging around 10 000 facial images as well.

4.6 Conclusion

The study's experiments results show the Local Binary Pattern and Local Directional Pattern classified by a Voting Classification Algorithm where kNearest Neighbour has the dominant weight (seventy percent) improves facial emotions accuracy. Although higher facial expression classification rates were achieved by the proposed methods, there is still some issues which should be furthered addressed such as finding the optimal threshold for each database automatically, or applying the proposed algorithm with other features, in the purpose of improving the recognition rate. Therefore, the hybrid LDP+ELBP Feature Extractor with a Voting Classifier operators presented in this study gave improved accuracy for facial emotion recognition compared to a uniform LBP or LDP on their own. Happiness in all the cases received the highest classification accuracy for both google dataset and the CK+ dataset as well ranging from around 90–100%.

References

1. Aung, M.S., et al.: The auto- matic detection of chronic pain-related expression: requirements, challenges and a multimodal dataset. Trans. Affect. Comput. (2015)
2. Pavithra, P., Ganesh, A.B.: Detection of human facial behavioral expression using image processing. ICTACT J. Image Video Process. **1**(3), February 2011
3. Nurzynska, K., Smolka, B.: Smiling and neutral facial display recognition with the local binary patterns operator. J. Med. Imaging Health Inform. **5**(6), 1374–1382 (2015)
4. Viola, P., Jones, M.J.: Robust real-time face detection. Int J. Comput. Vis. **57**, 137–154 (2004)
5. Blom, P.M., Bakkes, S., Tan, C.T., Whiteson, S., Roijers, D., Valenti, R., Gevers, T.: Towards personalised gaming via facial expression recognition. In: AAAI Press, Palo (2014)
6. Gernoth, T., Goossen, A., Grigat, R.-R.: Face recognition under pose variations using shape-adapted texture features. In: 2010 17th IEEE International Conference on Image Processing (ICIP), pp. 4525–4528, September 2010
7. Longmore, C., Liu, C., Young, A.: The importance of internal features in learning new faces. Q. J. Exp. Psychol. **68**(2), 249–260 (2015)
8. Zhao, X., Zhang, S.: Facial expression recognition based on local binary patterns and kernel discriminant isomap. Sensors **11**(10), 9573–9588 (2011)
9. Ramirez Rivera, A., Castillo, R., Chae, O.: Local directional number pattern for face analysis: face and expression recognition. IEEE Trans. Image Process. **22**, 1740–1752 (2013)
10. Alizadeh, S., Fazel, A.: Convolutional Neural Networks for Facial Expression Recognition. arXiv:1704.06756 (2017)
11. Chen, L., Xi, M.: Local binary pattern network: a deep learning approach for face recognition. In: 2016 IEEE International Conference on Image Processing (ICIP) (2016)
12. Chengeta, K., Viriri, S.: A survey on facial recognition based on local directional and local binary patterns. In: 2018 Conference on Information Communications Technology and Society (ICTAS) (2018)

13. Uddin, M.Z., Khaksar, W., Torresen, J.: Facial expression recognition using salient features and convolutional neural network. IEEE Access **5**, 26146–26161 (2017). https://doi.org/10.1109/ACCESS.2017.2777003
14. Vupputuri, A., Meher, S.: Facial Expression recognition using Local Binary Patterns and Kullback Leibler divergence. In: 2015 International Conference on Communications and Signal Processing (ICCSP), Melmaruvathur, pp. 0349–0353 (2015). https://doi.org/10.1109/ICCSP.2015.7322904
15. Lucey, P., Cohn, J.F., Kanade, T., Saragih, J., Ambadar, Z., Matthews, I.: The Extended Cohn-Kanade dataset (CK+): a complete expression dataset for action unit and emotion-specified expression. In: CVPR4HB 2010, San Francisco, USA, pp. 94–101 (2010)

Energy-Based Centroid Identification and Cluster Propagation with Noise Detection

Alexander Krassovitskiy$^{(\boxtimes)}$ and Rustam Mussabayev

Institute of Information and Computational Technologies,
Pushkin str. 125, Almaty, Kazakhstan
akrassovitskiy@gmail.com

Abstract. Clustering algorithms are used to partition an existing set of objects into groups according to similarity of their attributes. Parametric algorithms for determining initial points (centroids) and subsequent cluster propagation are proposed. The principle of competitive growth of clusters due to the absorption of boundary (contiguous) objects is used. The object is absorbed by that cluster or transferred from an adjacent cluster if it maximizes the total energy of cluster. The remaining objects that have not been clustered are classified as noise. Then the parameter identification problem for the algorithm is considered. Preliminary results on clustering and parameter identification are obtained on several public test data sets.

Keywords: Centroid identification · Energy based clustering
Noise detection · Parameter identification · Data analysis

1 Introduction

Clustering algorithms gain significant popularity as one of the efficient instruments for data analysis by determining the non-intersected object groups (clusters) by certain criteria. A number of clustering algorithms have been implemented recently: as general purpose techniques as well as ones directed to the broad scope of application [8]; some of them gained popularity due to their efficiency and clarity [6,13]. Each of the existing algorithms has its own advantages as well as drawbacks. The aim of our work is to construct an algorithm that can provide high level of flexibility for clustering of different data sets with various and complex topological configurations.

This work is inspired by the idea of handling energy values assigned to data points, so that their accumulating cluster energy growths faster in regions with higher local energy. In our case the energy is the quantitative property that is transferred between objects in order to perform work on their merging to clusters.

A number of different approaches of using energy in clustering have been proposed recently: In [3] a function of energy inspired by Lyapunov energy function

© Springer Nature Switzerland AG 2018
N. T. Nguyen et al. (Eds.): ICCCI 2018, LNAI 11055, pp. 523–533, 2018.
https://doi.org/10.1007/978-3-319-98443-8_48

is considered. Clustering quality is measured in terms of spatial and temporal characteristics for streamed data sets. The motivation behind this model is to make possible extracting temporal evolution of data streams so the behavior and tendency changes in the clustering can be captured. In [14] an energy-based optimization method is introduced helping to visualize arbitrary graph in an efficient way. Sampled edge segments of an input graph are filtered and hierarchically clustered by means of energy function. The latter one gives quantitative representation of the graph congestion and is used as a criterion to select the most cluttered edges to be reformed. In [10] edge repulsion/attraction model to find a proper graph separation is used. The introduced energy function captures properties of vertex/edge degrees and their positions so an unbiased decision can be made on extraction of a subgraph as a distinctive cluster. In [9] an image segmentation problem is considered by taking into account several known segmentation approaches. In order to combine different conflicting and complementary criteria (the region-based variation of information and the contour-based F-measure), a fusion global consistency criterion of energy is introduced. Hence, the image segmentation is processed by k-means algorithm with different color spaces which is followed by energy maximization to solve consensus clustering problem.

In the above publications it is shown that using the energy concept a certain aggregated state of a system can be achieved and measured during the clustering process. This opens a possibility to treat clustering problem by means of optimization of a single energy criterion. Also, since in general the nature of physical laws are non-linear, certain empirical heuristics can be added in order to achieve better quality of the clustering by means of adjusting their energy incoming criteria. Moreover, it is better to reason about clustering in terms of aggregated property such as energy.

In our work a related approach is used to measure local properties of data points being clustered with the help of a special energy function that reflects spatial clustering characteristics. However, we are also using the term of energy, it differs from the energy in the previous mentioned works.

We have constructed the clustering algorithm that tries to yield a maximum of energy for every cluster being assembled, and the total energy of the whole clustering.

Specifically, the data objects are considered as n-dimensional real-value vector points. Introducing the metric distance on data points and closeness relations can be done in a few different ways as suggested in [1]. Since our basic knowledge on relation between data set attributes somehow limited and for the reason of general purpose we use the standard approach of the Euclidean distance metric. A distance measure function is defined on these data points as a criterion of closeness or data likelihood. For this we use the Euclidean distance d which is normalized by the number of attributes n: $d(i,j) = \frac{\|s_i - s_j\|}{\sqrt{n}}$; $s_i, s_j \in \mathbb{R}^n$. While the measure of energy transfer between data points depends on a nonlinear function $p_q : \mathbb{R} \to \mathbb{R}$. p_q is an empirically-based energy function of parameter $-1 \le q \le 1$. Thus, for two points s_i and $s_j \in S$, $p_q(d(i,j))$ can be considered as an energy transfer from s_i to s_j and vice versa, that is a function of distance $d(i,j)$ between s_i and s_j. In our work it is used for merging data points into a cluster according to the local energy maximization.

The main assumption we make is that each clustering point s has a quantum of energy that is continuously emitted outward. We assume that the emitted energy is maximal for the closest neighborhood of s and fades out with distance non-linearly.

The algorithm has a number of heuristic parameters (e.g., p_q) by using which it is possible to control the clustering process for adjusting to the data set peculiarities. However, the adjusting process requires the application of parameter identification methods. The identification problem in turn is solved by encapsulating the clustering algorithm into a "black box", so its parameters are used as input, while we get the precision in output as a quality criterion.

We have compared results of our algorithm with well known clustering algorithms DBScan [6], k-means, ELM k-means [4,5], Mercer kernel method [7] on several common used test data sets from UCI [2]: Iris, Wisconsin Prognostic Breast Cancer (WPBC), Wine. Thus certain preliminary results on clustering and identification problems are obtained.

The presented algorithm consists of two parts: finding centroids, as initial points of cluster propagation, and the cluster propagation itself. For the first part we realize an algorithm that for a given data set points finds such a subset that corresponds to the center points of data highest local energy. In our case the centroid is the most representative data point of some cluster and the most suitable point for starting cluster growing process.

For centroids identification we make an assumption that centroids are those points that have the highest local energy (computed by an aggregated local energy function). In the first part we mark all these centroid points as clusters, each having exactly one element.

In the second part, we realize the approach of cluster propagation by including data points into one of the clusters in such a way that it "fits best" to the energy of those clusters. In particularly, we sort out the points and the clusters, after that we find the pair that gives maximum energy gain according to the function p_q and merge the point to this cluster. In our algorithm the data points may transfer (jump) from one cluster to another if corresponding energy increases. This procedure is repeated until no data points can be merged or jump. We solve the parameter identification problem by iterative search of the best input parameters, *i.e.*, as an optimization problem regarding precision of the clustering. The precision is calculated by comparing the clusters we compute with known distribution by clusters from the corresponding data set.

2   Energy Based Heuristic Functions

Let $s_{i,j}$ be a real value j-th attribute of i-th data point of the considered data set S. Normalization is done separately for each attribute. In details, for *j-th* attribute first we compute max_i and min_i values, then the proper normalized values are:

$$s^{norm}_{i,j} = \frac{s_{i,j} - min_i(s_{i,j})}{max_i(s_{i,j}) - min_i(s_{i,j})},$$

We define S as a set $\{s_i = (s^{norm}_{i,1}, ..., s^{norm}_{i,n}) | i = 1, ..., |S|\}$, from now on we omit the index of attribute j and $norm$.

We define the empirically-based energy function $p_q : [0, 1] \rightarrow [0, 1]$ with the following properties

$$p_q(0) = 1, p_q(1) = 0,$$

$p_q(\cdot)$ is a continues monotonic function on $[0, 1]$, q in $p_q(d(i, j))$; is a parameter that specifies how intensively a data point propagates its information in local vicinity, where $s_i, s_j \in S$. In certain cases we also consider

$$p_q(1) \approx 0, p_q(1) \geq 0. \tag{1}$$

The parameter q is one of the main parameters we adjust for different data sets to achieve best clustering algorithm precision. As a possible function can be used the circle arc with its center on the median line and starting from point $(0, 1)$ to point $(1, 0)$ (see Fig. 1 left). The second variant of function p_q we chose a heuristic function

$$p_q(x) = \begin{cases} [(1 - x^2)(1 - q) + (1 - x)(1 - q)]K, q \in [-1, 0), \\ \left[e^{-\frac{x^2}{\sigma^2}}(1 - q) + (1 - x)(1 - q) \right] K, q \in [0, 1]; \end{cases}$$

where $\sigma = 0.2$ has fixed value, $0 \leq x \leq 1$. For a given parameter q we compute K such that the condition (1) is fulfilled. The right Fig. 1. shows the empirical function p_q we use in our simulation, where $q = 0.9$. By changing parameter q

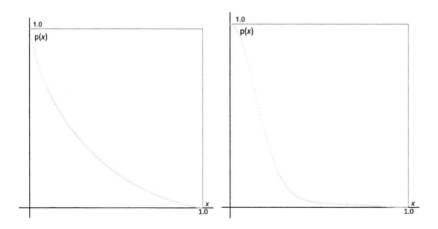

Fig. 1. Empirical energy functions that satisfy (1).

we obtain the modified shape of p_q which reflects in the corresponding changes of peculiarities of the clustering process and can be used to control it. This kind of empirically defined function represents best the idea that at a certain local radial distance the data points are independent, while nearby points are closely related. Note, that we introduce the function p_q as a measure of the local energy transfer.

3   Implementation of the Clustering Algorithm

Here we introduce our main algorithm of the clustering. It is composed of two parts: identification of centroids and the following competitive growth of clusters from these centroids. We use the previously defined parametric energy function p_q on normalized data set (*i.e.* attributes are scaled in [0,1] intervals).

3.1   Identification of Centroids

At this step the identification of centroids is performed for future propagation of clusters. A point s_i from the data set S is selected as a centroid point for s_j if $e_i p_q(d(i,j)/R_2)$ is a maximum for every $i = 1,...,N; N = |S|$, where $e_i = \sum_{k=1}^{N} p_q(d(i,k)/R_1)$. Let's discuss the meaning of the above defined parameters: R_1 and R_2 are used for finding (initial) centroid points. Both R_1 and R_2 define scaling factors for the neighborhood energy influence on the data set. e_i represents local energy coefficient of i-th data point. Thus the index j of data point s_j for the initial clustering are selected by the given procedure as sequent centroid if it is a local maximum of the aggregated energy function:

$$j = arg_j max_i [\ p_q(\frac{d(i,j)}{R_2}) \cdot \sum_{k=1}^{N} p_q(\frac{d(i,k)}{R_1})\ ], \tag{2}$$

where $i, j = 1,...,N$. It means R_1 and R_2 define radial distances at witch one data point may transfer energy to the neighborhood ones. For a compact cluster which has a maximal neighborhood measure ε (intercluster distance among data points), R_1 and R_2 can be chosen, e.g., as follows: $R_{1,2} = \varepsilon/(p_q^{-1}(1/2))$.

Algorithm:
   Input: S is the set of normalized data points.
   R_1, R_2, q are adjustable parameters of the algorithm.
   Build: $\{e_i\}$ is a set of potential energy values of data points S, L is set of indexes of the clusters that represent maximum energy points.
   Output: C is a set of centroid data points from S.
   for $i \in \{1,...,N\}$ **do**
   $$e_i \leftarrow \sum_{j=1}^{N} p_q(d(i,j)/R_1)$$
   end for

for $i \in \{1, ..., N\}$ **do**

    $L_i \leftarrow 0,\ PMax \leftarrow 0$

    for $j \in \{1, ..., N\}$ **do**

        $P \leftarrow e_j p_q(d(i,j)/R_2)$

        if $P > PMax$ **then**

            $PMax \leftarrow P,\ L_i \leftarrow j$

        end if

    end for

end for

Remove duplicates in L

Assign to C all data points which indexes are in L, *i.e.* for each $s_i \in S$ add $\{s_i\}$ to C if $i \in L$.

Now C contains such centroid data points for which local maximum of energy is obtained.

3.2   Cluster Propagation

On the second part of the algorithm the previously obtained clusters C can be extended by including not yet clustered data points or by transferring clustered ones from another clusters. In both cases it is realized by maximizing their contribution to the overall energy of clusters. Here, the main cycle is executed while possible, *i.e.*, while each step changes at least one cluster.

Algorithm:

    Input: S is the data set, C is obtained from the previous section.

    Output: Resulted clusters C_l and remaining noise.

    R_3, q are adjustable parameters of the algorithm.

    Define C_l for $l \in \{1, ..., |C|\}$ as singleton clusters of data points from C;

$$\bigcup_l C_l = C.$$

$E_{i,l} = \sum_{s_j \in C_l \setminus s_i} p_q(d(i,j)/R_3)$ is energy contribution of point s_i to the cluster C_l.

repeat

    $Changed \leftarrow false$

    for $i \in \{1, ..., N\}$ **do**

        for all C_l **do**

            Compute $E_{i,l}$

        end for

    end for

    Range $E_{i,l}$ for all C_l, $s_i \in S$ in descending order: $E_{i_1,l_1} \geq E_{i_2,l_2} \geq \cdots \geq E_{i_N,l_N}$.

    Find maximum $E_{i',l'}$ such that $s_{i'} \notin C_l$ for all C_l s.t. $l < l'$.

    if $E_{i',l'} > 0$ and $s_{i'} \notin C_k$, for all k **then**

        Add $s_{i'}$ to l'-th cluster $C_{l'}$

        $Changed = true$

    else

if $E_{i',l'} = 0$ **then**
  $Exit \leftarrow true$
else
    There is C_k such that $s_{i'} \in C_k$ and $E_{i',l'} > E_{i',k}$
    Replace $s_{i'}$ from cluster C_k to cluster C_l
    $Changed = true$
  end if
 end if
until $Changed = false$ or $Exit = true$
Output: all C_l as clusters,
   $S \setminus \bigcup_l C_l$ as noise.

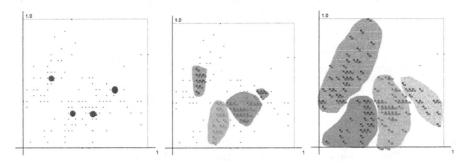

Fig. 2. Example of clustering on Iris data set. Projection on the first and second attributes. Initial clustering (left), clustering at step num. 50 (middle), stabilized clustering after step num. 125 (right).

The Fig. 2 demonstrates the clustering process simulation on Iris data set. There are three data points on the right side of the Figure that have not been clustered, and by the algorithm they are classified as noise.

3.3   Estimation on Computational Difficulty

It is easy to show that the given algorithm converges. Let us estimate the number of steps at which the algorithm stops. It is clear there is a maximum value by which the summary of cluster energies is limited (as an upper bound a constant $N(N-1)/2$ can be taken).

Since the algorithm uses $n \leq N$ data points as initial clustering and each step replaces/merges exactly one point the worst time estimation can be taken as (e.g., number of points × number of initial clusters = 150 × n, for Iris data set). Having the distance matrix $d(i,j)$ and in addition the function of energy $p_q(d(i,j))$ being precomputed for all $s_i, s_j \in S$, we get the worst case of cubic time difficulty with respect to the number of data points N.

At every step the algorithm of the cluster propagation requires to recompute the values $E_{i,l}$, for $i \in \{1, ..., N\}$, and for all obtained clusters C_l. These values are hardly used on the next steps. Also, each algorithmic step finds only a single point that is added to its proper cluster or it is replaced from one cluster to another. Hence, it is likely a further optimization of the algorithm is possible.

3.4   Clustering Precision Estimation

For the clustering precision estimation we use the original data sets from UCI [2] which are listed in Table 1. The right distribution by clusters for given data sets is known. The clustering precision is calculated on the following basis: consider each cluster as an equivalence relation, *i.e.* two data points s_i and s_j are equivalent if they are in the same cluster $s_i \sim s_j$, and they are not equivalent, if they are from different clusters. Similarly, we define equivalence relation $\overline{s_i} \sim \overline{s_j}$, for s_i, s_j in the data set according to their original clusters. The precision is calculated as

$$Pr = \frac{V}{(N-1)N},$$

where N is the number of data points; and

$$V = \sum_{s_i, s_j \in S; s_i \neq s_j} (s_i \sim s_j) \circ (\overline{s_i} \sim \overline{s_j}).$$

Operator $\circ$ means the equivalence function: $a \circ b = 1$, if $a = b$, and 0 otherwise. This technique of precision estimation allows to calculate difference in variants of clusterings for the same data set without considering the problem of cluster renaming (permutation of indexes).

Table 1. Simulation of our method on UCI [2] data sets

Data set	Num of data points	Num of attributes	Algorithm simulation
Iris	150	4	More than 1500 runs, discreet step: 1/50
WPBC	683	9	More than 2000 runs, discreet step: 1/200
Wine	178	13	More than 1500 runs, discreet step: 1/200

3.5   Parameter Identification Problem

We treat the problem by iterative search method using Pr as criterion for optimization. In order to process with the optimal parameter search we define value limits for every parameter $R_1 \in (0, 1], R_2 \in (0, 2.5], R_3 \in (0, 1], q \in (-1.1)$, with various relatively small discreet steps. By this approach the clustering algorithm is encapsulated into a "black box": the parameters are used as its initial/input data and we receive the precision as its result output $Pr(R_1, R_2, R_3, q) : \mathbb{R}^4 \to \mathbb{R}$.

The searching is realized independently for each attribute with subsequent fixation of the best resulted parameter value according to maximum of Pr. The results of the parameter identification are given in Table 2, the discreet step values is shown in Table 1. We have chosen this method due to its simplicity. The parameter identification problem can be solved by other optimization methods.

Table 2. The best parameter identification results (regarding Pr)

Data set	Precision (Pr)	$R1$	$R2$	$R3$	q
Iris	0.9574	0.5300	1.0193	0.2388	0.9862
WPBC	0.9140	0.6989	1.6511	0.2192	0.9836
Wine	0.9575	0.4724	0.9173	0.2378	0.9907

The algorithm was optimized so that it can run in feasible time. E.g., it runs approximately. 0.4 s. on Iris data set per algorithmic convergence for the known parameters.

Table 3. Comparing our algorithm with k-means, ELM k-means, kernel algorithms [12]

Data set	Our algorithm	k-means	ELM k-means	Kernel
Iris	0.9574	0.891	0.960	0.947
WPBC	0.9140	0.961	0.979	0.970
Wine	0.9575	0.693	0.702	0.696

4   Result Analysis and Discussion

Preliminary results on clustering and parameter identification are obtained on several public data sets. We consider the advantages as well as disadvantages of our solution comparing to several known algorithms [11] applied on Iris, WPBC and Wine from UCI [2]; see Table 1. Comparing our algorithm with the known ones we mention that its quality is comparative and even better for some cases, see Table 3 for more details. Unfortunately, the current realization lacks its efficiency in terms of execution time. Generally speaking, we improve the quality of clustering by processing redundant amount of information on spacial data layout. This may be considered as a trade-off between time efficiency and clustering quality.

Positive features of the algorithm: The algorithm does not depend on the number of attributes and number of data set points. It uses a physically motivated and representative approach based on energetic "stability" of samples inside their clusters. It determines automatically the number of clusters. The rest of data points that have not been included into clusters is considered as noise. Also, it can be used in practical tasks by a simple technique:

1. Determine the algorithm parameters on an existing data subset with known classification.
2. Proceed with new data points with unknown classification inside the same model.

Negative features of the algorithm: Relatively slow, especially at the beginning. It is partially dependent on the distribution properties of input data set.

5   Conclusion

We note that our energy based clustering approach is considered as an optimization problem with respect to heuristically defined energy function. It is also good to mention that our algorithm has a natural "stopping" property, *i.e.*, the algorithms necessary completes each execution run in a finite number of steps. Moreover, it yields a definite number of clusters corresponding to the input parameters of energy function.

Another advantage is that our approach simplifies the noise identification problem as it easy detects those data points that avoid clustering on different initial parameters.

On the other hand, our computational experiments also show basic limits of the above technique, e.g. we conclude empirically a highly discreet nature of optimization problem regarding Pr, see Sect. 3. We realize a computationally stable algorithm for identifying parameters that give a proper clustering. Our intention is to use this computational model in real problem oriented clustering, yet its use in concrete applications may require further investigation. What is essential about our approach is the possibility to adapt it to a range of clustering data sets and achieve the suitable precision.

Our numerical results show high diversity among input parameters. The variability of these parameters give additional control and flexibility in the clustering process. It appears even in the cases Pr has their values in the close radial distance the corresponding initial parameters may vary in broad intervals.

Remarks about further directions. The procedure of competitive growth of clusters from centroids can be implemented in several more optimal ways. For example, by incorporating parallelism in the clustering algorithm. In case we reuse energy contribution $E_{i,l}$ of point s_i to the cluster C_l on different independent clusters we obtain simultaneous growth of the clusters and hence reduce the number of computations.

One of the main contributions of this work is that the first part of the algorithm for centroid identification (Sect. 3.1) can be used independently for initialization of other algorithms that require the setting of initial centroids (e.g., k-means).

We also note that the current algorithm considers all data points in a sparse distance that have not been clustered as noise. In case it is preliminary known that the clustering contains no noise, an improvement to the algorithm is possible(noise free mode). In this case the data points in question are placed into their nearest clusters. This improvement does not increase the computational degree of our method, yet it may yield an improvement to the resulting precision.

Acknowledgments. This research is conducted within the framework of the grant num. BR05236839 "Development of information technologies and systems for stimulation of personality's sustainable development as one of the bases of development of digital Kazakhstan".

References

1. Comparing different clustering algorithms on toy datasets. http://scikit-learn.org/stable/auto_examples/cluster/plot_cluster_comparison.html
2. Online clustering data sets UCI. https://archive.ics.uci.edu/ml/datasets.html
3. Albertini, M.K., de Mello, R.F.: Energy-based function to evaluate data stream clustering. Adv. Data Anal. Classif. **7**(4), 435–464 (2013). https://doi.org/10.1007/s11634-013-0145-3
4. Bagirov, A.M.: Modified global k-means algorithm for minimum sum-of-squares clustering problems. Pattern Recognit. **41**(10), 3192–3199 (2008). https://doi.org/10.1016/j.patcog.2008.04.004. http://www.sciencedirect.com/science/article/pii/S0031320308001362
5. Carrizosa, E., Alguwaizani, A., Hansen, P., Mladenović, N.: New heuristic for harmonic means clustering. J. Global Optim. **63**(3), 427–443 (2015). https://doi.org/10.1007/s10898-014-0175-1
6. Ester, M., Kriegel, H.P., Sander, J., Xu, X.: A density-based algorithm for discovering clusters in large spatial databases with noise, pp. 226–231. AAAI Press (1996)
7. Girolami, M.: Mercer kernel-based clustering in feature space. IEEE Trans. Neural Netw. **13**, 780–784 (2002)
8. Karlsson, C. (ed.): Handbook of Research on Cluster Theory. Edward Elgar Publishing, Cheltenham (2010)
9. Khelifi, L., Mignotte, M.: EFA-BMFM: a multi-criteria framework for the fusion of colour image segmentation. Inf. Fusion **38**, 104–121 (2017)
10. Noack, A.: Energy-based clustering of graphs with nonuniform degrees. In: Healy, P., Nikolov, N.S. (eds.) GD 2005. LNCS, vol. 3843, pp. 309–320. Springer, Heidelberg (2006). https://doi.org/10.1007/11618058_28
11. Wu, X., et al.: Top 10 algorithms in datamining. Knowl. Inf. Syst. **14**(1), 1–37 (2008). https://doi.org/10.1007/s10115-007-0114-2
12. Xie, L., Lu, C., Mei, Y., Du, H., Man, Z.: An optimal method for data clustering. Neural Comput. Appl. **27**(2), 283–289 (2016). https://doi.org/10.1007/s00521-014-1818-3
13. Yue, Z., Chuansheng, Z.: A new clustering algorithm based on probability. In: Pan, J.-S., Snasel, V., Corchado, E.S., Abraham, A., Wang, S.-L. (eds.) Intelligent Data analysis and its Applications, Volume II. AISC, vol. 298, pp. 119–126. Springer, Cham (2014). https://doi.org/10.1007/978-3-319-07773-4_12
14. Zhou, H., Yuan, X., Cui, W., Qu, H., Chen, B.: Energy-based hierarchical edge clustering of graphs. In: PacificVIS 2008, Visualization Symposium, pp. 55–61 (2008)

An Approach to Property Valuation Based on Market Segmentation with Crisp and Fuzzy Clustering

Adrian Malinowski[1], Mateusz Piwowarczyk[1], Zbigniew Telec[1],
Bogdan Trawiński[1(✉)], Olgierd Kempa[2], and Tadeusz Lasota[2]

[1] Faculty of Computer Science and Management, Wrocław University
of Science and Technology, Wrocław, Poland
apmalinowski@gmail.com, {mateusz.piwowarczyk,
zbigniew.telec,trawinski}@pwr.edu.pl
[2] Department of Spatial Management, Wrocław University of Environmental
and Life Sciences, Wrocław, Poland
{olgierd.kempa,tadeusz.lasota}@up.wroc.pl

Abstract. Property valuation is a complex and time-consuming process which is carried out by qualified real estate appraisers. Number of properties and number of purchase-sale transactions grows year by year. Mass real estate appraisal appears as another big problem. These issues are connected with deficiency of human and time resources. Therefore, numerous studies are carried out on computer systems which can support the real estate appraisers. Automated property valuation systems are also developed. A method utilizing clustering algorithms to automate property valuation according to sales comparison approach was proposed in this paper. A crisp and fuzzy clustering algorithms were employed to divide the properties located in a given city into a number of clusters. These clusters established the basis for property valuation process. The effectiveness of the proposed method was examined and compared with the real estate appraisal based on the spatial partition of an area of the city into cadastral regions and expert zones.

Keywords: Property valuation · Mass appraisal · Sales comparison approach
Expert algorithms · K-means · C-means · Submarket segmentation

1 Introduction

Diverse methods and models are designed for automated computer systems to assist with valuation of individual premises as well as mass appraisal of properties located in a given area. They extend from multiple regression analysis [1] to intelligent techniques such as decision trees [2], fuzzy systems [3], neural networks [4], and hybrid methods [5]. The sales comparison approach is employed to generate majority of data-driven real estate appraisal models. They are applied in a specific category of computer systems named Automated Valuation Models (*AVM*) and Computer Assisted Mass Appraisal (*CAMA*) [6–8].

© Springer Nature Switzerland AG 2018
N. T. Nguyen et al. (Eds.): ICCCI 2018, LNAI 11055, pp. 534–548, 2018.
https://doi.org/10.1007/978-3-319-98443-8_49

Due to the heterogeneous nature of data utilized to create property valuation models the segmentation of real estate markets into more homogeneous submarkets is needed. Such segmentation can result in more accurate predictions of real estate values. Numerous researchers claimed that geographic and administratively determined boundaries were not effective, therefore various statistical methods such as clustering were proposed to segment the properties in a given market. The most popular the K-Means crisp clustering was successfully applied by Goodman and Thibodeau [9], Bourassa et al. [10], and Chen et al. [11]. In turn, Kauko et al. utilized neural network techniques, namely the self-organising map (*SOM*) and the learning vector quantisation (*LVQ*) [12]. The former technique was employed also by Kontrimas and Verikas [13]. Shi et al. [14] adapted the *Fuzzy C-Means* algorithm and applied to the market segmentation. Geostatistical approach was explored by Hayles [15] and Wu and Sharma [16]. Bourassa et al. [17] compared alternative methods for considering spatial dependence in house price prediction and concluded that geostatistical models with disaggregated submarket variables revealed the best performance.

The authors of the paper explored different the state-of-the-art machine learning methods and employed them to build data-driven property valuation models. They were ensemble techniques [18–20], hybrid approaches embracing evolutionary fuzzy systems [21] and evolving fuzzy systems [22]. We proposed several techniques for creating regression models for real estate appraisal using ensembles of genetic fuzzy systems [23, 24] and evolving fuzzy systems [25, 26]. We utilized ensembles of genetic fuzzy systems and neural networks to predict from a data stream of flat purchase and sale transactions [27, 28]. We worked out methods for merging different areas of a city into uniform zones to obtain homogenous segments containing a greater number of records necessary for constructing well-fitting real estate appraisal models [29, 30].

Our further research into expert algorithms developed for an automated system to assist with real estate appraisal, which we proposed in [31, 32], is presented in this paper. The expert algorithms were created according to the sales comparison approach. They estimated the price of a flat using a number of the latest transactions of similar flats encompassed by the same market segment in which the appraised flat was located. Two kinds of market areas were considered in [31, 32], namely administrative cadastral regions of a city and zones with uniform price behaviour devised by an expert. In this paper we extend the segmentation methods of real estate market to include statistical techniques of crisp and fuzzy clustering. Two crisp clustering algorithms: *K-Means* [33] and *OPTICS* [34] as well as two fuzzy clustering algorithms *C-Means* [35] and *FAC* (*Fuzzy Adaptive Clustering*) [36] were applied to segment the city area into more homogeneous submarkets. The expert algorithms were adjusted to predict the prices of flats based on a number of clusters. All six algorithms were experimentally compared in terms of accuracy based on real-world data derived from a cadastral system and registry of real estate transactions exploited is one of big Polish cities.

2   Expert Algorithms for Property Valuation

Six variants of expert algorithms for residential property valuation implementing the sales comparison approach are considered in the paper. To estimate the price of a flat, all algorithms compute the mean price of a number of similar flats to the one being appraised located in the same segment of market. The first step of each algorithm is aimed at determining similar properties to a given flat. The experts defined 21 classes of similarity by selecting five key features of flats and dividing their values by ranges. There are following key features of flats: usable area of a flat (*Area*), year of a building construction (*Year*), number of storeys in a building (*Storeys*), distance from the centre of the city (*Centre*), and distance from the nearest shopping centre (*Shopping*). The values of individual parameters were partitioned into 3, 5, 3, 5, and 5 ranges, respectively, as shown in Table 1. Two flats are regarded as similar if the values of all their five key features belong to the same similarity classes.

Table 1. Similarity classes of flats within a Polish city determined by the experts

Feature	Class	Values
Area	1-small	under 40 m^2
	2-medium	40–60 m^2
	3-large	over 60 m^2
Year	1-very old	before 1918
	2-old	1919–1945
	3-medium	1946–1975
	4-new	1976–1995
	5-newest	after 1995
Storeys	1-low-rise	1–2
	2-mid-rise	3–5
	3-high-rise	6 and more
Centre	1-very close	less than 2 km
	2-close	2–4 km
	3-medium	4.1–6 km
	4-far	6.1–9 km
	5-veryfar	more than 9 km
Shopping	1-very close	less than 1 km
	2-close	1–2 km
	3-medium	2.1–3 km
	4-far	3.1–4 km
	5-veryfar	more than 4 km

Each variant of the expert algorithm is based on a different partition of the city area into submarkets. Six types of market segmentation were investigated in the paper, namely cadastral regions of a city (*REG*) and expert zones with uniform price behaviour (*ZON*), two crisp clustering algorithms: *K-Means* [33] (*KME*) and *OPTICS* [34]

Table 2. Expert algorithms *REG/ZON: N-Latest Transactions in a cadastral region/expert zone*

Step	Action
1	Compute the average price per square metre for each region/zone Z_j: $AvgP_j$ for $j=1,2,..,NZ$
2	Take the next flat X_i to estimate its price
3	Give the number N of flats taken to estimate the price of the flat X_i
4	Determine all flats similar to X_i included in the same similarity classes
5	If the number of flats similar to X_i is less than N, then omit X_i and go to *Step 2*
6	Determine the region/zone Z_i including the flat to estimate its price
7	If the number of flats similar to X_i is greater or equal to N, then select the N latest transactions and go to *Step 11*
8	Repeat Steps 9-10 until N transactions are found
9	Find the next region/zone Z_k where the difference $|avgP_k-avgP_i|$ is the smallest
10	Select the m latest transactions from Z_k to complement the number N
11	Compute the estimated price of X_i as the average price per square metre of N transactions found

Table 3. Algorithms *KME/OPT: N-Latest Transactions in a crisp cluster (K-Means/OPTICS)*

Step	Action
1	Compute the average price per square metre for each crisp cluster C_j: $AvgP_j$ for $j=1,2,..,NC$
2	Take the next flat X_i to estimate its price
3	Give the number N of flats taken to estimate the price of the flat X_i
4	Determine all flats similar to X_i included in the same similarity classes
5	If the number of flats similar to X_i is less than N, then omit X_i and go to *Step 2*
6	Determine distances from X_i to centroids of all crisp clusters C_j
7	Determine the crisp cluster C_j including the flat to estimate its price, i.e. the cluster with the nearest centroid to X_i
8	If the number of flats similar to X_i is greater or equal to N, then select the N latest transactions and go to *Step 12*
9	Repeat Steps 10-11 until N transactions are found
10	Find the next crisp cluster C_k where the difference $|avgP_k-avgP_i|$ is the smallest
11	Select the m latest transactions from C_k to complement the number N
12	Compute the estimated price of X_i as the average price per square metre of N transactions found

(*OPT*) and two fuzzy clustering algorithms *C-Means* [35] (*CME*) and *Fuzzy Adaptive Clustering* [36] (*FAC*). All six algorithms considered in the paper have the same names as the methods of market segmentation they utilize. The *REG* and *ZON* algorithms are identical with the *LTA: N-Latest Transactions in an Area* proposed by the authors in [32]. It computes the predicted price as an average price of the N-latest transactions of similar flats within the same area, i.e. cadastral region or expert zone. The pseudocode of the *REG* and *ZON* algorithms is shown in Table 2. In turn, *KME, OPT, FCM*, and *FAC* are the extensions of *LTA* which was adopted to predict the prices of flats based on a number of clusters. The pseudocode of the *KME* and *OPT* algorithms using crisp clustering segmentation is presented in Table 3. In turn, the pseudocode of *FCM* and *FAC* algorithms employing fuzzy clustering segmentation is given in Table 4.

Table 4. Algorithms *FCM/FAC: N-Transactions in a fuzzy cluster (C-Means/FAC)*

Step	Action
1	Take the next flat X_i to estimate its price
2	Give the number N of flats taken to estimate the price of the flat X_i
3	Determine all flats similar to X_i included in the same similarity classes
4	If the number of flats similar to X_i is less than N, then omit X_i and go to *Step 1*
5	Determine the membership values of X_i to each fuzzy cluster F_j
6	Determine the fuzzy cluster F_j with the largest membership value of X_i
7	Select N flats with the largest degree of membership to F_j
8	Compute the estimated price of X_i as the average price per square metre of N transactions selected

3 Experimental Setup

The main goal of evaluation experiments was to compare the performance of expert algorithms utilizing the segmentation of the city area into cadastral regions and expert zones with the versions of algorithms based on the crisp and fuzzy clustering of the data on flat sale transactions. Moreover, the parameter N, i.e. the number of flats taken to compute the price of a flat being appraised, was examined.

The experiments were performed using real-world data about sale purchase transactions drawn from a cadastral system and registry of real estate transactions exploited in one of Polish big cities. The unrefined dataset consisted of 37,791 records of flat sales made within nine years from 2007 to 2015. The records with incomplete and erroneous data as well as outliers were removed during the cleansing process. In order to compensate fluctuations of real estate prices over time, the transactional prices were updated for the last day of 2015 employing trend functions. The final dataset counted 17,407 samples which were afterwards split randomly into 70% training set and 30% test set. Hence, the training set contained 12,185 records and the test set was composed of 5,222 samples.

Two crisp and two fuzzy clustering algorithms were implemented in the *MATLAB* environment and employed to partition data of flat sales accomplished in the city except for cadastral regions and expert zones:

K-Means – one of the simplest and most popular unsupervised learning algorithms devised by MacQueen in 1967 [33],

OPTICS – a density-based algorithm invented by Ankrest et al. in 1999 [34],

C-Means – the most well-known and used algorithm for fuzzy clustering in which data points can belong to multiple clusters, developed by Dunn in 1973 and improved by Bezdek in 1981 [35],

FAC (*Fuzzy Adaptive Clustering*) – a modification of C-Means proposed by Krishnapuram and Keller and enhanced by Young-Jun Lee [36].

The spread of sold properties over 69 cadastral regions and 23 expert zones is shown in Figs. 1 and 2, respectively. In turn, the spread of sold properties over 44 crisp clusters generated using *K-Means* and over 36 fuzzy clusters created with *C-Means* is depicted in Figs. 3 and 4, respectively. The boundaries of cadastral regions and expert zones were left in Figs. 3 and 4 only for illustration purposes because both crisp and fuzzy clusters do not have any spatial boundaries. It could be seen that crisp and fuzzy clustering produced different real estate market segmentation compared to regions and zones.

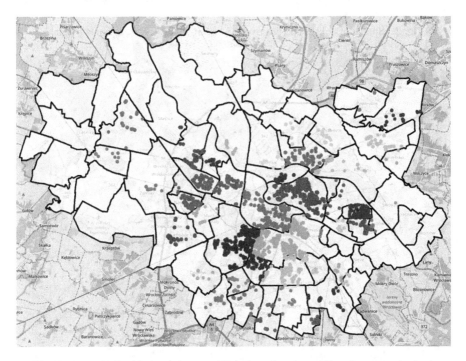

Fig. 1. Spread of sold properties over 69 cadastral regions (48 regions contain data)

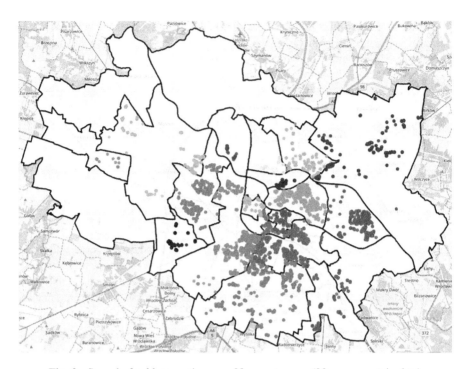

Fig. 2. Spread of sold properties over 23 expert zones (20 zones contain data)

Fig. 3. Spread of sold properties clustered with *K-Means* over the city area

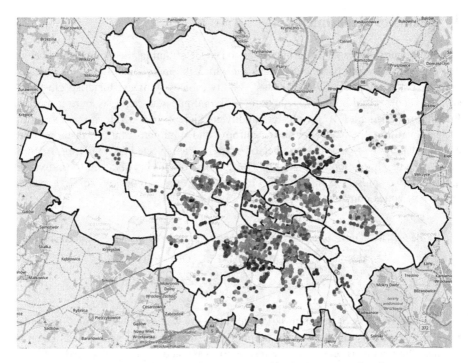

Fig. 4. Spread of sold properties clustered with *C-Means* over the city area

The following procedure was implemented to compare the performance of *REG, ZON, KME, OPT, FCM,* and *FAC* algorithms. The prices of 100 flats randomly selected from the test set were computed with individual algorithms. Then, the performance measure *MAE* (*mean absolute error*) was determined for each run in accordance with Formula 1, where P_i^a and P_i^p denote the actual and predicted prices respectively and n stands for the number of flats being estimated in each run.

$$MAE = \frac{1}{n} * \sum_{i=1}^{n} \left| P_i^p - P_i^a \right| \tag{1}$$

This schema was repeated 120 times yielding 120 values of *MAE* for each algorithm which enabled us to conduct the tests for statistical significance.

Each algorithm was tested for seven values of the parameter *N,* i.e. the number of similar flats taken to compute the price of the flat being evaluated, namely $N = 3, 5, 10, 20, 30, 40,$ and 50. Then, the algorithms with optimal values of *N* for which individual algorithms revealed the best performance were selected for final evaluation. They were compared in terms of *MAE* and the tests for statistical significance of differences between the algorithms were conducted.

4 Results of Experiments

The preliminary series of experiments aimed at determining the optimal number of clusters and clustering parameter values for which the individual algorithms revealed the best performance. For *KME* cluster validity indices designed for crisp clustering such as Davies–Bouldin index, silhouette [37], and gap statistics [38] were inspected. In turn, for *FCM*, and *FAC* the cluster validity coefficients intended for fuzzy clustering such as partition coefficient, partition entropy, and minimum centroid distance were examined [39]. For *OPT* the number of clusters was selected by applying different values of a neighbourhood radius [40]. The numbers of clusters obtained for individual algorithms was presented in Table 5. For illustration the number of cadastral regions and expert zones containing data was also given.

Table 5. The number of clusters, regions, and zones utilized to market segmentation

Algorithm	KME	FCM	FAC	OPT	REG	ZON
Clusters/regions	44	36	38	27	48	20

The effectiveness of determining cluster numbers was examined using 20 samples for 100 transactions taken from the training set. The number of similar flats taken to estimate the price was set to 40. The results for *KME* and *FCM* are depicted in Fig. 5.

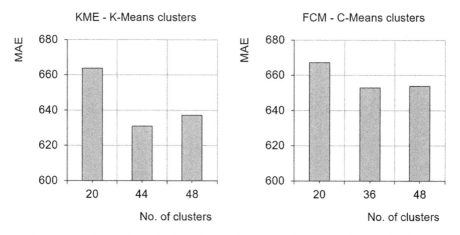

Fig. 5. Selection of the optimal number of clusters for the *KME* and *FCM* algorithms

The main part of experiments consisted in the examination of the accuracy of all algorithms depending on the number of similar flats utilized to estimate the price of the flat being evaluated. Each algorithm was run 120 times according to the schema described in the Sect. 3, for *N* = 3, 5, 10, 20, 30, 40, and 50 separately. The median of *MAE* values obtained for individual algorithms is shown in Fig. 6. It is clearly seen that better results can be achieved using the partition of the city area into cadastral regions (*REG*) and expert zones (*ZON*). Moreover, the accuracy of all algorithms is lower for smaller values of N equal to 3, 5, and 10. The final selection of the N values yielding the best performance was based on the results of the Friedman test. Average rank positions of *MAE* for individual algorithms for different values of N delivered by the Friedman test are presented in Table 6, where the lower rank value the better version of an algorithm. For further comparative analysis the following values of N were chosen: 20, 20, 40, 50, 40, and 40 for *REG, ZON, KME, OPT, FCM*, and *FAC* algorithms, respectively. The algorithms with selected values of N were denoted as follows: *REG (20), ZON (20), KME (40), OPT (50), FCM (40), and FAC (40)*.

Table 6. Rank positions of algorithms with different values of N produced by the Friedman test

	1st	2nd	3rd	4th	5th	6th	7th
REG	REG (20)	REG (40)	REG (10)	REG (30)	REG (50)	REG (5)	REG (3)
Rank	2.93	3.24	3.30	3.49	3.52	5.04	6.47
ZON	ZON (20)	ZON (40)	ZON (30)	ZON (50)	ZON (10)	ZON (5)	ZON (3)
Rank	2.88	2.88	3.39	3.39	3.67	5.28	6.51
KME	KME (40)	KME (50)	KME (20)	KME (30)	KME (10)	KME (5)	KME (3)
Rank	2.52	2.76	2.97	3.42	4.11	5.43	6.80
OPT	OPT (50)	OPT (40)	OPT (30)	OPT (20)	OPT (10)	OPT (5)	OPT (3)
Rank	2.43	2.50	3.34	3.54	4.18	5.27	6.74
FCM	FCM (40)	FCM (50)	FCM (20)	FCM (30)	FCM (10)	FCM (5)	FCM (3)
Rank	2.43	2.65	3.21	3.30	4.08	5.75	6.57
FAC	FAC (40)	FAC (50)	FAC (30)	FAC (20)	FAC (10)	FAC (5)	FAC (3)
Rank	2.55	2.88	3.11	3.29	4.29	5.44	6.43

The median of *MAE* in terms of Polish currency *PLN* provided by the algorithms with optimal values of N was contrasted in Fig. 7. The lowest values of *MAE,* namely 630, 633, and 638 were obtained for cadastral regions, expert zones, and *K-Means* clusters, respectively. In turn, *OPTICS* and Fuzzy Adaptive Clusters produced the worst outcome. The Shapiro-Wilk tests revealed that some results provided by algorithms being examined were not distributed normally. Therefore, non-parametric Friedman and Wilcoxon tests were applied. The Friedman test carried out for 120 test samples detected that there were significant disparity among the algorithms. Average rank positions of algorithm accuracy in terms of *MAE* provided by the Friedman test are given in Table 7, where the lower rank the more precise algorithm. In order to judge the significant discrepancy in performance between pairs of individual algorithms the nonparametric Wilcoxon test was applied. The null hypothesis assumed that there were no significant differences in accuracy, in terms of *MAE,* between individual pairs of results. The results of the Wilcoxon test for all pairs of algorithms are placed in Table 8. In Table 8 the sign + indicates that the algorithm in the row surpassed significantly the one in the corresponding column, the sign – denotes that the algorithm in the row revealed significantly worse accuracy than the one in the corresponding

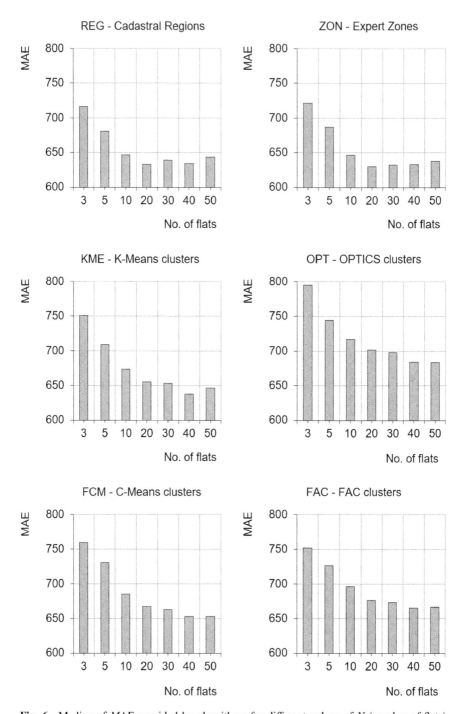

Fig. 6. Median of *MAE* provided by algorithms for different values of *N* (number of flats)

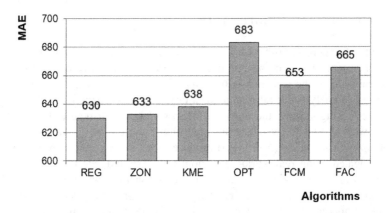

Fig. 7. Median of *MAE* provided by individual algorithms with optimal values of *N*, i.e. *REG (20), ZON (20), KME (40), OPT (50), FCM (40), FAC (40)*

column. In turn, ≈ means that no sufficient evidence was found to reject the null hypothesis. The significance level of 0.05 was taken for discarding the null hypothesis. Admittedly, *ZON (20)* and *REG (20)* took the lowest positions in the ranking, but the average rank positions of *KME (40)* and *FCM (40)* were only slightly higher. This was confirmed by the Wilcoxon paired test which did not detect statistically significant differences in accuracy between *ZON (20), REG (20), KME (40),* and *FCM (40)* algorithms. All these algorithms outperformed significantly *OPT (50)* and *FAC (40)* but *FCM (40)* compared to *FAC (40)*. In turn, *OPT (50)* showed significantly worse performance than any other algorithm.

Table 7. Rank positions of individual algorithms provided by the Friedman test

	1st	2nd	3rd	4th	5th	6th
Algorithm	ZON (20)	REG (20)	KME (40)	FCM (40)	FAC (40)	OPT (50)
Avg. rank	2.46	2.83	3.12	3.33	4.23	5.03

Table 8. Wilcoxon test results for pairwise comparison of individual algorithms

	ZON (20)	REG (20)	KME (40)	FCM (40)	FAC (40)	OPT (50)
ZON (20)		≈	≈	≈	+	+
REG (20)	≈		≈	≈	+	+
KME (40)	≈	≈		≈	+	+
FCM (40)	≈	≈	≈		≈	+
FAC (40)	–	–	–	≈		+
OPT (50)	–	–	–	–	–	

5 Conclusions and Future Work

Six variants of expert algorithms for residential property valuation adopting the sales comparison approach were compared in terms of accuracy. Each variant was based on a different partition of the city area into submarkets. Six types of market segmentation were applied, namely cadastral regions of a city (*REG*) and expert zones (*ZON*), two crisp clustering algorithms: the classic *K-Means* (*KME*) and density-based *OPTICS* (*OPT*) as well as two fuzzy clustering algorithms: the classic *C-Means* (*CME*) and *Fuzzy Adaptive Clustering* (*FAC*). Performance of all six algorithms was experimentally investigated using real-world data about purchase and sale transactions derived from a cadastral system and registry of real estate transactions which is exploited in one of big Polish cities. Mean absolute error (*MAE*) was applied as the measure of accuracy. The nonparametric Friedman and Wilcoxon tests were employed to analyse statistically significant differences among the algorithms.

ZON and *REG* algorithms revealed the best performance. However, there were no statistically significant differences in accuracy between *ZON, REG, KME,* and *FCM* algorithms. Thus, the main conclusion is as follows. Administrative cadastral regions of a city and expert zones can be used as effective submarkets in a city where an administrative partition was made in accordance with "natural boundaries" and individual segments were established by the boundaries of housing estates, forests, parks as well as by roads, railway lines, and rivers, etc. The same applies to expert zones which in our case were created by merging adjacent regions marked by similar behaviour of property prices. On the other hand, crisp and fuzzy clustering may turn out to be an effective way to delineate the boundaries of homogeneous submarkets. Clustering might be applied especially when an administrative partition was made according to artificial dividing lines or when an appraiser has no segmentation into submarkets at this disposal. All considered algorithms could be applied in an automated system to support professional valuators in real estate appraisal.

Examination of the accuracy of the algorithms depending on the number of similar flats (*N*) utilized to estimate the price of the flat being evaluated resulted in the conclusion that all algorithms performed worse for smaller values of *N* equal to 3, 5, and 10 compared to the larger values of *N* equal to 20, 30, 40, and 50.

Further study is planned to examine the performance of expert algorithms with different market areas of a city. Moreover, machine learning techniques will be applied to create various property valuation models over submarkets ranging from single models obtained with multiple linear regression and decision trees to compound models like random forests as well as bagging and boosting ensembles of neural networks.

References

1. Zurada, J., Levitan, A.S., Guan, J.: A comparison of regression and artificial intelligence methods in a mass appraisal context. J. Real Estate Res. **33**(3), 349–388 (2011)
2. Antipov, E.A., Pokryshevskaya, E.B.: Mass appraisal of residential apartments: An application of random forest for valuation and a CART-based approach for model diagnostics. Expert Syst. Appl. **39**, 1772–1778 (2012)

3. Kusan, H., Aytekin, O., Özdemir, I.: The use of fuzzy logic in predicting house selling price. Expert Syst. Appl. **37**(3), 1808–1813 (2010)
4. Peterson, S., Flangan, A.B.: Neural network hedonic pricing models in mass real estate appraisal. J. Real Estate Res. **31**(2), 147–164 (2009)
5. Musa, A.G., Daramola, O., Owoloko, A., Olugbara, O.: A neural-CBR system for real property valuation. J. Emerg. Trends Comput. Inf. Sci. **4**(8), 611–622 (2013)
6. Jahanshiri, E., Buyong, T., Shariff, A.R.M.: A review of property mass valuation models. Pertanika J. Sci. Technol. **19**(S), 23–30 (2011)
7. McCluskey, W.J., McCord, M., Davis, P.T., Haran, M., McIlhatton, D.: Prediction accuracy in mass appraisal: a comparison of modern approaches. J. Prop. Res. **30**(4), 239–265 (2013)
8. d'Amato, M., Kauko, T. (eds.): Advances in Automated Valuation Modeling AVM After the Non-agency Mortgage Crisis. Studies in Systems, Decision and Control, vol. 86. Springer, Cham (2017). https://doi.org/10.1007/978-3-319-49746-4
9. Goodman, A.C., Thibodeau, T.G.: Housing market segmentation and hedonic prediction accuracy. J. Hous. Econ. **12**(3), 181–201 (2003)
10. Bourassa, S.C., Hoesli, M., Peng, V.S.: Do housing submarkets really matter? J. Hous. Econ. **12**, 12–28 (2003)
11. Chen, Z., Cho, S.-H., Poudyal, N., Roberts, R.K.: Forecasting housing prices under different submarket assumptions. Urban Stud. **46**(1), 67–87 (2009)
12. Kauko, T., Hooimeijer, P., Hakfoort, J.: Capturing housing market segmentation: an alternative approach based on neural network modelling. Hous. Stud. **17**(6), 875–894 (2002)
13. Kontrimas, V., Verikas, A.: The mass appraisal of the real estate by computational intelligence. Appl. Soft Comput. **11**(1), 443–448 (2011)
14. Shi, D., Guan, J., Zurada, J., Levitan, A.S.: An innovative clustering approach to market segmentation for improved price prediction. J. Int. Technol. Inf. Manag. **24**(1), 15–32 (2015)
15. Hayles, K.: The use of GIS and cluster analysis to enhance property valuation modelling in Rural Victoria. J. Spat. Sci. **51**(2), 19–31 (2010)
16. Wu, C., Sharma, R.: Housing submarket classification: the role of spatial contiguity. Appl. Geogr. **32**, 746–756 (2012)
17. Bourassa, S.C., Cantoni, E., Hoesli, M.: Predicting house prices with spatial dependence: a comparison of alternative methods. J. Real Estate Res. **32**(2), 139–159 (2010)
18. Woźniak, M., Graña, M., Corchado, E.: A survey of multiple classifier systems as hybrid systems. Inf. Fusion **16**, 3–17 (2014)
19. Krawczyk, B., Woźniak, M., Cyganek, B.: Clustering-based ensembles for one-class classification. Inf. Sci. **264**, 182–195 (2014)
20. Burduk, R., Walkowiak, K.: Static classifier selection with interval weights of base classifiers. In: Nguyen, N.T., Trawiński, B., Kosala, R. (eds.) ACIIDS 2015. LNCS (LNAI), vol. 9011, pp. 494–502. Springer, Cham (2015). https://doi.org/10.1007/978-3-319-15702-3_48
21. Fernández, A., López, V., José del Jesus, M., Herrera, F.: Revisiting evolutionary fuzzy systems: taxonomy, applications, new trends and challenges. Knowl. Based Syst. **80**, 109–121 (2015)
22. Lughofer, E.: Evolving Fuzzy Systems – Methodologies, Advanced Concepts and Applications. STUDFUZZ. Springer, Heidelberg (2011). https://doi.org/10.1007/978-3-642-18087-3
23. Lasota, T., Telec, Z., Trawiński, B., Trawiński, K.: Exploration of bagging ensembles comprising genetic fuzzy models to assist with real estate appraisals. In: Corchado, E., Yin, H. (eds.) IDEAL 2009. LNCS, vol. 5788, pp. 554–561. Springer, Heidelberg (2009). https://doi.org/10.1007/978-3-642-04394-9_67

24. Krzystanek, M., Lasota, T., Telec, Z., Trawiński, B.: Analysis of bagging ensembles of fuzzy models for premises valuation. In: Nguyen, N.T., Le, M.T., Świątek, J. (eds.) ACIIDS 2010. LNCS (LNAI), vol. 5991, pp. 330–339. Springer, Heidelberg (2010). https://doi.org/10. 1007/978-3-642-12101-2_34

25. Lasota, T., Telec, Z., Trawiński, B., Trawiński, K.: Investigation of the eTS evolving fuzzy systems applied to real estate appraisal. J. Multiple-Valued Log. Soft Comput. **17**(2–3), 229–253 (2011)

26. Lughofer, E., Trawiński, B., Trawiński, K., Kempa, O., Lasota, T.: On employing fuzzy modeling algorithms for the valuation of residential premises. Inf. Sci. **181**, 5123–5142 (2011)

27. Trawiński, B.: Evolutionary fuzzy system ensemble approach to model real estate market based on data stream exploration. J. Univ. Comput. Sci. **19**(4), 539–562 (2013)

28. Telec, Z., Trawiński, B., Lasota, T., Trawiński, G.: Evaluation of neural network ensemble approach to predict from a data stream. In: Hwang, D., Jung, Jason J., Nguyen, N.-T. (eds.) ICCCI 2014. LNCS (LNAI), vol. 8733, pp. 472–482. Springer, Cham (2014). https://doi. org/10.1007/978-3-319-11289-3_48

29. Lasota, T., Sawiłow, E., Trawiński, B., Roman, M., Marczuk, P., Popowicz, P.: A method for merging similar zones to improve intelligent models for real estate appraisal. In: Nguyen, N.T., Trawiński, B., Kosala, R. (eds.) ACIIDS 2015. LNCS (LNAI), vol. 9011, pp. 472–483. Springer, Cham (2015). https://doi.org/10.1007/978-3-319-15702-3_46

30. Lasota, T., et al.: Enhancing intelligent property valuation models by merging similar cadastral regions of a municipality. In: Núñez, M., Nguyen, N.T., Camacho, D., Trawiński, B. (eds.) ICCCI 2015. LNCS (LNAI), vol. 9330, pp. 566–577. Springer, Cham (2015). https://doi.org/10.1007/978-3-319-24306-1_55

31. Trawiński, B., et al.: Comparison of expert algorithms with machine learning models for a real estate appraisal system. In: The 2017 IEEE International Conference on INnovations in Intelligent SysTems and Applications (INISTA 2017). IEEE (2017)

32. Trawiński, B., Lasota, T., Kempa, O., Telec, Z., Kutrzyński, M.: Comparison of ensemble learning models with expert algorithms designed for a property valuation system. In: Nguyen, N.T., Papadopoulos, George A., Jędrzejowicz, P., Trawiński, B., Vossen, G. (eds.) ICCCI 2017. LNCS (LNAI), vol. 10448, pp. 317–327. Springer, Cham (2017). https://doi. org/10.1007/978-3-319-67074-4_31

33. Hartigan, J.A., Wong, M.A.: A k-means clustering algorithm. J. R. Stat. Soc. Ser. C (Appl. Stat.) **28**(1), 100–108 (1979)

34. Ankrest, M., Breunig, M., Kriegel, H., Sander, J.: OPTICS: Ordering points to identify the clustering structure. In: Proceedings of the 1999 ACM SIGMOD International Conference on Management of Data, SIGMOD 1999, pp. 49–60, Philadelphia PA (1999)

35. Bezdek, J.C., Ehrlich, R., Full, W.: FCM: the fuzzy c-means clustering algorithm. Comput. Geosci. **10**(2–3), 191–203 (1984)

36. Cox, E.: Fuzzy Modeling and Genetic Algorithms for Data Mining and Exploration. Elsevier, Boston (2005)

37. Vendramin, L., Campello, R.J.G.B., Hruschka, E.R.: Relative clustering validity criteria: a comparative overview. Stat. Anal. Data Min. **3**(4), 209–235 (2010)

38. Tibshirani, R., Walther, G., Hastie., T.: Estimating the number of clusters in a data set via the gap statistic. J. R. Stat. Soc. Ser. B **63**(2), 411–423 (2001)

39. Wu, K.-L., Yang, M.-S.: A cluster validity index for fuzzy clustering. Pattern Recogn. Lett. **26**, 1275–1291 (2005)

40. Halkidi, M., Batistakis, Y., Vazirgiannis, M.: On clustering validation techniques. Intell. Inf. Syst. J. **17**(2–3), 107–145 (2001)

Predicting Solar Intensity Using Cluster Analysis

Waseem Ahmad[1(✉)], Sahil Sahil[1], and Aftab Mughal[2]

[1] Toi Ohomai Institute of Technology, Rotorua, New Zealand
Waseem.ahmad@toiohomai.ac.nz, schahal96@gmail.com
[2] AGI Education Limited, Auckland, New Zealand
aftab@agi.ac.nz

Abstract. A key goal of smart grid initiatives is significantly increasing the fraction of grid energy contributed by renewable sources and especially from solar power. One challenge with integrating solar power into the grid is that its power generation is stochastic and depends on various environmental factors. Thus, predicting future energy generation is important to moderate the overall energy requirements. In recent years, the use of machine learning approaches to solar power forecasting is becoming very popular. In this paper, a clustering based data segmentation approach is used to find natural subgrouping in the data. These subgroups are then used to construct forecasting models using various machine learning algorithms. The effectiveness of the approach is demonstrated by comparing the accuracy of clustering based forecasting to the standard forecasting models. The experimental results demonstrate that the proposed clustering based models produce more accurate models.

Keywords: Clustering · EM algorithm · Solar power · Machine learning

1 Introduction

In recent years, power infrastructure has been changed significantly due to the awareness of harnessing the powers from renewable energy sources. This paradigm shift has introduced the term 'Smart Grid'. The Smart Grid (SG) initiatives aimed at introducing new technologies and services in power systems to make the electrical networks more reliable, efficient and distributed [1]. Moreover, renewable energy has introduced significant economic and environmental interests in public such as the drive to reduce CO_2 emissions and creating new employment opportunities.

The main objective of SG is to exploit the renewable energy sources such as wind and solar energy. One of the challenges of such energy sources is their intermittent and stochastic nature as the power is dependent on weather characteristics. In solar power generation, Photovoltaic (PV) cells are the basic technology that converts solar energy into electric power. The installation of PV systems has become very popular at both domestic and commercial levels due to the cost reduction and the increased efficiency of both PV panels and converters. The power generated by a PV power system varies according to the solar radiation on the Earth's surface, which mainly depends on the installation site and the weather conditions. While the dependence on the specific

© Springer Nature Switzerland AG 2018
N. T. Nguyen et al. (Eds.): ICCCI 2018, LNAI 11055, pp. 549–560, 2018.
https://doi.org/10.1007/978-3-319-98443-8_50

location can be essentially predicted on a deterministic way, the atmospheric conditions (such as cloud cover, ambient temperature, relative humidity, etc.) are the main causes of the randomness of the solar radiation. Therefore, there is a need to develop tools and techniques to get optimal benefits from these sources. In tackling these issues, modelling and forecasting is a fundamental task for an efficient utilization of the available distributed energy resources.

The problem of forecasting Photovoltaic power generation lies in the domain of time series data analysis and forecasting. In time series, data points are indexed and organised in time order. A common goal of time series analysis is to extrapolate past behaviour into the future. In recent years, machine learning techniques have been extensively used in forecasting solar radiation. Raza et al. [26] divided the power forecasting into three categories: long-term, medium-term and short-term power forecasting. Long-term power forecasting (1 year to 10 years ahead) is used for future planning of power systems and provide guidance for devising various power consumption and generation policies. Medium-term power forecasting (1 month to 1 year ahead) is used to manage the effective use of the power systems. Short term power forecasting (1 h to 1 day ahead) is used to efficiently manage the flow of power on the grid and optimum unit commitment by the users. The machine learning forecasting techniques discussed in the literature mainly cover short term to medium term power forecasting.

Machine learning is a subfield of artificial intelligence, where various algorithms are developed to make computers learn without being explicitly programmed. Machine learning techniques have been successfully applied in anomaly detection, classification, automating processes, financial trading, fraud detection, healthcare, etc. [2–4]. In the literature, machine learning algorithms have been extensively used to model and forecast the PV data [5–9].

In model building and forecasting, data is divided into two parts: training and test datasets. The training data is used to construct the models using machine learning algorithm and test data is used to evaluate the forecasting capabilities of the model. In existing literature, two methods are used to present data to machine learning algorithms, namely, full data presentation and segmented training presentation. The full data presentation method is simple, where complete training data is used to create the forecasting model. In segmented training data presentation, the training data is subdivided into a number of groups and forecasting model on each sub-group is constructed separately. The examples of these segmentation criteria are splitting the data based on seasonal variations (i.e. summer, spring, autumn and winter) or weather conditions (i.e. sunny, cloudy and rainy days). It has been argued that as the segmented data has more specialised subset of data and therefore, the model is bound to have more robustness and better forecasting accuracy.

This paper presents a clustering based data partitioning method. Clustering is an unsupervised machine learning domain where data is portioned into similar subgroups with a constraint that group members within a cluster must have maximum similarity and there must be maximum dissimilarity across different clusters. In this paper, training data is divided into specified number of clusters and then forecasting models are constructed on each cluster separately. The aim of this research is to present a novel data segmentation method that is not based on any prior knowledge such as seasons in a year or weather conditions. Furthermore, this paper compares the model forecasting results on training data, seasonal segmented training data and clustered training data.

2 Literature Review

In the literature, solar power forecasting field of research has been divided into three main categories, namely, time series statistical methods, physical methods and ensemble methods [8]. In the time series statistical method, statistical and machine learning algorithms are used to construct predictive models on the past data consisting of meteorological and solar radiance data [5]. Artificial neural networks, support vector machines, autoregressive models and multiple regressions are the examples of time series statistical methods [8]. In the physical methods, sky imaging and satellite imaging are used to construct physical satellite models for solar irradiance prediction [10]. Ensemble methods were developed to reduce the limitations of the individual methods. Bagging and Stacking are two main approaches in ensemble model construction. In ensemble method, multiple solutions are generated by sampling the feature space, changing the parameter settings or using various statistical or machine learning algorithms [11]. In this paper, further literature review of the time series statistical models will be presented. An extensive review of the all three methods on solar power forecasting is provided in [8].

Mandal *et al.* [12] proposed a hybrid method with the combination of Neural Network (NN) and Wavelet Transform (WT) to predict short-term solar PV power forecast. WT filtered the spike and chaotic changes in solar PV power time-series data whereas, RBFNN captured the non-linear solar PV power fluctuation in a substantial way. To minimize the number of input data while keeping the accuracy maximum, the authors analyzed the historical data and attempted to include various time-lag inputs into proposed forecasting model such as PV power (PV), solar radiation (R), and temperature (T). The authors observed the MAPE value of the prediction models to achieve the best combination. Two days were chosen randomly from each season of the year 2011 (D1 & D2). The MAPE values obtained from the BPNN and RBFNN models were in the range of 15%–32% and 5%–35% respectively whereas when WT was combined, the MAPE values improved significantly and they were recorded in the range of 4%–14% and 4%–13% for WT+BPNN and WT+RBFNN respectively. Also, the proposed model recorded lower MAE and RMSE values demonstrating the higher degree of accuracy.

Sharma *et al.* [5] proposed prediction models to predict solar generation from weather forecasts representing both observational and forecast weather matrices as a time-series data due to the high dimensionality. They used eight months training data as input to generate each model which includes Solar Intensity readings as class. They tested their models' accuracy on 2 months' data. Their research outcomes underlined that the prediction models are unable to predict changes in the weather patterns in advance but the models generated using historical data automatically predict solar intensity. Researchers have concluded that increasing the accuracy of these prediction models is a promising area and plan on using their prediction models to better match renewable generation to consumption in both smart homes and data centres.

Researchers proposed three different models, models were generated using Linear Least Square Regression, Support Vector Machines and Eliminating Redundant Information. The values of RMS-Error for both cross-validation and prediction were recorded to compare the different models. The Table 1 shows the recorded values of RMSE.

Table 1. Residual mean squared error values obtained using various algorithms [5]

Model	Cross-Validation W/m^2	Prediction W/m^2
Linear least square regression	165	130
Support vector machines (linear and polynomial kernel)	201	228
Support vector machines (RBF kernel)	164	163
Eliminating redundant information (RBF SVM)	159	128

Takahashi and Mori [13] also used ANN and studied 3 different models for 30 min ahead predictions: A Radial Basis Function Network and Generalized Radial Basis Function Networks with clustering via k-means and Deterministic Annealing. They used actual and lagged values of Power (P) and cell temperature (T) as inputs. The k-means method outperformed MLP by reducing 12% of the average and 28% of the maximum error. The DA method succeeded in reducing approximately 34% of the average and maximum error of MLP. Results showed that their Deterministic Annealing model outperformed the other models, having a standard deviation of errors 88% of the MLP.

Rana *et al.* [14] worked with a machine learning approach using SVR. They introduced the 2D interval forecasts for predicting intra-hour and intra-day. 2D interval forecast can be defined as an interval forecast for a range of future values, what is a continuous probabilistic forecast for a defined future period. They also studied different sets of inputs depending on how frequently data was sampled. Results for the 30-min forecast horizon showed a MRE of 6.47% and an Internal Coverage Probability (ICP) of 65.58%, which represented big improvements with respect to baselines.

Yona *et al.* [15] also proposed the power output forecasting for PV system based on insolation prediction using NN. The proposed technique for application of NN was trained by only weather data and tested for the target term. They also suggested that by using only meteorological data, the forecast time can be shortened at the time of insolation forecasting. During their research, they studied three different models: FFNN, RBFNN and RNN. They chose RBFNN for its structural simplicity and universal approximation properties. RNN was chosen as it is known as a good tool for time-series data forecasting. Their results indicated that the RBFNN and RNN outperform the results of FFNN in some months. The validity of their proposed NN model was confirmed using one-day-ahead 24 h forecasting simulation.

Ding *et al.* [16] proposed an ANN-based approach for forecasting the power output of photovoltaic system through an effective analysis of influencing factor on PV power output, combined with the improved back-propagation learning algorithm and similar day selection algorithm. The research underlined that the proposed model does not require complex modelling and complicated calculation. The PV forecast can be carried out under different weather types using only historical data. The forecasting accuracy of the proposed model on a sunny and a rainy day was validated using actual value and forecasted value. The validation results of a sunny and a rainy day were recorded as MAP-Error values were 10.06% and 18.89% respectively. As a result of very less fluctuation in weather on a sunny day as compared to a rainy day, the forecasting results were more accurate on a sunny day.

3 Methodology

The dataset used in this paper is from a weather station based in Martinborough, New Zealand. The data was acquired for years 2015, 2016 and 2017. The first two years of data was used to generate the model and the year 2017 data was used to measure the accuracy of the model. The data has following features:

- Day of the Year
- Wind Direction (points in degrees)
- Air Temperature (average in °C)
- Humidity (average percentage relative humidity)
- Air Pressure (average hectopascals – hPa)
- Rainfall (points in millimetre – mm)
- Wind Speed (maximum in kph)
- Wind Speed (average in kph)
- Light Intensity (average in W/m^2).

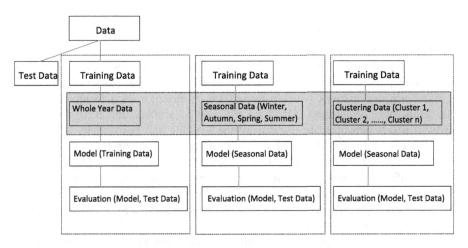

Fig. 1. Research methodology used in this paper

In this paper, three approaches are used to construct models from the data and then test data is used to compare the performance of each approach. The data is normalised between zero and one to incorporate uniform feature weightage. In the first approach

(Fig. 1-Left), training data is used to generate the machine learning models and then test data is used to evaluate the model accuracy. In the second approach (Fig. 1-Middle), training data is split into segments based on their seasons (winter, autumn, spring and summer) and four separate models are constructed (one for each season). The test data is also divided into seasons and each seasons' test data is passed on to its respective model for evaluation. Similarly, segmentation on the basis of other criterion can be achieved such as splitting the data based on weather condition (i.e. rainy and not rainy days; or sunny, cloudy and rainy days). In the third approach (Fig. 1-Right), hierarchical clustering is implemented on the training data and then models are constructed on each cluster separately. The models are later evaluated on the test data. The training data without PV intensity variable is used to create clustering and similarly, cluster membership of each test data instance is assigned based on the proximity of the test instance to each cluster.

In this paper, the six models, namely, linear regression, radial basis function (RBF) regressor, multi-layer perceptron (MLP) regressor, M5 rules, forest tree and ensemble RBF regressor are used to demonstrate the applicability of the proposed approach. First four algorithms in the list are single model classifiers (or standard classifiers) whereas, last two algorithms are ensemble-based classifiers. Ensemble-based classifiers are developed on the idea that instead of creating a single model, it is better to generate multiple models and then forecasts are made by considering the outcome of all the models. It has been demonstrated in the literature in a number of cases that ensemble methods have outperformed single models [17, 18].

Linear regression is a statistical method that looks into the relationship of two or more input variables and an output variable by fitting a linear equation to the observed data [19]. Formally, the model for multiple linear regression, given n observations, is

$$y_i = w_0 + w_1 x_1 + w_2 x_2 \ldots \ldots \ldots + w_n x_n$$

where w_0 is a constant and n is the number of correlated attributes.

RBF regressor implements Gaussian radial basis function networks for regression using some optimiser that minimises the squared error. Radial basis function network is an artificial neural network which uses radial basis functions as activation functions [20]. Multi-layer perceptrons (MLP) are the class of artificial neural networks [21]. Artificial neural networks are inspired by the working and connectivity of the biological neural networks. Artificial neurons are typically connected to each other through weights and these weights are adjusted to carryout learning. The application areas of artificial neural networks are pattern recognition, anomaly detection, computer vision, video games and medical diagnosis [22]. In classification, a model is built to best differential different classes of the data and in regression, a MLP model is built to approximate a function that fit the real value output.

In the M5 Rules approach, repeatedly decision trees using J48 algorithms are created and then the best rule is selected from the tree at each iteration. This approach produces rule with good accuracy and high degree of coverage of the data space [23]. The random forest is an ensemble-based approach, where multiple models are created from the data by changing the parameter settings. In random forest approach, multiple decision trees are built by randomly selecting the subset of features. The accumulated results from these forest of decision trees would form an ensemble (random forest) [24]. The forecasting error on all these algorithms are calculated on the basis of the mean absolute error and the formula is as below:

$$MAE = \frac{1}{N} \sum |y - x|$$

Where x and y are the actual solar intensity value and forecasted solar intensity value, respectively. The N is the number of observations in the data.

Other methods of error calculations are root mean squared error, relative absolute error and root relative squared error. These methods compare the difference between the estimated and actual values. In the literature researchers have used different error measuring criteria, however, in this paper, MAE is selected due to its simplicity and effectiveness.

The clustering algorithm used in this paper is EM algorithm. This method is very similar to the K-means clustering method. In K-means clustering, given a fixed number of clusters, observations are assigned to the centroids of the clusters so that cluster centres are as far from each other as possible. In EM algorithm, instead of assigning samples to the cluster centroids, probabilities of the cluster memberships are computed based on one or more probability distributions. The final objective of the algorithm then is to maximise the overall probability or likelihood of the data, given the final clusters [25]. The hierarchical EM algorithm is used in this paper, where number of clusters is set to be in the range of two to eight. On all of these clustering solutions, supervised learning models are built using 10-fold cross-validation. The clustering scheme with lowest MAE value is selected as an optimal number of clusters for the forecasting model.

4   Experimental Results

The raw data acquired from the source had missing values, false values and variating frequency. On a typical day, the weather readings were recorded at an hourly interval but in some cases, the readings were recorded at interval of 30 min, 15 min and 5 min. The readings are averaged to daily weather readings. The training data has 724 instances which were daily records of the year 2015 and 2016 and the test data has 271 instances of the year 2017.

The training and test errors of the models generated on the six algorithms used in this paper are presented in Table 2. It is noticeable that Forest Tree and Linear regression have produced the best and worst results, respectively, on the training and test data.

One possible reason of the better performance of the Forest tree is that it is an ensemble based model. In ensemble multiple models are created using subset of the training feature space and majority vote is considered on the test data. The ensemble RBF regressor was also used and its results are almost similar to the single RBF regressor model.

Table 2. MAE values obtained from six algorithms, where first two years data was used as training data and third year as test data

Algorithms	Training error (MAE)	Test error (MAE)
Linear regression	0.135	0.127
RBF regressor	0.127	0.103
MLP regressor	0.125	0.117
M5 rules	0.101	0.097
Forest tree	0.098	0.079
Ensemble RBF regressor	0.114	0.100

According to the model proposed by Yona *et al.* [15] shortened training and forecasting time data can produce better results. Thus, we planned to produce four prediction models as per four different seasons i.e. autumn, spring, summer and winter. The training datasets were built on years 2015 and 2016 data based on autumn, spring, summer and winter; the test data sets were built using the year 2017 data based on autumn, spring, summer and winter. There were 181, 178, 184 and 181 training instances in summer, spring, autumn and winter seasonal segmentations, respectively. Moreover, there were 55, 31, 92 and 92 test instances in summer, spring, autumn and winter segmentation, respectively.

In Table 3 four models against each season are generated separately and tested on the respective 2017 seasonal data. The average of each algorithm across four models is also presented in Table 3. There are a number of things to notice. Firstly, the average training and test errors on all the algorithms except Random forest have reduced when compared to the results of Table 2, where single model for data was generated. Secondly, winter season model has the lowest training and testing error, whereas, summer season has the highest error across all models. Autumn and spring seasons have similar testing errors. However, spring models have higher training (1- fold cross validation) errors than autumn. This information is important in terms of knowing which segments produce more accurate models. Thirdly, both RBF regressor and ensemble RBF regressor have lowest cross-validation error (0.092 MAE) while ensemble RBF regressor model has lower testing error (0.84 MAE) than single model RBF regressor (0.089 MAE).

Table 3. Four seasonal models generated using multiple algorithms, and their individual, and average errors (MAE) on training and test data sets

	Data set	Summer	Spring	Autumn	Winter	Average
Linear regression	Training	0.151	0.109	0.080	0.053	**0.098**
	Test	0.169	0.078	0.086	0.047	**0.095**
RBF regressor	Training	0.149	0.104	0.068	0.048	**0.092**
	Test	0.154	0.072	0.077	0.052	**0.089**
MLP regressor	Training	0.161	0.114	0.079	0.048	**0.101**
	Test	0.175	0.075	0.087	0.050	**0.097**

(*continued*)

Table 3. (*continued*)

	Data set	Summer	Spring	Autumn	Winter	Average
M5 rules	Training	0.151	0.105	0.080	0.056	**0.098**
	Test	0.169	0.078	0.074	0.047	**0.092**
Random forest	Training	0.162	0.103	0.071	0.050	0.096
	Test	0.150	0.068	0.067	0.046	0.083
Ensemble RBF regressor	Training	0.151	0.103	0.069	0.046	**0.092**
	Test	0.150	0.066	0.071	0.047	**0.084**

The data is separated on the basis of rain and no-rain due to the high fluctuations in weather on a rainy day. There were 650 and 74 not rainy and rainy training instances and the number of rainy and not rainy test instances were 41 and 230, respectively in the data. The MAE values for the 10-fold cross-validation and testing are provided in Table 4. The average MAE in linear regression, RBF regressor and ensemble RBF regressor are marginally better than results of Table 2, where full training data was used to create the model. However, results obtained using rainy and not rainy segmentation are poor in accuracy than the results of seasonal model averages in Table 3.

Table 4. Two segment (Rainy and not rainy) models generated using multiple algorithms, and their individual, and average errors (MAE) on training and test data sets

Algorithms	Data set	Rainy	Not rainy	Average
Linear regression	Training	0.124	0.136	**0.130**
	Test	0.115	0.129	**0.122**
RBF regressor	Training	0.123	0.120	**0.121**
	Test	0.086	0.100	**0.093**
MLP regressor	Training	0.154	0.109	0.131
	Test	0.115	0.091	0.103
M5 rules	Training	0.125	0.099	0.112
	Test	0.103	0.097	0.100
Forest tree	Training	0.102	0.097	0.099
	Test	0.091	0.083	0.087
Ensemble RBF regressor	Training	0.116	0.112	**0.114**
	Test	0.092	0.103	**0.098**

In Tables 3 and 4, data is divided into the groups based on either state of rain or the four seasons. It has been seen that the model accuracy increases by dividing the data into subgroups and creating models on each subgroup separately. Another way to group the data is on the basis of finding natural groups in the data rather than dividing the data based on any pre-defined criteria (i.e. seasons, rainy/not rainy weather condition). An expectation maximization (EM) clustering algorithm was used to extract natural grouping in the data. The data is split into up to eight clusters and models are generated on training and tested on test data set. The Fig. 2 shows the cross-validation

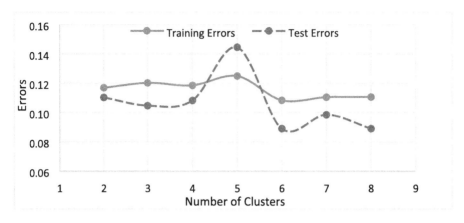

Fig. 2. Mean absolute errors (y-axis) against number of clusters (2 to 8) along x-axis on training and test data sets using linear regression method. Six clusters are selected as it has the minimum 10-fold cross validation error on training data.

error (on training data) and testing error for minimum two and maximum eight number of clusters using linear regression method. The optimal clustering is found at index 6 which has minimum cross-validation error.

10-fold cross-validation and test errors on all six algorithms are mentioned in Table 5 along with optimal number of clusters. Results have shown that four algorithms have found four clusters in the data one algorithm each has found three and six clusters. It is evident that all algorithms have produced better or similar errors results than the single model results in Table 2. The clustering based average model accuracies (test data) of linear regression, RBF regressor and MLP regressors were better than the four seasonal models in Table 3.

Table 5. 10-fold cross-validation and test errors (MAE) using EM clustering algorithm

Algorithms	Dataset	Error
Linear regression (6 clusters)	Training	**0.108**
	Test	**0.089**
RBF regressor (4 clusters)	Training	**0.099**
	Test	**0.087**
MLP regressor (3 clusters)	Training	**0.104**
	Test	**0.094**
M5 rules (4 clusters)	Training	**0.108**
	Test	**0.097**
Forest tree (4 clusters)	Training	0.097
	Test	0.093
Ensemble RBF regressor (4 clusters)	Training	**0.096**
	Test	**0.088**

5   Conclusion

In this paper, a novel clustering-based PV forecasting approach has been proposed. The mean absolute error on 10-fold cross validation method was used to select the optimal number of clusters. It has been demonstrated with experimental results that cluster based forecasting method produced superior forecasting accuracy as compared to forecasting models constructed using full training data. Moreover, it was presented in Tables 3 and 5 that seasonal forecasting models and cluster-based forecasting models' results were comparable. It was found that different forecasting algorithms had found different number of clusters in the data. For example, linear regression method found six clusters and MLP regressor found three optimal clusters in the data. In this paper, EM clustering method was used to find natural clusters in the data. It will be interesting to perform clustering using other state of the art clustering methods (i.e. K-means and Hierarchical clustering) and compare the forecasting results on PV data. In the future, a novel ensemble-based PV forecasting approach can be developed by considering various clustering data.

References

1. Wan, C., et al.: Photovoltaic and solar power forecasting for smart grid energy management. CSEE J. Power Energy Syst. **1**(4), 38–46 (2015)
2. Hsu, K., Gupta, H.V., Sorooshian, S.: Artificial neural network modeling of the rainfall runoff process. Water Resour. Res. **31**(10), 2517–2530 (1995)
3. Burges, C.: A tutorial on support vector machines for pattern recognition. Data Min. Knowl. Discov. **2**(2), 121–167 (1998)
4. Bishop, C.M.: Pattern Recognition and Machine Learning. Information Science and Statistics. Springer, New York (2006)
5. Sharma, N., et al.: Predicting solar generation from weather forecasts using machine learning (2011)
6. Järventausta, P., et al.: Smart grid power system control in distributed generation environment. Annu. Rev. Control **34**(2), 277–286 (2010)
7. Haque, A., Nehrir, M.H., Mandal, P.: Solar PV power generation forecast using a hybrid intelligent approach, pp. 1–5 (2013)
8. Sobri, S., Koohi-Kamali, S., Rahim, N.A.: Solar photovoltaic generation forecasting methods: a review. Energy Convers. Manag. **156**, 459–497 (2018)
9. Yona, A., et al.: Determination method of insolation prediction with fuzzy and applying neural network for long-term ahead PV power output correction. IEEE Trans. Sustain. Energy **4**(2), 527–533 (2013)
10. Tuohy, A., et al.: Solar forecasting: methods, challenges, and performance. IEEE Power Energy Mag. **13**(6), 50–59 (2015)
11. Ren, Y., Suganthan, P.N., Srikanth, N.: Ensemble methods for wind and solar power forecasting—a state-of-the-art review. Renew. Sustain. Energy Rev. **50**, 82–91 (2015)
12. Mandal, P., et al.: Forecasting power output of solar photovoltaic system using wavelet transform and artificial intelligence techniques. Procedia Comput. Sci. **12**, 332–337 (2012)
13. Takahashi, M., Mori, H.: A hybrid intelligent system approach to forecasting of pv generation output. J. Int. Counc. Electr. Eng. **3**(4), 295–299 (2013)

14. Rana, M., Koprinska, I., Agelidis, V.G.: 2D-interval forecasts for solar power production. Sol. Energy **122**, 191–203 (2015)
15. Yona, A., et al. Application of neural network to one-day-ahead 24 hours generating power forecasting for photovoltaic system. In: 2007 International Conference on Intelligent Systems Applications to Power Systems (2007)
16. Ding, M., Wang, L., Bi, R.: An ANN-based approach for forecasting the power output of photovoltaic system. Procedia Environ. Sci. **11**, 1308–1315 (2011)
17. Cunningham, P., Carney, J., Jacob, S.: Stability problems with artificial neural networks and the ensemble solution. Artif. Intell. Med. **20**(3), 217–225 (2000)
18. Bauer, E., Kohavi, R.: An empirical comparison of voting classification algorithms: bagging, boosting, and variants. Mach. Learn. **36**(1–2), 105–139 (1999)
19. Grégoire, G.: Multiple linear regression. Eur. Astron. Soc. Publ. Ser. **66**, 45–72 (2014)
20. Chen, S., Cowan, C.F., Grant, P.M.: Orthogonal least squares learning algorithm for radial basis function networks. Neural Netw. **2**(2), 302–309 (1991)
21. Jain, A.K., Mao, J., Mohiuddin, K.M.: Artificial neural networks: a tutorial. IEEE Comput. **29**, 31–44 (1996)
22. Zhang, G.: Neural networks for classification: a survey. IEEE Trans. Syst. Man Cybern. Part C **30**(4), 451–462 (2000)
23. Murthy, S.K.: Automatic construction of decision trees from data: a multi-disciplinary survey. Data Mining Knowl. Discov. **2**(4), 345–389 (1998)
24. Breiman, L.: Random Forests. Mach. Learn. **45**(1), 5–32 (2001)
25. Dempster, A.P., Laird, N.M., Rubin, D.B.: Maximum likelihood from incomplete data via the EM algorithm. J. Roy. Stat. Soc. Ser. B (Methodological) **39**(1), 1–38 (1977)
26. Raza, M.Q., Nadarajah, M., Ekanayake, C.: Review on recent advances in PV output power forecast. Sol. Energy **136**, 125–144 (2017)

Author Index